Control and Estimation of Distributed Parameter Systems

International Conference in Vorau (Austria), July 14–20, 1996

Edited by

W. Desch
F. Kappel
K. Kunisch

Birkhäuser Verlag
Basel · Boston · Berlin

Editors:

W. Desch, F. Kappel and K. Kunisch
Universität Graz
Institut für Mathematik
Heinrichstraße 36
A-8010 Graz

e-mails: desch@bkfug.kfunigraz.ac.at
franz.kappel@kfunigraz.ac.at
karl.kunisch@kfunigraz.ac.at

1991 Mathematics Subject Classification 93-06, 49-06, 35-06

Library of Congress Cataloging-in-Publication Data

Control and estimation of distributed parameter systems :
international conference in Vorau (Austria), July 14–20, 1996 / edited by W. Desch, F. Kappel, K. Kunisch.
p. cm. — (International series of numerical mathematics ; vol. 126)
Papers from the International Conference on Control and Estimation of Distributed Parameter Systems held July 14–20, 1996, in Vorau, Austria.
Includes bibliographical references.
ISBN 3-7643-5835-1 (acid-free paper). — ISBN 0-8176-5835-1 (Boston : acid-free paper)
1. Control theory – Congresses. 2. Distributed parameter systems – Congresses. 3. Estimation theory – Congresses. I. Desch, W. (Wolfgang), 1953– . II. Kappel, F. III. Kunisch, K. (Karl). 1952– . IV. International Conference on Control and Estimation of Distributed Parameter Systems (1996: Vorau, Styria, Austria)
V. Series: International series of numerical mathematics ; v. 126.
QA402.3.C6262 1998
003',78–dc21

Deutsche Bibliothek Cataloging-in-Publication Data

Control and estimation of distributed parameter systems :
international conference in Vorau (Austria), July 14–20, 1996 / ed.
by W. Desch ... – Basel ; Boston ; Berlin : Birkhäuser, 1998
(International series of numerical mathematics ; Vol. 126)
ISBN 3-7643-5835-1 (Basel ...)
ISBN 0-8176-5835-1 (Boston)

Printed on acid-free paper produced of chlorine-free pulp. TCF ∞
Cover design: Heinz Hiltbrunner, Basel
Printed in Germany
ISBN 3-7643-5835-1
ISBN 0-8176-5835-1

9 8 7 6 5 4 3 2 1

Preface

The International Conference on Control and Estimation of Distributed Parameter Systems took place from July 14–20, 1996, at the Bildungshaus Chorherrenstift Vorau in Vorau (Austria). It was the seventh in a series of conferences that begun in 1982. 51 researchers from 11 states contributed to draw a broad and diverse picture of the recent developments in optimal control and parameter identification of partial differential equations, both from a theoretical and numerical viewpoint. We thank them all for their contributions to an enjoyable and interesting conference.

We address our thanks to the whole staff of the Bildungshaus Chorherrenstift Vorau. The pleasant atmosphere at the Bildungshaus has been a key ingredient to the success of the meeting and the stimulating interaction between the participants. We are particularly indebted to Mrs. L. Reiß, who helped us omnipresently with all the everyday issues of a conference like this one.

This meeting was facilitated by funding from the following organizations:

Amt der Steiermärkischen Landesregierung,
Bundesministerium für Wissenschaft und Verkehr,
Christian Doppler Laboratorium für Parameter Identifikation
und Inverse Probleme,
European Research Office of the U.S. Army,
Spezialforschungsbereich F003 "Optimierung und Kontrolle",
Stadt Graz,
U.S. Air Force European Office of Aerospace Research and Development.

It is our pleasure to acknowledge the generous support by these institutions.

Once again, the friendly and supportive team of Birkhäuser, in particular Dr. T. Hintermann and Mrs. S. Lotrovsky, have provided an optimal opportunity to publish our proceedings. Our special thanks go to Mrs. G. Krois. Her enthusiasm, skill and workpower have been the backbone of the organization of the conference and the preparation of the TEX manuscript of the proceedings you are presently reading.

Graz, July 1997

W. Desch, F. Kappel, K. Kunisch

List of Participants

H. T. BANKS	North Carolina State University
M. BERGOUNIOUX	Université d'Orléans
F. BONNANS	INRIA
A. BRIANI	Università di Pisa
M. BROKATE	Universität Kiel
J. A. BURNS	Virginia Tech
P. CANNARSA	Università di Roma "Tor Vergata"
S. J. COX	Rice University
W. DESCH	Karl–Franzens–Universität Graz
H. EGGHART	US Army European Research Office
M. FALCONE	Università di Roma "La Sapienza"
H. O. FATTORINI	University of California
E. FERNÁNDEZ-CARA	Universidad de Sevilla
A. V. FURSIKOV	Moscow State University
J. HASLINGER	KFK MFF UK
M. HEINKENSCHLOSS	Rice University
M. HINZE	TU Berlin
K. ITO	North Carolina State University
F. KAPPEL	Karl–Franzens–Universität Graz
A. KAUFFMANN	TU Berlin
C. T. KELLEY	North Carolina State University
V. KOMORNIK	Université Louis Pasteur
M. KROLLER	Karl–Franzens–Universität Graz
K. KUNISCH	Karl–Franzens–Universität Graz
A. V. KUNTSEVICH	Academy of Sciences of the Ukraine
G. LEUGERING	Universität Bayreuth
J. MALINEN	Helsinki University of Technology
S. MICU	Universidad Complutense de Madrid

F. Mignot	Université de Paris-Sud
B. S. Mordukhovich	Wayne State University
G. Peichl	Karl–Franzens–Universität Graz
W. Prager	Karl–Franzens–Universität Graz
G. Propst	Karl–Franzens–Universität Graz
J.-P. Puel	CMAP
A.M. Ramos	Universidad Complutense de Madrid
J.-P. Raymond	Université Paul Sabatier
W. Ring	Karl–Franzens–Universität Graz
A. Rösch	TU Chemnitz-Zwickau
T. Slawig	TU Berlin
R. C. Smith	Iowa State University
O. J. Staffans	Åbo Akademi University
D. Tataru	Princeton University
M. E. Tessitore	Università di Roma "Tor Vergata"
S. Thaller	Karl–Franzens–Universität Graz
R. Tichatschke	Universität Trier
M. Tucsnak	CMAP
J. Turi	University of Texas at Dallas
C. Wang	University of Southern California
J.-P. Yvon	Université de Technologie de Compiègne
Bing-Yu Zhang	University of Cincinnati
E. Zuazua	Universidad Complutense de Madrid

Contents

International Series of Numerical Mathematics
Vol. 126,

Approximation Results for Parameter Estimation in Nonlinear Elastomers

H.T. BANKS AND GABRIELLA A. PINTER

Center for Research in Scientific Computation
North Carolina State University

Department of Mathematics
Texas Tech University

ABSTRACT. In this paper we present an approximation framework and theoretical convergence results for a class of parameter estimation problems for general abstract nonlinear hyperbolic systems. These systems include as a special case those modeling a large class of nonlinear elastomers.

1991 *Mathematics Subject Classification.* 35R30

Key words and phrases. Abstract nonlinear hyperbolic systems, parameter estimation, finite dimensional approximation.

1. Introduction

We consider the following class of abstract nonlinear damped parameter dependent hyperbolic systems evolving in a complex separable Hilbert space H:

$$w_{tt} + \mathcal{A}_1(q)w + \mathcal{A}_2(q)w_t + \mathcal{N}^* g(q)(\mathcal{N}w) = f(t;q) \tag{1.1}$$

$$w(0) = \varphi_0 \tag{1.2}$$

$$w_t(0) = \varphi_1. \tag{1.3}$$

Here $\mathcal{A}_1(q), \mathcal{A}_2(q)$ are unbounded operators depending on some parameter q, $g(q)$ is a parameter dependent nonlinear operator in H, $\mathcal{N}$ is an unbounded operator, and f is a parameter dependent forcing term. Precise conditions on these operators are given below.

This class of systems was introduced in [BGS, BLMY] and further studied in [BLGMY] as a model for the behavior of nonlinear elastomers. These materials, which are used in the development of active and passive vibration devices, are rubber or polymer based composites that involve complex viscoelastic materials. Their behavior cannot be adequately modelled using the theory of linear elasticity. Indeed, they exhibit nonlinearities in material and geometric properties so that there is a nonlinear relationship between stress and strain even for small strains. We illustrate with a simple example that takes into account these nonlinearities, and describe the associated general parameter estimation problems. (For detailed discussions of this and other models see [BLMY, BGS, BLGMY, BL].)

Consider an isotropic, incompressible rubber-like rod under simple elongation with a finite applied stress in the principal axis direction $x_1 = x$. Let w denote deformation in the x direction. Following the derivation of the model in [BLMY, BGS, BL] we arrive at the partial differential equation

$$\rho A \frac{\partial^2 w}{\partial t^2} - \frac{\partial}{\partial x}\left(\frac{EA}{3}\frac{\partial w}{\partial x}\right) - \frac{\partial}{\partial x}\left(\frac{EA}{3}\tilde{g}\left(\frac{\partial w}{\partial x}\right)\right) = F, \tag{1.4}$$

where ρ is mass density, E is the *generalized* modulus of elasticity, A is the cross sectional area, and F is an applied external force. If one assumes that the rod is composed of a neo-Hookean material (see [BL]), then the nonlinearity $\tilde{g}$ in (1.4) is given by $\tilde{g}(\xi) = 1 - \frac{1}{(1+\xi)^2}$ for $-1 < \xi < 1$. Assuming that we have a slender rod of length ℓ that satisfies $w(t,0) = w(t,\ell) = 0$, and defining $V = H_0^1(0,\ell)$ and $H = L^2(0,\ell)$, we obtain the usual Gelfand triple $V \hookrightarrow H \approx H^* \hookrightarrow V^*$ where $V^* = H^{-1}(0,\ell)$. Then equation (1.4) with the specified boundary conditions can be written in the variational form:

$$\rho A w_{tt} + \mathcal{A}_1 w + D^*\tilde{g}(Dw) = F \quad \text{in } V^*, \tag{1.5}$$

where $\mathcal{A}_1 \in \mathcal{L}(V, V^*)$ is given by

$$\langle \mathcal{A}_1\varphi, \psi\rangle_{V^*,V} = \left\langle \frac{EA}{3} D\varphi, D\psi \right\rangle_H$$

and $D = \frac{\partial}{\partial x} \in \mathcal{L}(V, H)$ is the spatial differentiation operator. For a realistic model we also must include some type of material damping which is known to be present in elastomers (indeed in *all* materials). Here we assume an internal damping of the form $\mathcal{A}_2 w_t$, where $\mathcal{A}_2 \in \mathcal{L}(V_2, V_2^*)$ and $V \hookrightarrow V_2 \hookrightarrow H$. In the case of Kelvin-Voigt damping we define $V_2 = V = H_0^1(0,\ell)$ and

$$\langle \mathcal{A}_2\varphi, \psi\rangle_{V_2^*,V_2} = \langle c_D D\varphi, D\psi\rangle_H ,$$

where $c_D \in L_\infty(0,\ell)$. (We remark that the exact form of the damping mechanism in elastomers is not known and, indeed, is the subject of current research.) With the damping included, we find that our model in variational form for the neo-Hookean elastomer rod is given by

$$\rho A w_{tt} + \mathcal{A}_1 w + \mathcal{A}_2 w_t + D^*\tilde{g}(Dw) = F \quad \text{in } V^*. \tag{1.6}$$

If this model is to be used for simulation or control of the behavior of the elastomer rod we need values for ρ, E, A, c_D, F, ℓ. Some of these can be given or measured explicitly (e.g., A, ℓ, F), or can be found from manufacturers specifications (so-called "book-values"). However, some parameters (e.g., E, c_D) cannot be measured or obtained this way. Also, the "book-values" can vary considerably between samples. Thus we need a method to estimate these "unknown" parameters by dynamic experiments with the sample itself. Moreover, the nonlinearity $\tilde{g}$ is in general unknown and must be estimated (the neo-Hookean assumption is only a first approximation to actual material properties) or chosen from a general class of admissible nonlinearities.

In one general parameter estimation formulation equation (1.6) takes the form (1.1)–(1.3) where the structural operators $\mathcal{A}_1, \mathcal{A}_2$, the nonlinearity g and the input f have all been parameterized by a vector (possibly infinite dimensional) parameter

q that must be estimated. Here the parameter q takes values from an admissible parameter set Q. Suppose that we have a set of measured observations $z = \{z_i\}_{i=1}^K$ corresponding to measurements (e.g., displacements, velocities) taken at time t_i. In a general least squares parameter estimation problem, we seek to minimize the least squares output functional

$$J(q, z) = \left|\tilde{C}_2 \left\{\tilde{C}_1\{w(t_i, \cdot; q)\} - \{z_i\}\right\}\right|^2,$$

over $q \in Q$, where $\{w(t_i, \cdot; q)\}$ are the parameter dependent solutions of (1.1)–(1.3) evaluated at time $t_i, i = 1, 2, \ldots K$, and $|\cdot|$ is an appropriately chosen Euclidean norm. The operators $\tilde{C}_1, \tilde{C}_2$ depend on the type of the collected data. For example, if z_i is time domain displacement, velocity or acceleration at a point x, then $\tilde{C}_1$ involves differentiation (0,1 or 2 times, respectively) with respect to time followed by pointwise evaluation in t and x. The operator $\tilde{C}_2$ is the identity in the case of time domain identification, while it is related to the Fourier transform if we consider fitting the data in the frequency domain (see Chapter 5 of [BSW] for details).

In this formulation the minimization problem involves an infinite dimensional state space and (in general) an infinite dimensional admissible parameter set Q. To overcome this difficulty and to obtain a computationally tractable method, we use the general ideas described in [BSW]. Namely, let H^N be finite dimensional subspaces of H, and Q^M be a sequence of finite dimensional sets approximating the parameter set Q. Denote the orthogonal projections of H onto H^N by P^N. One can formulate a family of approximating estimation problems with finite dimensional state spaces and finite dimensional parameter sets in the following way: find $q \in Q^M$ which minimizes

$$J^N(q, z) = \left|\tilde{C}_2 \left\{\tilde{C}_1\{w^N(t_i, \cdot; q)\} - \{z_i\}\right\}\right|^2, \tag{1.7}$$

where $w^N(t; q) \in H^N$ is the solution to the finite dimensional approximation of (1.1)–(1.3) given by:

$$\begin{aligned}\langle w_{tt}^N, \phi\rangle_{V^*,V} + \langle \mathcal{A}_1(q) w^N, \phi\rangle_{V^*,V} + \langle \mathcal{A}_2 w_t^N, \phi\rangle_{V_2^*,V_2} + \langle g(q)(\mathcal{N} w^N), \mathcal{N}\phi\rangle \\ = \langle f(t; q), \phi\rangle_{V_2^*,V_2}\end{aligned} \tag{1.8}$$

$$w^N(0) = P^N\varphi_0, \; w_t^N(0) = P^N\varphi_1 \tag{1.9}$$

for all $\varphi \in H^N$.

Solution of these approximate estimation problems (1.7)–(1.9) provides one with a sequence of parameter estimates $\{\bar{q}^{N,M}\}$. The crucial question is when one can guarantee that this sequence (or some subsequence) converges to a solution of the original infinite dimensional parameter estimation problem. Under certain suitable assumptions on the approximating spaces H^N and approximating sets Q^M this question is answered in [BSW] for linear systems and here we extend these ideas to include nonlinear systems.

To permit use of the method outlined above we must be certain that the above systems (1.1)–(1.3) and (1.8)–(1.9) have solutions in some sense for each $q \in Q$. This well-posedness problem (without considering the parameter dependent case) was solved in the recent paper [BGS]. In the following section we summarize these results and give precise conditions under which (1.1)–(1.3) has a unique weak solution for each $q \in Q$. Then in Section 3 we give assumptions on the general parameter estimation problem that we shall use to prove convergence. We also recall Theorem 5.1 from [BSW] that provides a sufficient condition for the convergence of the solutions $\{\bar{q}^{N,M}\}$ of the approximate estimation problems to a solution of the original parameter estimation problem. Then in Section 4 we show that this condition is satisfied in our case under natural assumptions on the parameter dependence of $\mathcal{A}_1, \mathcal{A}_2, g$ and f.

2. Formulation of the Problem

We assume that there is a sequence of separable Hilbert spaces V, V_2, H, V^*, V_2^* forming a Gelfand quintuple [BIW, Wl] satisfying

$$V \hookrightarrow V_2 \hookrightarrow H \hookrightarrow V_2^* \hookrightarrow V^*, \tag{2.1}$$

where we assume that the embedding $V \hookrightarrow V_2$ is dense and continuous with $|\varphi|_{V_2} \leq c|\varphi|_V$ for $\varphi \in V$ and $V_2 \hookrightarrow H$ is a dense compact embedding with $|\varphi| \leq \tilde{c}|\varphi|_{V_2}$. The norm in H will be denoted by $|\cdot|$ while those in V, V_2 etc. will carry an appropriate subscript. We denote by $\langle\ ,\ \rangle_{V^*,V}$, etc., the usual duality products [Wl]. These duality products are the extensions by continuity of the inner product in H, denoted by $\langle\ ,\ \rangle$ throughout. Let Q be an infinite dimensional parameter set. The operators $\mathcal{A}_1(q)$ and $\mathcal{A}_2(q)$ are defined in terms of their sesquilinear forms $\sigma_1(q) : V \times V \to \mathbb{C}$ and $\sigma_2(q) : V_2 \times V_2 \to \mathbb{C}$. That is, $\mathcal{A}_1(q) \in \mathcal{L}(V, V^*), \mathcal{A}_2(q) \in \mathcal{L}(V_2, V_2^*)$ and $\langle \mathcal{A}_1(q)\varphi, \psi\rangle_{V^*,V} = \sigma_1(q)(\varphi, \psi), \langle \mathcal{A}_2(q)\varphi, \psi\rangle_{V_2^*,V_2} = \sigma_2(q)(\varphi, \psi)$.

Let $\mathcal{L}_T$ denote the space of functions $w : [0,T] \to H$ such that

$$w \in C_W([0,T], V_2) \cap L^\infty([0,T], V)$$

(the subscript W denotes weak continuity), and

$$w_t \in C_W([0,T], H) \cap L^2([0,T], V_2),$$

where the time derivative w_t is understood in the sense of distributions with values in a Hilbert Space (see, e.g., [Li1]). The space $\mathcal{L}_T$ is equipped with the norm

$$|w|_{\mathcal{L}_T} = \operatorname*{ess\ sup}_{t\in[0,T]} \left(|w_t(t)| + |w(t)|_V\right) + \left(\int_0^T |w_t(t)|_{V_2}^2 dt\right)^{1/2}. \tag{2.2}$$

Definition 2.1. *We say that $w \in \mathcal{L}_T$ is a* ***weak solution*** *of the problem (1.1)–(1.3) if it satisfies the equation:*

$$\begin{aligned}\int_0^t \Big[&- \langle w_\tau(\tau), \eta_\tau(\tau)\rangle + \sigma_1(q)\left(w(\tau), \eta(\tau)\right) + \sigma_2(q)\left(w_\tau(\tau), \eta(\tau)\right) \\ &+ \langle g(q)\left(\mathcal{N}w(\tau)\right), \mathcal{N}\eta(\tau)\rangle\Big] d\tau + \langle w_t(t), \eta(t)\rangle \\ = \langle \varphi_1, \eta(0)\rangle &+ \int_0^t \langle f(\tau; q), \eta(\tau)\rangle_{V_2^*,V_2} d\tau,\end{aligned} \tag{2.3}$$

for any $t \in [0,T]$ and any $\eta \in \mathcal{L}_T$, as well as the initial condition

$$w(0) = \varphi_0. \tag{2.4}$$

Equivalently, w is a weak solution if

$$(2.5)\quad \langle w_{tt},\eta\rangle_{V^*,V}+\sigma_1(q)(w,\eta)+\sigma_2(q)(w_t,\eta)+\langle g(q)(\mathcal{N}w),\mathcal{N}\eta\rangle=\langle f(q),\eta\rangle_{V_2^*,V_2}$$

is satisfied for all $\eta \in \mathcal{L}_T$ and for almost all $t \in [0,T]$.

To establish our parameter estimation convergence results, we first make the following assumptions (these assumptions are the same as in [BGS] except that here we require them to be satisfied uniformly for all $q \in Q$) which will guarantee well-posedness for all $q \in Q$.

A1) The form $\sigma_1(q)$ is a Hermitian sesquilinear form: for $\varphi, \psi \in V$

$$(2.6)\qquad \sigma_1(q)(\varphi,\psi) = \overline{\sigma_1(q)(\psi,\varphi)} \ \text{ for every } \ q \in Q.$$

A2) The form $\sigma_1(q)$ is V bounded: for $\varphi, \psi \in V$

$$(2.7)\qquad |\sigma_1(q)(\varphi,\psi) \le c_1|\varphi|_V|\psi|_V \ \text{ for every } \ q \in Q.$$

A3) The form $\sigma_1(q)$ is strictly V coercive: for $\varphi \in V$

$$(2.8)\qquad Re\,\sigma_1(q)(\varphi,\varphi) = \sigma_1(q)(\varphi,\varphi) \ge k_1|\varphi|_V^2, \quad k_1 > 0$$

for every $q \in Q$.

A4) The form $\sigma_2(q)$ is V_2 bounded: for $\varphi, \psi \in V_2$

$$(2.9)\qquad |\sigma_2(q)(\varphi,\psi)| \le c_2|\varphi|_{V_2}|\psi|_{V_2} \ \text{ for every } \ q \in Q.$$

A5) The real part of $\sigma_2(q)$ is V_2 coercive and is symmetric:

$$(2.10)\qquad Re\,\sigma_2(q)(\varphi,\varphi) + \lambda_0|\varphi|^2 \ge k_2|\varphi|_{V_2}^2 \quad k_2 > 0, \lambda_0 \ge 0$$

$$(2.11)\qquad Re\,\sigma_2(q)(\varphi,\psi) = Re\,\sigma_2(q)(\psi,\varphi), \ \text{ for any } \ \varphi,\psi \in V_2, q \in Q.$$

A6) The forcing term $f(q)$ satisfies $f \in L^2([0,T],V_2^*)$ for every $q \in Q$.

A7) The operator $\mathcal{N}$ satisfies

$$(2.12)\qquad \mathcal{N} \in \mathcal{L}(V_2,H) \ \text{ with } |\mathcal{N}\varphi| \le \sqrt{\tilde{k}}\,|\varphi|_{V_2}$$

and the range of $\mathcal{N}$ on V is dense in H.

Note that (2.12) and $V \hookrightarrow V_2$ implies

$$(2.13)\qquad \mathcal{N} \in \mathcal{L}(V,H) \ \text{ with } \ |\mathcal{N}\varphi| \le \sqrt{k}\,|\varphi|_V$$

with $k = c^2\tilde{k}$.

A8) The nonlinear function $g(q) : H \to H$ is a continuous nonlinear mapping of real gradient (or potential) type. This means that there exists a continuous Frechet-differentiable nonlinear functional $G(q) : H \to \mathbb{R}^1$, whose Frechet derivative $G'(q)(\varphi) \in \mathcal{L}(H,\mathbb{R}^1)$ at any $\varphi \in H$ can be represented in the form

$$(2.14)\qquad G'(q)(\varphi)\psi = Re\langle g(q)(\varphi),\psi\rangle \quad \text{for any } \psi \in H.$$

We also require that there are constants C_1, C_2, C_3 and $\varepsilon > 0$ such that

$$-\frac{1}{2}k^{-1}(k_1-\varepsilon)|\varphi|^2 - C_1 \le G(q)(\varphi) \le C_2|\varphi|^2 + C_3, \tag{2.15}$$

for every $q \in Q$, where k is from (2.13) and k_1 from (2.8).

A9) The nonlinear function $g(q)$ also satisfies

$$|g(q)(\varphi)| \le \tilde{C}_1|\varphi| + \tilde{C}_2, \quad \varphi \in H, \tag{2.16}$$

for every $q \in Q$, for some constants $\tilde{C}_1$, $\tilde{C}_2$.

An additional condition is necessary for uniqueness of solutions.

A10) For any $\varphi \in H$ the Frechet derivative of $g(q)$ exists and satisfies

$$g'(q)(\varphi) \in \mathcal{L}(H,H) \text{ with } |g'(q)(\varphi)|_{\mathcal{L}(H,H)} \le \tilde{C}_3 \text{ for every } q \in Q. \tag{2.17}$$

A11) We assume that for any $u, v \in \mathcal{L}_T$, the following inequality is satisfied for any $t \in [0,T], q \in Q$:

$$\begin{aligned} &\int_0^t \Big\{ Re\langle g(q)(\mathcal{N}u(\tau)) - g(q)(\mathcal{N}v(\tau)), \mathcal{N}u(\tau) - \mathcal{N}v(\tau)\rangle \\ &\quad + k_1k^{-1}|\mathcal{N}u(\tau) - \mathcal{N}v(\tau)|^2\Big\}\,dt \\ &\quad + a\left(\left(\int_0^t |u(\tau)-v(\tau)|^2\,dt\right)^{1/2}\right) \ge 0, \end{aligned} \tag{2.18}$$

where $a(\xi) \ge 0$ is a continuous function in $\xi \ge 0$ such that
i) $a(0) = 0$,
ii) there exists a first derivative such that $a'(0) = 0$.
Note that (2.18) is satisfied if, for example,

$$Re\langle g(q)(\varphi) - g(q)(\psi), \varphi - \psi\rangle + k_1k^{-1}|\varphi - \psi|^2 \ge 0 \tag{2.19}$$

for any $\varphi, \psi \in H$, where k and k_1 are the constants in (2.8) and (2.13). Thus if $H = L^2(\Omega)$, $\Omega \subset \mathbb{R}^m$, so that $g(q) : \mathbb{R} \to \mathbb{R}$, then a sufficient condition for (2.19) is that $g'(q)(\xi) \ge -l_1$ for some $l_1 > 0$.

In [BGS] it is shown that

Theorem 2.1. *Under conditions A1)–A11) the system (1.1)–(1.3) has a unique weak solution $w \in \mathcal{L}_T$ for every initial condition $(\varphi_0, \varphi_1) \in V \times H$. The weak solution satisfies*

$$\langle w_{tt}, \eta\rangle_{V^*,V} + \sigma_1(q)(w,\eta) + \sigma_2(q)(w_t,\eta) + \langle g(q)(\mathcal{N}w), \mathcal{N}\eta\rangle = \langle f(q), \eta\rangle_{V_2^*,V_2} \tag{2.20}$$

for all $\eta \in \mathcal{L}_T, q \in Q$ and for almost all $t \in [0,T]$. Also, $w \in C_W([0,T],V_2)$, $w_t \in C_W([0,T],H)$ and the weak solution depends continuously on initial conditions.

3. The General Parameter Estimation Problem

Assume that we have a set of observations $z = \{z_i\}_{i=1}^K$ corresponding to measurements taken at time t_i. As stated in the Introduction we would like to find a solution for the least squares minimization problem, i.e., find $q \in Q$ that minimizes

$$J(q,z) = \left|\tilde{C}_2\left\{\tilde{C}_1\{w(t_i,\cdot;q)\} - \{z_i\}\right\}\right|^2, \tag{3.1}$$

where $\{w(t_i,\cdot;q)\}$ are the parameter dependent solutions of (1.1)–(1.3) evaluated at time $t_i, i = 1,2,\dots K$. To this end, we consider Galerkin type approximations to (1.1)–(1.3) and define a family of approximating parameter estimation problems.

Let H^N be finite dimensional subspaces of H and let Q^M be finite dimensional sets approximating (in a sense to be made precise below) the parameter set Q. Let P^N denote the orthogonal projections of H onto H^N. Then the approximate parameter estimation problems can be stated in the following way: find $q \in Q^M$ that minimizes

$$J^N(q,z) = \left|\tilde{C}_2\left\{\tilde{C}_1\{w^N(t_i,\cdot;q)\} - \{z_i\}\right\}\right|^2, \tag{3.2}$$

where $w^N(t;q) \in H^N$ is the solution to the finite dimensional approximation of (1.1)–(1.3) given by (1.8)–(1.9).

We make the following assumptions for the spaces H^N and H and the sets Q^M and Q (see [BSW]).

B1) The sets Q and Q^M lie in a metric space $\tilde{Q}$ with metric d. We assume that Q and Q^M are compact in this metric and there is a mapping $i^M : Q \to Q^M$ such that $Q^M = i^M(Q)$. Also, for each $q \in Q$, $i^M(q) \to q$ in $\tilde{Q}$ with the convergence uniform in $q \in Q$.

B2) The finite dimensional subspaces H^N satisfy $H^N \subset V$.

B3) For each $\psi \in V$, $|\psi - P^N\psi|_V \to 0$ as $N \to \infty$.

B4) For each $\psi \in V_2$, $|\psi - P^N\psi|_{V_2} \to 0$ as $N \to \infty$.

We also assume that $\mathcal{A}_1, \mathcal{A}_2, g, f$ depend continuously on the parameter $q \in Q$, i.e., they satisfy the following conditions:

C1) $|\sigma_1(q)(\phi,\psi) - \sigma_1(\tilde{q})(\phi,\psi)| \le \gamma_1 d(q,\tilde{q})|\phi|_V|\psi|_V$, for every $\phi,\psi \in V$.

C2) $|\sigma_2(q)(\xi,\eta) - \sigma_2(\tilde{q})(\xi,\eta)| \le \gamma_2 d(q,\tilde{q})|\xi|_{V_2}|\eta|_{V_2}$, for every $\xi,\eta \in V_2$.

C3) $|g(q)(\phi) - g(\tilde{q})(\phi)| \le \gamma_3 d(q,\tilde{q})|\phi|$ for all $\phi \in H$.

C4) The mapping $q \to f(\cdot;q)$ is continuous from Q to $L^2([0,T],V_2^*)$.

Under conditions A1)–A11), B1), C1)–C4) we know that a solution $\{\bar{q}^{N,M}\}$ to the approximate parameter estimation problem (1.7)–(1.9) and a solution $\bar{q}$ to the original parameter estimation problem for (3.1) exist. A general sufficient condition for the convergence of $\{\bar{q}^{N,M}\}$ to $\bar{q}$ is given in Theorem 5.1 of [BSW] (see also [BK]):

Theorem 3.1. *To obtain convergence of at least a subsequence of $\{\bar{q}^{N,M}\}$ to a solution $\bar{q}$ of minimizing (3.1) subject to (1.1)–(1.3), it suffices, under assumption B1), to argue that for arbitrary sequences $\{q^{N,M}\}$ in Q^M with $q^{N,M} \to q \in Q$, we have*

$$\tilde{C}_2\tilde{C}_1 w^N(t;q^{N,M}) \to \tilde{C}_2\tilde{C}_1 w(t;q). \tag{3.3}$$

4. Convergence Results

In this section we show that under our general conditions given above, the convergence criteria (3.3) of Theorem 3.1 holds for a reasonable class of observation operators $\tilde{C}_1, \tilde{C}_2$.

Theorem 4.1. *Suppose that assumptions A1)–A11), B1)–B4) and C1)–C4) are satisfied. Let q^N be arbitrary in Q^N such that $q^N \to q \in Q$ as $N \to \infty$. Then we have*

$$w^N(t, q^N) \to w(t, q) \text{ in } V, \text{ for all } t \geq 0$$

and

$$w_t^N(t, q^N) \to w_t(t, q) \text{ in } H \text{ for all } t \geq 0 \text{ and in } V_2 \text{ for almost all } t \geq 0,$$

where w^N, w_t^N satisfy

$$\begin{aligned} &\langle w_{tt}^N(t), \phi\rangle_{V^*,V} + \sigma_2(q^N)(w_t^N(t), \phi) + \sigma_1(q^N)(w^N(t), \phi) + \langle g(q^N)(\mathcal{N}w^N), \mathcal{N}\phi\rangle \\ &= \langle f(t, q^N), \phi\rangle_{V_2^*,V_2} \\ (4.1)\quad &w^N(0) = P^N\varphi_0\ , \ w_t^N(0) = P^N\varphi_1 \end{aligned}$$

for all $\phi \in H^N$, and w, w_t satisfy

$$\begin{aligned} &\langle w_{tt}(t), \phi\rangle_{V^*,V} + \sigma_2(q)(w_t(t), \phi) + \sigma_1(q)(w(t), \phi) + \langle g(q)(\mathcal{N}w), \mathcal{N}\phi\rangle \\ &= \langle f(t, q), \phi\rangle_{V_2^*,V_2} \\ (4.2)\qquad &w(0) = \varphi_0\ , \ w_t(0) = \varphi_1 \end{aligned}$$

for all $\phi \in V$.

Proof: We know that $w(t) \in V$, $w_t(t) \in H$ for all $t \geq 0$ and $w_t(t) \in V_2$ for almost all $t \geq 0$. By the triangle inequality

$$|w^N(t, q^N) - w(t, q)|_V \leq |w^N(t, q^N) - P^N w(t, q)|_V + |P^N w(t, q) - w(t, q)|_V.$$

By assumption B3) the second term on the right side goes to 0 as $N \to \infty$. So to prove our statement about $w^N(t, q^N)$ it is enough to show that the first term on the right side also goes to 0 as $N \to \infty$. Similarly,

$$|w_t^N(t, q^N) - w_t(t, q)|_{V_2} \leq |w_t^N(t, q^N) - P^N w_t(t, q)|_{V_2} + |P^N w_t(t, q) - w_t(t, q)|_{V_2}.$$

The last term again goes to 0 by B4), so to prove our statement it is enough to show that the first term also converges to zero.

Let us introduce the following notation:

$$w^N = w^N(t, q^N),\ w = w(t, q) \quad \text{and} \quad \Delta^N = w^N(t, q^N) - P^N w(t, q).$$

Then

$$\Delta_t^N = w_t^N - \frac{d}{dt}P^N w = w_t^N - P^N w_t,$$

$$\Delta_{tt}^N = w_{tt}^N - \frac{d^2}{dt^2}P^N w$$

since $w_t \in L^2([0,t],V_2)$.

From (4.1) and (4.2) we have that for every $\psi \in H^N$:

$$\begin{aligned}
&\langle \Delta_{tt}^N, \psi\rangle_{V^*,V} = \langle w_{tt}^N - w_{tt} + w_{tt} - \frac{d^2}{dt^2}P^N w, \psi\rangle_{V^*,V} \\
&= \langle f(q^N), \psi\rangle_{V_2^*,V_2} - \sigma_2(q^N)(w_t^N,\psi) - \sigma_1(q^N)(w^N,\psi) \\
&\quad - \langle g(q^N)\mathcal{N}w^N, \mathcal{N}\psi\rangle - \langle f(q),\psi\rangle_{V_2^*,V_2} + \sigma_2(q)(w_t,\psi) + \sigma_1(q)(w,\psi) \\
&\quad + \langle g(q)\mathcal{N}w, \mathcal{N}\psi\rangle + \langle w_{tt} - \frac{d^2}{dt^2}P^N w, \psi\rangle_{V^*,V}.
\end{aligned} \tag{4.3}$$

By adding and subtracting we obtain for all $\psi \in H^N$

$$\begin{aligned}
&\langle \Delta_{tt}^N, \psi\rangle_{V^*,V} + \sigma_1(q^N)(\Delta^N,\psi) = \langle w_{tt} - \frac{d^2}{dt^2}P^N w, \psi\rangle_{V^*,V} \\
&\quad - \langle f(q) - f(q^N), \psi\rangle_{V_2^*,V_2} \\
&\quad + \sigma_2(q^N)(w_t - P^N w_t, \psi) + \sigma_2(q)(w_t,\psi) - \sigma_2(q^N)(w_t,\psi) \\
&\quad + \sigma_1(q^N)(w - P^N w, \psi) + \sigma_1(q)(w,\psi) - \sigma_1(q^N)(w,\psi) \\
&\quad - \sigma_2(q^N)(\Delta_t^N,\psi) + \langle g(q)\mathcal{N}w, \mathcal{N}\psi\rangle - \langle g(q^N)\mathcal{N}w^N, \mathcal{N}\psi\rangle.
\end{aligned} \tag{4.4}$$

We may choose $\psi = \Delta_t^N$ since $\Delta_t^N \in H^N$. Then $\langle \Delta_{tt}^N, \Delta_t^N\rangle_{V^*,V} = \frac{1}{2}\frac{d}{dt}|\Delta_t^N|_V^2$. As in [BSW] we find:

$$\begin{aligned}
&\frac{d}{dt}\Big(\langle|\Delta_t^N|_V^2 + \sigma_1(q^N)(\Delta^N,\Delta^N)\Big) = 2\,Re\Big\{\langle w_{tt} - \frac{d^2}{dt^2}P^N w, \Delta_t^N\rangle_{V^*,V} \\
&\quad - \langle f(q) - f(q^N), \Delta_t^N\rangle_{V_2^*,V_2} \\
&\quad + \sigma_2(q^N)(w_t - P^N w_t, \Delta_t^N) + \sigma_2(q)(w_t,\Delta_t^N) - \sigma_2(q^N)(w_t,\Delta_t^N) \\
&\quad + \sigma_1(q^N)(w - P^N w, \Delta_t^N) + \sigma_1(q)(w,\Delta_t^N) - \sigma_1(q^N)(w,\Delta_t^N) \\
&\quad - \sigma_2(q^N)(\Delta_t^N,\Delta_t^N) + \langle g(q)\mathcal{N}w, \mathcal{N}\Delta_t^N\rangle - \langle g(q^N)\mathcal{N}w^N, \mathcal{N}\Delta_t^N\rangle\Big\}.
\end{aligned} \tag{4.5}$$

We denote the left side of (4.5) by $L(t)$ and the right side by $R(t)$. Integrating L from 0 to t, using initial conditions

$$\Delta^N(0) = w^N(0) - P^N w(0) = w^N(0) - P^N\varphi_0 = 0$$

and

$$\Delta_t^N(0) = w_t^N(0) - P^N w_t(0) = w_t^N(0) - P^N\varphi_1 = 0,$$

along with A3), we have

$$\int_0^t L(s)ds \geq |\Delta_t^N|_V^2 + k_1|\Delta^N|_V^2. \tag{4.6}$$

We next argue that

$$\int_0^t R(s)ds \le \mu_1 \tilde{\delta}^N(t) + \mu_2 \int_0^t |\Delta_t^N|_V^2 + k_1 |\Delta^N|_V^2 ds, \tag{4.7}$$

where $\tilde{\delta}^N(t) \to 0$ as $N \to \infty$ and μ_1, μ_2 are positive constants. Then by Gronwall's inequality we obtain that

$$k_1 |\Delta^N(t)|_V^2 + |\Delta_t^N(t)|_V^2 \to 0 \text{ as } N \to \infty \text{ for each } t \ge 0,$$

which implies the desired results.

Proceeding as in [BSW], we have that

$$\begin{aligned} \int_0^t R(s)ds \le \nu_1 \delta^N(t) + \nu_2 \int_0^t |\Delta_t^N|_V^2 + k_1 |\Delta^N|_V^2 ds \\ + 2\,Re \int_0^t \left(\langle g(q)\mathcal{N}w, \mathcal{N}\Delta_t^N\rangle - \langle g(q^N)\mathcal{N}w^N, \mathcal{N}\Delta_t^N\rangle \right) ds, \end{aligned} \tag{4.8}$$

where

$$\begin{aligned} \delta^N(t) = \int_0^t Re \Big\{ \langle w_{tt} - \frac{d^2}{dt^2} P^N w, \Delta_t^N \rangle_{V^*,V} + |f(q) - f(q^N)|_{V_2^*}^2 \\ + c_2^2 |w_t - P^N w_t|_{V_2}^2 + \gamma_2^2 d^2(q, q^N) |w_t|_{V_2}^2 \\ + c_2 |w - P^N w|_V^2 + \gamma_1^2 d^2(q, q^N) |w|_V^2 \Big\} ds, \end{aligned} \tag{4.9}$$

and $\delta^N(t) \to 0$ as $N \to \infty$ by (5.18) of [BSW], B3), B4), properties of w, w_t and assumptions of the theorem. Finally, we need to show that the last integral containing the terms involving g can also be estimated from above by an expression similar to the right side of (4.7). We may argue

$$\begin{aligned} &|\int_0^t \langle g(q)\mathcal{N}w, \mathcal{N}\Delta_t^N\rangle - \langle g(q^N)\mathcal{N}w^N, \mathcal{N}\Delta_t^N\rangle ds| \\ &\le |\int_0^t \langle g(q)\mathcal{N}w - g(q^N)\mathcal{N}w, \mathcal{N}\Delta_t^N\rangle + \langle g(q^N)\mathcal{N}w - g(q^N)\mathcal{N}w^N, \mathcal{N}\Delta_t^N\rangle ds| \\ &\le \frac{1}{2} \int_0^t (\gamma_3^2 d^2(q, q^N) k |w|_V^2 + k |\Delta_t^N|_V^2) ds \\ &\quad + |\int_0^t \langle g(q^N)\mathcal{N}w - g(q^N)\mathcal{N}P^N w, \mathcal{N}\Delta_t^N\rangle ds| \\ &\quad + |\int_0^t \langle g(q^N)\mathcal{N}P^N w - g(q^N)\mathcal{N}w^N, \mathcal{N}\Delta_t^N\rangle ds|. \end{aligned} \tag{4.10}$$

Now the first integral on the right is dominated by the right side of (4.7) (with suitably chosen constants). To estimate the last two integrals we use the same method as in

[BGS]. We have

$$
\begin{aligned}
&|\int_0^t \langle g(q^N)\mathcal{N}w - g(q^N)\mathcal{N}P^N w, \mathcal{N}\Delta_t^N\rangle ds| \\
&= |\int_0^t \langle \int_0^1 g'(\theta\mathcal{N}w(s) + (1-\theta)\mathcal{N}P^N w(s))[\mathcal{N}w(s) - \mathcal{N}P^N w(s)]d\theta, \mathcal{N}\Delta_t^N\rangle ds| \\
&\le \int_0^t \tilde{C}_3|\mathcal{N}w - \mathcal{N}P^N w||\mathcal{N}\Delta_t^N|ds \le \frac{1}{2}\int_0^t (\tilde{C}_3^2|\mathcal{N}w - \mathcal{N}P^N w|^2 + |\mathcal{N}\Delta_t^N|^2)ds \\
&\le \frac{1}{2}k\int_0^t \tilde{C}_3^2|w - P^N w|_V^2 ds + \frac{1}{2}k\int_0^t |\Delta_t^N|_V^2 ds.
\end{aligned} \tag{4.11}
$$

Here the first term in the last expression goes to 0 as $N \to \infty$ by B3) and properties of w, while the second is dominated by second term on the right side of (4.7). Similarly,

$$
\begin{aligned}
&|\int_0^t \langle g(q^N)\mathcal{N}P^N w - g(q^N)\mathcal{N}w^N, \mathcal{N}\Delta_t^N\rangle ds| \\
&\le \frac{1}{2}k\int_0^t \tilde{C}_3^2|P^N w - w^N|_V^2 + |\Delta_t^N|_V^2 ds \le \frac{1}{2}k\int_0^t \tilde{C}_3^2|\Delta^N|_V^2 + |\Delta_t^N|_V^2 ds,
\end{aligned} \tag{4.12}
$$

which again is dominated by the right side of (4.7). This completes the required arguments. □

We note that the above theorem gives a computationally tractable method to solve the parameter estimation problem involving (3.1) in case the data collected consists of displacement or velocity measurements, i.e., $\tilde{C}_1$ is either the identity or differentiation with respect to time once followed by evaluation in t and x. However, the case of accelerometer data is more complicated, since then Theorem 3.1 requires $w_{tt}^N(t;q^N) \to w_{tt}(t;q)$ in V^* for $t \in [0,T]$. We will now give conditions under which this convergence can be obtained.

Let us suppose that $V_2 = V$, i.e., we have strong damping, such as Kelvin-Voigt damping in the example given in the Introduction. We can formulate the system (1.1)–(1.3) in variational form (4.2) and rewrite it in first order vector form on $\mathcal{H} = V \times H$ in the coordinates

$$
z = \begin{pmatrix} z_1 \\ z_2 \end{pmatrix} = \begin{pmatrix} w \\ w_t \end{pmatrix}.
$$

We define $\mathcal{V} = V \times V$ and $\sigma(q) : \mathcal{V} \times \mathcal{V} \to \mathbb{C}$ by

$$
\sigma(q)\left(\begin{pmatrix} \xi \\ \eta \end{pmatrix}, \begin{pmatrix} \phi \\ \psi \end{pmatrix}\right) = -\langle \eta, \phi\rangle_V + \sigma_1(q)(\xi, \psi) + \sigma_2(q)(\eta, \psi).
$$

Then (4.2) can be rewritten as

$$
\langle z_t, \Phi\rangle + \sigma(q)(z, \Phi) = \langle F(q), \Phi\rangle \text{ for all } \Phi \in \mathcal{V} \tag{4.13}
$$

$$
z(0) = \begin{pmatrix} \varphi_0 \\ \varphi_1 \end{pmatrix}, \tag{4.14}
$$

where

$$
F(q) = \begin{pmatrix} 0 \\ f(q) - \mathcal{N}^* g(q)(\mathcal{N}z_1) \end{pmatrix}.
$$

We can also write this in the equivalent operator form (not distinguishing between row and column vectors):

$$z_t = A(q)z + F(q) \tag{4.15}$$

$$z(0) = \begin{pmatrix} \varphi_0 \\ \varphi_1 \end{pmatrix} \tag{4.16}$$

where $\sigma(q)(\Phi, \Psi) = \langle -A(q)\Phi, \Psi \rangle_{\mathcal{V}^*, V}$ with

$$A(q) = \begin{bmatrix} 0 & I \\ -\mathcal{A}_1(q) & -\mathcal{A}_2(q) \end{bmatrix}.$$

It is shown in [BIW, BSW] that if $V = V_2$, then $A(q)$ generates an analytic semigroup $S(t; q)$ on $\mathcal{V}^* = V \times V^*$. Then the weak solution of (1.1)–(1.3) can be represented in the form:

$$z(t;q) = \begin{pmatrix} w(t;q) \\ w_t(t;q) \end{pmatrix} = S(t;q)\begin{pmatrix} \varphi_0 \\ \varphi_1 \end{pmatrix} + \int_0^t S(t-\tau;q)F(\tau;q)d\tau. \tag{4.17}$$

Letting $\mathcal{H}^N = H^N \times H^N$, we can restrict $\sigma(q)$ to $\mathcal{H}^N \times \mathcal{H}^N$, denote this restriction by σ^N and define $A^N(q) : \mathcal{H}^N \to \mathcal{H}^N$ by $\langle A^N(q)\Phi, \Psi \rangle = \sigma^N(\Phi, \Psi)$ for all $\Phi, \Psi \in H^N \times H^N$. Then A^N generates an analytic semigroup $S^N(t; q)$ on $\mathcal{H}^N$. Solutions of (1.8)–(1.9) can then be represented as:

$$z^N(t;q) = \begin{pmatrix} w^N(t;q) \\ w_t^N(t;q) \end{pmatrix} = S^N(t;q)\begin{pmatrix} P^N\varphi_0 \\ P^N\varphi_1 \end{pmatrix} + \int_0^t S^N(t-\tau;q)P^N F(\tau;q)d\tau, \tag{4.18}$$

where $P^N F$ is understood to mean $\begin{pmatrix} P^N F_1 \\ P^N F_2 \end{pmatrix}$ if F_i are the components of F and $P^N : V^* \to H^N$ is the generalized projection (in the sense of the duality product). We can then use the theory developed in [BR] to obtain $z_t^N(t; q^N) \to z_t(t; q)$ when $q^N \to q$. Note that this will guarantee that $w_{tt}^N(t; q^N) \to w_{tt}(t; q)$ in V^* for $t \in [0, T]$ (which is what we desired). According to Theorem 3.1, 3.2 in [BR] and the remarks following them, this convergence is guaranteed if we can argue convergence of (4.18) to (4.17) after differentiation of these terms with respect to time. Since we have analytic semigroups we obtain this property if $F(q) \in \tilde{L}^\infty([0, T], \mathcal{V}^*)$ (i.e., pointwise defined and bounded $\mathcal{V}^*$ valued functions) and $P^N\varphi \to \varphi$ in V^*. Thus, we can state the following theorem:

Theorem 4.2. *Let $V = V_2$ and $f(q) \in \tilde{L}^\infty([0, T], V^*)$ in the system (1.1)-(1.3). Let A1)–A11), B1)–B4), C1)–C4) hold. Morover, assume that $P^N\varphi \to \varphi$ in V^* for $\varphi \in V^*$. Then for any $q^N \to q \in Q$ we have $w_{tt}^N(t; q^N) \to w_{tt}(t; q)$ in V^* for $t \in [0, T]$.*

Proof: Using the arguments in [BR] with (4.17), 4.18 we only need to argue that $f(\cdot; q) - \mathcal{N}^* g(q)(\mathcal{N} z_1) \in \tilde{L}^\infty([0, T], V^*)$. But this follows from the fact that $f(q) \in \tilde{L}^\infty([0, T], V^*)$, $z = w \in C_W([0, T], V_2)$ and $\mathcal{N} \in \mathcal{L}(V_2, H)$. □

Acknowledgement: This research was supported in part by the Air Force Office of Scientific Research under grant AFOSR F49620-95-1-0236.

References

[BGS] H.T. Banks, D.S. Gilliam, and V.I. Shubov, Global solvability for damped nonlinear hyperbolic systems, *CRSC-TR95-25*, August, 1995; *Differential and Integral Equations* **10** (1997), pp. 309–332.

[BIW] H.T. Banks, K. Ito, and Y. Wang, Well-posedness for damped second order systems with unbounded input operators, *CRSC-TR93-10*, June 1993; *Differential and Integral Equations*, **8**(1995), pp. 587–606.

[BK] H.T. Banks and K. Kunisch, *Estimation Techniques for Distributed Parameter Systems*, Birkhäuser, Boston, 1989.

[BL] H.T. Banks and N.J. Lybeck, Modeling methodology for elastomer dynamics, CRSC-TR96-29, September, 1996; *Systems and Control in the 21st Century*, (ed. by C. Byrnes, B. Datta, D. Gilliam and C. Martin), Birkhäuser, Boston, 1996, pp. 37–50.

[BLGMY] H.T. Banks, N.J. Lybeck, M.J. Gaitens, B.C. Munoz, and L.C. Yanyo, Modeling the dynamic mechanical behavior of elastomers, CRSC-TR96-26, September, 1996; *Rubber Chem and Tech.*, submitted.

[BLMY] H.T. Banks, N.J. Lybeck, B.C. Munoz, and L.C. Yanyo, Nonlinear elastomers: Modeling and estimation, *CRSC-TR95-19*, May, 1995; Proc. 3rd IEEE Mediterranean Symp. on New Directions in Control, Limassol, Cyprus, July, 1995, Vol. 1. pp. 1–7.

[BR] H.T. Banks and D.A. Rebnord, Analytic semigroups: Applications to inverse problems for flexible structures, in *Differential Equations with Applications in Biology, Physics and Engineering*, Marcel Dekker Inc., New York, 1991, pp. 21–35.

[BSW] H.T. Banks, R. C. Smith and Y. Wang, *Smart Material Structures, Modeling, Estimation and Control*, Masson/Wiley, Paris/Chichester, 1996.

[Li1] J.L. Lions, *Optimal Control of Systems Governed by Partial Differential Equations*, Springer-Verlag, New York, 1971.

[P] A. Pazy, *Semigroups of Linear Operators and Applications to Partial Differential Equations*, Springer-Verlag, New York, 1983.

[Wl] J. Wloka, *Partial Differential Equations*, Cambridge Univ. Press, 1992.

H.T. Banks
Center for Research in Scientific Computation
North Carolina State University
Raleigh, NC 27695-8205, USA

Gabriella A. Pinter
Department of Mathematics
Texas Tech University
Lubbock, TX 79409, USA

International Series of Numerical Mathematics
Vol. 126, © 1998 Birkhäuser Verlag, Basel

Preconditioners for Karush–Kuhn–Tucker Matrices Arising in the Optimal Control of Distributed Systems

A. BATTERMANN AND M. HEINKENSCHLOSS*

FB IV–Mathematik
Universität Trier

Department of Computational and Applied Mathematics
Rice University

ABSTRACT. In this paper preconditioners for linear systems arising in interior–point methods for the solution of distributed control problems are derived and analyzed. The matrices K in these systems have a block structure with blocks obtained from the discretization of the objective function and the governing differential equation. The preconditioners have a block structure with blocks being composed of preconditioners for the subblocks of the system matrix K. The effectiveness of the preconditioners is analyzed and numerical examples for an elliptic model problem are shown.

1991 *Mathematics Subject Classification.* 49M30, 49N10, 90C06, 90C20

Key words and phrases. Preconditioners, iterative methods, interior point methods, linear quadratic optimal control problems.

1. Introduction

The discretization of distributed linear quadratic optimal control problems with bound constraints on the controls and on the states leads to large scale quadratic programming problems. Because of their complexity and convergence properties, interior point methods are attractive solvers for such problems. They are iterative methods which in each iteration generate approximations to solutions that are strictly feasible with respect to the bound constraints. Within each iteration, large indefinite linear systems have to be solved. If interior point methods are applied to linear quadratic control problems governed by partial differential equations, then iterative techniques usually have to be applied to solve these linear systems. To make interior point methods efficient, it is important to solve these linear systems efficiently. Krylov subspace methods are iterative linear system solvers, which are very suitable in this context. They do not require the system matrix in explicit form, but only require matrix vector multiplications. This is very useful since for the problems under investigation the system matrices have a block structure in which blocks are related to discretized differential equations. The convergence of Krylov subspace methods depends on the distribution of the eigenvalues of the system matrix. Roughly speak-

*This author was supported by the NSF DMS–9403699, AFOSR F49620–93–1–0280, and in part by the DoE DE–FG03–95ER25257.

ing, their convergence is the better the more the eigenvalues of the system matrix are clustered and the smaller the clusters are. Ill–conditioning of the matrix, i.e. a large quotient of largest absolute eigenvalue divided by smallest absolute eigenvalue, typically corresponds to a poor convergence of Krylov subspace methods. To improve the convergence of these methods nonsingular matrices are constructed so that the similarity transformation with these matrices leads to a system with better clustered eigenvalues. These matrices are called preconditioners. The purpose of this paper is the construction of such preconditioners for systems arising in interior–point methods for certain distributed control problems.

To illustrate the issues, we consider the following elliptic model problem.

$$\min \frac{1}{2}\int_{\Omega}(y(x)-y_d(x))^2+\frac{\gamma}{2}\int_{\partial\Omega}u^2(x)ds \tag{1.1}$$

over all (y,u) satisfying the state equation

$$\begin{aligned} -\Delta y(x)+y(x) &= f(x) \qquad x\in\Omega, \\ \frac{\partial}{\partial n}y(x) &= u(x) \qquad x\in\partial\Omega \end{aligned} \tag{1.2}$$

and the bound constraints

$$\begin{aligned} y_{low} \le y(x) \le y_{upp} \quad \text{a.e.}, \\ u_{low} \le u(x) \le u_{upp} \quad \text{a.e.} \end{aligned} \tag{1.3}$$

A discretization of the problem with, say, finite elements, leads to a quadratic programming problem of the form

$$\min \frac{1}{2}y_h^T M_y y_h + \frac{\gamma}{2}u_h^T M_u u_h + c^T y_h + d^T u_h \tag{1.4}$$

subject to

$$Ay_h + Bu_h = b, \tag{1.5}$$

$$\begin{aligned} y_{h,low} \le y_h \le y_{h,upp}, \\ u_{h,low} \le u_h \le u_{h,upp}. \end{aligned} \tag{1.6}$$

Here h indicates the mesh size of the discretization and $u_h \in \mathbb{R}^{n_u}$, $y_h \in \mathbb{R}^{n_y}$ represent the discretized controls and states, respectively. The matrices $M_u \in \mathbb{R}^{n_u\times n_u}$ and $M_y \in \mathbb{R}^{n_y\times n_y}$ are positive definite. The vectors $y_{h,low},\dots,u_{h,upp}$ are obtained from the bound constraints (1.3) in a straightforward way.

There are various classes of interior point methods. They all (after possible transformations) require the solution of linear systems with system matrices

$$K=\begin{pmatrix} H_y & 0 & A^T \\ 0 & H_u & B^T \\ A & B & 0 \end{pmatrix}, \tag{1.7}$$

where

$$H_y = M_y + D_y, \quad H_u = \gamma M_u + D_u, \tag{1.8}$$

with some positive semidefinite diagonal matrices D_y and D_u. Since the matrix K is related to matrices arising in the Karush–Kuhn–Tucker optimality conditions, we call K a Karush–Kuhn–Tucker (KKT) matrix.

Even though the exact form of the diagonal matrices D_y and D_u differs from interior point method to interior point method, they all have in common that diagonals of D_y and D_u grow unbounded if the corresponding components of y_h or u_h converge towards a bound.

The matrices K arising in interior–point methods for the solution of problems like (1.1)–(1.3) are usually ill–conditioned. There are at least two sources for the ill-conditioning. One source is the discretization of the infinite dimensional problem. Typically, the eigenvalues of K spread out towards zero if the discretization is refined. The second source are the large diagonals in D_y and D_u that arise if variables approach the bound. This source is due to the interior–point method. Ill–conditioning also arises if the original infinite dimensional problem is ill–posed. The preconditioners derived in this paper are designed to remedy the ill–conditioning arising from the first two sources. They use the block structure of K and are composed of preconditioners for the blocks M_y, M_u, and A of K. This allows the use of known preconditioners for the governing differential equations. Moreover, computationally expensive parts of the preconditioner have to be computed only once during the interior–point method, since only the diagonal contributions D_y and D_u change from one interior–point iteration to another.

Preconditioners for problems related to this one are investigated in other papers. There are several papers, e.g. [5], [15], [17], investigating preconditioners for systems arising in the numerical solution of partial differential equations such as the Stokes equations, or the biharmonic equation. These systems can also be viewed as KKT systems. However, the blocks in those matrices are different and, therefore, the preconditioners for those problems are different than the ones introduced here. In fact, if the governing equations would be the Stokes equations, or the biharmonic equation, then the preconditioners in the papers cited above could be used as blocks in the preconditioners introduced here. Some of the tools provided in those papers, in particular a result from [15], cf. Lemma 5.1, are heavily used in our analysis. Preconditioners for interior–point methods for linear programs (LP) are investigated in [9], [10]. Those preconditioners are for general LPs and are based on sparse matrix factorizations or on the SOR method. Since no particular structure is assumed, those papers do not contain any theoretical result on the quality of the preconditioner.

This paper is organized as follows. In the first part we study the QP problem. Section 2 investigates the problem (1.1)–(1.3) and its discretization. The Sections 3 and 4 discuss the optimality conditions for the QP (1.4)–(1.6) and some aspects of interior–point methods relevant for the construction of preconditioners. Section 5 contains some essential results about the Krylov subspace methods MINRES and SYMMLQ. The preconditioners are introduced and analyzed in Section 6. This section also contains some numerical tests demonstrating the quality of the preconditioners.

2. The Control Problem

As noted in the introduction, one source of ill–conditioning in the KKT matrix is the discretization of the infinite dimensional problem. This section provides some results needed to address this aspect of the problem. These results can be proven for a general class of problems, which include the model problem (1.1)–(1.3) as a special case. In this section we do not consider the control or the state constraints.

2.1. The Abstract Problem. Let $\mathcal{Y}$ and $\mathcal{U}$ be Hilbert spaces. These spaces play the role of the state and the control space, respectively. Moreover, let a, b be continuous bilinear forms on $\mathcal{Y} \times \mathcal{Y}$ and $\mathcal{U} \times \mathcal{Y}$, respectively. In addition, we assume that a is $\mathcal{Y}$–elliptic. In particular, there exist constants $\alpha > 0$ and $\beta > 0$ with

$$\alpha \|y\|_{\mathcal{Y}}^2 \leq a(y,y), \quad b(u,y) \leq \beta \|u\|_{\mathcal{U}} \|y\|_{\mathcal{Y}}, \quad \forall y \in \mathcal{Y},\ u \in \mathcal{U}.$$

Furthermore, let $\mathcal{Z}$ be a Hilbert space and $\mathcal{C} \in \mathcal{L}(\mathcal{Y}, \mathcal{Z})$. In particular there exists $\zeta > 0$ such that

$$\|\mathcal{C}y\|_{\mathcal{Z}} \leq \zeta \|y\|_{\mathcal{Y}} \quad \forall y \in \mathcal{Y}.$$

With some linear functional l on $\mathcal{Y}$ we consider the problem

$$\text{(2.1)} \qquad \min \quad \frac{1}{2}\|\mathcal{C}y - z_d\|_{\mathcal{Z}}^2 + \frac{\gamma}{2}\|u\|_{\mathcal{U}}^2,$$

$$\text{(2.2)} \qquad \text{s.t.} \quad a(y,v) + b(u,v) = l(v) \quad \forall v \in \mathcal{Y}.$$

Results on the existence of solutions for problems like (2.1), (2.2) are given e.g. in [1], [13] and we refer to those books.

We consider the following discretizations. Let

$$\mathcal{Y}_h = \text{span}\{\phi_1, \ldots, \phi_{n_y}\} \subset \mathcal{Y}, \quad \mathcal{U}_h = \text{span}\{\psi_1, \ldots, \psi_{n_u}\} \subset \mathcal{U},$$

and define matrices $A \in \mathbb{R}^{n_y \times n_y}$ and $B \in \mathbb{R}^{n_y \times n_u}$ by

$$\begin{aligned} A_{ij} &= a(\phi_j, \phi_i), \quad i,j = 1, \ldots, n_y, \\ B_{ij} &= b(\psi_j, \phi_i), \quad j = 1, \ldots, n_u, \quad i = 1, \ldots, n_y, \end{aligned}$$

and matrices $M_y \in \mathbb{R}^{n_y \times n_y}$ and $M_u \in \mathbb{R}^{n_u \times n_u}$ by

$$\begin{aligned} (M_y)_{ij} &= \langle \mathcal{C}\phi_j, \mathcal{C}\phi_i \rangle_{\mathcal{Z}}, \quad i,j = 1, \ldots, n_y, \\ (M_u)_{ij} &= \langle \psi_j, \psi_i \rangle_{\mathcal{U}}, \quad i,j = 1, \ldots, n_u. \end{aligned}$$

Obviously,

$$y_h^T M_y y_h = \|\sum_{i=1}^{n_y} y_{h,i} \mathcal{C}\phi_i\|_{\mathcal{Z}}^2, \quad u_h^T M_u u_h = \|\sum_{i=1}^{n_u} u_{h,i} \psi_i\|_{\mathcal{U}}^2.$$

In particular, the matrix M_u is positive definite and the matrix M_y is positive semidefinite.

By $\|\cdot\|$ we denote the Euclidean norm in $\mathbb{R}^k$ for some k. We can show the following simple, but important result.

Lemma 2.1. *There exists a constant $c > 0$, independent of the discretization parameter h, such that*

$$\|M_y^{1/2} A^{-1} B M_u^{-1/2}\| \leq c.$$

Proof. Let $u_h \neq 0$ be arbitrary and set $y_h = A^{-1}BM_u^{-1/2}u_h$, $\tilde{u}_h = M_u^{-1/2}u_h$. Define $y = \sum_{i=1}^{n_y} y_{h,i}\phi_i$ and $\tilde{u} = \sum_{i=1}^{n_u} \tilde{u}_{h,i}\psi_i$. By definition of A and B, $a(y,\phi_i) = b(\tilde{u},\phi_i)$, $i = 1,\dots,n_y$. Hence,

$$\alpha\|y\|_{\mathcal{Y}}^2 \leq a(y,y) = b(\tilde{u},y) \leq \beta\|\tilde{u}\|_{\mathcal{U}}\|y\|_{\mathcal{Y}}.$$

This implies

$$\begin{aligned}\frac{\|M_y^{1/2}A^{-1}BM_u^{-1/2}u_h\|^2}{\|u_h\|^2} &= \frac{\|M_y^{1/2}y_h\|^2}{\|u_h\|^2} = \frac{\|\mathcal{C}y\|_{\mathcal{Z}}^2}{\|u_h\|^2}\\ &\leq \frac{\zeta^2\|y\|_{\mathcal{Y}}^2}{\|M_u^{1/2}\tilde{u}_h\|^2} = \frac{\zeta^2\|y\|_{\mathcal{Y}}^2}{\|\tilde{u}\|_{\mathcal{U}}^2} \leq \frac{\zeta^2\beta^2}{\alpha^2}.\end{aligned}$$

□

2.2. The Model Problem. The model problem (1.1)–(1.3) fits into the above framework, if we use the weak formulation of (1.2). The Hilbert spaces are $\mathcal{Y} = H^1(\Omega)$, $\mathcal{U} = L^2(\partial\Omega)$, and $\mathcal{Z} = L^2(\Omega)$. The bilinear forms and the functional are $a(y,v) = \int_\Omega \nabla y(x)\nabla v(x) + y(x)v(x)dx$, $b(u,v) = -\int_{\partial\Omega} u(x)v(x)dx$, and $l(v) = \int_\Omega f(x)v(x)dx$. The operator $\mathcal{C}$ is the imbedding operator. For our discretization we use a finite element discretization with piecewise linear functions over triangles. In our numerical experiments we use $\Omega = (0,1)^2$ and we construct the triangulation as follows: The $x-$ and $y-$ intervals are subdivided into d_x and d_y subintervals. The resulting rectangles are subdivided into two triangles by connecting the lower left corner and the upper right corner of the rectangle. Since piecewise linear approximations are used, the number of state variables is $n_y = (d_x+1)(d_y+1)$ and the number of controls is $n_u = 2(d_x + d_y)$.

3. The Quadratic Programming Problem

We consider the following quadratic programming problem (QP) in standard form:

$$\text{Minimize } \frac{1}{2}\begin{pmatrix} y \\ u \end{pmatrix}^T \begin{pmatrix} M_{yy} & M_{yu} \\ M_{uy} & M_{uu} \end{pmatrix}\begin{pmatrix} y \\ u \end{pmatrix} + \begin{pmatrix} c \\ d \end{pmatrix}^T\begin{pmatrix} y \\ u \end{pmatrix} \tag{3.1}$$

subject to

$$Ay + Bu = b, \tag{3.2}$$

$$y \geq 0,\ u \geq 0. \tag{3.3}$$

In this section the origin of the QP is not important and we omit the subscript h. Moreover, we absorb γ into M_{uu}. The standard form (3.1)–(3.3) is considered to reduce the complexity of notation. Using straightforward extensions, bound constraints of the form (1.6) can be handled as well. Throughout this section we use the notation

$$M = \begin{pmatrix} M_{yy} & M_{yu} \\ M_{uy} & M_{uu} \end{pmatrix},\quad g = \begin{pmatrix} c \\ d \end{pmatrix},\quad C = (A \mid B),\quad x = \begin{pmatrix} y \\ u \end{pmatrix},\quad q = \begin{pmatrix} q_y \\ q_u \end{pmatrix}.$$

We limit our discussion to convex problems and assume that M is positive semidefinite. The existence of solutions of the QP (3.1)–(3.3) is guaranteed if the objective function is bounded from below on the set of feasible points. More precisely, we have the following well-known result (e.g. [6, § 12.3]):

Theorem 3.1 (Necessary and Sufficient Optimality Conditions). *If M is positive semidefinite and if $q(x) = \frac{1}{2}x^T Mx + g^T x$ is bounded from below on the set of feasible points $\{(y,u) | Ay + Bu = b,\ y \geq 0, u \geq 0\}$, then the QP (3.1)–(3.3) admits a solution x_*. If M is positive definite, the QP admits a unique solution.*

The vector (y, u) is a solution of (3.1)–(3.3) if and only if there exist $p \in \mathbb{R}^{n_y}$, $q_y \in \mathbb{R}^{n_y}$, and $q_u \in \mathbb{R}^{n_u}$ such that the Karush–Kuhn–Tucker (KKT) conditions

$$
\begin{aligned}
M_{yy}y + M_{yu}u + A^T p - q_y &= -d, \\
M_{uy}y + M_{uu}u + B^T p - q_u &= -c, \\
Ay + Bu &= b, \\
y^T q_y + u^T q_u &= 0, \\
q_y, q_u &\geq 0, \\
y, u &\geq 0
\end{aligned}
\tag{3.4}
$$

are satisfied.

To learn more about the QP and the optimality system (3.4) it will be helpful to distinguish three cases. This discussion will also help us to relate the results in this paper to the results on the solution of KKT systems in interior–point methods for linear programming that can be found in the literature, see e.g. [10].

Throughout this subsection we assume that A is nonsingular and that the QP has a solution. As a consequence, the matrix $C = (A \mid B)$ has full row rank and the KKT system (3.4) has a solution.

Bound constraints for u and y. Let (y_*, u_*) be a solution of the QP. Furthermore, let $\{l_1^u, \dots, l_{k_u}^u\}$ and $\{l_1^y, \dots, l_{k_y}^y\}$ denote the set of active indices for u_* and y_*, respectively,

$$\{l_1^u, \dots, l_{k_u}^u\} = \{i \mid (u_*)_i = 0\}, \quad \{l_1^y, \dots, l_{k_y}^y\} = \{i \mid (y_*)_i = 0\}.$$

The Lagrange multipliers at the solution satisfy

$$(q_y)_i = 0,\ i \notin \{l_1^y, \dots, l_{k_y}^y\} \quad \text{and} \quad (q_u)_i = 0,\ i \notin \{l_1^u, \dots, l_{k_u}^u\}.$$

If we define the matrices $I(y_*) \in \mathbb{R}^{k_y \times n_y}$, $I(u_*) \in \mathbb{R}^{k_u \times n_u}$ by

$$(I(y_*))_{ij} = \begin{cases} 1 & \text{if } j = l_i^y, \\ 0 & \text{otherwise,} \end{cases} \qquad \text{and} \qquad (I(u_*))_{ij} = \begin{cases} 1 & \text{if } j = l_i^u, \\ 0 & \text{otherwise,} \end{cases}$$

then the KKT conditions (3.4) are equivalent to

$$
\begin{pmatrix}
M_{yy} & M_{yu} & A^T & I(y_*)^T & 0 \\
M_{uy} & M_{uu} & B^T & 0 & I(u_*)^T \\
A & B & 0 & 0 & 0 \\
I(y_*) & 0 & 0 & 0 & 0 \\
0 & I(u_*) & 0 & 0 & 0
\end{pmatrix}
\begin{pmatrix} y \\ u \\ p \\ q_y^a \\ q_u^a \end{pmatrix}
=
\begin{pmatrix} -c \\ -d \\ b \\ 0 \\ 0 \end{pmatrix},
\tag{3.5}
$$

where q_y^a, q_u^a denote the Lagrange multipliers corresponding to the active indices.

Let l denote the number of positive components in the solution (y_*, u_*) of the QP. The assumption that A is nonsingular is not sufficient to guarantee that the matrix

$$\widehat{C} = \begin{pmatrix} A & B \\ I(y_*) & 0 \\ 0 & I(u_*) \end{pmatrix} \in \mathbb{R}^{(n_y+(n_y+n_u-l))\times(n_y+n_u)} \tag{3.6}$$

has full row rank. If $\widehat{C}$ does not have full rank, then the system (3.5) does not have a unique solution, even if the QP has a unique solution (y_*, u_*). It is not difficult to see that in this case the Lagrange multipliers (p, q_y^a, q_u^a) are not uniquely determined.

If $M = 0$, then the QP reduces to an LP. In this case the solution of the optimization problem can be found in a vertex (y_*, u_*). Recall that a feasible point (y, u) is called a vertex if the columns of $C = (A \mid B)$ corresponding to the positive components are linearly independent, see e.g. [6, § 2]. If (y_*, u_*) is a vertex, at most n_y components of (y_*, u_*) can be positive and the columns of $C = (A \mid B)$ corresponding to the positive components of the vertex (y_*, u_*) are linearly independent. If less than n_y components of (y_*, u_*) are positive, the vertex is called *degenerate*, see e.g. [6, § 2]. In the nondegenerate case, i.e. if $l = n_y$ components of (y_*, u_*) are positive, then the matrix $\widehat{C}$ has full row rank. In the degenerate case, however, $l < n_y$ components of (y_*, u_*) are positive. Thus, $2n_y + n_u - l > n_y + n_u$ and the matrix $\widehat{C}$ cannot have full row rank. Hence, the solution is degenerate if and only if $\widehat{C}$ does not have full row rank.

Bound constraints for u. Let (y_*, u_*) be a solution of the QP and suppose that no bound constraints are imposed on y_* or that the bound constraints for y_* are not active. In this case,

$$\widehat{C} = \begin{pmatrix} A & B \\ 0 & I(u_*) \end{pmatrix}.$$

Since A is nonsingular, $\widehat{C}$ has full row rank. Therefore, the system (3.4) is uniquely solvable if the matrix M is positive definite on the null-space of $\widehat{C}$.

In the LP case, i.e. $M = 0$, the solution can be found in a vertex (y_*, u_*). Since, by assumption, $y_* > 0$ and A is nonsingular, we can conclude that $u_* = 0$. Consequently, $I(u_*) \in \mathbb{R}^{n_u \times n_u}$ is the identity matrix. In the language of linear programming, y_* are the basis variables and u_* are the nonbasis variables. Thus, this case always corresponds to the *nondegenerate case* in linear programming.

No bound constraints. If the bound constraints are not active, then the Lagrange multipliers q_y and q_u are zero and the KKT conditions (3.4) are equivalent to the system (3.5) with the last two row and column blocks of the system matrix removed. If the matrix M is positive definite on the null–space of C, the system (3.4) has a unique solution.

4. Interior–Point Methods for the Solution of the Quadratic Programming Problem

It is not the purpose of this section to give an overview of interior point methods. We primarily address the structure of the linear systems arising in these methods to provide the necessary background for the construction of preconditioners. Because of space limitations, we focus on primal–dual interior–point methods. However, matrices with similar structure also arise in barrier methods, see e.g. [19] and [8], and certain affine–scaling methods, see e.g. [18].

We continue to use the notation of Section 3 and we will employ the notation common in interior point methods: For a given vector x, the diagonal matrix with diagonal entries equal to the entries of x is denoted by X. Moreover, e denotes the vector of ones, $e = (1, \dots, 1)^T$.

The construction of primal–dual interior–point methods is based on the so–called perturbed KKT conditions corresponding to (3.4), which are given by

$$\begin{aligned} Mx + C^T p - q &= -g, \\ Cx &= b, \\ XQe &= \theta e, \end{aligned} \tag{4.1}$$

and $x, q > 0$, where $\theta > 0$. To move from a current iterate (x, p, q) with $x, q > 0$ to the next iterate (x_+, p_+, q_+), primal–dual Newton interior–point methods compute the Newton step $(\Delta x, \Delta p, \Delta q)$ for the perturbed KKT conditions (4.1) and set

$$(x_+, p_+, q_+) = (x + \alpha_x \Delta x, p + \alpha_p \Delta p, q + \alpha_q \Delta q),$$

where the step sizes $\alpha_x, \alpha_p, \alpha_q \in (0, 1]$ are chosen so that $x_+, q_+ > 0$. Then the perturbation parameter θ is updated based on $x_+^T q_+$ and the previous step is repeated. We refer to the literature, e.g. [20] for details.

The Newton system for the perturbed KKT conditions (4.1) is given by

$$\begin{pmatrix} M & C^T & -I \\ C & & \\ Q & & X \end{pmatrix} \begin{pmatrix} \Delta x \\ \Delta p \\ \Delta q \end{pmatrix} = - \begin{pmatrix} Mx + C^T p - q + g \\ Cx - b \\ XQe - \theta e \end{pmatrix}. \tag{4.2}$$

The nonsymmetric system (4.2) can be reduced to a symmetric system. If we use the last equation in (4.2) to eliminate Δq,

$$\Delta q = -X^{-1} Q \Delta x - Qe + \theta X^{-1} e, \tag{4.3}$$

then we arrive at the system

$$\begin{pmatrix} M + X^{-1} Q & C^T \\ C & \end{pmatrix} \begin{pmatrix} \Delta x \\ \Delta p \end{pmatrix} = - \begin{pmatrix} Mx + C^T p + g - \theta X^{-1} e \\ Cx - b \end{pmatrix}. \tag{4.4}$$

If $M_{yu} = 0, M_{uy} = 0$, the system (4.4) is of the form (1.7). As variables y_j or u_i approach the bound, i.e. approach zero, large quantities are added to the diagonals (j, j) or (i, i), respectively.

In actual computations more care must be taken during the reduction of the system (4.2) to avoid cancellation in the reduction process due to very large elements in X^{-1}, see e.g. [9]. A stable reduction of the system (4.2) is discussed in [9]. The unknowns and the right hand side in that reduced system differ from those in (4.4). However, the system matrix in the stable reduction is equal to the system matrix in (4.4). For our purposes it is therefore not necessary to present the lengthier stable reduction and we refer to [9] for details.

The influence of inexact solutions of the linear systems (4.4) onto the convergence behavior of the primal–dual interior–point method and the control of the inexactness is studied in [4], [12].

Before we continue, we briefly discuss the three cases explored in Section 3.

No bound constraints. In this case the diagonal contributions D_y and D_u coming from the interior point method will be zero or close to zero. Since in our case the matrix M is positive definite, the system (3.4) has a unique solution. The ill-conditioning in the matrix K in this case is purely due to the discretization of the infinite dimensional control problem.

Bound constraints for u. It has been observed, e.g. [10], that in the nondegenerate case the KKT systems in barrier methods for linear programming can be preconditioned effectively. This will also be true in our case. If only bounds on u are active, efficient preconditioners can be constructed for the problems investigated in this paper. However, in our applications, ill–conditioning also arises from the matrices A. Although proven to be nonsingular, the matrices A arising in our applications have a wide spectrum which causes a large spread in the spectrum of the KKT matrix K. This will be investigated in more detail in Section 6.

Bound constraints for u **and** y. For the construction of preconditioners in barrier methods for linear programming the degenerate case is the difficult one. For example, the preconditioners discussed in [10] are far less effective in reducing the condition number of the KKT matrix in the degenerate case than they are in the nondegenerate case, cf. Tables 1 and 2 in [10]. This will also be the case in our situation. If bounds are only imposed on the controls u, efficient and rather general preconditioners can be derived. However, if state constraints, i.e. bounds on y, are present and active, then the QP (1.4)–(1.6) is very often degenerate and the design of preconditioners is much more difficult.

5. Solution of the Linear System

5.1. MINRES and SYMMLQ. Two Krylov subspace methods for the solution of indefinite linear systems, MINRES and SYMMLQ, have been introduced in [14]. These methods have been successfully used for problems like the one studied in this paper and are used for the solution of our systems.

We set $x = (y_h, u_h, p_h)^T$. Suppose the system to be solved is $Kx = b$. Given an initial iterate x_0 we set $r_0 = b - Kx_0$. The Krylov subspace $\mathcal{K}_j(K, r_0)$ is defined by

$$\mathcal{K}_j(K, r_0) = \text{span}\{r_0, Kr_0, \dots, K^{j-1}r_0\}. \tag{5.1}$$

In iteration j, $j = 0, 1, \dots$, the minimum residual method MINRES computes

$$x_j \in \mathcal{K}_j(K, r)$$

such that x_j solves

$$\min_{x \in \mathcal{K}_j(K, r_0)} \|r_0 - Kx\|.$$

In iteration j, $j = 0, 1, \dots$, SYMMLQ computes the iterate

$$x_j \in \mathcal{K}_j(K, r_0)$$

such that x_j solves

$$(r_0 - Kx_j)^T v = 0 \quad \forall v \in \mathcal{K}_j(K, r_0).$$

Since K is indefinite, such an x_j may not exist. If it does not exist, SYMMLQ generates an iterate using information obtained from the Lanczos tridiagonalization. See [14].

The representation of Krylov subspaces (5.1) show that $x_j \in \mathcal{K}_j(K, r_0)$ if and only if

$$x_j = p_{j-1}(K)\, r_0,$$

where p_{j-1} is a polynomial of degree less or equal to $j-1$. This yields an upper bound for the residuals in MINRES:

$$\|r_0 - Kx_j\| = \|p_j^1(K) r_0\| \le \min_{p \in \Pi_j^1} \max_{\lambda \in \Lambda(K)} |p^1(\lambda)| \quad \|r_0\|. \tag{5.2}$$

Here $\Lambda(K)$ denotes the spectrum of K and Π_j^1 denotes the set of all polynomials p of degree less or equal to j which satisfy $p(0) = 1$. From (5.2) one can derive error estimates, see e.g. [16]. For example, using Chebyshev polynomials, one can show the following convergence estimate for MINRES:

$$\|r_0 - Kx_j\| \le 2 \left(\frac{\kappa - 1}{\kappa + 1} \right)^{\lfloor j/2 \rfloor} \|r_0\|,$$

where $\kappa = \bar{\lambda}/\underline{\lambda}$ is the condition number of K with $\underline{\lambda} = \min_{\lambda \in \Lambda(K)} |\lambda|$, $\bar{\lambda} = \max_{\lambda \in \Lambda(K)} |\lambda|$, and $\lfloor j/2 \rfloor$ is the largest integer less or equal to $j/2$.

If the matrix K has an unfavorable eigenvalue distribution, one constructs a nonsingular matrix P such that $K = P^{-1}KP^{-T}$ has a smaller condition number and better clustered eigenvalues. Instead of $Kx = b$ one solves the preconditioned system $\tilde{K}\tilde{x} = \tilde{b}$, where $\tilde{K} = P^{-1}KP^{-T}$, $\tilde{x} = P^T x$, and $\tilde{b} = P^{-1}b$. Of course, the preconditioner P has to be constructed so that matrix–vector multiplications with P^{-1} and P^{-T} can be done efficiently and so that the eigenvalue distribution of $P^{-1}KP^{-T}$ is improved.

For more details on MINRES and SYMMLQ we refer to [14], [2], and [3]. Those references also contain some details of the implementation. Complete listings of the preconditioned MINRES and SYMMLQ algorithms are given in [3]. We have implemented MINRES and SYMMLQ in Matlab.[1] Recently a version of the QMR algorithm has been developed in [9] to solve symmetric indefinite linear systems. These allow the application of indefinite preconditioners. If the preconditioner is positive definite, as in our case, then this QMR based method is equivalent to MINRES.

5.2. Eigenvalue Estimates. If A is invertible and if H_y and H_u are positive definite, then the matrix K defined by (1.7) has $n_y + n_u$ positive eigenvalues and n_y negative eigenvalues. More information on the eigenvalue distribution of K is provided by the following result, which is proven in [15]:

[1] A Fortran implementation of SYMMLQ written by M. Saunders is available from Netlib. See `linalg/symmlq` at `http://www.netlib.org/linalg/index.html`.

Lemma 5.1 (Rusten/Winther). *Suppose that H_y and H_u are positive definite and that $(A \mid B)$ has rank n_y. Let $\mu_1 \geq \mu_2 \geq \ldots \geq \mu_{n_y+n_u} > 0$ be the combined eigenvalues of H_y and H_u and let $\sigma_1 \geq \sigma_2 \geq \ldots \geq \sigma_{n_y} > 0$ be the singular values of $(A \mid B)^T$. The eigenvalues $\lambda_1 \geq \ldots \geq \lambda_{n_y+n_u} > 0 > \lambda_{n_y+n_u+1} \geq \ldots \geq \lambda_{2n_y+n_u}$ of K obey*

$$\lambda_{2n_y+n_u} \geq \frac{1}{2}\Big(\mu_{n_y+n_u} - \sqrt{\mu_{n_y+n_u}^2 + 4\sigma_1^2}\Big), \tag{5.3}$$

$$\lambda_{n_y+n_u+1} \leq \frac{1}{2}\Big(\mu_1 - \sqrt{\mu_1^2 + 4\sigma_{n_y}^2}\Big), \tag{5.4}$$

$$\lambda_{n_y+n_u} \geq \mu_{n_y+n_u}, \tag{5.5}$$

$$\lambda_1 \leq \frac{1}{2}\Big(\mu_1 + \sqrt{\mu_1^2 + 4\sigma_1^2}\Big). \tag{5.6}$$

6. The Preconditioners

We now turn to the preconditioners for the matrix K in (1.7). We assume that $H_y \in \mathbb{R}^{n_y \times n_y}$, $H_u \in \mathbb{R}^{n_u \times n_u}$ are symmetric positive definite and that $A \in \mathbb{R}^{n_y \times n_y}$ is nonsingular.

In the following P_y and P_u are preconditioners of H_y and H_u, respectively, i.e. P_y and P_u are nonsingular matrices such that

$$P_y^{-1} H_y P_y^{-T} \approx I, \quad \text{and} \quad P_u^{-1} H_u P_u^{-T} \approx I. \tag{6.1}$$

By $\tilde{A}^{-1}$ we denote an approximate inverse of A,

$$\tilde{A}^{-1} A \approx I. \tag{6.2}$$

In our numerical tests we use $P_u = [\mathrm{diag}(H_u)]^{1/2}$, $P_y = [\mathrm{diag}(H_y)]^{1/2}$, and $\tilde{A} = A$. Since the diagonals of the mass matrices M_u and M_y are very good preconditioners for these matrices, these choices for the preconditioners P_u, P_y are efficient and satisfy (6.1).

In our computations we use K derived from the model problem and the finite element discretization outlined in Section 2.2. In all computations we use $d_x = d_y$. MINRES and SYMMLQ were used with starting value $x_0 = (y_h, u_h, p_h) = 0$ and the iterations were stopped when $\|P^{-1}b - P^{-1}KP^{-T}\tilde{x}_j\| < 10^{-5}$. We do not test our preconditioners within an interior–point method, but simulate the matrices K in (1.7) that would arise in an interior point method by adding diagonal matrices D_y and D_u. All computations are done in Matlab.

In the analysis of the preconditioners it will be helpful to distinguish four cases.

Case 1 ($\gamma = 1$, $D_y = 0$, $D_u = 0$): In this case we can reduce the condition number of the systems under consideration considerably. By preconditioning we reduce the iterations required by MINRES and SYMMLQ to a number which appears to be independent of the grid size.

Case 2 ($\gamma \ll 1$, $D_y = 0$, $D_u = 0$): In this case, the spectrum of H_u moves towards the origin, and while the conditioning of H_u itself is not changed, the condition number of K increases significantly. In this situation, ill–conditioning of K is induced by ill–posedness of the original problem. As γ decreases, the system with K becomes hard to solve, and for sufficiently small values of γ MINRES and SYMMLQ need an unacceptably large number of iterations. The performance of MINRES and SYMMLQ improves on the preconditioned systems.

Case 3 ($\gamma = 1$, $D_y = 0$, $D_u \gg I$): If bound constraints for u are active, corresponding diagonal entries in D_u increase. We write $D_u \gg I$ and mean this to be understood component wise. Large entries in D_u can be shown to affect the conditioning of the preconditioned system only to a moderate amount. In fact, they can even help to neutralize a small parameter γ or large entries in D_y. In this case our preconditioners are very effective.

Case 4 ($\gamma = 1$, $D_y \gg I$, $D_u = 0$): This case corresponds to the situation where bound constraints on y are active. As mentioned in Sections 3 and 4 the solution may be degenerate and this case may correspond to the degenerate case in linear programming. Often, a large diagonal in H_y unfavorably affects the performance of MINRES and SYMMLQ on the preconditioned systems. While the preconditioners introduced in the following lead to some improvement, their effectiveness in this case is much smaller than in the Cases 1 and 3. We point out that in our applications the number n_y of states is much larger than the number n_u of controls. Hence if more than n_u states are active at the solution, then the matrix $\widehat{C}$ in (3.6) can not have full row rank. In our numerical tests for Case 4 we set $D_y = 10^4 I$. This simulates the worst case in the sense that this corresponds to the case where all states approach the bounds. Our numerical tests always correspond to the degenerate case, which is the hard case.

6.1. The First Preconditioner. The first preconditioner is given by

$$P_1^{-1} = \begin{pmatrix} P_y^{-1} & 0 & 0 \\ 0 & P_u^{-1} & 0 \\ 0 & 0 & P_y^T \tilde{A}^{-1} \end{pmatrix}.$$

The preconditioned KKT matrix is

$$P_1^{-1} K P_1^{-T} = \begin{pmatrix} P_y^{-1} H_y P_y^{-T} & 0 & P_y^{-1} \tilde{A}^{-T} A^T P_y \\ 0 & P_u^{-1} H_u P_u^{-T} & P_u^{-1} B^T \tilde{A}^{-T} P_y \\ P_y^T \tilde{A}^{-1} A P_y^{-T} & P_y^T \tilde{A}^{-1} B P_u^{-T} & 0 \end{pmatrix} \tag{6.3}$$

and we expect that

$$P_1^{-1} K P_1^{-T} = \begin{pmatrix} \tilde{I}_{n_y} & 0 & \tilde{I}_{n_y} \\ 0 & \tilde{I}_{n_u} & P_u^{-1} B^T \tilde{A}^{-T} P_y \\ \tilde{I}_{n_y} & P_y^T \tilde{A}^{-1} B P_u^{-T} & 0 \end{pmatrix}, \tag{6.4}$$

where $\tilde{I}$ is an approximate identity matrix. The preconditioned system still has the structure allowing us to estimate its spectrum using Lemma 5.1. The derivation of the general form of our first preconditioner is motivated by the assumption that for preconditioners P_y, P_u of H_y, H_u and for an approximate inverse $\tilde{A}^{-1}$ of A the singular values of

$$\tilde{B} = P_y^T \tilde{A}^{-1} B P_u^{-T} \tag{6.5}$$

are of moderate size. If $P_y = M_y^{1/2}$, $P_u = M_u^{1/2}$, and $\tilde{A} = A$, this is guaranteed in the situation of Section 2.1. See Lemma 2.1.

Lemma 6.1. *Let $\tilde{B} \in \mathbb{R}^{n_y \times n_u}$. The singular values σ_i of $(I_m|\tilde{B})$ are given by*

$$\sigma_i = \sqrt{1 + \sigma_i^2(\tilde{B})}, \quad i = 1, \dots, n_y,$$

where $\sigma_i(\tilde{B})$ are the singular values of $\tilde{B}$. If $n_y \geq n_u$, $\tilde{B}$ has n_u singular values, and we set $\sigma_i(\tilde{B}) = 0$ for $i = n_u + 1, \dots, n_y$.

Proof. The proof follows immediately from the fact that the squares of the singular values of a matrix B are the eigenvalues of BB^T. □

In the situation of Section 2.1 the estimate in Lemma 2.1 shows that

$$\sigma_i(\tilde{B}) \leq \|M_y^{1/2} A^{-1} B M_u^{-1/2}\| \leq c, \quad i = 1, \dots, n_y, \tag{6.6}$$

for a constant c independent of h. Thus in Case 1 ($H_y = M_y$ and $H_u = M_u$) we expect that, for preconditioners P_u, P_y and $\tilde{A}$ neutralizing the dependency of H_y, H_u and A on the mesh constant h, we can similarly bound the singular values of $P_y^T \tilde{A}^{-1} B P_u^{-T}$ such that

$$\sigma_i(\tilde{B}) \leq \|P_y^T \tilde{A}^{-1} B P_u^{-T}\| \leq c_P, \tag{6.7}$$

where c_P is a constant independent of h.

Assuming that (6.7) is valid we discuss the expected performance of the first preconditioner in the four cases defined earlier. By $\sigma_i^{(l)} = \sigma_i^{(l)}(P_y^T \tilde{A}^{-1} B P_u^{-T})$, $l = 1, 2, 3, 4$, we denote the singular values of $P_y^T \tilde{A}^{-1} B P_u^{-T}$ in Case $l = 1, 2, 3, 4$.

Case 1 ($\gamma = 1$, $D_y = 0$, $D_u = 0$): If $\gamma = 1$, (6.7) shows that there exists a constant upper bound for the singular values $\sigma^{(1)}(H_y^{1/2} A^{-1} B H_u^{-1/2})$. The preconditioner P_1 can be expected to perform well if the preconditioning matrices P_y, P_u and $\tilde{A}$ neutralize the influence of the mesh size h on the submatrices and thus on the system, and if the singular values of $P_y^T \tilde{A}^{-1} B P_u^{-T}$ are bounded by a small constant c_P. If the eigenvalues of $P_y^{-1} H_y P_y^{-1}$ and $P_u^{-T} H_u P_u^{-1}$ are close to one and if $\sigma_{min}^{(1)} \ll 1$, where $\sigma_i^{(1)}$ denote the singular values of $(P_y^T \tilde{A}^{-1} B P_u^{-T})$, we can deduce

$$\lambda_{n_y+n_u} \approx 1, \quad \lambda_{n_y+n_u+1} \approx \frac{1}{2}\left(1 - \sqrt{5}\right),$$

so that the eigenvalues of the preconditioned system are bounded away from zero. If in addition $\sigma_{max}^{(1)}$, i.e. the constant c_P in (6.7) is of moderate size, Lemma 5.1 guarantees that the condition number of the preconditioned system $P_1^{-1} K P_1^{-T}$ is small. MINRES and SYMMLQ will perform very well on the preconditioned system. This is confirmed by our numerical tests. See Table 1.

The preconditioner will perform poorly if the singular values of $P_y^T \tilde{A}^{-1} B P_u^{-T}$ are not small. This happens in two of the remaining three cases.

Case 2 ($\gamma \ll 1$, $D_y = 0$, $D_u = 0$): If a small parameter γ determines the size of the eigenvalues of the matrix M_u, we must expect that bounds on the norm

$\|H_y^{1/2}A^{-1}BH_u^{-1/2}\|$ grow with the reciprocal of $\sqrt{\gamma}$. Denoting by $\sigma_i^{(2)}$ the singular values of $H_y^{1/2}A^{-1}BH_u^{-1/2}$, we have the relationship

$$\sigma_i^{(2)} = \frac{1}{\sqrt{\gamma}}\sigma_i^{(1)}.$$

For decreasing values of γ the spectrum of $P_y^T\tilde{A}^{-1}BP_u^{-T}$ expands and the conditioning of the preconditioned system deteriorates.

Case 3 ($\gamma = 1$, $D_y = 0$, $D_u \gg I$): In this case $H_u = \gamma M_u + D_u$, where $D_u \gg I$, i.e. some diagonal entries may become very large. Analogously we write $P_u = \gamma P_O + P_D$, where P_D stands for the (large) diagonal entries and P_O for the off–diagonal entries that are generally of moderate size. By $\sigma_i^{(3)}$ we denote the singular values of $P_y^T\tilde{A}^{-1}BP_u^{-T}$. We obtain the estimate

$$\begin{aligned}
\sigma_i^{(3)} &= \sigma_i^{(3)}(P_y^T\tilde{A}^{-1}BP_u^{-T}) = \sigma_i^{(3)}(P_y^T\tilde{A}^{-1}B(\gamma P_O + P_D)^{-T}) \\
&= \sigma_i^{(3)}(P_y^T\tilde{A}^{-1}BP_D^{-T}(\gamma P_D^{-1}P_O + I)^{-T}) \\
&\leq \|P_y^T\tilde{A}^{-1}B\|\,\|P_D^{-T}\|\,\|\gamma(P_D^{-1}P_O + I)^{-T}\| \leq \|P_y^T\tilde{A}^{-1}B\|\,\|P_D^{-T}\|\,\frac{1}{1-\|\gamma P_O^T P_D^{-T}\|}.
\end{aligned}$$

If D_u dominates the matrix H_u, $\|\gamma P_O P_D^{-T}\|$ will be of negligible size. If additionally $\gamma \ll 1$, this contributes to reducing the factor $1/(1 - \|\gamma P_O P_D^{-1}\|)$ to a constant close to one. The norm $\|P_y^T\tilde{A}^{-1}B\|$ can be expected to be of moderate size, while $\|P_D^{-1}\|$ will be very small. The singular values $\sigma^{(3)}$ converge to zero as the entries in the diagonal D_u, and with it in P_D, grow. In the case of large diagonal entries in H_u we can expect a good performance of the solvers on the preconditioned system, due to a small condition number of $P_1^{-1}KP_1^{-T}$ which is in turn induced by small singular values of $P_y^T\tilde{A}^{-1}BP_u^{-T}$. The performance of MINRES and SYMMLQ on the preconditioned system is documented in Table 2.

Case 4 ($\gamma = 1$, $D_y \gg I$, $D_u = 0$): If we denote by P_y the preconditioner for H_y and by P_O, P_D its off-diagonal part and its diagonal part, respectively, then we see that the matrix $P_y^T\tilde{A}^{-1}BP_u^{-T}$ will have very large singular values. This is indicated by the estimates ($M = \tilde{A}^{-1}BP_u^{-T}P_u^{-1}B^T\tilde{A}^{-T}$)

$$\lambda_{max}((P_O+P_D)^TM(P_O+P_D)) \geq \lambda_{max}(P_D^TMP_D)+\lambda_{min}(P_O^TMP_O+P_O^TMP_D+P_D^TMP_O)$$

and

$$\lambda_{min}((P_O+P_D)^TM(P_O+P_D)) \leq \lambda_{min}(P_O^TMP_D+P_D^TMP_O+P_D^TMP_D)+\lambda_{max}(P_O^TMP_O).$$

For the estimates see [11, p. 411]. While the preconditioner yields a considerable improvement over the unpreconditioned system, the improvement is less than in Cases 1 and 3. See Table 3. However, the improvement is expected to decrease as the diagonals in D_y become larger.

6.2. The Second Preconditioner. We have seen that the effectiveness of preconditioner P_1 depends on the size of the singular values of the matrix $\tilde{B}$ defined in (6.5). The preconditioner P_2 is designed to isolate the effect of $\tilde{B}$. In order to make the action of the second preconditioner transparent, we consider the ideal version of P_2, denoted by P_2^*, i.e. we choose $P_u = H_u^{1/2}$, $P_y = H_y^{1/2}$, and $\tilde{A} = A$. For the general form of the preconditioner, which is used in the computations, we refer to [3].

The ideal preconditioner P_2^* is given by its inverse as

$$(P_2^*)^{-1} = \begin{pmatrix} H_y^{-1/2} & 0 & 0 \\ 0 & H_u^{-1/2} & 0 \\ -H_y^{-1/2} & -H_y^{1/2}A^{-1}BH_u^{-1} & H_y^{1/2}A^{-1} \end{pmatrix}.$$

The ideal preconditioned system is

$$P_2^{*-1}KP_2^{*-T} = \begin{pmatrix} I_{n_y} & 0 & 0 \\ 0 & I_{n_u} & 0 \\ 0 & 0 & -(I_{n_y} + \tilde{B}\tilde{B}^T) \end{pmatrix},$$

where $\tilde{B}$ is defined by (6.5) with $P_u = H_u^{1/2}$, $P_y = H_y^{1/2}$, and $\tilde{A} = A$.

The application of the preconditioner P_2 is roughly as expensive as the application of the preconditioner P_1. The performance of P_2 is slightly inferior to the performance of P_1. See Tables 1–3. The eigenvalue distribution of the preconditioned system, i.e. the eigenvalue distribution of $(I_{n_y} + \tilde{B}\tilde{B}^T)$, can be analyzed analogously to the previous case.

Table 1
Iterations of MINRES and SYMMLQ on K with $\gamma = 1$, $D_y = 0$, $D_u = 0$.

grid size d_x	5	10	15	20	25	30
dimension	92	282	572	962	1452	2042
Without Preconditioning						
MINRES	47	185	431	784	1070	1483
SYMMLQ	47	179	407	647	902	1209
Preconditioner P_1						
MINRES	23	25	24	21	21	19
SYMMLQ	23	24	22	21	19	19
Preconditioner P_2						
MINRES	24	35	37	37	35	35
SYMMLQ	24	35	36	35	35	33
Preconditioner P_3						
MINRES	7	6	5	5	5	4
SYMMLQ	7	6	5	5	5	4

6.3. The Third Preconditioner. A third preconditioner is derived from reductions performed to solve QP subproblems in sequential quadratic programming methods, see e.g. [7]. As before we use the ideal form for the presentation of the preconditioner. The general form of the preconditioner, see [3], is used in the computations. The ideal

Table 2
Iterations of MINRES and SYMMLQ for K with $\gamma = 1$ and $D_u = 10^4 \cdot I$, $D_y = 0$.

grid size d_x	5	10	15	20	25	30
dimension	92	282	572	962	1452	2042
Without Preconditioning						
MINRES	54	173	349	589	857	1183
SYMMLQ	54	173	349	579	848	1165
Preconditioner P_1						
MINRES	16	18	18	18	18	16
SYMMLQ	16	18	18	18	17	16
Preconditioner P_2						
MINRES	21	33	35	37	35	35
SYMMLQ	21	33	35	35	33	33
Preconditioner P_3						
MINRES	5	4	4	4	4	4
SYMMLQ	5	4	4	4	4	4

Table 3
Iterations of MINRES and SYMMLQ for K with $\gamma = 1$ and $D_y = 10^4 \cdot I$, $D_u = 0$.

grid size d_x	5	10	15	20	25	30
dimension	92	282	572	962	1452	2042
Without Preconditioning						
MINRES	73	282	572	962	1452	2042
SYMMLQ	73	282	572	962	1452	2042
Preconditioner P_1						
MINRES	50	98	194	289	449	530
SYMMLQ	50	98	187	283	410	524
Preconditioner P_2						
MINRES	61	146	235	323	453	583
SYMMLQ	61	143	233	323	447	547
Preconditioner P_3						
MINRES	44	67	120	203	275	366
SYMMLQ	44	56	120	167	286	355

preconditioner P_3^*, given by its inverse as

$$(P_3^*)^{-1} = \begin{pmatrix} I_{n_y} & 0 & -1/2\, H_y A^{-1} \\ 0 & 0 & A^{-1} \\ -(A^{-1}B)^T & I_{n_u} & (A^{-1}B)^T\, H_y A^{-1} \end{pmatrix},$$

transforms K into the preconditioned system

$$(P_3^*)^{-1}K(P_3^*)^{-T} = \begin{pmatrix} 0 & I_{n_y} & 0 \\ I_{n_y} & 0 & 0 \\ 0 & 0 & W^T H W \end{pmatrix},$$

where

$$W = \begin{pmatrix} -A^{-1}B \\ I_{n_u} \end{pmatrix}, \quad H - \begin{pmatrix} H_y & 0 \\ 0 & H_u \end{pmatrix}.$$

The matrix W is a representation for the nullspace of $C = (A|B)$. The matrix $W^T H W$ is given by

$$W^T H W = B^T A^{-T} H_y A^{-1} B + H_u = H_u^{1/2}\Big(\tilde{B}^T \tilde{B} + I_{n_u}\Big) H_u^{1/2},$$

where $\tilde{B}$ is defined by (6.5) with $P_u = H_u^{1/2}$, $P_y = H_y^{1/2}$, and $\tilde{A} = A$. Note that the partitioning of the blocks in the preconditioned system has changed.

The preconditioner P_3 is the most effective in reducing the number of iterations. See Tables 1–3. However, the application of the general preconditioner P_3 is roughly twice as expensive as the application of the preconditioners P_1 and P_2. See [3]. The eigenvalue distribution of $W^T H W$ can be analyzed analogously to the preconditioned system with P_1.

7. Conclusions

In this paper we have derived preconditioners for matrices K arising in the numerical solution of certain distributed linear quadratic control problems by interior–point methods. The preconditioners are in block form, with blocks composed of preconditioners for the individual blocks of the matrix K. This allows the incorporation of known preconditioners for the governing equations of the original problem and it allows to reuse computationally expensive information within all interior–point iterations. The effectiveness of the preconditioners was analyzed using the properties of the control problem and its discretization, the block structure of the matrix K, and information from the optimality conditions. Numerical results supporting the theoretical analysis were given.

Acknowledgments. The work in this paper was done while A. Battermann and M. Heinkenschloss were student and faculty, respectively, in the Department of Mathematics and the Interdisciplinary Center for Applied Mathematics at the Virginia Polytechnic Institute and State University. The authors would like to thank their friends and colleagues at those places for their support.

References

1. H. T. Banks and K. Kunisch, *Estimation Techniques for Distributed Parameter Systems*, Systems & Control: Foundations & Applications, Birkhäuser-Verlag, Boston, Basel, Berlin, 1989.
2. R. Barrett, M. Berry, T. F. Chan, J. D. J. Donato, J. Dongarra, V. Eijkhout, R. Pozo, C. Romine, and H. van der Vorst, *Templates for the Solution of Linear Systems: Building Blocks for Iterative Methods*, SIAM, Philadelphia, 1993.

3. A. Battermann, *Preconditioners for Karush–Kuhn–Tucker systems arising in optimal control*, Master's thesis, Department of Mathematics, Virginia Polytechnic Institute and State University, Blacksburg, Virginia, 1996. Available electronically from `http://scholar.lib.vt.edu/theses/`.
4. J. Bonnans, C. Pola, and R. Rebai, *Perturbed path following interior point algorithm*, Tech. Rep. No. 2745, INRIA, Domaine de Voluceau, 78153 Rocquencourt, France, 1995.
5. D. Braess and P. Peisker, *On the numerical solution of the biharmonic equation and the role of squaring matrices*, IMA J. Numer. Anal, 6 (1986), pp. 393–404.
6. L. Collatz and W. Wetterling, *Optimization Problems*, Springer-Verlag, Berlin, Heidelberg, New York, 1975.
7. J. E. Dennis, M. Heinkenschloss, and L. N. Vicente, *Trust–region interior–point algorithms for a class of nonlinear programming problems*, Tech. Rep. TR94–45, Department of Computational and Applied Mathematics, Rice University, Houston, TX 77005–1892, 1994. Available electronically at `http://www.caam.rice.edu/~trice/trice_soft.html`.
8. A. S. El-Bakry, R. A. Tapia, T. Tsuchiya, and Y. Zhang, *On the formulation and theory of the primal–dual Newton interior–point method for nonlinear programming*, Journal of Optimization Theory and Applications, 89 (1996), pp. 507–541.
9. R. W. Freund and F. Jarre, *A qmr-based interior-point method for solving linear programs*, Mathematical Programming, Series B, (to appear).
10. P. E. Gill, W. Murray, D. B. Ponceleón, and M. A. Saunders, *Preconditioners for indefinite systems arising in optimization*, SIAM J. Matrix Anal. Appl., 13 (1992), pp. 292–311.
11. G. Golub and C. F. van Loan, *Matrix Computations*, John Hopkins University Press, Baltimore, London, 1989.
12. S. Ito, C. T. Kelley, and E. W. Sachs, *Inexact primal–dual interior–point iteration for linear programs in function spaces*, Computational Optimization and Applications, 4 (1995), pp. 189–201.
13. J. L. Lions, *Optimal Control of Systems Governed by Partial Differential Equations*, Springer-Verlag, Berlin, Heidelberg, New York, 1971.
14. C. C. Paige and M. A. Saunders, *Solution of sparse indefinite systems of linear equations*, SIAM J. Numer. Anal., 12 (1975), pp. 617–629.
15. T. Rusten and R. Winther, *A preconditioned iterative method for saddlepoint problems*, SIAM J. Matrix Anal. Appl., 13 (1992), pp. 887–904.
16. J. Stoer, *Solution of large linear systems of equations by conjugate gradient type methods*, in Mathematical Programming, The State of The Art, A. Bachem, M. Grötschel, and B. Korte, eds., Springer-Verlag, Berlin, Heidelberg, New-York, 1983, pp. 540–565.
17. D. Sylvester and A. Wathen, *Fast iterative solution of stabilized Stokes systems part II: using general block preconditioners*, SIAM J. Numer. Anal., 31 (1994), pp. 1352–1367.
18. L. N. Vicente, *On interior–point Newton algorithms for discretized optimal control problems with state constraints*, tech. rep., Departamento de Matemática, Universidade de Coimbra, 3000 Coimbra, Portugal, 1996.
19. M. H. Wright, *Interior point methods for constrained optimization*, in Acta Numerica 1992, A. Iserles, ed., Cambridge University Press, Cambridge, London, New York, 1992, pp. 341–407.
20. S. J. Wright, *Primal–Dual Interior–Point Methods*, SIAM, Philadelphia, PA, 1996.

A. Battermann
Universität Trier
FB IV – Mathematik
D–54286 Trier
Federal Republic of Germany

M. Heinkenschloss
Department of Computational
and Applied Mathematics
Rice University
Houston, TX 77005–1892, USA
e-mail: heinken@caam.rice.edu

International Series of Numerical Mathematics
Vol. 126, © 1998 Birkhäuser Verlag, Basel

Augmented Lagrangian Algorithms for State Constrained Optimal Control Problems

M. BERGOUNIOUX AND K. KUNISCH

URA-CNRS 1803
Université d'Orléans

Fachbereich Mathematik
Technische Universität Berlin

ABSTRACT. We investigate augmented Lagrangian algorithms to solve state and control constrained optimal control problems. We augment both the state-equation and the non-smooth state and control constraints. We present the method with the example of linear optimal control problem with a boundary control function but the proposed algorithms are general and can be adapted to a much wider class of problems.

1991 *Mathematics Subject Classification.* 49J20, 49M29

Key words and phrases. State and control constrained optimal control problems, augmented Lagrangian, elliptic equations.

1. Setting of the Problem

Let Ω be an open, bounded subset of $\mathbb{R}^n$, n ≤ 3 , with a smooth boundary Γ. We consider the following optimal control problem:

$$(\mathcal{P}) \qquad \min \; J(y,u) = \frac{1}{2}\int_\Omega (y-z_d)^2 \, dx + \frac{\alpha}{2}\int_\Gamma (u-u_d)^2 \, d\sigma$$

$$(1.1) \qquad Ay = f \text{ in } \Omega \; , \; y = u \text{ on } \Gamma \; ,$$

$$(1.2) \qquad \Lambda_1 y \in K \; , \; u \in U \; ,$$

where

- $f, \; z_d \in L^2(\Omega), \; u, \; u_d \in L^2(\Gamma)$ and either $\alpha > 0$ or U is bounded in $L^2(\Gamma)$,
- L is a finite dimensional (Hilbert) space and $\Lambda_1 \in \mathcal{L}(W, L)$, ($W$ is defined just below).
- K and U are nonempty, closed, convex subsets of L and $L^2(\Gamma)$ respectively.
- A is an elliptic operator defined by:

$$(1.3) \qquad \begin{cases} Ay = -\sum\limits_{i,j=1}^{n} \partial_{x_i}(a_{ij}(x)\partial_{x_j} y) + a_0(x)y \;\; \text{with} \\ a_{ij}, a_0 \in \mathcal{C}^2(\bar\Omega) \; for \; i,j = 1, \dots, n, \; \inf\{a_0(x) \mid x \in \bar\Omega\} > 0 \\ \sum\limits_{ij=1}^{n} a_{ij}(x)\xi_i\xi_j \geq \delta \sum\limits_{i=1}^{n} \xi_i^2 \; , \forall x \in \bar\Omega, \forall \xi \in \mathbb{R}^n, \delta > 0 \, . \end{cases}$$

System (1.1) is well-posed: for every $(u,f) \in L^2(\Gamma) \times L^2(\Omega)$ there exists a unique solution $y = \mathcal{T}(u,f)$ in W, where

$$W = \{ \ y \in L^2(\Omega) \mid Ay \in L^2(\Omega) \ , \ y_{|\Gamma} \in L^2(\Gamma) \ \}.$$

Moreover $\mathcal{T}$ is continuous from $L^2(\Gamma) \times L^2(\Omega)$ to W, when W is endowed with the graph norm:

$$|y|_W^2 = |y|_\Omega^2 + |Ay|_\Omega^2 + |y_{|\Gamma}|_\Gamma^2 \ .$$

From now on, when H is an Hilbert-space, we denote by $(\ , \ )_H$ (resp. $(\ , \ )_\Omega$ and $(\ , \ )_\Gamma$) the H (resp. $L^2(\Omega)$ and $L^2(\Gamma)$) inner products and by $| \ |_H$, $| \ |_\Omega$, $| \ |_\Gamma$, the H, $L^2(\Omega)$ and $L^2(\Gamma)$-norms, respectively. Moreover, we define $\Lambda : W \times L^2(\Gamma) \to L \times L^2(\Gamma)$ by $\Lambda(y,v) = (\Lambda_1(y), v)$ and we assume that the feasible domain

$$\mathcal{D} \ = \ \{ \ (y,u) \in W \times L^2(\Gamma) \mid Ay \ = \ f \ \text{ in } \Omega, \ y = u \text{ on } \Gamma, \ \Lambda(y,u) \in K \times U \ \} \ ,$$

is nonempty. It is easy to see that problem ($\mathcal{P}$) has a unique solution $(\bar{y}, \bar{u})$ since the functional J is strictly convex and coercive and $\mathcal{D}$ is convex, closed and nonempty. Our main purpose is to present new augmented Lagrangian algorithms to solve numerically optimal control problems of the above type. Usually such algorithms use the augmentation of the "smooth" part of the constraints, that is the state-equation. This has been done in Fortin-Glowinski [3] and adapted to the present example in Bergounioux [1]. Here we use a different point of view, since we use a Lagrangian function where **both** the state-equation and the nonsmooth constraints "$\Lambda_1 y \in K$, $u \in U$" are augmented. These last constraints are augmented using a method developed in Ito and Kunisch [4].

2. Optimality Conditions

In this section we recall a result which is crucial to interpret the forthcoming algorithms and to give convergence results.

Theorem 2.1. *Let $(\bar{y}, \bar{u})$ be the optimal solution of ($\mathcal{P}$) and assume the following qualification condition*

$$(\mathcal{H}) \begin{cases} \textit{There exists a bounded (in } L^2(\Omega) \times L^2(\Gamma)) \textit{ subset } \mathcal{M} \textit{ of } W \times L^2(\Gamma) \textit{ such} \\ \textit{that } \Lambda(\mathcal{M}) \subset K \times U \textit{and } 0 \in Int_2(\mathcal{V}(\mathcal{M})) \ , \\ \textit{where } Int_2 \textit{ denotes the interior with respect to the } L^2(\Omega) \times L^2(\Gamma)\textit{-topology} \\ \textit{and } \mathcal{V}(y,u) = (Ay - f, y|_\Gamma - u). \end{cases}$$

Then there exists $(\bar{q}, \bar{r}) \in L^2(\Omega) \times L^2(\Gamma)$ and $(\bar{\mu}_1, \bar{\mu}_2) \in \Lambda_1(W) \times L^2(\Gamma)$ such that:

$$A\bar{y} = f \quad \text{in } \Omega \ , \ \bar{y} = \bar{u} \quad \text{on } \Gamma \ , \tag{2.1}$$

$$(\bar{y} - z_d, y)_\Omega + (\bar{q}, Ay)_\Omega + (\bar{r}, y)_\Gamma + (\bar{\mu}_1, \Lambda_1 y)_L = 0 \quad \text{for all } \ y \in W \ , \tag{2.2}$$

$$\alpha \ (\bar{u} - u_d) = \bar{r} - \bar{\mu}_2 \in L^2(\Gamma) \ , \tag{2.3}$$

$$\begin{aligned} &(\bar{\mu}_1, \Lambda_1(y - \bar{y}))_L \leq 0 \text{ for all } y \text{ such that } \Lambda_1 y \in K \ , \\ &(\bar{\mu}_2, u - \bar{u})_\Gamma \leq 0 \text{ for all } u \in U \ . \end{aligned} \tag{2.4}$$

Proof. This is a particular case of a more general result that can be found in [2]. ■

As a specific example, L can be chosen as the set of linear finite elements with respect to a triangulation of Ω and $\Lambda_1 : W \to L$ can be the L^2-projection.

3. Lagrangian Algorithms

In this section we turn to the numerical realization of the constrained optimal control problem $(\mathcal{P})$. We shall combine the techniques from [1] and [4] augmenting the state equation as well as the constraints characterizing the feasible set D, to obtain well performing algorithms.

3.1. Augmentation of the State Equation. First we recall an augmented Lagrangian algorithm based on the penalization of the state equation (see [3], [1] and the references therein).

Algorithm $\mathcal{A}_o$

- Step 1. Initialization: Set $n = 0$, and choose $\gamma > 0$, $q_o \in L^2(\Omega)$, $r_o \in L^2(\Gamma)$.
- Step 2. Compute

$$(y_n, u_n) \quad = \quad \begin{array}{l}\text{Arg min } L_\gamma(y, v, q_n, r_n)\\ \Lambda(y,u) \in K \times U\end{array}$$

where

$$L_\gamma(y, u, q, r) = J(y, u) + (q, Ay)_\Omega + (r, y-u)_\Gamma + \frac{\gamma}{2}|Ay - f|^2_\Omega + \frac{\gamma}{2}|y-u|^2_\Gamma$$

is the augmented Lagrangian with respect to the state equation constraint.
- Step 3. Set

$$q_{n+1} = q_n + \rho_1\,(Ay_n - f) \quad \text{where } \rho_1 \in (0, 2\gamma\,]\,,$$
$$r_{n+1} = r_n + \rho_2\,(y_{n|\Gamma} - u_n) \quad \text{where } \rho_2 \in (0, 2\gamma\,]\,.$$

The analysis of this algorithm is rather standard, see [1] and the references there. For the convenience of the reader we provide a precise convergence result (which appears to slightly generalize the existing ones) and give a concise proof.

Theorem 3.1. *Let $(\bar y, \bar u)$ be the solution to $(\mathcal{P})$ and suppose that $(\mathcal{H})$ holds. Then the iterates of Algorithm $\mathcal{A}_o$ satisfy*

$$\text{(3.1)}\quad |y_n - \bar y|^2_\Omega + \alpha|u_n - \bar u|^2_\Gamma + \frac{1}{2\rho_1}|q_{n+1} - \bar q|^2_\Omega + \frac{1}{2\rho_2}|r_{n+1} - \bar r|^2_\Gamma$$
$$+ (\gamma - \frac{\rho_1}{2})|Ay_n - f|^2_\Omega + (\gamma - \frac{\rho_2}{2})|y_n - u_n|^2_\Gamma \le \frac{1}{2\rho_1}|q_n - \bar q|^2_\Omega + \frac{1}{2\rho_2}|r_n - \bar r|^2_\Gamma$$

for all $n = 0, 1, 2, \ldots$. This implies

$$\text{(3.2)}\quad \sum_{n=0}^{\infty}|y_n - \bar y|^2_\Omega + \alpha\sum_{n=0}^{\infty}|u_n - \bar u|^2_\Gamma + (\gamma - \frac{\rho_1}{2})\sum_{n=0}^{\infty}|Ay_n - f|^2_\Omega$$
$$+ (\gamma - \frac{\rho_2}{2})\sum_{n=0}^{\infty}|y_n - u_n|^2_\Gamma \le \frac{1}{2\rho_1}|q_0 - \bar q|^2_\Omega + \frac{1}{2\rho_2}|r_0 - \bar r|^2_\Gamma\,,$$

and in particular strong convergence of $(y_n, u_n) \to (\bar y, \bar u)$ in $L^2(\Omega) \times L^2(\Gamma)$, and boundedness of $\{(q_n, r_n)\}$. If moreover $\rho_1 < 2\gamma$ and $\rho_2 < 2\gamma$ then $(y_n, u_n) \to (\bar y, \bar u)$

in $W \times L^2(\Gamma)$, and every weak limit $(\tilde{q}, \tilde{r})$ of (q_n, r_n) has the property that $(\bar{y}, \bar{u}, \tilde{q}, \tilde{r})$ satisfies (3.3)*.*

Proof. From Theorem 2.1 we obtain

$$\text{(3.3)} \quad \Big(J'(\bar{y}, \bar{u}), (y, u) - (\bar{y}, \bar{u})\Big)_{\Omega\times\Gamma} + \Big(\bar{q}, A(y - \bar{y})\Big)_{\Omega} + \Big(\bar{r}, y - \bar{y} - (u - \bar{u})\Big)_{\Gamma} \geq 0$$

for all $\Lambda(y, u) \in K \times U$. The solutions (y_n, u_n) of Step 2 are characterized by

$$\text{(3.4)} \quad \begin{aligned} &\Big(J'(y_n, u_n), (y, u) - (y_n, u_n)\Big)_{\Omega\times\Gamma} + \Big(q_{n+1}, A(y - y_n)\Big)_{\Omega} \\ &+\Big(r_{n+1}, y - y_n - (u - u_n)\Big)_{\Gamma} + (\gamma - \rho_1)\Big(Ay_n - f, A(y - y_n)\Big)_{\Omega} \\ &+(\gamma - \rho_2)\Big(y_n - u_n, y - y_n - (u - u_n)\Big)_{\Gamma} \qquad \geq 0 \end{aligned}$$

for all $\Lambda(y, u) \in K \times U$. Adding (3.3) with $(y, u) = (y_n, u_n)$ to (3.4) with $(y, u) = (\bar{y}, \bar{u})$ one obtains

$$\text{(3.5)} \quad \begin{aligned} &\Big(J'(y_n, u_n) - J'(\bar{y}, \bar{u}), (y_n, u_n) - (\bar{y}, \bar{u})\Big)_{\Omega\times\Gamma} + \Big(q_{n+1} - \bar{q}, Ay_n - A\bar{y}\Big)_{\Omega} \\ &+ (r_{n+1} - \bar{r}, y_n - u_n)_{\Gamma} + (\gamma - \rho_1)|Ay_n - f|^2_{\Omega} + (\gamma - \rho_2)|y_n - u_n|^2_{\Gamma} \leq 0. \end{aligned}$$

Let us note the following equality

$$\text{(3.6)} \qquad (a + \rho b, b)_H = \frac{1}{2\rho}|a + \rho b|^2_H - \frac{1}{2\rho}|a|^2_H + \frac{\rho}{2}|b|^2_H,$$

for all elements a,b of a real Hilbert space H and all $\rho \in \mathbb{R}$.
Due to (3.6) we find

$$\Big(q_{n+1} - \bar{q}, A(y_n - \bar{y})\Big)_{\Omega} = \frac{1}{2\rho_1}|q_{n+1} - \bar{q}|^2_{\Omega} - \frac{1}{2\rho_1}|q_n - \bar{q}|^2_{\Omega} + \frac{\rho_1}{2}|Ay_n - A\bar{y}|^2_{\Omega}$$

and

$$\Big(r_{n+1} - \bar{r}, y_n - u_n\Big)_{\Gamma} = \frac{1}{2\rho_2}|r_{n+1} - \bar{r}|^2_{\Gamma} - \frac{1}{2\rho_2}|r_n - \bar{r}|^2_{\Gamma} + \frac{\rho_2}{2}|y_n - u_n|^2_{\Gamma}\ .$$

Inserting these equalities into (3.5) we obtain

$$\begin{aligned} &|y_n - \bar{y}|^2_{\Omega} + \alpha|u_n - \bar{u}|^2_{\Gamma} + \frac{1}{2\rho_1}|q_{n+1} - \bar{q}|^2_{\Omega} + \frac{1}{2\rho_2}|r_{n+1} - \bar{r}|^2_{\Gamma} \\ &\quad + \frac{\rho_1}{2}|Ay_n - f|^2_{\Omega} + \frac{\rho_2}{2}|y_n - u_n|^2_{\Gamma} + (\gamma - \rho_1)|Ay_n - f|^2_{\Omega} + (\gamma - \rho_2)|y_n - u_n|^2_{\Gamma} \\ &\leq \frac{1}{2\rho_1}|q_n - \bar{q}|^2_{\Omega} + \frac{1}{2\rho_2}|r_n - \bar{r}|^2_{\Gamma}\ , \end{aligned}$$

and (3.1) follows. Using a telescoping argument (3.2) is implied by (3.1). The ascertained convergence properties follow from (3.2), (3.4) and uniqueness of $(\bar{y}, \bar{u}, \bar{q}, \bar{r})$.

■

3.2. Augmentation of the Non Smooth Constraints. The main remaining problem is the resolution of the auxiliary problem of Step 2 in Algorithm $\mathcal{A}_o$. This auxiliary problem can be written as:

$$\begin{array}{rcl} (y_n, u_n) & = & \text{Arg min } L_\gamma(y,u) \\ & & \Lambda(y,u) \in D \ . \end{array}$$

To simplify the notation we omit to indicate the dependence of L_γ on q and r. During Step 2 these functions are fixed. We set $H = L \times L^2(\Gamma)$. Let φ be the characteristic function of the convex set D. Then, following [4], we define (for any $c > 0$) the function $\varphi_c : H \times H \to \mathbb{R}$ by:

$$\varphi_c(x,\lambda) = \inf_{\xi \in H} \{ \varphi(x-\xi) + (\lambda,\xi)_H + \frac{c}{2}|\xi|_H^2 \} \ , \tag{3.7}$$

where $x = (y,u)$. Here $(\ ,\ )_H$ (denoting the H-inner product) is given by $(\lambda,\xi)_H = (\lambda_1,\xi_1)_\Omega + (\lambda_2,\xi_2)_\Gamma$, with $\lambda = (\lambda_1,\lambda_2)$ and $\xi = (\xi_1,\xi_2)$.
We recall some properties of the function φ_c (for more details one can refer to [4, 2]):

Proposition 3.1. *For all $x = (y,u) \in H$ and $\lambda = (\lambda_1,\lambda_2) \in H$*

$$\begin{array}{rcl} \varphi_c(x,\lambda) & = & \frac{c}{2}|y - P_K(y+\frac{\lambda_1}{c})|_L^2 + \left(\lambda_1, y - P_K(y+\frac{\lambda_1}{c})\right)_L \\ & & +\frac{c}{2}|u - P_U(u+\frac{\lambda_2}{c})|_\Gamma^2 + \left(\lambda_2, u - P_U(u+\frac{\lambda_2}{c})\right)_\Gamma \ , \end{array} \tag{3.8}$$

$$\varphi_c'(x,\lambda) = c\left(y + \frac{\lambda_1}{c} - P_K(y+\frac{\lambda_1}{c}), u + \frac{\lambda_2}{c} - P_U(u+\frac{\lambda_2}{c})\right) \ , \tag{3.9}$$

where P_K (resp. P_U, P_D) is the L (resp. $L^2(\Gamma)$, H) projection on K (resp. on U, D).

Proof. See [2]. ∎

We are going to use the following algorithm and a splitting variant to solve the auxiliary problem:

Algorithm $\mathcal{A}_1$

- Step 1. Initialization: Choose $\lambda^o \in H$ and $c > 0$.
- Step 2. Compute

$$\begin{array}{rcl} (y^j, u^j) & = & \text{Arg min } L_\gamma(y,u) + \varphi_c(\Lambda(y,u), \lambda^j) \\ & & \Lambda(y,u) \in W \times L^2(\Gamma) \ , \end{array}$$

  where φ_c has been defined in the previous section.
- Step 3. Set

$$\lambda^{j+1} = \varphi_c'(\Lambda(y^j,u^j),\lambda^j) = c\Big(\Lambda_1 y^j + \frac{\lambda_1^j}{c} - P_K(\Lambda_1 y^j + \frac{\lambda_1^j}{c}), u^j + \frac{\lambda_2^j}{c} - P_U(u^j + \frac{\lambda_2^j}{c})\Big),$$

  (see 3.9).

The convergence of this algorithm under the assumption that L is finite dimensional follows from result in [4].

3.3. Final Algorithm. We now write the version where Algorithm $\mathcal{A}_1$ appears as an inner loop in algorithm $\mathcal{A}_o$:

Algorithm $\mathcal{A}$

- Step 1. Initialization: Set $n = 0$, and choose $\gamma > 0,\ c > 0$.
  Choose $(q_o, r_o) \in L^2(\Omega) \times L^2(\Gamma)$ and $\lambda_o = (\lambda_{o1}, \lambda_{o2}) \in L \times L^2(\Gamma)$.
- Step 2. Choose $k_n \in \mathbb{N}$, set $\lambda_n^o = \lambda_n$, and for $j = 0, \ldots, k_n$:

$$\begin{cases} (y_n^j, u_n^j) &= \text{Arg min } L_\gamma(y, u, q_n, r_n) + \varphi_c(\Lambda(y,u), \lambda_n^j) \\ & \quad (y,u) \in W \times L^2(\Gamma) , \\ \lambda_n^{j+1} &= (\lambda_{n,1}^{j+1}, \lambda_{n,2}^{j+1}) \ \text{ with} \\ \lambda_{n,1}^{j+1} &= c[\Lambda_1 y_n^j + \dfrac{\lambda_{n,1}^j}{c} - P_K(\Lambda_1 y_n^j + \dfrac{\lambda_{n,1}^j}{c})] , \\ \lambda_{n,2}^{j+1} &= c[u_n^j + \dfrac{\lambda_{n,2}^j}{c} - P_U(u_n^j + \dfrac{\lambda_{n,2}^j}{c})] . \end{cases}$$

  End of the inner loop:

$$\lambda_{n+1} = \lambda_n^{k_n+1},\ y_n = y_n^{k_n},\ u_n = u_n^{k_n} .$$

- Step 3.

$$q_{n+1} = q_n + \rho_1 \left(\frac{1}{k_n+1} \sum_{j=0}^{k_n} A y_n^j - f\right) \quad \text{where } \rho_1 \in (0, 2\gamma\ ] ,$$

$$r_{n+1} = r_n + \rho_2 \left(\frac{1}{k_n+1} \sum_{j=0}^{k_n} (y_n^j|_\Gamma - u_n^j)\right) \qquad \rho_2 \in (0, 2\gamma\ ] .$$

Theorem 3.2. *Let $(\bar y, \bar u)$ be the solution to $(\mathcal{P})$ and suppose that $(\mathcal{H})$ holds. Let $(\bar q, \bar r, \bar\mu) \in L^2(\Omega) \times L^2(\Gamma) \times L \times L^2(\Gamma)$ be an associated Lagrange multiplier. Then the iterates of Algorithm $\mathcal{A}$ satisfy*

$$\begin{aligned} &|y_n - \bar y|_\Omega^2 + \alpha |u_n - \bar u|_\Gamma^2 + \frac{k_n+1}{2\rho_1}|q_{n+1} - \bar q|_\Omega^2 + \frac{k_n+1}{2\rho_2}|r_{n+1} - \bar r|_\Gamma^2 \\ &\quad +(\gamma - \frac{\rho_1}{2})|Ay_n - f|_\Omega^2 + (\gamma - \frac{\rho_2}{2})|u_n - y_n|_\Gamma^2 + \frac{1}{2c}|\lambda_{n+1} - \bar\mu|_{L\times L^2(\Gamma)}^2 \\ &\le \frac{k_n+1}{2\rho_1}|q_n - \bar q|_\Omega^2 + \frac{k_n+1}{2\rho_2}|r_n - \bar r|_\Gamma^2 + \frac{1}{2c}|\lambda_n - \bar\mu|_{L\times L^2(\Gamma)}^2 \end{aligned} \tag{3.10}$$

for all $n = 0, 1, 2, \ldots$. If k_n is nonincreasing this implies

$$\begin{aligned} &\sum_{n=0}^{\infty} |y_n - \bar y|_\Omega^2 + \alpha \sum_{n=0}^{\infty} |u_n - \bar u|_\Gamma^2 \\ &\quad +(\gamma - \frac{\rho_1}{2}) \sum_{n=0}^{\infty} |Ay_n - f|_\Omega^2 + (\gamma - \frac{\rho_2}{2}) \sum_{n=0}^{\infty} |u_n - y_n|_\Gamma^2 \\ &\le \frac{k_0+1}{2\rho_1}|q_0 - \bar q|_\Omega^2 + \frac{k_0+1}{2\rho_2}|r_0 - \bar r|_\Gamma^2 + \frac{1}{2c}|\lambda_0 - \bar\mu|_{L\times L^2(\Gamma)}^2 \end{aligned} \tag{3.11}$$

and the strong convergence of $(y_n, u_n) \to (\bar{y}, \bar{u})$ *in* $L^2(\Omega) \times L^2(\Gamma)$, *and boundedness of* $\{(q_n, r_n, \lambda_n)\}$. *If moreover* $\rho_1 < 2\gamma$ *and* $\rho_2 < 2\gamma$ *then* $(y_n, u_n) \to (\bar{y}, \bar{u})$ *in* $W \times L^2(\Gamma)$ *and every weak limit* $(\tilde{q}, \tilde{r}, \tilde{\lambda})$ *of* $\{(q_n, r_n, \lambda_n)\}$ *has the property that* $(\bar{y}, \bar{u}, \tilde{q}, \tilde{r}, \tilde{\lambda})$ *satisfies (2.2), (2.3).*

Proof. From (2.2), (2.3) it follows that

$$\left(J'(\bar{y},\bar{u}),(y,u)\right)_{\Omega\times\Gamma} + (\bar{q}, Ay)_\Omega + (\bar{r}, y-u)_\Gamma + \left(\bar{\mu}, \Lambda(y,u)\right)_{L\times L^2(\Gamma)} = 0 \tag{3.12}$$

for all $(y,u) \in W \times L^2(\Gamma)$. The solutions (y_n^j, u_n^j) of Step 2 satisfy

$$\begin{aligned}
&\left(J'(y_n^j,u_n^j),(y,u)\right)_{\Omega\times\Gamma} + \left(q_n + \frac{\rho_1}{k_n+1}(Ay_n^j - f), Ay\right)_\Omega \\
&+(\gamma - \frac{\rho_1}{k_n+1})(Ay_n^j - f, Ay)_\Omega + \left(r_n + \frac{\rho_2}{k_n+1}(y_n^j - u_n^j), y-u\right)_\Gamma \\
&+(\gamma - \frac{\rho_2}{k_n+1})(y_n^j - u_n^j, y-u)_\Gamma + \left(\varphi_c'(\Lambda(y_n^j,u_n^j),\lambda_n^j),\Lambda(y,u)\right)_{L\times L^2(\Gamma)} = 0
\end{aligned} \tag{3.13}$$

for all $(y,u) \in W \times L^2(\Gamma)$. Let us denote $q_n^{-1} := q_n$, $r_n^{-1} := r_n$ and for $j = 0, \ldots, k_n$, $n \in \mathbb{N}$

$$q_n^j = q_n + \frac{\rho_1}{k_n+1}\sum_{i=0}^{j}(Ay_n^i - f), \quad r_n^j = r_n + \frac{\rho_2}{k_n+1}\sum_{i=0}^{j}(y_n^i - u_n^i)\ .$$

From now, for convenience, we omit the indication of the norm since there is no possible confusion. From [4] it follows that $\bar{\mu} = \varphi_c'(\Lambda(\bar{y},\bar{u}),\bar{\mu})$. From (3.12) and (3.13) we deduce for $j = 0, 1, \ldots$

$$\begin{aligned}
&\left(J'(y_n^j,u_n^j) - J'(\bar{y},\bar{u}),((y_n^j,u_n^j)-(\bar{y},\bar{u})\right) + \left(q_n^{j-1} + \frac{\rho_1}{k_n+1}(Ay_n^j - f) - \bar{q}, Ay_n^j - f\right) \\
&+\left(r_n^{j-1} + \frac{\rho_2}{k_n+1}(y_n^j - u_n^j) - \bar{r}, y_n^j - u_n^j\right) + (\gamma - \frac{\rho_1}{k_n+1})|Ay_n^j - f|^2 \\
&+(\gamma - \frac{\rho_2}{k_n+1})|y_n^j - u_n^j|^2 - \frac{\rho_1}{k_n+1}\sum_{i=0}^{j-1}(Ay_n^i - f, Ay_n^j - f) \\
&-\frac{\rho_2}{k_n+1}\sum_{i=0}^{j-1}(y_n^i - u_n^i, y_n^j - u_n^j) \\
&+\left(\varphi_c'(\Lambda(y_n^j,u_n^j),\lambda_n^j) - \varphi_c'(\Lambda(\bar{y},\bar{u}),\bar{\mu}), \Lambda(y_n^j,u_n^j) - \Lambda(\bar{y},\bar{u})\right) = 0\ ,
\end{aligned}$$

where $\sum_{i=0}^{-1} := 0$. From [4] it is known that

$$\begin{aligned}
&\left(\varphi_c'(\Lambda(y_n^j,u_n^j),\lambda_n^j) - \varphi_c'(\Lambda(\bar{y},\bar{u}),\bar{\mu}), \Lambda(y_n^j,u_n^j) - \Lambda(\bar{y},\bar{u})\right) \\
&\geq \frac{1}{2c}|\lambda_n^{j+1} - \bar{\mu}|^2 - \frac{1}{2c}|\lambda_n^j - \bar{\mu}|^2\ ,
\end{aligned} \tag{3.14}$$

for $j = 0, 1, \ldots, k_n$. Inserting this inequality in the above equality and using (3.6) we obtain

$$
\begin{aligned}
&|y_n^j - \bar{y}|^2 + \alpha |u_n^j - \bar{u}|^2 + \frac{k_n+1}{2\rho_1}|q_n^j - \bar{q}|^2 - \frac{k_n+1}{2\rho_1}|q_n^{j-1} - \bar{q}|^2 \\
&+(\gamma - \frac{\rho_1}{2(k_n+1)})|Ay_n^j - f|^2 + \frac{k_n+1}{2\rho_2}|r_n^j - \bar{r}|^2 - \frac{k_n+1}{2\rho_2}|r_n^{j-1} - \bar{r}|^2 \\
&+(\gamma - \frac{\rho_2}{2(k_n+1)})|y_n^j - u_n^j|^2 - \frac{\rho_1}{k_n+1}\sum_{i=0}^{j-1}(Ay_n^i - f, Ay_n^j - f) \\
&-\frac{\rho_2}{k_n+1}\sum_{i=0}^{j-1}(y_n^i - u_n^i, y_n^j - u_n^j) + \frac{1}{2c}(|\lambda_n^{j+1} - \bar{\mu}|^2 - |\lambda_n^j - \bar{\mu}|^2) \le 0 \,,
\end{aligned}
$$

for $j = 0, 1, \ldots, k_n$. Summing the above inequality over j and using the fact that

$$
\sum_{j=1}^{k_n}\sum_{i=0}^{j-1}(a_i, a_j)_H \le \frac{k_n}{2}\sum_{j=0}^{k_n}(a_j^2)
$$

we arrive at

$$
\begin{aligned}
&\sum_{j=0}^{k_n}(|y_n^j - \bar{y}|^2 + \alpha|u_n^j - \bar{u}|^2) + \frac{k_n+1}{2\rho_1}|q_n^{k_n} - \bar{q}|^2 + \frac{k_n+1}{2\rho_2}|r_n^{k_n} - \bar{r}|^2 \\
&\quad +(\gamma - \frac{\rho_1}{2})\sum_{j=0}^{k_n}|Ay_n^j - f|^2 + (\gamma - \frac{\rho_2}{2})\sum_{j=0}^{k_n}|y_n^j - u_n^j|^2 + \frac{1}{2c}|\lambda_n^{k_n+1} - \bar{\mu}|^2 \\
&\le \frac{k_n+1}{2\rho_1}|q_n - \bar{q}|^2 + \frac{k_n+1}{2\rho_2}|r_n - \bar{r}|^2 + \frac{1}{2c}|\lambda_n^0 - \bar{\mu}|^2 \,.
\end{aligned}
$$

Since $(y_n, u_n) = (y_n^{k_n}, u_n^{k_n})$, $(q_{n+1}, r_{n+1}) = (q_n^{k_n}, r_n^{k_n})$ and $2\gamma \ge \rho_1$, $2\gamma \ge \rho_2$ this implies

$$
\begin{aligned}
&|y_n - \bar{y}|^2 + \alpha|u_n - \bar{u}|^2 + \frac{k_n+1}{2\rho_1}|q_{n+1} - \bar{q}|^2 + \frac{k_n+1}{2\rho_2}|r_{n+1} - \bar{r}|^2 \\
&\quad +(\gamma - \frac{\rho_1}{2})|Ay_n - f|^2 + (\gamma - \frac{\rho_2}{2})|y_n - u_n|^2 + \frac{1}{2c}|\lambda_{n+1} - \bar{\mu}|^2 \\
&\qquad \le \frac{k_n+1}{2\rho_1}|q_n - \bar{q}|^2 + \frac{k_n+1}{2\rho_2}|r_n - \bar{r}|^2 + \frac{1}{2c}|\lambda_n - \bar{\mu}|^2 \,,
\end{aligned}
$$

which is the desired estimate (3.10). Summation over n implies (3.2). This ends the proof. ■

3.4. Adding Gauss-Seidel Spitting. Our final goal is the analysis of Gauss-Seidel splitting techniques to solve the auxiliary problems. A similar approach was taken in [1]. The splitting avoids the minimization of the auxiliary problem with respect to y and u simultaneously. The new algorithm is:

Algorithm $\mathcal{A}_o^{GS}$

- Step 1. Initialization: Set $n = 0$, choose $\gamma > 0$, $q_o \in L^2(\Omega)$, $r_o \in L^2(\Gamma)$, $u_{-1} \in U$.
- Step 2.

$$y_n = \operatorname*{Arg\ min}_{\Lambda_1 y \in K} L_\gamma(y, u_{n-1}, q_n, r_n)$$

$$u_n = \operatorname*{Arg\ min}_{u \in U} L_\gamma(y_n, u, q_n, r_n)$$

- Step 3.

$$q_{n+1} = q_n + \rho_1 \; (Ay_n - f) \;\text{ where }\; \rho_1 \in (0, 2\gamma] \; ,$$

$$r_{n+1} = r_n + \rho_2 \; (y_{n|\Gamma} - u_n) \;\text{ where }\; \rho_2 \in (0, \gamma] \; .$$

Theorem 3.3. *Under the assumptions of Theorem 3.1 the iterates (y_n, u_n, q_n, r_n) of Algorithm $\mathcal{A}_o^{GS}$ satisfy*

$$\begin{aligned}
&|y_n - \bar{y}|_\Omega^2 + \alpha|u_n - \bar{u}|_\Gamma^2 + \frac{1}{2\rho_1}|q_{n+1} - \bar{q}|_\Omega^2 + \frac{1}{2\rho_2}|r_{n+1} - \bar{r}|_\Gamma^2 \\
&\quad + (\gamma - \frac{\rho_1}{2})|Ay_n - f|_\Omega^2 + (\gamma - \frac{\rho_2}{2})|u_n - y_n|_\Gamma^2 \\
&\quad + (\alpha + \frac{\rho_2}{2})|u_n - u_{n-1}|_\Gamma^2 + \frac{\gamma}{2}|u_n - \bar{u}|_\Gamma^2 \\
&\le \frac{1}{2\rho_1}|q_n - \bar{q}|_\Omega^2 + \frac{1}{2\rho_2}|r_n - \bar{r}|_\Gamma^2 + (\frac{\gamma - \rho_2}{2})|u_{n-1} - y_{n-1}|_\Gamma^2 + \frac{\gamma}{2}|u_{n-1} - \bar{u}|_\Gamma^2
\end{aligned} \tag{3.15}$$

for all $n = 0, 1, 2, \dots$. This implies

$$\begin{aligned}
&\sum_{n=0}^{\infty} |y_n - \bar{y}|_\Omega^2 + \alpha \sum_{n=0}^{\infty} |u_n - \bar{u}|_\Gamma^2 + (\gamma - \frac{\rho_1}{2}) \sum_{n=0}^{\infty} |Ay_n - f|_\Omega^2 \\
&\quad + \frac{\gamma}{2} \sum_{n=0}^{\infty} |u_n - y_n|_\Gamma^2 + (\alpha + \frac{\rho_2}{2}) \sum_{n=0}^{\infty} |u_n - u_{n-1}|_\Gamma^2 \\
&\le \frac{1}{2\rho_1}|q_o - \bar{q}|_\Omega^2 + \frac{1}{2\rho_2}|r_o - \bar{r}|_\Gamma^2 + \frac{(\gamma - \rho_2)}{2}|u_{-1} - y_{-1}|_\Gamma^2 + \frac{\gamma}{2}|u_{-1} - \bar{u}|_\Gamma^2 \; .
\end{aligned} \tag{3.16}$$

Proof. The optimality conditions for the two auxiliary problems of Step 2 give

$$\begin{aligned}
&\Big(J_y'(y_n, u_{n-1}), y - y_n\Big)_\Omega + \Big(q_n, A(y - y_n)\Big)_\Omega + \Big(r_n, y - y_n\Big)_\Gamma \\
&+\gamma\Big(Ay_n - f, A(y - y_n)\Big)_\Omega + \gamma\Big(y_n - u_{n-1}, y - y_n\Big)_\Gamma \ge 0 \;\text{ for all } \Lambda_1 y \in K \; ,
\end{aligned}$$

and

$$\begin{aligned}
&\Big(J_u'(y_n, u_n), u - u_n\Big)_\Gamma - \Big(r_n, u - u_n\Big)_\Gamma \\
&\qquad -\gamma\Big(y_n - u_n, u - u_n\Big)_\Gamma \ge 0 \;\text{ for all } u \in U \; .
\end{aligned} \tag{3.17}$$

Adding these two inequalities and combining with (3.3) we obtain

$$\begin{aligned}
&\Big(J'(y_n,u_n)-J'(\bar y,\bar u),(y_n,u_n)-(\bar y,\bar u)\Big)_{\Omega\times\Gamma}\\
&+\Big(q_{n+1}-\bar q, A(y_n-\bar y)\Big)_\Omega+\Big(r_{n+1}-\bar r, y_n-u_n\Big)_\Gamma \qquad (3.18)\\
&+(\gamma-\rho_1)|Ay_n-f|^2_\Omega+(\gamma-\rho_2)|y_n-u_n|^2_\Gamma+\gamma\Big(u_{n-1}-u_n,\bar y-y_n\Big)_\Gamma\le 0\,.
\end{aligned}$$

This expression coincides with (3.5) except for the last term on the left hand side of (3.18). Proceeding as in the proof of Theorem 4.1 we therefore obtain

$$\begin{aligned}
&|y_n-\bar y|^2_\Omega+\alpha|u_n-\bar u|^2_\Gamma+\frac{1}{2\rho_1}|q_{n+1}-\bar q|^2_\Omega+\frac{1}{2\rho_2}|r_{n+1}-\bar r|^2_\Gamma\\
&\quad+(\gamma-\frac{\rho_1}{2})|Ay_n-f|^2_\Omega+(\gamma-\frac{\rho_2}{2})|y_n-u_n|^2_\Gamma+\gamma\big(u_{n-1}-u_n,\bar y-y_n\big)_\Gamma \qquad (3.19)\\
&\le\frac{1}{2\rho_1}|q_n-\bar q|^2_\Omega+\frac{1}{2\rho_2}|r_n-\bar r|^2_\Gamma\,.
\end{aligned}$$

The method of estimation of $\Big(u_{n-1}-u_n,\bar y-y_n\Big)_\Gamma$ is standard [3], but is given for the sake of completeness. First we note that

$$\begin{aligned}
\Big(u_{n-1}-u_n,\bar y-y_n\Big)_\Gamma=&\Big(u_n-u_{n-1},y_n-y_{n-1}\Big)_\Gamma\\
&+\Big(u_n-u_{n-1},y_{n-1}-\bar y-u_{n-1}+\bar u\Big)_\Gamma \qquad (3.20)\\
&+\frac{1}{2}(|u_n-\bar u|^2_\Gamma-|u_{n-1}-\bar u|^2_\Gamma-|u_n-u_{n-1}|^2_\Gamma)\,.
\end{aligned}$$

Using the optimality condition (3.17) for n and $n-1$, with $u=u_n$ and $u=u_{n-1}$ respectively and adding the two resulting inequalities one arrives at

$$-\alpha|u_n-u_{n-1}|^2_\Gamma+\Big(r_n-r_{n-1},u_n-u_{n-1}\Big)_\Gamma-\gamma\Big(y_n-y_{n-1}-(u_n-u_{n-1}),u_{n-1}-u_n\Big)_\Gamma\ge 0\,,$$

which in turn implies

$$\begin{aligned}
&\gamma\Big(y_n-y_{n-1},u_n-u_{n-1}\Big)_\Gamma \qquad (3.21)\\
&\qquad\ge(\gamma+\alpha)|u_n-u_{n-1}|^2_\Gamma-\rho_2\Big(u_n-u_{n-1},y_{n-1}-u_{n-1}\Big)_\Gamma\,,
\end{aligned}$$

for $n=1,2,\ldots$. Inserting (3.21) into (3.20) implies that

$$\begin{aligned}
\gamma\Big(u_{n-1}-u_n,\bar y-y_n\Big)_\Gamma\ \ge\ &(\alpha+\frac{\gamma}{2})|u_n-u_{n-1}|^2_\Gamma\\
&+(\gamma-\rho_2)\Big(u_n-u_{n-1},y_{n-1}-u_{n-1}\Big)_\Gamma\\
&+\frac{\gamma}{2}(|u_n-\bar u|^2_\Gamma-|u_{n-1}-\bar u|^2_\Gamma)
\end{aligned}$$

$$\begin{aligned} &\geq \frac{\rho_2-\gamma}{2}|y_{n-1}-u_{n-1}|^2_\Gamma + (\alpha+\frac{\rho_2}{2})|u_n-u_{n-1}|^2_\Gamma \\ &+\frac{\gamma}{2}(|u_n-\bar u|^2_\Gamma - |u_{n-1}-\bar u|^2_\Gamma)\ , \end{aligned} \tag{3.22}$$

where $\rho_2 \leq \gamma$ is used. Inserting (3.22) into (3.19) the desired inequality (3.15) follows. The second inequality in the claim of the theorem follows by a telescoping argument from the first one. ■

Once again, we may use algorithm $\mathcal{A}_1$ to solve the first sub-problem of Step 2. The second one is easily solved directly, see Remark 3.1 below. For convenience we shall henceforth delete the index 1 in the notation of the state component of the multiplier.

Algorithm $\mathcal{A}^{GS}$

- Step 1. Initialization: Set $n=0$ and choose $\gamma>0,\ c>0$.
  Choose $(q_o, r_o) \in L \times L^2(\Gamma),\ \lambda_o \in L^2(\Omega)$ and $u_{-1} \in L^2(\Gamma)$.
- Step 2. Choose $k_n \in \mathbb{N}$, set $\lambda_n^o = \lambda_n,\ u_n^{-1} = u_{n-1}$ and for $j=0,\dots,k_n$

$$\left[\begin{array}{lll} y_n^j &=& \underset{y\in W}{\text{Arg min}}\ L_\gamma(y,u_n^{j-1},q_n,r_n)+\varphi_c(\Lambda(y,u_n^{j-1}),(\lambda_n^j,0)) \\ \\ \lambda_n^{j+1} &=& c[\Lambda_1 y_n^j + \dfrac{\lambda_n^j}{c} - P_K(\Lambda_1 y_n^j + \dfrac{\lambda_n^j}{c})]\ , \\ \\ u_n^j &=& \underset{u\in U}{\text{Arg min}}\ L_\gamma(y_n^j,u,q_n,r_n) \end{array}\right.$$

  End of the inner loop: $\lambda_{n+1} = \lambda_n^{k_n+1},\ y_n = y_n^{k_n}\ , u_n = u_n^{k_n}$.
- Step 3.

$$q_{n+1} = q_n + \frac{\rho_1}{k_n+1}\sum_{j=0}^{k_n}(Ay_n^j - f),\quad \text{where } \rho_1 \in (0,2\gamma]\ ,$$

$$r_{n+1} = r_n + \frac{\rho_2}{k_n+1}\sum_{j=0}^{k_n}(y^j_{n|\Gamma} - u_n^j),\quad \text{where } \rho_2 \in (0,\gamma]\ .$$

Remark 3.1. *The second minimization problem is indeed equivalent to*

$$u_n^j = \underset{u\in U\ ,}{\text{Arg min}} \left| u - \frac{\alpha u_d + r_n + \gamma y_n^j}{\alpha+\gamma}\right|_\Gamma$$

that is u_n^j is the $L^2(\Gamma)$-projection of $\dfrac{\alpha u_d + r_n + \gamma y_n^j}{\alpha+\gamma}$ *on U.*

We may now end this section with a convergence result for Algorithm $\mathcal{A}^{GS}$.

Theorem 3.4. *Let $(\bar y, \bar u)$ be the solution to $(\mathcal{P})$ and suppose that $(\mathcal{H})$ holds. Let $(\bar q, \bar r, \bar\mu) \in L^2(\Omega) \times L^2(\Gamma) \times L \times L^2(\Gamma)$ be a Lagrange multiplier associated to the* *state equation and the state constraint. Then the iterates (y_n, u_n, q_n, r_n) of Algorithm $\mathcal{A}^{GS}$ satisfy*

$$
\begin{aligned}
&|y_n - \bar{y}|^2_\Omega + (\alpha + \frac{\gamma}{2})|u_n - \bar{u}|^2_\Gamma + \frac{k_n+1}{2\rho_1}|q_{n+1} - \bar{q}|^2_\Omega + \frac{k_n+1}{2\rho_2}|r_{n+1} - \bar{r}|^2_\Gamma \\
&\quad + (\gamma - \frac{\rho_1}{2})|Ay_n - f|^2_\Omega + \frac{\gamma - \rho_2}{2}|u_n - y_n|^2_\Gamma + \frac{1}{2c}|\lambda_{n+1} - \bar{\mu}|^2_L \\
&\qquad \leq \frac{k_n+1}{2\rho_1}|q_n - \bar{q}|^2_\Omega + \frac{k_n+1}{2\rho_2}|r_n - \bar{r}|^2_\Gamma \\
&\quad + \frac{1}{2c}|\lambda_n - \bar{\mu}|^2_L + \frac{\gamma - \rho_2}{2}|u_{n-1} - y_{n-1}|^2_\Gamma + \frac{\gamma}{2}|u_{n-1} - \bar{u}|^2_\Gamma
\end{aligned}
\tag{3.23}
$$

for all $n = 1, 2, \ldots$. If k_n is nonincreasing this implies

$$
\begin{aligned}
&\sum_{n=1}^{\infty}\left(|y_n - \bar{y}|^2_\Omega + \alpha|u_n - \bar{u}|^2_\Gamma + (\gamma - \frac{\rho_1}{2})|Ay_n - f|^2_\Omega + \frac{\gamma}{2}|u_n - y_n|^2_\Gamma\right) \leq \\
&\frac{k_1+1}{2\rho_1}|q_1 - \bar{q}|^2_\Omega + \frac{k_1+1}{2\rho_2}|r_1 - \bar{r}|^2_\Gamma + \frac{1}{2c}|\lambda_1 - \bar{\mu}|^2_L + \frac{\gamma - \rho_2}{2}|y_o - u_o|^2_\Gamma + \frac{\gamma}{2}|u_o - \bar{u}|^2_\Gamma \; .
\end{aligned}
$$

Proof. We combine the techniques used in the proofs of Theorems 3.2 and 3.3 which allows to omit some details. Once again we use the optimality conditions issued from Step 2 of Algorithm $\mathcal{A}^{GS}$. The iterates (y^j_n, u^j_n) of Step 2 satisfy, for $j = 0, \ldots, k_n$, for all $y \in W$

$$
\begin{aligned}
&\left(J'_y(y^j_n, u^{j-1}_n), y)\right)_\Omega + \left(q_n + \frac{\rho_1}{k_n+1}(Ay^j_n - f), Ay\right)_\Omega \\
&+ (\gamma - \frac{\rho_1}{k_n+1})(Ay^j_n - f, Ay)_\Omega + \left(r_n + \frac{\rho_2}{k_n+1}(y^j_n - u^{j-1}_n), y\right)_\Gamma \\
&+ (\gamma - \frac{\rho_2}{k_n+1})(y^j_n - u^{j-1}_n, y)_\Gamma + \left(\varphi'_{c,1}(\Lambda_1 y^j_n, \lambda^j_n), \Lambda_1 y)\right)_L = 0,
\end{aligned}
\tag{3.24}
$$

and for all $u \in U$

$$
\begin{aligned}
&\left(J'_u(y^j_n, u^j_n), u - u^j_n)\right)_\Gamma - \left(r_n + \frac{\rho_2}{k_n+1}(y^j_n - u^j_n), u - u^j_n\right)_\Gamma \\
&\qquad - (\gamma - \frac{\rho_2}{k_n+1})(y^j_n - u^j_n, u - u^j_n)_\Gamma \geq 0 \; .
\end{aligned}
\tag{3.25}
$$

Relation (3.12) implies

$$
\begin{aligned}
&\left(J'(\bar{y}, \bar{u}), (y, u - \bar{u})\right)_{\Omega\times\Gamma} + (\bar{q}, Ay)_\Omega \\
&\qquad + (\bar{r}, y - (u - \bar{u}))_\Gamma + \left(\bar{\mu}, \Lambda_1 y\right)_{L\times L^2(\Gamma)} \geq 0
\end{aligned}
\tag{3.26}
$$

for all $(y,u) \in W \times U$. Combining (3.24)–(3.26) and (3.14) of the proof of Theorem 3.2 implies

$$\begin{aligned}
&|y_n^j - \bar y|_\Omega^2 + \alpha |u_n^j - \bar u|_\Gamma^2 + \frac{k_n+1}{2\rho_1}|q_n^j - \bar q|_\Omega^2 - \frac{k_n+1}{2\rho_1}|q_n^{j-1} - \bar q|_\Omega^2 \\
&+(\gamma - \frac{\rho_1}{2(k_n+1)})|Ay_n^j - f|_\Omega^2 + \frac{k_n+1}{2\rho_2}|r_n^j - \bar r|_\Gamma^2 - \frac{k_n+1}{2\rho_2}|r_n^{j-1} - \bar r|_\Gamma^2 \\
&+(\gamma - \frac{\rho_2}{2(k_n+1)})|y_n^j - u_n^j|_\Gamma^2 - \frac{\rho_1}{k_n+1}\sum_{i=0}^{j-1}(Ay_n^i - f, Ay_n^j - f)_\Omega \\
&- \frac{\rho_2}{k_n+1}\sum_{i=0}^{j-1}(y_n^i - u_n^i, y_n^j - u_n^j)_\Gamma + \frac{1}{2c}(|\lambda_n^{j+1} - \bar\mu|_L^2 - |\lambda_n^j - \bar\mu|_L^2) \\
&+\gamma\Big(u_n^j - u_n^{j-1}, y_n^j - \bar y\Big)_\Gamma \le 0\,,
\end{aligned}$$

for $n, j = 0, 1, \ldots, k_n$. Summing the above inequality over j we arrive at

$$\begin{aligned}
&\sum_{j=0}^{k_n}(|y_n^j - \bar y|_\Omega^2 + \alpha|u_n^j - \bar u|_\Gamma^2) + \frac{k_n+1}{2\rho_1}|q_n^{k_n} - \bar q|_\Omega^2 + \frac{k_n+1}{2\rho_2}|r_n^{k_n} - \bar r|_\Gamma^2 + \frac{1}{2c}|\lambda_{n+1} - \bar\mu|_L^2 \\
&\quad +(\gamma - \frac{\rho_1}{2})\sum_{j=0}^{k_n}|Ay_n^j - f|_\Omega^2 + (\gamma - \frac{\rho_2}{2})\sum_{j=0}^{k_n}|y_n^j - u_n^j|_\Gamma^2 + \gamma\sum_{j=0}^{k_n}\Big(u_n^j - u_n^{j-1}, y_n^j - \bar y\Big)_\Gamma \\
&\le \frac{k_n+1}{2\rho_1}|q_n - \bar q|_\Omega^2 + \frac{k_n+1}{2\rho_2}|r_n - \bar r|_\Gamma^2 + \frac{1}{2c}|\lambda_n^0 - \bar\mu|_L^2\,.
\end{aligned}$$

Now we estimate $\Big(u_n^j - u_n^{j-1}, y_n^j - \bar y\Big)_\Gamma$ as in the proof of Theorem 3.3. We obtain, for $j = 1, 2, \ldots$ and $n = 0, 1, \ldots$

$$\gamma\Big(u_n^{j-1} - u_n^j, \bar y - y_n^j\Big)_\Gamma \ge \alpha|u_n^j - u_n^{j-1}|_\Gamma^2 - \frac{\gamma}{2}(|y_n^{j-1} - u_n^{j-1}|_\Gamma^2 + |u_n^{j-1} - \bar u|_\Gamma^2 - |u_n^j - \bar u|_\Gamma^2)\,.$$

A similar calculus provides the estimation of $\Big(u_n^o - u_n^{-1}, y_n^o - \bar y\Big)_\Gamma$ for $n = 1, 2, \ldots$

$$\begin{aligned}
\gamma\Big(u_n^{-1} - u_n^o, \bar y - y_n^o\Big)_\Gamma \ge\ & (\alpha + \frac{\rho_2}{2})|u_n^o - u_n^{-1}|_\Gamma^2 + \frac{\rho_2 - \gamma}{2}|y_{n-1} - u_n^{-1}|_\Gamma^2 \\
& + \frac{\gamma}{2}|u_n^o - \bar u|_\Gamma^2 - \frac{\gamma}{2}|u_n^{-1} - \bar u|_\Gamma^2\,.
\end{aligned} \tag{3.27}$$

We henceforth assume $n \ge 1$. We obtain

$$\begin{aligned}
\gamma\sum_{j=0}^{k_n}\Big(u_n^j - u_n^{j-1}, y_n^j - \bar y\Big)_\Gamma \ge & (\alpha + \frac{\rho_2}{2})|u_n^o - u_{n-1}|_\Gamma^2 + \frac{\rho_2 - \gamma}{2}|y_{n-1} - u_{n-1}|_\Gamma^2 \\
& + \frac{\gamma}{2}|u_n^o - \bar u|_\Gamma^2 - \frac{\gamma}{2}|u_{n-1} - \bar u|_\Gamma^2 + \alpha\sum_{j=1}^{k_n}|u_n^j - u_n^{j-1}|_\Gamma^2 \\
& - \frac{\gamma}{2}\sum_{j=1}^{k_n}(|y_n^{j-1} - u_n^{j-1}|_\Gamma^2 + |u_n^{j-1} - \bar u|_\Gamma^2 - |u_n^j - \bar u|_\Gamma^2)\,.
\end{aligned} \tag{3.28}$$

We finally get for $k_n \geq 1$

$$
\begin{aligned}
&\sum_{j=0}^{k_n}(|y_n^j - \bar{y}|_\Omega^2 + \alpha|u_n^j - \bar{u}|_\Gamma^2) + \frac{k_n+1}{2\rho_1}|q_{n+1} - \bar{q}|_\Omega^2 + \frac{k_n+1}{2\rho_2}|r_{n+1} - \bar{r}|_\Gamma^2 \\
&\quad + \frac{1}{2c}|\lambda_{n+1} - \bar{\mu}|_L^2 + (\gamma - \frac{\rho_1}{2})\sum_{j=0}^{k_n}|Ay_n^j - f|_\Omega^2 + \frac{\gamma-\rho_2}{2}\sum_{j=0}^{k_n}|y_n^j - u_n^j|_\Gamma^2 + \frac{\gamma}{2}|u_n - \bar{u}|_\Gamma^2 \\
&\leq \frac{k_n+1}{2\rho_1}|q_n - \bar{q}|_\Omega^2 + \frac{k_n+1}{2\rho_2}|r_n - \bar{r}|_\Gamma^2 \\
&\quad + \frac{1}{2c}|\lambda_n - \bar{\mu}|_L^2 + \frac{\gamma-\rho_2}{2}|y_n - y_{n-1}|_\Omega^2 + \frac{\gamma}{2}|u_{n-1} - \bar{u}|_\Gamma^2 .
\end{aligned}
$$

Since $\rho_2 \leq \gamma$ we deduce that if $k_n \geq 1$

$$
\begin{aligned}
&|y_n - \bar{y}|_\Omega^2 + \alpha|u_n - \bar{u}|_\Gamma^2 + \frac{k_n+1}{2\rho_1}|q_{n+1} - \bar{q}|_\Omega^2 + \frac{k_n+1}{2\rho_2}|r_{n+1} - \bar{r}|_\Gamma^2 \\
&\quad + (\gamma - \frac{\rho_1}{2})|Ay_n - f|_\Omega^2 + \frac{\gamma-\rho_2}{2}|y_n - u_n|_\Gamma^2 + \frac{1}{2c}|\lambda_{n+1} - \bar{\mu}|_\Omega^2 + \frac{\gamma}{2}|u_n - \bar{u}|_\Gamma^2 \\
&\leq \frac{k_n+1}{2\rho_1}|q_n - \bar{q}|_\Omega^2 + \frac{k_n+1}{2\rho_2}|r_n - \bar{r}|_\Gamma^2 + \frac{1}{2c}|\lambda_n - \bar{\mu}|_\Omega^2 \\
&\quad + \frac{\gamma-\rho_2}{2}|y_{n-1} - u_{n-1}|_\Gamma^2 + \frac{\gamma}{2}|u_{n-1} - \bar{u}|_\Gamma^2 .
\end{aligned}
$$

Using (3.27) the same estimate follows for $k_n = 0$. The final claim again follows with a telescoping argument. ∎

4. Numerical Experiments

4.1. Implementation. Numerical experiments were carried out for one and two dimensional problems but we present only a 1D-example. Since Algorithm $\mathcal{A}^{GS}$ is the simplest for implementation we have used it for our tests. The discretization of the problem was done with finite-differences discretization schemes. The size of the grid was $\frac{1}{N}$ so that $L = \mathbb{R}^{N+1}$. Λ was chosen as the discretization operator with respect to the given equidistant grid.

The main difficulty that remains in applying Algorithm $\mathcal{A}^{GS}$ is given by the (unconstrained) minimization with respect to y. This was done via the adjoint state equation and results, for fixed u, q and r in the resolution of

$$
(4.1) \qquad \begin{cases} A^*p = y - z_d + c\,[y + \frac{\lambda}{c} - P_K(y + \frac{\lambda}{c})] \text{ in } \Omega,\ p = 0 \text{ on } \Gamma , \\ Ay = f - \frac{q+p}{\gamma} \text{ in } \Omega ,\ y = u - \frac{r}{\gamma} + \frac{1}{\gamma}\frac{\partial p}{\partial \nu_{A^*}} \quad \text{on } \Gamma , \end{cases}
$$

for p and y. Here $\frac{\partial p}{\partial \nu_{A^*}}$ denotes the conormal derivative of p with respect of A^* (which is the adjoint operator of A). The control function was computed using the L^∞-projection of $\frac{r + \alpha u_d + \gamma y}{\alpha + \gamma}$ on U.

All numerical tests were carried out on an HP workstation using the MATLAB© package. The required accuracy and stopping criteria were set to 10^{-6}.

4.2. 1D-Example. In this example we chose $\Omega =]0,1[$ and $N = 30$, $A = -\Delta$ and $f(x) = -(x+2)\exp(x)$, $z_d \equiv -1$, $\alpha = 0.1$, $u_d(0) = -2$, $u_d(1) = 1$, $U = [-3,3]$ and $K = \{ Y \in L \mid -1.1 \leq Y \leq 1 \}$.
Note that z_d is quite close to the boundary of K.
In fact, as can be seen from Figure 1, the lower bound on the state is active. The active set is a singleton. In view of the fact that the influence of the boundary control at $x = 0$ and $x = 1$ is restricted to the superposition of straight lines to the uncontrolled state, this is not surprising.

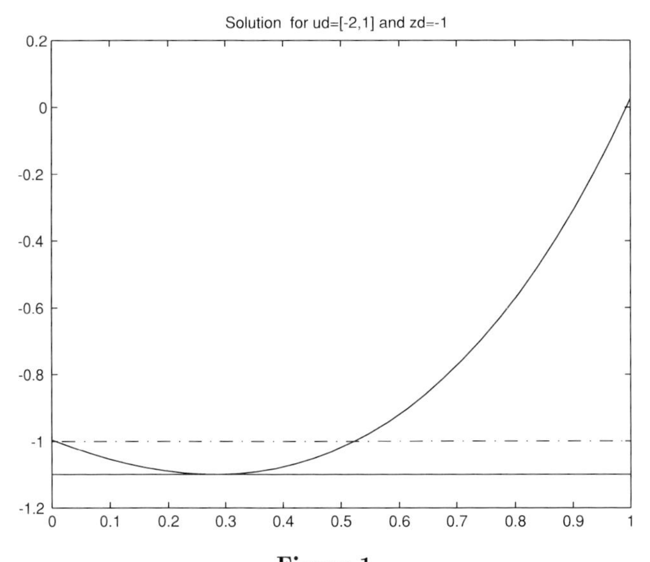

Figure 1

c	γ	k_n (constant)	$\|\Delta y + f\|_\infty$	$\|y - v\|_\infty$	n	CPU units	min[y-(-1.1)]
10	1	10	4.8 e-07	4. e-07	58	1	4. e-10
10	1	1	9.3 e-07	6. e-07	154	2.17	2.5 e-06
10	1	100	2.2 e-07	5. e-07	13	1.35	-2 e-09
100	10	10	6.2 e-07	9. e-07	95	1.01	-1.3 e-11

Table 1

One of the main questions concerning the class of algorithms that we analyzed is the choice of the parameters ρ_i, c and γ. From Table 1 we conclude that while the choice of the parameters certainly has an influence on the convergence properties of the algorithm, there is a wide range of parameters values for which convergence is achieved numerically, for this and other examples that we tested. In all calculations we chose $\rho_i = 1$. Some tests shows that the ratio $\dfrac{\gamma}{c} = \dfrac{1}{10}$ is a good one. For $(c,\gamma) = (1,1)$, $(c,\gamma) = (100,50)$, $(c,\gamma) = (1,0.5)$ (all with $k_n = 10$ for all n), convergence is achieved but it is slower than for those pairs that are presented in Table 1. From that Table, as well as from other tests, it can also be seen that the auxiliary problem should be solved sufficiently accurately, before the Lagrange-multipliers (q,r) for the state equation and the boundary condition are updated (see $k_n \equiv 1$). The values $(c,\gamma) = (10, 0.1)$ still with $\rho_i = 1$ led to divergence. This is not unexpected in view of the result of Theorem 4.5 which requires $\rho_2 \leq \gamma$.

The numerical values for J and the control at the minimum are:

$$J = 1.5862\ 10^{-1} \text{ and } \bar{u}(0) = -9.9573\ 10^{-1},\ \bar{u}(1) = 2.6314\ 10^{-2}\ .$$

4.3. Conclusion. The augmented Lagrangian algorithms with splitting into state and control variable can effectively be used to solve state and control constrained optimization problems. For the first order methods that are presented in this paper the auxiliary problems in the inner-loop must be solved sufficiently accurately before the Lagrange-multipliers of state equation and boundary condition are updated. Appropriate choices for the penalty parameters (here c and γ) and the step lengths ρ_i for the dual variables are easily determined since the algorithm are not particularly sensitive to them.

References

1. BERGOUNIOUX M.: *On Boundary State Constrained Control Problems,* Numerical Functional Analysis and Optimization, Vol.14 , pp. 515–543, 1993.
2. BERGOUNIOUX M., KUNISCH K.: *Augmented Lagrangian Techniques for Elliptic State Constrained Optimal Control Problems*, to appear in SIAM Journal on Control and Optimization.
3. FORTIN M., GLOWINSKI R.: *Méthodes de Lagrangien Augmenté - Applications à la Résolution de Problèmes aux Limites*, Méthodes Mathématiques pour l'Informatique, Dunod, Paris, France, 1982.
4. ITO K., KUNISCH K.: *Augmented Lagrangian Methods for Nonsmooth Convex Optimization in Hilbert Spaces*, preprint, 1994.

M. Bergounioux
URA-CNRS 1803
Université d'Orléans
B.P. 6759
F-45067 Orléans Cedex 2, France
e-mail: maitine@univ-orleans.fr

K. Kunisch
Fachbereich Mathematik
Technische Universität Berlin
Str. des 17 Juni 136
D-10623 Berlin, Germany
e-mail: kunisch@math.tu-berlin.de

International Series of Numerical Mathematics
Vol. 126, © 1998 Birkhäuser Verlag, Basel

A Priori Estimates for the Approximation of a Parabolic Boundary Control Problem

ARIELA BRIANI AND MAURIZIO FALCONE
Dipartimento di Matematica
Università di Pisa
Dipartimento di Matematica
Università di Roma

ABSTRACT. We study an approximation of the boundary control problem for the heat equation over a finite horizon. Our goal is to obtain an approximation of the value function and of the corresponding "locally optimal" trajectories. We examine here a time discretization also proving some a priori estimates of convergence for the value function of the time–discrete problem. Some hints are also given for the construction of a fully discrete scheme.

1991 *Mathematics Subject Classification.* Primary 65K10, 49L10 Secondary 65M12, 93C20

Key words and phrases. Optimal control, parabolic equations, dynamic programming, approximations schemes.

1. Introduction

We deal with the finite horizon control problem for a system governed by the heat equation focusing our attention on its numerical approximation. In particular, we will examine in detail the case of boundary controls. We refer to Lions [17] for the theoretical framework of the continuos problem.
It is well known that via the Pontryagin's maximum principle, one can characterize the open–loop solution in terms of a coupled system involving the state $y(u)$ and the adjoint state $p(u)$ (u is the control). In this formulation the adjoint state satisfies a backward heat equation in $[0, T]$ having $y(x, T; u)$ as initial condition so that the numerical solution of the coupled system requires a long iterative process. This approach has been followed by Hackbush ([13], [14]) who has applied the multigrid technique to obtain the solution of the system describing the necessary conditions for optimality. Note that the approximation of the system of necessary conditions requires a good initial guess for the multipliers (the adjoint states) to start the iterative procedure and to guarantee a local convergence to the solution. Searching for the initial guess is often one of the more difficult tasks in that approximation method.
Having in mind those limitations, we consider here a different approach where Dynamic Programming plays a role also in the construction of the approximation scheme. The main goal is to reduce the global complexity of the algorithm and to avoid the cumbersome solution of a Hamilton–Jacobi type equation in the whole space of initial data (which in our case would require the solution of a nonlinear PDE in $L^2(\Omega)$).

This work has been partially supported by the Ministry for University and Scientific Research (MURST Project "Analisi Numerica e Matematica Computazionale").

Our method is based on a backward–forward procedure where we try to compute a trajectory minimizing the pay–off in a neighbourhood of a reference trajectory (which is assumed to be given). From this point of view this paper is close to the numerical experiments presented for finite dimensional control problems in [1].
We will establish convergence for a time–discrete scheme and we will also give some hints for the construction of the fully discrete algorithm based on a space discretization (by finite elements or finite differences). A similar approach has been followed by Ferretti for distributed and boundary control problems ([8], [10]) without state constraints and localization around a trajectory. Among the huge amount of literature on the control of parabolic type equations we quote the paper [2] where the approximation of the linear regulator problem is considered and [16] where abstract convergence results for approximation methods have been given mainly for the linear quadratic problem (see also the reference therein and [15]). More recently Banks and Ito [3] have studied the numerical aproximation of boundary control problems by means of augmented lagrangian techniques.

The paper is organized as follows. In Section 2 we set up the problem, introduce our notations and prove some useful properties of the value function for the continuous control problem. Those properties are rather important for the construction of the discretization scheme and also play a role in the proof of the dynamic programming principle which is given in the Appendix. The time discretization is treated in Section 3 where we prove a discrete Dynamic Programming Principle which will be used to obtain the approximation scheme and to prove our convergence result. Section 4 is devoted to the presentation of the basic ideas for the numerical algorithm.

2. Some preliminary results for the continuous problem

Let Ω be un open subset of $\mathbb{R}^n$, $\Gamma = \partial\Omega$ and t_0, T be two real numbers such that $T > t_0$. We set $Q = \Omega \times (t_0, T)$ and $\Sigma = \Gamma \times (t_0, T)$.
We will say that Γ is regular if it is piecewise C^1 in the sense given by the definition of [18, p. 25].

Let E be an Hilbert space, we will use the following spaces

$$L^2(t_0, T; E) = \{h : [t_0, T] \longrightarrow E \text{ such that } \int_{t_0}^{T} \| h(s) \|_E^2 \, ds < +\infty\}$$

$$W(t_0, T) = \{h \in L^2(t_0, T; E) \text{ such that } \frac{\partial h}{\partial t} \in L^2(t_0, T; E')\}$$

where the derivative $\partial h/\partial t$ should be understood in the distributional sense. We will consider a boundary control problem with Neumann condition for the heat equation:

$$\text{(2.1)} \qquad \begin{cases} \frac{\partial}{\partial t} y(x,t) - \Delta y(x,t) = g(x,t) & \text{in } Q \\ \frac{\partial}{\partial \nu} y(x,t) = u(x,t) & \text{on } \Sigma \\ y(x,t_0) = \eta(x) & \text{in } \Omega \end{cases}$$

where η is given in $L^2(\Omega)$, $g \in L^2(t_0, T; H^1(\Omega))$, and the controls u belong to $U \equiv L^2(\Sigma)$, i.e. we start investigating the problem without constraints. Later on in this section we will also introduce some constraints on the control space.

We will always consider solutions of (2.1) in the weak sense. We will denote that solution by $y_\eta(x, t; u)$ and when it will be possible without ambiguities we will also use the short notation $y_\eta(t; u)$ or $y_\eta(t)$.

In order to have existence and uniqueness for the solution of (2.1) for every fixed $u \in U$ one can apply the standard results for the heat equation (see f.e. Theorem 1.2 p.102 in [17]). The solution depends continuously on the data and lives in $W(t_0, T)$. Sometimes we will refer to $L^2(\Omega)$ as the space of observations, i.e. the space where the solution lives for every fixed t in $[t_0, T]$.

In order to simplify our notations, let us define

$$\| u(t) \|_b^2 \equiv \int_\Gamma | u(x,t) |^2 \, dx ,$$

$$\| y(t;u) \|^2 \equiv \int_\Omega | y_\eta(x,t;u) |^2 \, dx .$$

The final time T being fixed, for every initial condition (η, t_0) and for any control u we define the pay–off:

$$J(\eta, t_0, u) = \int_{t_0}^{T} f(y_\eta(t;u), u(t)) \, \mathrm{e}^{-\lambda(t-t_0)} dt + \psi(y_\eta(T;u)), \; \mathrm{e}^{-\lambda(T-t_0)} \tag{2.2}$$

where $f : L^2(\Omega) \times L^2(\Gamma) \to \mathbb{R}$, $\psi : L^2(\Omega) \to \mathbb{R}$ and $\lambda > 0$. Note that the discount factor $e^{-\lambda(t-t_0)}$ appearing in the functional (2.2) is usually included to up–date all the cost at the initial time (when the decision has to be made) and to decrease the costs over long time intervals.

We will make the following assumptions:

(a) there exists a constant B_f such that

$$\left| f(y,u) \right| \le B_f, \quad \text{for any } (y,u) \in L^2(\Omega) \times L^2(\Gamma) , \tag{2.3}$$

$$f \text{ is continuous with respect to } (y,u); \tag{2.4}$$

(b) there exists a constant $C_f > 0$ such that for any $z, w \in L^2(\Omega)$ and for any $u \in L^2(\Gamma)$,

$$\left| f(z,u) - f(w,u) \right| \le C_f \, \| z - w \|, \tag{2.5}$$

$$f \text{ is convex with respect to } (y,u); \tag{2.6}$$

(c) there exists a constant B_ψ such that

$$\left| \psi(y) \right| \le B_\psi, \quad \text{for any } y \in L^2(\Omega); \tag{2.7}$$

(d) there exists a constant $C_\psi > 0$ such that for any $z, w \in L^2(\Omega)$

$$\left| \psi(z) - \psi(w) \right| \le C_\psi \, \| z - w \|, \tag{2.8}$$

$$\psi \text{ is convex.} \tag{2.9}$$

We want to minimize J, i.e. we want to find an optimal control $u \in U$ such that

$$J(\eta, t_0, u) = \inf_{v \in U} J(\eta, t_0, v). \tag{2.10}$$

The value function for our problem is defined as

$$v(\eta, t_0) = \inf_{u \in U} J(\eta, t_0, u). \tag{2.11}$$

Definition 2.1. *The system (2.1) is said to be controllable if for $u \in U$, the observation $y(T; u)$ spans a dense subspace in the space of observations.*

For the proof that our system is controllable when $U = L^2(\Sigma)$ see [17, p. 207]. It is clear that, in general, the existence of an optimal control attaining the minimum of J is not guaranteed. In order to prove it we need some additional assumptions.

Theorem 2.2. *Let (2.3), (2.4), (2.5), (2.6), (2.7), (2.8), (2.9) be satisfied. Let $U_{ad} = \{u \in U : \| u \|_{L^2(\Sigma)} \leq R\}$ and assume that $f(y, \cdot)$ is continuous for any $y \in L^2(\Omega)$.*
Then, there exists a control $u \in U_{ad}$ such that

$$J(\eta, t_0, u) = \inf_{v \in U_{ad}} J(\eta, t_0, v). \tag{2.12}$$

Proof. Let us define $\phi : U_{ad} \to \mathbb{R}$ as $\phi(u) = J(\eta, t_0, u)$. In order to get the result we will apply Corollary III.20 in [6] (for readers convenience its exact statement has been included in the Appendix) so let us check that all the assumptions are satisfied. It is easy to see that U_{ad} is a closed, bounded, convex subset of the reflexive Banach space $L^2(\Sigma)$. By (2.3) and (2.7) we have that $\phi \not\equiv +\infty$.
To prove the convexity and the continuity of ϕ we first observe that for any $\alpha \in (0, 1)$ and for any $u, v \in U_{ad}$,

$$y_\eta(x, t; \alpha u + (1 - \alpha)v) = \alpha y_\eta(x, t; u) + (1 - \alpha) y_\eta(x, t; v). \tag{2.13}$$

In fact, we can write $y_\eta(x, t; \alpha u + (1 - \alpha)v) = a(x, t) + b_{\alpha u + (1-\alpha)v}(x, t)$ where $a(x, t)$ is solution of (2.1) for $u = 0$, i.e.

$$\begin{cases} \frac{\partial}{\partial t} y(x, t) - \Delta y(x, t) = g(x, t) & \text{in } Q \\ \frac{\partial}{\partial \nu} y(x, t) = 0 & \text{on } \Sigma \\ y(x, t_0) = \eta(x) & \text{in } \Omega \end{cases} \tag{2.14}$$

and $b_{\alpha u + (1-\alpha)v}(x, t)$ is solution of (2.1) for $g = 0$ and $\eta = 0$ with the control $\alpha u + (1 - \alpha)v$, i.e.

$$\begin{cases} \frac{\partial}{\partial t} y(x, t) - \Delta y(x, t) = 0 & \text{in } Q \\ \frac{\partial}{\partial \nu} y(x, t) = \alpha u + (1 - \alpha)v & \text{on } \Sigma \\ y(x, t_0) = 0 & \text{in } \Omega\,. \end{cases} \tag{2.15}$$

By linearity of the equation we have that $b_{\alpha u + (1-\alpha)v}(x, t) = \alpha b_u(x, t) + (1 - \alpha) b_v(x, t)$ and this give us (2.13).

By the convexity of ψ and f and by (2.13) we can easily prove that ϕ is convex. Let us prove that ϕ is lower semicontinuous.

Let $u_n \in U_{ad}$ be a sequence converging to u in $L^2(\Sigma)$. By the weak formulation of the heat equation and exploiting the linearity of the dynamics we get the following estimate

$$\| y_\eta(t;u) - y_\eta(t;u_n) \| \leq \| u - u_n \|_{L^2(\Sigma)} \ . \tag{2.16}$$

The continuity of $\psi(y_\eta(T;u))\mathrm{e}^{-\lambda(T-t_0)}$ with respect to u then follows by (2.8) and (2.16).

Let us examine the integral term appearing in the definition of ϕ.

$$\begin{aligned}
&\int_{t_0}^{T} f(y_\eta(t;u),u(t))\mathrm{e}^{-\lambda(t-t_0)}dt = \\
&\quad = \int_{t_0}^{T} [f(y_\eta(t;u),u(t)) - f(y_\eta(t;0),u(t))]\mathrm{e}^{-\lambda(t-t_0)}dt + \\
&\quad + \int_{t_0}^{T} [f(y_\eta(t;0),u(t)) - f(y_\eta(t;0),0)]\mathrm{e}^{-\lambda(t-t_0)}dt + \\
&\quad + \int_{t_0}^{T} f(y_\eta(t;0),0)]\mathrm{e}^{-\lambda(t-t_0)}dt \ .
\end{aligned} \tag{2.17}$$

By (2.16) and the Lipschitz continuity of $f(\cdot,u)$, we get

$$\int_{t_0}^{T} [f(y_\eta(t;u),u(t)) - f(y_\eta(t;0),u(t))]\mathrm{e}^{-\lambda(t-t_0)}dt \leq C_f \| u \|_{L^2(\Sigma)} \ .$$

Note that the continuity assumption on $f(y,\cdot)$ implies, by the Fatou's Lemma, that

$$I(u) \equiv \int_{t_0}^{T} f(y_\eta(t;u),\cdot)\mathrm{e}^{-\lambda(t-t_0)}dt$$

is a lower semicontinuous function over U_{ad}. Then, we can conclude that ϕ is lower semicontinuous over U_{ad} since the last term in (2.17) is constant with respect to u. The proof can be completed simply applying Corollary III.20 in [6]. □

Assuming that f and ψ are strictly convex, we can conclude that the optimal control is unique. Note that hypothesis (2.6) is fullfilled when the running cost f has the form $f(y,u) = f_1(y) + f_2(u)$ with f_1 and f_2 convex.

The Lipschitz continuity of the cost function with respect to the initial data is established in the next theorem.

Theorem 2.3. *Let Ω be a bounded open set with a regular boundary and let the assumptions (2.5) and (2.8) be satisfied. Then, there exists a constant $C > 0$ such that for any u in U*

$$\Big| J(\eta,t_0,u) - J(\mu,t_0,u) \Big| \leq C \| \eta - \mu \| \,, \qquad \forall \eta,\mu \in L^2(\Omega). \tag{2.18}$$

Proof. We can write $y_\eta(x,t;u) = a_\eta(x,t) + b(x,t,u)$ where $a_\eta(x,t)$ is the solution of (2.1) for $g = 0$ and $u = 0$, and $b(x,t,u)$ is solution of (2.1) for $\eta = 0$.
The map $a_\eta(x,t) = A(t)\eta(x)$ is linear in η by linearity of the equation. Theorem 7.2-2 p.161 in [18] gives for $g = 0$, the continuous dependence of the solution with

respect to the initial data. We can conclude that it exists a constant $C_a > 0$ such that

$$\| A(t)\eta(x) \| \leq C_a \| \eta \| \ , \forall t \in [t_0, T] \ . \tag{2.19}$$

Let us denote by $y_\eta(x, t; u)$ the solution of (2.1) and by $y_\mu(x, t; u)$ the solution of the same system with initial data $\mu(x)$.
The linearity of A and the inequality (2.19) imply

$$\| y_\eta(t; u) - y_\mu(t; u) \| = \| A(t)(\eta - \mu) \| \leq C_a \| \eta - \mu \| \ , \qquad \forall t \in [t_0, T]. \tag{2.20}$$

Recalling the definition of the pay-off, by (2.5) and (2.8) we get

$$\begin{aligned} | J(\eta, t_0, u) - J(\mu, t_0, u) | \leq & \int_{t_0}^{T} |f(y_\eta(t; u), u(t)) - f(y_\mu(t; u), u(t))| \mathrm{e}^{-\lambda(t-t_0)} dt + \\ & + |\psi(y_\eta(T)) - \psi(y_\mu(T))| e^{-\lambda(T-t_0)} \leq \\ \leq & \int_{t_0}^{T} C_f \| y_\eta(t) - y_\mu(t) \| \, dt + C_\psi \| y_\eta(T) - y_\mu(T) \| \ . \end{aligned}$$

By (2.20), we can conclude that there exists a constant independent from u, $C = C_f C_a T + C_\psi C_a$, such that

$$| J(\eta, t_0, u) - J(\mu, t_0, u) | \leq C \| \eta - \mu \| \ . \tag{2.21}$$

□

As an easy corollary we get the Lipschitz continuity of the value function with respect to the initial data.

Corollary 2.4. *In the same hypotheses of Lemma* 2.3 *there exists* $C > 0$ *such that for any* $\eta, \mu \in L^2(\Omega)$ *we have*

$$\left| v(\eta, t_0) - v(\mu, t_0) \right| \leq C \| \eta - \mu \| \ . \tag{2.22}$$

We state now the Dynamic Programming Principle which will give a characterization of the value function also useful for numerical purposes (see the Appendix for the proof).

Theorem 2.5. *Let the value function be defined as in* (2.11)*. Then*

$$v(\eta, t_0) = \inf_{u \in U} \left\{ \int_{t_0}^{\tau} f(y_\eta(t; u), u(t)) e^{-\lambda(t-t_0)} dt + v(y_\eta(\tau), \tau) e^{-\lambda(\tau - t_0)} \right\} \tag{2.23}$$

for every τ, $t_0 \leq \tau \leq T$.

It is well known that the Dynamic Programming Principle gives a characterization of the value function where the value at the initial condition $v(\eta, t_0)$ depends on its value at the point $y_\eta(\tau)$ belonging to the optimal trajectory. This is the basic principle of optimality for the trajectories. This principle is also useful to derive the Hamilton–Jacobi-Bellman equation giving a characterization of v in terms of a partial differential equation. In what follows we will use a discrete version of this principle (Theorem 3.1) to get a semi–discrete approximation scheme for the value function. Moreover, (2.23) and its discrete version will play an important role when proving the convergence of that scheme to the value function of our problem.

For the numerical approximation, we are interested to a particular choice of the cost function (2.2). Note that an explicit dependence on t can be easily included in the running cost. We will take then

$$f(t, y_\eta, u) = \| u(t) \|_b^2 + \| y_\eta(t; u) - \zeta(t) \|^2 \tag{2.24}$$

where ζ is a given function from $L^2(Q)$ in $\mathbb{R}$ and

$$\psi(y_\eta(T)) = \| y_\eta(T) - z_T \|^2 \tag{2.25}$$

where z_T is a given function from $L^2(\Omega)$ in $\mathbb{R}$.

Moreover, we also want to restrict the space of admissible controls and of the initial conditions assuming that

$$u \in U_{ad} = \{u \in U \text{ such that } \| u(\cdot, t) \|_{L^2(\Gamma)} \leq R \text{ for any } t \in [t_0, T]\}, \tag{2.26}$$

$$\eta \in K = \{\mu \in L^\infty(\Omega) \text{ such that } \| \mu \|_\infty \leq M\}. \tag{2.27}$$

Note that the restriction on the controls which appear replacing U by U_{ad} may imply the loss of the controllability property for our dynamics. However, under the assumptions of Theorem 2.2 there exists a minimum for the pay-off. In practice, we can think that $\zeta(t)$ is a trajectory starting at η and reaching a neighbourhood of our terminal state z_T obtained by an analysis of the controllability problem or by some experiments. It is what we know about the problem *before* starting the optimization process. We want to minimize the pay–off given by

$$\int_{t_0}^{T} \| u(t) \|_b^2 + \| y_\eta(t; u) - \zeta(t) \|^2 \, \mathrm{e}^{-\lambda(t-t_0)} dt + \| y_\eta(T) - z_T \|^2 \, \mathrm{e}^{-\lambda(T-t_0)}$$

Note that the term

$$\int_{t_0}^{T} \| y_\eta(t; u) - \zeta(t) \|^2 \, \mathrm{e}^{-\lambda(t-t_0)} dt$$

has been added in order to penalize the L^2-distance from ζ in the interval $[t_0, T]$. We will come back to this point in the last section, where we will explain the algorithm giving more details.

In the following Lemma we will show that under some restrictive hypoteses we can apply Theorem 2.3 to the pay–off J corresponding to (2.24), (2.25) so that our value function is Lipschitz continuous with respect to the initial data.

Lemma 2.6. *Assume that Ω is a bounded open set with regular boundary, and that (2.26) and (2.27) hold true. Then (2.5) and (2.8) are satisfied.*

Proof. Let us start from the weak formulation of the heat equation. After some calculations one can obtain the following estimate

$$\| y(t; u) \|^2 \leq \| \eta \|^2 + (1/\alpha) \| g \|_{L^2(0,T;H^1)}^2 + \| u \|_{L^2(\Sigma)}^2 \tag{2.28}$$

for all t in $[t_0, T]$ ($\alpha > 0$). Then, (2.26) and (2.27) imply

$$\| y(t; u) \| \leq K \ , \text{ for all } t \in [t_0, T] \ , \tag{2.29}$$

where $K = (M^2 meas(\Omega) + (1/\alpha) \| g \|^2_{L^2(0,T;H^1)} + R^2)^{1/2}$.

Fix η and μ in $L^2(\Omega)$, t in $[t_0, T]$ and u in U_{ad}. By the definition (2.24) we get

$$f(y_\eta(t; u), u(t)) - f(y_\mu(t; u), u(t)) = \| y_\eta(t; u) - \zeta(t) \|^2 - \| y_\mu(t; u) - \zeta(t) \|^2 \leq$$

$$\leq \| y_\eta(t; u) - y_\mu(t; u) \| \Big\{ \| y_\eta(t; u) - y_\mu(t; u) \| + 2 \| y_\mu(t; u) - \zeta(t) \| \Big\}.$$

By (2.29) we have

$$f(y_\eta(t; u), u(t)) - f(y_\mu(t; u), u(t)) \leq [2K + 2(K + \| \zeta \|)] \| y_\eta(t; u) - y_\mu(t; u) \| \ .$$

Inverting the roles of μ and η, after some calculations we get

$$\Big| f(y_\eta(t; u), u(t)) - f(y_\mu(t; u), u(t)) \Big| \leq C_f \| y_\eta(t; u) - y_\mu(t; u) \|$$

where $C_f = [2K + 2(K + \| \zeta \|)]$. This proves (2.5).
The proof of (2.8) is similar so we skip the details. □

By Theorem 2.2, one can easily see that under the same hypoteses of Lemma 2.6 we have existence and uniqueness of the optimal control for the particular cost functional defined by (2.25) and (2.24). Note that the proof of the the Dynamic Programming Principle is still valid also when we assume (2.26) and (2.27) (see Appendix).

3. Semi–discretization and convergence

Let us introduce the discretization in time of our problem. For the sake of simplicity we will consider only the case when the time–step is constant, but the results can be extended to a variable time–step using standard arguments. Given $N \in \mathbb{N}$ we set $\Delta t = [T - t_0]/N$ and $t_n = t_0 + n\Delta t$ for $n = 0, \ldots, N$.

Let the set of controls U be replaced by the set of admissible discrete controls $\widehat{U} \subset U$. The set $\widehat{U}$ will play an important role in our discretization. Just to fix ideas, one can imagine that $\widehat{U}$ is some sort of finite representation of U more suitable for the construction of the algorithm (we will come back to this point in the next section).

We replace our dynamics (2.1) by a discrete time dynamics obtained, for example, by an explicit Euler scheme

$$\begin{cases} y(x, t_{n+1}) = y(x, t_n) + \Delta t \, [\Delta y(x, t_n) + g(x, t_n)] \\ \frac{\partial}{\partial \nu} y(x, t_n) = u(x, t_n) \\ y(x, t_0) = \eta(x) \ . \end{cases} \tag{3.1}$$

The solution will be denoted by $\widehat{y}_\eta(x, t_n; u)$ and, whenever is possible, we will use the short notations $\widehat{y}_\eta(t_n; u)$ and $\widehat{y}_\eta(t_n)$, for every $n = 0, \ldots, N$ and for every $u \in \widehat{U}$.

For any $n_0 \in \{0, \dots, N-1\}$ we define the pay–off corresponding to the discrete time dynamics (3.1) as

$$\widehat{J}(\eta, t_{n_0}, u) = \Delta t \sum_{n=n_0}^{N-1} f(\widehat{y}_\eta(t_n), u(t_n)) \mathrm{e}^{-\lambda(t_n - t_0)} + \psi(\widehat{y}_\eta(T; u)) \mathrm{e}^{-\lambda(T - t_{n_0})} \tag{3.2}$$

and for $n_0 = N$ we define,

$$\widehat{J}(\eta, t_N) = \psi(\eta). \tag{3.3}$$

The corresponding value function is

$$\widehat{v}(\eta, t_{n_0}) = \inf_{u \in \widehat{U}} \widehat{J}(\eta, t_{n_0}, u), \qquad n_0 \in \{0, \dots, N\}. \tag{3.4}$$

We will construct our scheme by means of the following discrete version of the dynamic programming principle.

Theorem 3.1. *Let the value function be defined by (3.4). Then,*

$$\widehat{v}(\eta, t_{n_0}) = \inf_{u \in \widehat{U}} \left\{ \Delta t \sum_{n=n_0}^{p-1} f(\widehat{y}_\eta(t_n), u(t_n)) e^{-\lambda(t_n - t_{n_0})} + \widehat{v}(\widehat{y}_\eta(t_p), t_p) e^{-\lambda(t_p - t_{n_0})} \right\} \tag{3.5}$$

for every integer p, $n_0 < p \le N$.

Proof. Let us denote by $\widehat{w}(\eta, t_{n_0})$ the right-hand side of (3.5).
We consider two cases.
1) Let $p = N$. We have

$$\widehat{w}(\eta, t_{n_0}) = \inf_{u \in \widehat{U}} \left\{ \Delta t \sum_{n=n_0}^{N-1} f(\widehat{y}_\eta(t_n), u(t_n)) \mathrm{e}^{-\lambda(t_n - t_0)} + \widehat{v}(\widehat{y}_\eta(T), T) \mathrm{e}^{-\lambda(T - t_{n_0})} \right\}. \tag{3.6}$$

By the definitions (3.4) and (3.2), we have

$$\widehat{v}(\widehat{y}_\eta(T), T) = \inf_{u \in \widehat{U}} \widehat{J}(\widehat{y}_\eta(T), T; u) = \inf_{u \in \widehat{U}} \psi(\widehat{y}_\eta(T)),$$

so that (3.6) coincides with the definition of value function.
2) Let $p < N$. We divide the proof into two parts.
a) $\widehat{v}(\eta, t_{n_0}) \ge \widehat{w}(\eta, t_{n_0})$.
Let us fix a control u, by (3.2) we have

$$\begin{aligned} \widehat{J}(\eta, t_{n_0}, u) = {} & \Delta t \sum_{n=n_0}^{p-1} f(\widehat{y}_\eta(t_n), u(t_n)) \mathrm{e}^{-\lambda(t_n - t_{n_0})} + \\ & + \Delta t \sum_{n=p}^{N-1} f(\widehat{y}_\eta(t_n), u(t_n)) \mathrm{e}^{-\lambda(T - t_{n_0})} + \psi(\widehat{y}_\eta(T)) \mathrm{e}^{-\lambda(T - t_{n_0})}. \end{aligned}$$

Defining $\mu = \widehat{y}_\eta(t_p; u)$, the uniqueness of the solution of (3.1) implies $\widehat{y}_\eta(x, t_n; u) = \widehat{y}_\mu(x, t_n; u)$ for every $t_n \geq t_p$. Then,

$$\begin{aligned}\widehat{J}(\eta, t_{n_0}, u) = \Delta t \sum_{n=n_0}^{p-1} f(\widehat{y}_\eta(t_n), u(t_n)) e^{-\lambda(t_n - t_{n_0})} + \\ + \Delta t \sum_{n=p}^{N-1} f(\widehat{y}_\mu(t_n), u(t_n)) e^{-\lambda(t_n - t_{n_0})} + \psi(\widehat{y}_\mu(T)) e^{-\lambda(T - t_{n_0})} = \\ = \Delta t \sum_{n=n_0}^{p-1} f(\widehat{y}_\eta(t_n), u(t_n)) e^{-\lambda(t_n - t_{n_0})} + \widehat{J}(\mu, t_p, u) e^{-\lambda(t_p - t_{n_0})} \geq \\ \geq \Delta t \sum_{n=n_0}^{p-1} f(\widehat{y}_\eta(t_n), u(t_n)) e^{-\lambda(t_n - t_{n_0})} + \widehat{v}(\mu, t_p) e^{-\lambda(t_p - t_{n_0})}\end{aligned}$$

by definition (3.2) and (3.4). Taking the infimum over $\widehat{U}$ we have

$$\widehat{v}(\eta, t_{n_0}) \geq \inf_{u \in \widehat{U}} \left\{ \Delta t \sum_{n=n_0}^{p-1} f(\widehat{y}_\eta(t_n), u(t_n)) e^{-\lambda(t_n - t_{n_0})} + \widehat{v}(\mu, t_p) e^{-\lambda(t_p - t_{n_0})} \right\} .$$

b) $\widehat{v}(\eta, t_{n_0}) \leq \widehat{w}(\eta, t_{n_0})$.
Fix $\tilde{u} \in \widehat{U}$ let $\mu = \widehat{y}_\eta(t_p; \tilde{u})$. For any fixed $\varepsilon > 0$ there exists a control u_ε such that

$$\widehat{v}(\mu, t_p) + \varepsilon > \widehat{J}(\mu, t_p, u_\varepsilon) . \tag{3.7}$$

Let us define the control

$$\overline{u}(x, t_n) = \begin{cases} \tilde{u}(x, t_n) & t_{n_0} \leq t_n \leq t_{p-1} \\ u_\varepsilon(x, t_n) & t_p \leq t_n \leq T . \end{cases}$$

We first note that the uniqueness of the solution of (3.1) implies

$$\widehat{y}_\eta(x, t_n; \overline{u}) = \begin{cases} \widehat{y}_\eta(x, t_n; \tilde{u}) & t_{n_0} \leq t_n \leq t_{p-1} , \\ \widehat{y}_\mu(x, t_n; u_\varepsilon) & t_p \leq t_n \leq T . \end{cases} \tag{3.8}$$

Recalling the definitions $\overline{u}$, (3.2), (3.8) and the inequality (3.7) we have

$$\begin{aligned}\widehat{v}(\eta, t_{n_0}) \;\leq\; \widehat{J}(\eta, t_{n_0}, \overline{u}) = \Delta t \sum_{n=n_0}^{p-1} f(\widehat{y}_\eta(t; \overline{u}), \overline{u}(t_n)) e^{-\lambda(t_n - t_{n_0})} + \\ + \Delta t \sum_{n=p}^{N-1} f(\widehat{y}_\eta(t; \overline{u}), \overline{u}(t_n)) e^{-\lambda(t_n - t_{n_0})} + \psi(\widehat{y}_\eta(T)) e^{-\lambda(T - t_{n_0})} =\end{aligned}$$

$$
\begin{aligned}
&= \Delta t \sum_{n=n_0}^{p-1} f(\widehat{y}_\eta(t;\tilde{u}), \tilde{u}(t_n)) e^{-\lambda(t_n - t_{n_0})} + \\
&\quad + \Delta t \sum_{n=p}^{N-1} f(\widehat{y}_\mu(t;u_\varepsilon), u_\varepsilon(t_n)) e^{-\lambda(t_n - t_{n_0})} + \psi(\widehat{y}_\mu(T)) e^{-\lambda(T - t_{n_0})} = \\
&= \Delta t \sum_{n=n_0}^{p-1} f(\widehat{y}_\eta(t;\tilde{u}), \tilde{u}(t_n)) e^{-\lambda(t_n - t_{n_0})} + \widehat{J}(\mu, t_p, u_\varepsilon) e^{-\lambda(t_p - t_{n_0})} < \\
&\quad < \Delta t \sum_{n=n_0}^{p-1} f(\widehat{y}_\eta(t;\tilde{u}), \tilde{u}(t_n)) e^{-\lambda(t_n - t_{n_0})} + (\widehat{v}(\mu, t_p) + \varepsilon)\ e^{-\lambda(t_p - t_{n_0})}.
\end{aligned}
$$

For ε tending to 0 and taking the infimum over $\tilde{u} \in \widehat{U}$ we end the proof. □

Let us turn now to the proof of convergence. The basic idea is simple: coupling an Euler discretization scheme for the dynamics with a quadrature formula (rectangles) for the cost we can get a reasonable approximation of the value function. Two main questions have to be clarified. Which conditions on the two discretization schemes guarantee the convergence to the value function of our approximation scheme ? How accurate is that discretization scheme ?

To obtain results in both directions we make the following assumptions:

$(H1)$ For any $\eta \in K$, $u \in U_{ad}$, $0 < \Delta t < T - t_0$ and $\xi \in [t_0, T]$ there exist $\widehat{u} \in \widehat{U}$ and two positive constants C_1 and C_2, such that

$$
\| y_\eta(\xi, u) - \widehat{y}_\eta(\xi, \widehat{u}) \| \leq C_1 (\Delta t)^2 \tag{3.9}
$$

$$
\left| \Delta t\, f(\widehat{y}_\eta(\xi), \widehat{u}(\xi)) e^{-\lambda(\xi - t_0)} - \int_\xi^{\xi + \Delta t} f(y_\eta(t), u(t)) e^{-\lambda(t - t_0)} dt \right| \leq C_2 (\Delta t)^2 \tag{3.10}
$$

$(H2)$ For any $\eta \in K$, $\widehat{u} \in \widehat{U}$, $0 < \Delta t < T - t_0$ and $\xi \in [t_0, T - \Delta t]$, there exist $u \in U_{ad}$ and two positive constants C_1, C_2 such that (3.9) and (3.10) hold.

Note that the constants C_1 and C_2 appearing in (3.9) and (3.10) are independent of any other variable, so that the above inequalities provide uniform estimates for the time discretization of the dynamics and of the cost functional.

The following result gives an estimate of the L^∞ error related to our time discretization. The proof follows the lines of Theorem 3.1 in Falcone-Ferretti [11] where a similar estimate is obtained for a finite dimensional control problem.

Theorem 3.2. *Let (H1), (H2) be satisfied. Then, for any $\eta \in K$, $\xi \in [t_0, T]$ and $0 < \Delta t < T - t_0$, there exists a constant $C > 0$ such that*

$$
\| v(\eta, \xi) - \widehat{v}(\eta, \xi) \|_\infty \leq C\, \Delta t\,. \tag{3.11}
$$

Proof. Let us assume that there exists a control $\widehat{u} \in \widehat{U}$ such that the minimum is attained in the discrete Dynamic Programming Principle (if not the same proof will work with slight modifications). Let $\overline{u}$ be an optimal control corresponding to $\widehat{u}$ such that (H2) holds and set $\beta = e^{-\lambda \Delta t}$. Note that in the following calculations we do not require neither $\widehat{u}$ nor $\overline{u}$ to be unique.

By applying Theorem 2.5 for $\tau = t_p$ and Theorem 3.1 we have

$$
\begin{aligned}
v(\eta,\xi) - \widehat{v}(\eta,\xi) \leq \Big| \int_{\xi}^{t_p} & f(y_\eta(t,\overline{u}),\overline{u}(t))\mathrm{e}^{-\lambda(t_p-t_0)}dt + \\
& - \Delta t \sum_{n=0}^{p-1} f(\widehat{y}_\eta(t_n),\widehat{u}(t_n))\mathrm{e}^{-\lambda(t_n-t_0)} \Big| + \\
& + \beta^p[v(y_\eta(t_p,\overline{u}),t_p) - \widehat{v}(\widehat{y}_\eta(t_p,\widehat{u}),t_p)].
\end{aligned}
\tag{3.12}
$$

The above inequality and (H2) (b) imply

$$
\begin{aligned}
v(\eta,\xi) - \widehat{v}(\eta,\xi) &\leq pC_2(\Delta t)^2 + \beta^p[v(y_\eta(t_p,\overline{u}),t_p) - v(y_\eta(t_p,\widehat{u}),t_p)] + \\
&+ \beta^p[v(y_\eta(t_p,\widehat{u}),t_p) - \widehat{v}(\widehat{y}_\eta(t_p,\widehat{u}),t_p)] \leq \\
&\leq pC_2(\Delta t)^2 + \beta^p C \parallel y_\eta(t_p,\overline{u}) - y_\eta(t_p,\widehat{u}) \parallel + \\
&+ \beta^p[v(y_\eta(t_p,\widehat{u}),t_p) - \widehat{v}(\widehat{y}_\eta(t_p,\widehat{u}),t_p)]
\end{aligned}
$$

for the Lipschitz continuity of the value function.
Then, assumption (H2)(a) implies

$$
v(\eta,\xi) - \widehat{v}(\eta,\xi) \leq pC_2(\Delta t)^2 + CC_1(\Delta t)^2\beta^p + \beta^p[v(y_\eta(t_p,\widehat{u}),t_p) - \widehat{v}(\widehat{y}_\eta(t_p,\widehat{u}),t_p)] \,.
$$

Then we can conclude that

$$
(1-\beta^p) \sup_{\substack{\eta\in L^\infty(\Omega)\\ t\in[t_0,T]}} \Big(v(\eta,t) - \widehat{v}(\eta,t)\Big) \leq pC_2(\Delta t)^2 + CC_1(\Delta t)^2\beta^p \,.
\tag{3.13}
$$

In the same way one can prove a similar inequality for $\sup_{\substack{\eta\in L^\infty(\Omega)\\ t\in[t_0,T]}} \Big(\widehat{v}(\eta,t) - v(\eta,t)\Big)$.

In conclusion, we get

$$
\parallel v(\eta,T-t) - \widehat{v}(\eta,T-t) \parallel_\infty \leq \frac{pC_2(\Delta t)^2 + CC_1(\Delta t)^2\beta^p}{1-\beta^p}
\tag{3.14}
$$

and since $1-\beta^p = O(p\Delta t)$ this ends the proof. □

The above conditions ($H1$) and ($H2$) can be interpreted as assumptions on the order of approximation of the time discretization for the dynamics and for the cost integral. In finite dimensional control problems one can also obtain sufficient conditions on the data guaranteeing ($H1$) and ($H2$) (see [11]). The inequality (3.9) is satisfied if the discrete dynamics is close enough to the continuous dynamics and this of course will depend on the accuracy of the approximation scheme and on the discretization of the control space (note that we are taking the control in a set $\widehat{U}$ which stands for a discretization of U_{ad}).

Let us examine in more detail the second inequality (3.10). In order to guarantee that (3.10) holds true for the approximation of our parabolic problem we need to know that the time derivative of the control u and of the solution of the heat equation exist. This can be obtained adding some regularity assumptions on the data of the problem (see f.e. Theorem 2 in [12, p. 144]).

For example, let us assume that f is Lipschitz continuous with respect to the couple (y,u) and that the space $\widehat{U}$ is such that for any $u \in U_{ad}$ there exists at least one control $\widehat{u} \in \widehat{U}$ guaranteeing

$$||u - \widehat{u}||_{L^2([0,T])} \leq C\Delta t$$

for some positive constant C. Then,

$$\begin{aligned}
&\int_{\xi}^{\xi+\Delta t} \left| f(\widehat{y}_\eta(\xi), \widehat{u}(\xi)) \mathrm{e}^{-\lambda(\xi - t_0)} - f(y_\eta(t), u(t)) \mathrm{e}^{-\lambda(t-t_0)} dt \right| \leq \\
&\leq \int_{\xi}^{\xi+\Delta t} \left| (f(\widehat{y}_\eta(\xi), \widehat{u}(\xi)) - f(y_\eta(t), u(t))) \mathrm{e}^{-\lambda(\xi-t_0)} \right| dt + \\
&+ \int_{\xi}^{\xi+\Delta t} |f(y_\eta(t), u(t)))| \, |\mathrm{e}^{-\lambda(t-t_0)} - \mathrm{e}^{-\lambda(\xi-t_0)}| dt \leq \\
&\leq \int_{\xi}^{\xi+\Delta t} \left[C_1 ||u - \widehat{u}||_{L^2([0,T])} + C_2 \Delta t \right] dt \leq C(\Delta t)^2 \, .
\end{aligned} \tag{3.15}$$

This tells us that it is important to built an accurate discretization of the control space (f.e. by means of piecewise polynomial functions of time) and couple this discretization with sufficiently accurate approximation schemes for the dynamics and the cost to get the error bound proved in Theorem 2.5.

4. Some hints for the algorithm

In order to solve numerically our boundary control problem we use a local version of Dynamic Programming trying to reduce the huge amount of computations usually needed by that approach. Let us assume that we want to compute the minimum over a subset of all the possible trajectories, f.e. we can imagine that there exists a trajectory ζ starting at our initial condition η and reaching a neighbourhood of the final state z_T.

The Bellman optimality principle gives the characterization of the value function for every initial condition $\eta \in H^1(\Omega)$ but, in order to have a feasible algorithm, we have to restrict ourselves to a compact set in that space. One possibility is to restrict the analysis to a neighbourhood of ζ enforcing some state constraints and to deal with the Hamilton–Jacobi equation associated to the infinite dimensional problem with state constraints. At present the theory and the numerical methods for such problems in infinite dimension seem to be rather incomplete and unsatisfactory so we prefer to attack the problem by means of a penalization method. In practice, we add to the pay–off a (penalization) term rapidly growing outside the tube around ζ.

Our algorithm to compute an approximate locally optimal trajectory will be divided into two parts. At first, in the backward procedure (from T to t_0), we compute a sequence of almost optimal controls and states guaranteeing the final condition *and* the state constraints. Then, in the forward procedure we actually solve our problem using the informations obtained in the backward steps.

Let N be a positive integer and let $\Delta t = (T - t_0)/N$, as in Section 3 we consider a discretization with time step Δt and we define $t_n = t_0 + n\Delta t$. In order to simplify let us assume that there are only m different controls, i.e. $\widehat{U} = \{u_1, \ldots, u_m\}$. Note that we can always construct a discretization of the control space U_{ad} leading to that

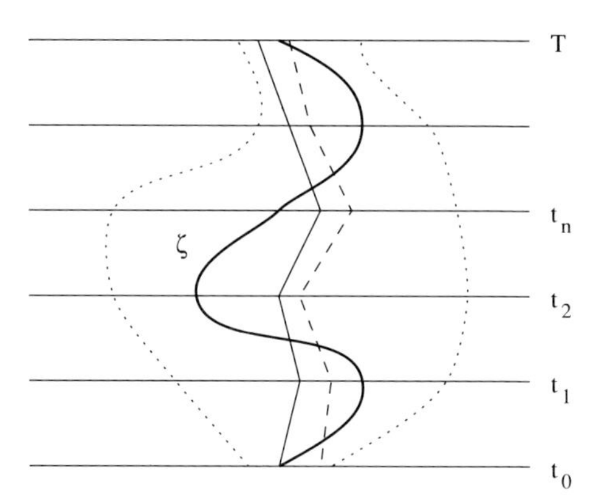

FIGURE 4.1. A sketch of the backward-forward procedure.

situation, f.e. using piecewice constant or piecewice linear functions to approximate an element in U_{ad}. The set $\widehat{U}$ is the finite dimensional representation of our original control space. A simple choice is to take controls which are constant on each time interval $I_n \equiv (t_n, t_{n+1})$.

Figure 1 gives a sketch of the algorithm. The thick solid line represents the trajectory ζ while the area contained between the two dotted lines represents the tube.

The *backward procedure* works as follows. Starting from the final condition z_T, we determine a new trajectory (possibly different from ζ) joining z_T to an initial state (possibly different from η) at time t_0. More precisely, starting from $g^N = z_T$, at each time step we take the solution g^{n+1} of the dynamics at time t_{n+1}, $n = N-1, .., 0$, as our target and we want to determine a state g^n and a control $\widehat{u}$ at time t_n such that we minimize the running cost related to that displacement. The sequence of all the intermediate targets g^n will give us the new "trajectory" (the dashed line in Figure 1) to be used in the forward procedure.

In fact, in the *forward procedure* starting from the initial condition η of the original problem we try to pass through the intermediate targets g^n, $n = 1, .., N$, always minimizing the costs over $\widehat{U}$. Since the cost functional depends continuosly on the initial conditions (see [17]), we will obtain a quasi-optimal control and a good approximation (the thin solid line in Figure 1) of the solution provided the discretization is sufficiently accurate.

The fully discrete algorithm requires a further discretization in space. This means that in the numerical solution of the boundary control problem one has to couple an approximation scheme for the dynamics (which can be a black-box solving the heat equation for any initial condition and piecewise constant boundary control) and an approximation scheme for the cost functional (which can also be a library routine for the numerical integration over Ω). In the backward procedure the black-box is used to compute the solutions corresponding to a finite number of initial conditions and controls, comparing the results of the integral on each of them one can get a couple (g^n, u^n) giving the optimal discrete value. The forward procedure uses the knowledge of the "intermediate targets" to solve a sequence of optimization problems in the intervals I_n. In each of this optimization problems one starts from the numerical solution at time t_n and tries to reach g^{n+1} minimizing the cost. Note that the algorithm computes local minima. So in general we will obtain an approximation of the optimal solution in the tube. Only if our tube contains the optimal solution for the unconstrained optimal control problem we would expect to converge to the globally optimal solution.

5. Appendix

1. We give for completeness the proof of the Dynamic Programming Principle (Theorem 2.5) for our problem.

Proof. Let us denote by $w(\eta, t_0)$ the right hand side of (2.23).
We consider two cases.
1) Let $\tau = T$. We first observe that, by definition,

$$v(y_\eta(T), T) = \inf_{u \in U} \psi(y_\eta(T)).$$

We have then

$$\begin{aligned} w(\eta, t_0) &= \inf_{u \in U} \left\{ \int_{t_0}^{T} f(y_\eta(t;u), u(t)) \mathrm{e}^{-\lambda(t-t_0)} dt + v(y_\eta(T), T) \mathrm{e}^{-\lambda(T-t_0)} \right\} = \\ &= \inf_{u \in U} \left\{ \int_{t_0}^{T} f(y_\eta(t;u), u(t)) \mathrm{e}^{-\lambda(t-t_0)} dt + \psi(y_\eta(T)) \mathrm{e}^{-\lambda(T-t_0)} \right\} \end{aligned}$$

which coincides with the definition of the value function.
2) Let $\tau < T$. We will show the two inequalities:
a) $v(\eta, t_0) \geq w(\eta, t_0)$.
For any fixed $u \in U$, we can write

$$\begin{aligned} J(\eta, t_0, u) = & \int_{t_0}^{\tau} f(y_\eta(t;u), u(t)) \mathrm{e}^{-\lambda(t-t_0)} dt + \\ & + \int_{\tau}^{T} f(y_\eta(t;u), u(t)) \mathrm{e}^{-\lambda(t-t_0)} dt + \psi(y_\eta(T)) \mathrm{e}^{\ \lambda(T-t_0)} . \end{aligned}$$

Since the solution of the heat equation is unique, setting $\mu = y_\eta(\tau; u)$ we have $y_\eta(t;u) = y_\mu(t;u)$, for every $t \geq \tau$. Then the definitions of the cost function and of the value function imply

$$\begin{aligned} J(\eta, t_0, u) = & \int_{t_0}^{\tau} f(y_\eta(t;u), u(t)) \mathrm{e}^{-\lambda(t-t_0)} dt + \\ & + \mathrm{e}^{-\lambda(\tau-t_0)} \Big\{ \int_{\tau}^{T} f(y_\mu(t;u), u(t)) \mathrm{e}^{-\lambda(t-t_0)} dt + \psi(y_\mu(T)) \mathrm{e}^{-\lambda(T-\tau)} \Big\} = \\ = & \int_{t_0}^{\tau} f(y_\eta(t;u), u(t)) \mathrm{e}^{-\lambda(t-t_0)} dt + \mathrm{e}^{-\lambda(\tau-t_0)} J(y_\mu(\tau), \tau, u) \geq \\ \geq & \int_{t_0}^{\tau} f(y_\eta(t;u), u(t)) \mathrm{e}^{-\lambda(t-t_0)} dt + \mathrm{e}^{-\lambda(\tau-t_0)} v(y_\mu(\tau), \tau). \end{aligned}$$

Taking the infimum over $u \in U$, we prove our first inequality.
b) $v(\eta, t_0) \leq w(\eta, t_0)$. For any $\varepsilon > 0$ there exists $u_\varepsilon \in U$ such that

$$v(\mu, \tau) + \varepsilon > J(\mu, \tau, u_\varepsilon) \tag{5.1}$$

where $\mu = y_\eta(\tau, u)$ and u is fixed.

Define

$$\overline{u}(x,t) = \begin{cases} u(x,t) & t_0 \le t \le \tau \\ u_\varepsilon(x,t) & \tau < t \le T. \end{cases}$$

Then, by definition,

$$\begin{aligned} v(\eta,t_0) \le J(\eta,t_0,\overline{u}) &= \int_{t_0}^{\tau} f(y_\eta(t;\overline{u}),\overline{u}(t))\mathrm{e}^{-\lambda(t-t_0)}dt + \\ &+ \int_{\tau}^{T} f(y_\eta(t;\overline{u}),\overline{u}(t))\mathrm{e}^{-\lambda(t-t_0)}dt + \psi(y_\eta(T))\mathrm{e}^{-\lambda(T-t_0)} . \end{aligned} \tag{5.2}$$

The uniqueness of the solution for (2.1) and the definition of $\overline{u}$, imply

$$y_\eta(x,t;\overline{u}) = \begin{cases} y_\eta(x,t;u) & t_0 \le t \le \tau \\ y_\mu(x,t;u_\varepsilon) & \tau < t \le T\,. \end{cases}$$

By substitution in (5.2), we get

$$\begin{aligned} v(\eta,t_0) \le& \int_{t_0}^{\tau} f(y_\eta(t;u),u(t))\mathrm{e}^{-\lambda(t-t_0)}dt + \\ &+ \int_{\tau}^{T} f(y_\mu(t;u_\varepsilon),u_\varepsilon(t))\mathrm{e}^{-\lambda(t-t_0)}dt + \psi(y_\mu(T))\mathrm{e}^{-\lambda(T-t_0)} = \\ &= \int_{t_0}^{\tau} f(y_\eta(t;u),u(t))\mathrm{e}^{-\lambda(t-t_0)}dt + \mathrm{e}^{-\lambda(\tau-t_0)}\{J(\mu,\tau,u_\varepsilon)\} < \\ &< \int_{t_0}^{\tau} f(y_\eta(t;u),u(t))\mathrm{e}^{-\lambda(t-t_0)}dt + \mathrm{e}^{-\lambda(\tau-t_0)}\left(v(\mu,\tau)+\varepsilon\right) . \end{aligned}$$

Taking the infimum over $u \in U$, by the arbitrariness of ε we get the reverse inequality. This ends the proof. □

Note that one of the crucial requirements for the proof is the fact that the set of controls is "closed by concatenation", i.e. if two controls u_1 and u_2 belong to U_{ad} then, for any $\tau \in [t_0,T]$, also the control

$$u(x,t) = \begin{cases} u_1(x,t) & t_0 \le t \le \tau \\ u_2(x,t) & \tau < t \le T \end{cases}$$

belongs to the same space U_{ad}. As a consequence the Dynamic Programming Principle holds also under the restrictions (2.26) and (2.27) on the controls and on the initial data.

2. Here is the statement in Corollary III.20 in [6]

Theorem 5.1. *Let E be a reflexive Banach space, $A \subset E$ be a closed, bounded, convex subset and the function $\phi : A \to]-\infty,+\infty]$ be convex and lower semicontinuous. Then, there exists $x_0 \in A$ such that $\phi(x_0) = \min_{x\in A}\phi(x)$.*

References

1. B. Alziary and P.L. Lions, *A grid refinement method for deterministic control and differential games*, Mathematical Models and Methods in Applied Sciences **4** (1994), 899–910.
2. H.T. Banks and K. Kunish, *The linear regulator problem for parabolic systems*, SIAM J. Control and Opt. **22** (1984), 684–698.
3. H.T. Banks and K. Ito, *Approximation in LQR problems for infinite dimensional systems with unbounded input operators*, J. Mathematical Systems, Estimation and Control, to appear.
4. V. Barbu, and G. Da Prato, Hamilton-Jacobi Equations in Hilbert Spaces, Research Notes in Mathematics, **86**, Pitman, Boston, 1983.
5. A. Bensoussan, G. Da Prato, M.C. Delfour, and S.K. Mitter, Representation and control of infinite dimensional systems, Birkhäuser, Boston, 1992.
6. H. Brezis, Analyse fonctionnelle: théorie et applications, Masson, Paris, 1983.
7. P. Cannarsa and M. E. Tessitore, *Cauchy problem for the dynamic programming equation of boundary control*, Proceedings IFIP Workshop on "Boundary Control and Boundary Variation", Marcel Dekker, 1993, 13–26.
8. R. Ferretti, *On a class of approximation schemes for linear boundary control problems*, in J.P. Zolesio (ed.), "Boundary Control and Variations", Lecture Notes in Pure and Applied Mathematics, 163, Marcel Dekker, 1994.
9. R. Ferretti, *Dynamic programming techniques in the approximation of optimal stopping time problems in Hilbert spaces*, in J.P. Zolesio (ed.), "Boundary Control and Variations", Lecture Notes in Pure and Applied Mathematics, Marcel Dekker, to appear.
10. R. Ferretti, *Internal approximation schemes for optimal control problems in Hilbert spaces*, Journal of Math. Sys. Est. Cont., to appear.
11. M. Falcone, and R. Ferretti, *Discrete time high-order schemes of Hamilton-Jacobi-Bellman equations*, Numerische Mathematik **67** (1994), 315–344.
12. A. Friedman, Partial differential equations of parabolic type, Prentice-Hall Inc., London, 1964.
13. W. Hackbusch, *On the fast solving of parabolic boundary control problems*, SIAM J. Control and Optimization **17** (1979), 231–244.
14. W. Hackbusch, *Multigrid Methods and Applications*, Springer series in Computational Mathematics 4, Springer-Verlag, (1985).
15. K. Ito and H.T. Tran, *Linear quadratic optimal control problems for linear systems with unbounded input and output operators: numerical approximations*, Inter. Series of Numerical Math., **91**, Birkhäuser Verlag (1989), 171–195.
16. I. Lasiecka and R. Triggiani, *Differential and Algebraic Riccati equations with application to boundary/point control problems: continuous theory and approximation theory*, Lecture notes in control and Information Sciences, 164, Springer-Verlag, Berlin, 1991.
17. J.L. Lions, *Optimal control of systems governed by partial differential equations*, Springer-Verlag, Berlin, 1971.
18. P.A. Raviart, and J.M. Thomas, *Introduction à l'analyse numérique des équations aux dérivées partielles*, Masson, Paris, 1988.

Ariela Briani
Dipartimento di Matematica
Università di Pisa
Via Buonarroti 2
I-56126 Pisa
e-mail:briani@dm.unipi.it

Maurizio Falcone
Dipartimento di Matematica
Università di Roma "La Sapienza"
P.le Aldo Moro 2
I-00185 Roma
e-mail:falcone@caspur.it

International Series of Numerical Mathematics
Vol. 126, © 1998 Birkhäuser Verlag, Basel

On the Wellposedness of the Chaboche Model

MARTIN BROKATE* AND PAVEL KREJČÍ†‡

Mathematisches Seminar
Universität Kiel

Institute of Mathematics
Academy of Sciences, Praha

ABSTRACT. We formally state and prove the wellposedness and the local Lipschitz continuity of the multisurface stress-strain law of nonlinear kinematic hardening type due to Chaboche within the space of time-dependent tensor-valued absolutely continuous functions. The results also include the more general case of a continuous family of auxiliary surfaces.

1991 *Mathematics Subject Classification.* 47H30, 73E05

Key words and phrases. Plasticity, Chaboche model, hysteresis operators, kinematic hardening.

1. Introduction

In rate independent plasticity, the Prandtl-Reuß model constitutes the basic model for the stress-strain law. Here, the elastic region Z is bounded by a yield surface ∂Z. Throughout this paper, we will assume the yield surface to be a sphere of radius r in the space of deviatoric stresses. If loading occurs while the stress deviator σ_d lies on the yield surface, there is plastic flow with a plastic strain rate $\dot{\varepsilon}^p$ proportional to the outer normal to ∂Z in σ_d. It has been known from experiments for a long time that for many materials the yield surface undergoes changes which depend upon the history of the loading process. In the Melan-Prager model which dates back to [12], [13], nowadays called linear kinematic hardening, the yield surface moves during plastic loading in the direction of the plastic strain rate. More sophisticated models have been developed to account for real material behaviour, in particular for the phenomenon called ratchetting. Among those, the Chaboche model [10], also called nonlinear kinematic hardening, enjoys a widespread popularity. In its standard form, it employs a finite family of auxiliary spherical surfaces. In the special case of a single auxiliary surface, assumed to be centered at 0 with radius R, the model is known as the Armstrong-Frederick model [1]; here, the center σ^b of the yield surface, also termed the backstress, moves according to the differential equation

$$\dot{\sigma}^b = \gamma \left(R\dot{\varepsilon}^p - \sigma^b |\dot{\varepsilon}^p| \right) , \tag{1.1}$$

* Supported by the BMBF, Grant No. 03-BR7KIE-9, within "Anwendungsorientierte Verbundprojekte auf dem Gebiet der Mathematik".

† Supported by the BMBF during his stay at Kiel.

‡ Partially supported by the Grant Agency of the Czech Republic under Grant No. 201/95/0568.

for some constant $\gamma > 0$, see Figure 1. (In the Melan-Prager model, the term $-\sigma^b|\dot{\varepsilon}^p|$ is omitted.)

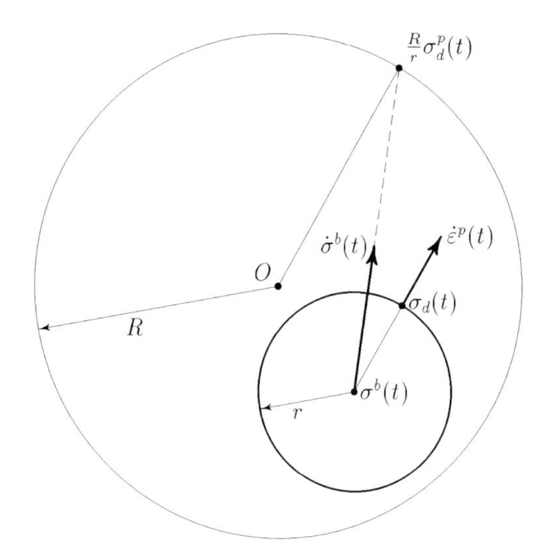

Figure 1: *The model of Armstrong and Frederick.*

In the Chaboche model, the backstress σ^b is decomposed into a sum

$$\sigma^b = \sum_{k \in I} \sigma_k^b , \tag{1.2}$$

where each constituent σ_k^b satisfies an equation of type (1.1), namely

$$\dot{\sigma}_k^b = \gamma(k) \left(R(k)\dot{\varepsilon}^p - \sigma_k^b|\dot{\varepsilon}^p| \right) , \quad k \in I . \tag{1.3}$$

In the standard Chaboche model, the index set I is finite; we will allow an arbitrary measure space and thus include the case of a continuous family of auxiliary surfaces.

Figure 2 shows the rheological structure of the model. It visualizes the relations between the various variables which occur in the model, stated formally in (2.5) - (2.12) below. The element $\mathcal{E}$ refers to the linear elastic part, $\mathcal{R}$ is called the rigid plastic element and represents the variational inequality, and $\mathcal{K}_k$ is the element defined by (1.3). The element $\mathcal{L}$ plays a special role; it stands for the linear element $\sigma^l = C^l\varepsilon^p$ of the Melan-Prager model. It may or may not be included within the Chaboche model, but its presence or absence influences the asymptotic behaviour (see e.g. [7]). If we remove all nonlinear elements $\mathcal{K}_k$ in Figure 2, we obtain the Melan-Prager model. If we moreover delete the element $\mathcal{L}$, we arrive at the Prandtl-Reuß model.

In this paper, we prove that the Chaboche model is well posed in the space $W^{1,1}$ both in the stress controlled and in the strain controlled case by proving that the defining equations and inequalities of the Chaboche model (see (2.5)–(2.12) below) lead to operators

$$\varepsilon = \mathcal{F}(\sigma)\,, \quad \sigma = \mathcal{G}(\varepsilon)\,, \tag{1.4}$$

which are well defined and Lipschitz continuous on their appropriate domains of definition. In doing this, we consider the stress-strain law in isolation, that is, we do not study the boundary value problems which arise from the coupling with the balance equations. For the proof we utilize the method of [2]. There we have introduced an auxiliary variable u in order to reformulate the model equations such that the unknown functions of Figure 2 appear only in terms of $|\dot{\varepsilon}^p|$ and σ_d^p. The analysis is based on the concept of hysteresis operators, that is, of operators which are rate-independent as well as causal, see e.g. [14], [8], [9], [3].

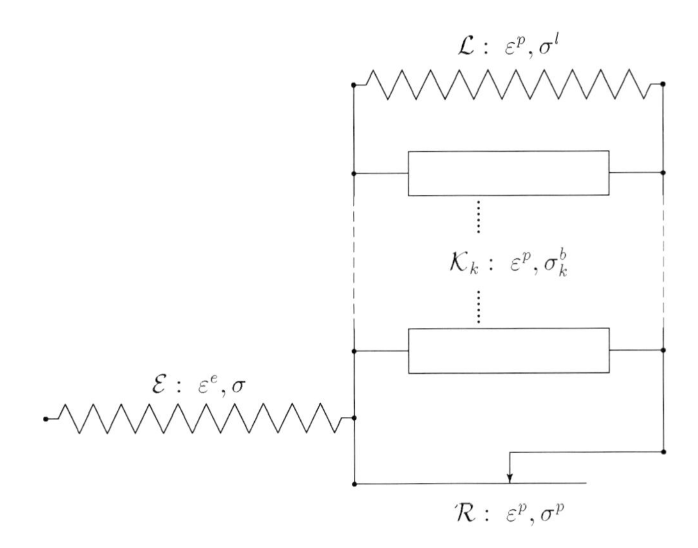

Figure 2: *The rheological structure of the Chaboche model.*

2. Model Formulation and Main Result

We first fix some basic tensor notation. By $\mathbb{T}$, we denote the space of symmetric $N \times N$ tensors endowed with the usual scalar product and the associated norm

$$\langle \tau, \eta \rangle = \sum_{i,j=1}^{N} \tau_{ij}\eta_{ij}\,, \quad |\tau| = \sqrt{\langle \tau, \tau \rangle}\,, \tag{2.1}$$

For $\tau \in \mathbb{T}$, we define its trace $\operatorname{Tr}\tau$ and its deviator τ_d by

$$\operatorname{Tr}\tau = \sum_{i=1}^{N} \tau_{ii} = \langle \tau, \delta \rangle\,, \quad \tau_d = \tau - \frac{\operatorname{Tr}\tau}{N}\delta\,, \tag{2.2}$$

where $\delta = (\delta_{ij})$ stands for the Kronecker symbol. We denote by

$$\mathbb{T}_d = \{\tau : \tau \in \mathbb{T}\,, \operatorname{Tr}\tau = 0\}\,, \quad \mathbb{T}_d^{\perp} = \{\tau : \tau = \lambda\delta\,, \lambda \in \mathbb{R}\}\,, \tag{2.3}$$

the space of all deviators respectively its orthogonal complement. We understand stress and strain as time-dependent tensor-valued functions which are absolutely continuous,

$$(2.4)\quad \sigma, \varepsilon \in W^{1,1}(t_0, t_1; \mathbb{T}) := \{\tau | \tau : [t_0, t_1] \to \mathbb{T}\,,\ \|\tau\|_{1,1} = |\tau(t_0)| + \int_{t_0}^{t_1} |\dot{\tau}(t)|\, dt < \infty\}\,.$$

As we study the stress-strain law in isolation, we do not consider the space dependence. In this terminology, the Chaboche model takes on the form

$$\sigma = \sigma^b + \sigma^p\,, \quad \varepsilon = \varepsilon^e + \varepsilon^p\,, \quad \varepsilon^p(t) \in \mathbb{T}_d \quad \forall\, t\,, \tag{2.5}$$

$$\langle \dot{\varepsilon}^p, \sigma^p_d - \tilde{\sigma} \rangle \geq 0\,, \quad \forall\, \tilde{\sigma} \in \mathbb{T}_d\,, |\tilde{\sigma}| \leq r\,, \tag{2.6}$$

$$|\sigma^p_d| \leq r\,, \tag{2.7}$$

$$\sigma = A\varepsilon^e\,, \tag{2.8}$$

$$\sigma^b(t) = \int_I \sigma^b_k(t)\, d\nu(k) + \nu^l \sigma^l(t) \quad \forall\, t\,, \tag{2.9}$$

$$\dot{\sigma}^b_k = \gamma(k) \left(R(k)\dot{\varepsilon}^p - \sigma^b_k |\dot{\varepsilon}^p| \right)\,, \quad \text{for all } k \in I\,, \tag{2.10}$$

$$\sigma^l = C^l \varepsilon^p\,, \tag{2.11}$$

$$\sigma^p(t_0) = \sigma^p_0\,, \quad \sigma^b_k(t_0) = \sigma^b_0(k)\,, \quad \text{for all } k \in I\,. \tag{2.12}$$

Throughout this paper, we assume the data to have the following properties.

Assumption 2.1.
(i) *I is a measure space, ν is a finite nonnegative measure on I, the numbers ν^l, C^l and functions $R \in L^1_\nu(I)$, $\gamma \in L^\infty_\nu(I)$ satisfy $\nu^l, R, \gamma \geq 0$, $C^l > 0$, $\int_I R(k) d\nu(k) > 0$ and*

$$0 < \gamma_{\min} \leq \gamma(k) \leq \gamma_{\max}\,, \quad \textit{for all } k \in I\,. \tag{2.13}$$

(ii) *The initial values in (2.12) satisfy*

$$\sigma^p_0 \in \mathbb{T}^p_r = \{\tau : \tau \in \mathbb{T}\,, |\tau_d| \leq r\}\,, \tag{2.14}$$

$$\sigma^b_0 \in \mathcal{T}^b = \{f | f \in L^1_\nu(I; \mathbb{T}_d)\,, |f(k)| \leq R(k)\ a.e.\}\,. \tag{2.15}$$

(iii) *$A : \mathbb{T} \to \mathbb{T}$ is linear, symmetric and positive definite.*

We also introduce the constants

$$\Gamma_i = \int_I \gamma(k)^i R(k)\, d\nu(k)\,, \quad i = 0, 1, 2, 3\,. \tag{2.16}$$

Remark 2.2.
(i) If the index set I is finite, say $I = \{1, \dots, K\}$, and if ν is chosen to be the counting measure, that is, $\nu(J)$ equals the number of elements in J for every subset J of I, then we obtain the standard formulation of the multisurface Chaboche model with K auxiliary (limiting) surfaces, namely

$$\sigma^b = \sum_{k=1}^{K} \sigma_k^b . \tag{2.17}$$

In this case, the model (2.5) - (2.12) is identical with the one discussed in ([10], Section 5.4.4), nonlinear kinematic case, if we change the notation according to

$$k \mathrel{\hat{=}} l, \quad \sigma_k^b \mathrel{\hat{=}} X_l, \quad \gamma(k) \mathrel{\hat{=}} \sqrt{\frac{2}{3}}\gamma_l, \quad \gamma(k)R(k) \mathrel{\hat{=}} \frac{2}{3}C_l . \tag{2.18}$$

(ii) If we have $K = 1$ in (i), or if we choose $\gamma(k) \equiv \gamma$ and $R(k) \equiv R/\nu(I)$ to be constant, the Chaboche model reduces to the model of Armstrong and Frederick [1]

$$\dot{\sigma}^b = \gamma(R\dot{\varepsilon}^p - \sigma^b|\dot{\varepsilon}^p|) . \tag{2.19}$$

(iii) If $d\nu(k) = g(k)d\lambda(k)$ for some function g, that is, if the measure ν has a density with respect to the Lebesgue measure λ, we obtain a version of the Chaboche model with a continuous one parameter family of backstresses respectively auxiliary surfaces.

We formulate our main results. For the strain controlled case, we assume Hooke's law for the linear elastic part, that is,

$$A\varepsilon = 2\mu\varepsilon + \lambda \mathrm{Tr}\,(\varepsilon)\delta , \tag{2.20}$$

where $\lambda, \mu > 0$ denote the Lamé constants.

Theorem 2.3 (Wellposedness, Strain Controlled Case).
Let Assumption 2.1 as well as (2.20) hold. Then the system (2.5)–(2.12) defines an operator

$$\sigma = \mathcal{G}(\varepsilon; \sigma_0^p, \sigma_0^b) , \tag{2.21}$$

$$\mathcal{G} : W^{1,1}(t_0, t_1; \mathbb{T}) \times \mathbb{T}_r^p \times \mathcal{T}^b \to W^{1,1}(t_0, t_1; \mathbb{T}) , \tag{2.22}$$

which satisfies the Lipschitz condition

$$\begin{aligned} \| \mathcal{G}(\varepsilon; \sigma_0^p, \sigma_0^b) - \mathcal{G}(\bar\varepsilon; \bar\sigma_0^p, \bar\sigma_0^b) \|_{1,1} & \\ \le L(K) \left(\| \varepsilon - \bar\varepsilon \|_{1,1} + |\sigma_0^p - \bar\sigma_0^p| + \|\sigma_0^b - \bar\sigma_0^b\|_{L^1_\nu(I;\mathbb{T}_d)} \right) , & \end{aligned} \tag{2.23}$$

where the Lipschitz constant is uniform over subsets $\{(\varepsilon, \sigma_0^p, \sigma_0^b) : \| \varepsilon \|_{1,1} \le K\}$ *of the domain of definition of* $\mathcal{G}$.

We now consider the stress controlled case. If $\nu^l = 0$, that is, if the Melan-Prager element is absent, our choice (2.12) of initial conditions restricts the initial value $\sigma(t_0)$ of the stress; on the other hand, there has to be an initial condition

$$\varepsilon^p(t_0) = \varepsilon_0^p \tag{2.24}$$

for the plastic strain. This setting also works for the case $\nu^l > 0$, the restriction being

$$\sigma(t_0) = \sigma_0^p + \int_I \sigma_0^b(k)\,d\nu(k) + \nu^l C^l \varepsilon_0^p \,. \tag{2.25}$$

In the case $\nu^l = 0$, the description of the domains where the Lipschitz constant is uniform involves the number

$$\beta = \frac{1}{\gamma_{\min}} + \frac{r\gamma_{\max}}{\Gamma_2} \,. \tag{2.26}$$

Theorem 2.4 (Wellposedness, Stress Controlled Case).
Let Assumption 2.1 *hold.*

(Case $\nu^l > 0$.***)*** *The system* (2.5)–(2.12), (2.24) *defines an operator*

$$\varepsilon = \mathcal{F}(\sigma; \sigma_0^p, \sigma_0^b, \varepsilon_0^p)\,, \quad \mathcal{F} : D_\sigma \to W^{1,1}(t_0, t_1; \mathbb{T})\,, \tag{2.27}$$

where $D_\sigma \subset W^{1,1}(t_0, t_1; \mathbb{T}) \times \mathbb{T}_r^p \times \mathcal{T}^b \times \mathbb{T}_d$ *is the subset of quadruples which satisfy* (2.25)*. Moreover,* $\mathcal{F}$ *satisfies on* D_σ *the Lipschitz condition*

$$\begin{aligned} &\| \mathcal{F}(\sigma; \sigma_0^p, \sigma_0^b, \varepsilon_0^p) - \mathcal{F}(\bar\sigma; \bar\sigma_0^p, \bar\sigma_0^b, \bar\varepsilon_0^p) \|_{1,1} \\ &\quad \le L(K) \left(\| \sigma - \bar\sigma \|_{1,1} + |\sigma_0^p - \bar\sigma_0^p| + \|\sigma_0^b - \bar\sigma_0^b\|_{L^1_\nu(I;\mathbb{T}_d)} + |\varepsilon_0^p - \bar\varepsilon_0^p| \right), \end{aligned} \tag{2.28}$$

where the Lipschitz constant is uniform over subsets $\{(\sigma, \sigma_0^p, \sigma_0^b, \varepsilon_0^p) : \| \sigma \|_{1,1} \le K\}$ *of the domain of definition of* $\mathcal{F}$.

(Case $\nu^l = 0$.***)*** *For every* $\kappa > 0$*, let* $D_{\sigma,\kappa}$ *be the subset of* D_σ *where the two conditions*

$$\left| \int_I \gamma(k)\sigma_0^b(k)\,d\nu(k) \right| \le \Gamma_1(1-\kappa)\,, \tag{2.29}$$

$$\| \sigma_d \|_\infty \le \Gamma_0 + r - \Gamma_1\beta\kappa\,, \tag{2.30}$$

hold, the number β *being defined in* (2.26)*. Then* $\mathcal{F}$ *has the properties as stated above on the domains* $D_{\sigma,\kappa}$ *instead of* D_σ*; in particular, the Lipschitz constant also depends on* κ.

A well known example (see [10], or Example 3.5 in [2]) shows that the bound $\| \sigma_d \|_\infty < \Gamma_0 + r$ in (2.30) cannot be improved.

The basic idea of the proof of the two theorems above is the same as in [2]. We replace the two unknown functions ε^p and σ_d^p by a single auxiliary function u, namely

$$u = C\varepsilon^p + \sigma_d^p\,, \tag{2.31}$$

where $C > 0$ is a suitably chosen constant. In fact, both functions ε^p and σ_d^p can be expressed as

$$\varepsilon^p = \frac{1}{C}\mathcal{P}(u\,;\sigma_{0d}^p)\,, \quad \sigma_d^p = \mathcal{S}(u\,;\sigma_{0d}^p)\,. \tag{2.32}$$

Here, the stop operator $\mathcal{S}$ represents the solution of the evolution variational inequality

$$|\sigma_d^p| \le r\,, \quad \langle \dot{u} - \dot{\sigma}_d^p, \sigma_d^p - \tilde{\sigma}\rangle \ge 0 \quad \text{a.e.} \quad \forall\, |\tilde{\sigma}| \le r\,, \tag{2.33}$$

with the initial condition

$$\sigma_d^p(t_0) = \sigma_{0d}^p\,, \tag{2.34}$$

and the play operator $\mathcal{P}$ is defined by

$$\mathcal{P}(u\,;\sigma_{0d}^p) = u - \mathcal{S}(u\,;\sigma_{0d}^p)\,. \tag{2.35}$$

We refer to [2] and [9] for more details. We now derive a differential equation for u where the internal variables $\sigma_k^b, \sigma^l, \varepsilon_d^e, \varepsilon^p$ appear only in terms of σ_d^p and $|\dot{\varepsilon}^p|$. In the stress controlled case, we set

$$C = \Gamma_1 + \nu^l C^l\,. \tag{2.36}$$

Using the model equations, we obtain

$$\begin{aligned} \dot{u} &= (\Gamma_1 + \nu^l C^l)\dot{\varepsilon}^p + \dot{\sigma}_d^p = (\Gamma_1 + \nu^l C^l)\dot{\varepsilon}^p + \dot{\sigma}_d - \dot{\sigma}^b \\ &= \dot{\sigma}_d + \int_I \gamma(k)\sigma_k^b\,d\nu(k)\,|\dot{\varepsilon}^p|\,. \end{aligned} \tag{2.37}$$

In the strain controlled case, where we have assumed Hooke's law (2.20) for the linear elastic part, the backstress σ^b satisfies

$$\sigma_d = 2\mu\varepsilon_d^e\,, \quad \sigma^b = 2\mu\varepsilon_d - (2\mu\varepsilon^p + \sigma_d^p)\,. \tag{2.38}$$

Here, we set the constant C in (2.31) to

$$C = 2\mu + \Gamma_1 + \nu^l C^l\,, \tag{2.39}$$

and obtain

$$\begin{aligned} \dot{u} &= (2\mu + \Gamma_1 + \nu^l C^l)\dot{\varepsilon}^p + \dot{\sigma}_d^p = 2\mu\dot{\varepsilon}^p + 2\mu\dot{\varepsilon}_d^e - \dot{\sigma}^b + (\Gamma_1 + \nu^l C^l)\dot{\varepsilon}^p \\ &= 2\mu\dot{\varepsilon}_d + \int_I \gamma(k)\sigma_k^b\,d\nu(k)\,|\dot{\varepsilon}^p|\,. \end{aligned} \tag{2.40}$$

As it is well known, one can easily eliminate the unknowns σ_k^b with the variations of constants formula. Using the basic identity

$$\dot{\varepsilon}^p = \frac{\sigma_d^p}{r}|\dot{\varepsilon}^p|\,, \tag{2.41}$$

the differential equation (2.10) for the backstresses becomes

$$\dot{\sigma}_k^b = \gamma(k)\left(\frac{R(k)}{r}\sigma_d^p - \sigma_k^b\right)|\dot{\varepsilon}^p|\,, \quad k \in I\,. \tag{2.42}$$

For later use, we will write down the solution formula in terms of the play and stop operator with the abbreviated notation

$$\xi = \mathcal{P}(u\,;\sigma_{0d}^p)\,, \quad x = \mathcal{S}(u\,;\sigma_{0d}^p)\,, \quad \xi, x : [t_0, t_1] \to \mathbb{T}_d\,. \tag{2.43}$$

The function

$$V(t) = \mathrm{Var}_{[t_0,t]}\xi \quad \left(= \int_{t_0}^{t} |\dot{\xi}(\tau)|\, d\tau\,, \quad \text{if}\, \xi \in W^{1,1}(t_0,t_1;\mathbb{T}_d)\right) \tag{2.44}$$

represents the accumulated plastic strain, scaled by a constant factor. If we set

$$W_k(t) = \exp\left(\frac{\gamma(k)}{C} V(t)\right), \tag{2.45}$$

the backstresses can be expressed as

$$\sigma_k^b(t) = \exp\left(-\frac{\gamma(k)}{C} V(t)\right)\left(\sigma_0^b(k) + \int_{t_0}^{t} \frac{R(k)}{r} x(\tau)\, dW_k(\tau)\right). \tag{2.46}$$

Thus, for the stress as well as for the strain controlled case, the auxiliary function u satisfies the equation

$$\dot{u} = \dot{\theta} + \mathcal{M}(u;\sigma_0^p,\sigma_0^b)|\dot{\xi}|\,, \tag{2.47}$$

where $\theta = \sigma_d$ respectively $\theta = 2\mu\varepsilon_d$,

$$\mathcal{M}(u;\sigma_0^p,\sigma_0^b)(t) = \frac{1}{C}\int_I \gamma(k)\sigma_k^b(t)\, d\nu(k)\,, \tag{2.48}$$

and (2.43) - (2.46) are used to express σ_k^b in terms of the arguments of $\mathcal{M}$. Equation (2.47) is complemented by the initial condition

$$u(t_0) = C\varepsilon^p(t_0) + \sigma_{0d}^p\,. \tag{2.49}$$

In the stress controlled case, $\varepsilon^p(t_0)$ is prescribed, whereas in the strain controlled case, it can be expressed in terms of the given data by (2.38).

Once the auxiliary equation (2.47) is solved, we can express the operators $\mathcal{F},\mathcal{G}$ in terms of u, namely

$$\varepsilon = \mathcal{F}(\sigma,\sigma_0^p,\sigma_0^b,\varepsilon_0^p) = \varepsilon^e + \varepsilon^p = A^{-1}\sigma + \frac{1}{C}\mathcal{P}(u\,;\sigma_{0d}^p)\,, \tag{2.50}$$

$$\sigma = \mathcal{G}(\varepsilon,\sigma_0^p,\sigma_0^b) = A(\varepsilon-\varepsilon^p) = A\varepsilon - \frac{2\mu}{C}\mathcal{P}(u\,;\sigma_{0d}^p)\,. \tag{2.51}$$

3. Proof of the Wellposedness

The wellposedness of the initial value problem

$$\dot{u}(t) = \dot{\theta}(t) + \mathcal{M}(u;\sigma_0^p,\sigma_0^b)(t)|\dot{\xi}(t)|\,, \quad \xi(t) = \mathcal{P}(u\,;\sigma_{0d}^p)(t)\,, \tag{3.1}$$

$$u(t_0) = u^0\,. \tag{3.2}$$

has been studied in [2] concerning the dependence on θ; the dependence on the initial conditions $(u^0,\sigma_0^p,\sigma_0^b)$ does not pose any new problems. For the convenience of the reader, we repeat the formulation of the existence theorem, adapted to the present case.

Theorem 3.1. *Let $\theta \in W^{1,1}(t_0,t_1;\mathbb{T}_d)$, $u^0 \in \mathbb{T}_d$ and an operator*

$$\mathcal{M} : C([t_0,t_1];\mathbb{T}_d) \times \mathbb{T}_r^p \times \mathcal{T}^b \to C([t_0,t_1];\mathbb{T}_d) \tag{3.3}$$

be given. Assume that $\mathcal{M}(\cdot;\sigma_0^p,\sigma_0^b)$ is causal and continuous with respect to the maximum norm for all $\sigma_0^p \in \mathbb{T}_r^p$ and $\sigma_0^b \in \mathcal{T}^b$, and that $\kappa > 0, \sigma_0^p \in \mathbb{T}_r^p, \sigma_0^b \in \mathcal{T}^b$ and $u^0 \in \mathbb{T}_d$ are given such that

$$\sup_{\tau\in[t_0,t]} |\mathcal{M}(u;\sigma_0^p,\sigma_0^b)| \le 1-\kappa \tag{3.4}$$

holds for all $t \in [t_0,t_1]$ and all $u \in W^{1,1}(t_0,t;\mathbb{T}_d)$ with $u(t_0) = u^0$ and

$$|\dot u(\tau)| \le \frac{1}{\kappa}|\dot\theta(\tau)|\,, \quad \text{a.e. in } (t_0,t)\,. \tag{3.5}$$

Then there exists a solution (u,ξ) of the Cauchy problem (3.1), (3.2) where the functions $u,\xi \in W^{1,1}(t_0,t_1;\mathbb{T}_d)$ fulfil (3.4) and (3.5). Moreover, every such solution which satisfies (3.4) also satisfies (3.5).

Proof. See [2], Theorem 3.2. □

Lemma 3.2. *The operator $\mathcal{M}(\cdot;\sigma_0^p,\sigma_0^b)$ as defined in (2.43) - (2.48) is causal and continuous on $C([t_0,t_1];\mathbb{T}_d)$ for all $\sigma_0^p \in \mathbb{T}_r^p$ and $\sigma_0^b \in \mathcal{T}^b$. The backstresses σ_k^b satisfy the a priori estimate*

$$|\sigma_k^b(t)| \le R(k)\,, \quad \text{a.e. in } (t_0,t_1)\,, \tag{3.6}$$

for all $k \in I$.

Proof. The estimate (3.6) follows from the variations of constants formula (2.46), since $|x(t)| \le r$ and $|\sigma_0^b(k)| \le R(k)$ hold for all t and k. Let now $u_n \in C([t_0,t_1];\mathbb{T}_d)$ converge uniformly to $u \in C([t_0,t_1];\mathbb{T}_d)$. It is known (see [9]) that

$$\xi_n = \mathcal{P}(u_n\,;\sigma_{0d}^p) \;\to\; \xi = \mathcal{P}(u\,;\sigma_{0d}^p)\,, \quad x_n = \mathcal{S}(u_n\,;\sigma_{0d}^p) \;\to\; x = \mathcal{S}(u\,;\sigma_{0d}^p)\,, \tag{3.7}$$

$$V_n(t) = \mathrm{Var}_{[t_0,t]}\xi_n \;\to\; V(t) = \mathrm{Var}_{[t_0,t]}\xi\,, \tag{3.8}$$

uniformly on $[t_0,t_1]$. An application of Lebesgue's dominated convergence theorem yields the assertion. □

We now discuss the boundedness property (3.4). By the definition of $\mathcal{M}$ in (2.48), the estimate (3.6) yields

$$\|\,\mathcal{M}(u;\sigma_0^p,\sigma_0^b)\,\|_\infty \le \frac{\Gamma_1}{C}\,, \tag{3.9}$$

so (3.4) holds for all arguments, regardless of (3.5), with

$$\kappa = \frac{\nu^l C^l}{C}\,, \quad \text{respectively} \quad \kappa = \frac{2\mu + \nu^l C^l}{C}\,, \tag{3.10}$$

in the stress respectively strain controlled case. Thus, the existence of a solution of (3.1), (3.2) follows for the strain controlled case and, if in addition $\nu^l > 0$, also for the stress controlled case.

Existence proof for the stress controlled case with $\nu^l = 0$. Let $\kappa > 0$. According to Theorem 2.4, we want to prove existence for initial conditions satisfying

$$\sigma(t_0) = \sigma_0^p + \int_I \sigma_0^b(k)d\nu(k)\,, \tag{3.11}$$

$$\left|\int_I \gamma(k)\sigma_0^b(k)d\nu(k)\right| \le \Gamma_1(1-\kappa)\,, \tag{3.12}$$

and for stress inputs $\sigma_d \mathrel{\hat=} \theta \in W^{1,1}(t_0,t_1;\mathbb{T}_d)$ satisfying

$$\|\,\sigma_d\,\|_\infty \le \Gamma_0 + r - \Gamma_1\beta\kappa\,, \quad \beta = \frac{1}{\gamma_{\min}} + \frac{r\gamma_{\max}}{\Gamma_2}\,. \tag{3.13}$$

Let such a σ_d be given, choose $\eta > 0$ small enough such that

$$\int_t^{t+\eta} |\dot\sigma_d(\tau)|\,d\tau \le \frac{\kappa^2\Gamma_1}{8\gamma_{\max}}\,, \quad \forall\, t \in [t_0, t_1-\eta]\,. \tag{3.14}$$

In the first step we will prove that, if we have a solution u of (3.1), (3.2) satisfying (3.4) on $[t_0,a]$, then it can be extended to $[a,a+\eta]$, and every such extension $\tilde u$ satisfies

$$\|\,\mathcal{M}(\tilde u;\sigma_0^p,\sigma_0^b)\,\|_\infty \le 1 - \frac{\kappa}{2} \tag{3.15}$$

on $[a,a+\eta]$, and

$$|\dot{\tilde u}(t)| \le \frac{2}{\kappa}|\dot\sigma_d(t)|\,, \qquad \text{a.e. on } (a,a+\eta)\,. \tag{3.16}$$

To this end, let $\tilde u \in W^{1,1}(a,a+\eta;\mathbb{T}_d)$ be an arbitrary function which satisfies (3.16) as well as $\tilde u(a) = u(a)$; setting $\tilde u = u$ on $[t_0,a]$ we may regard it as an element of $W^{1,1}(t_0,a+\eta;\mathbb{T}_d)$ as well. From the variation of constants formula (2.46), applied on the interval $[a,a+\eta]$, we obtain

$$\begin{aligned} &|\sigma_k^b(t) - \sigma_k^b(a)| \\ &\quad \le \left(1 - \exp\left(-\frac{\gamma(k)}{\Gamma_1}(V(t)-V(a))\right)\right)\cdot(|\sigma_k^b(a)| + R(k))\,, \quad t \in [a,a+\eta]\,, \end{aligned} \tag{3.17}$$

for the corresponding backstresses. Since

$$|V(t) - V(a)| \le \int_a^t |\dot{\tilde u}(\tau)|\,d\tau\,, \tag{3.18}$$

we get

$$\begin{aligned} |\sigma_k^b(t) - \sigma_k^b(a)| &\le 2R(k)\frac{\gamma_{\max}}{\Gamma_1}\int_a^t |\dot{\tilde u}(\tau)|\,d\tau \le \frac{4\gamma_{\max}}{\kappa C_1}R(k)\int_a^{a+\eta}|\dot\sigma_d(\tau)|\,d\tau \\ &\le \frac{\kappa}{2}R(k)\,, \end{aligned} \tag{3.19}$$

so

$$|\mathcal{M}(\tilde u;\sigma_0^p,\sigma_0^b)(t) - \mathcal{M}(\tilde u;\sigma_0^p,\sigma_0^b)(a)| \le \frac{\kappa}{2}\,. \tag{3.20}$$

Thus, the assumption $|\mathcal{M}(\tilde u;\sigma_0^p,\sigma_0^b)(a)| \le 1-\kappa$ implies that (3.15) holds if $\tilde u$ satisfies (3.16). We may therefore apply Theorem 3.1 on the interval $[a,a+\eta]$ to conclude the first step of the proof. In the second step, we use (3.13) to show that (3.15) can be improved to

$$|\mathcal{M}(\tilde{u};\sigma_0^p,\sigma_0^b)(t)| \le 1-\kappa\,, \quad \forall\, t\in[a,a+\eta]\,. \tag{3.21}$$

In fact, if (3.21) does not hold, then there must exist a $t\in(a,a+\eta)$ such that

$$|\mathcal{M}(\tilde{u};\sigma_0^p,\sigma_0^b)(t)| > 1-\kappa\,, \quad \frac{\mathrm{d}}{\mathrm{dt}}\left(|\mathcal{M}(\tilde{u};\sigma_0^p,\sigma_0^b)(t)|^2\right) > 0\,. \tag{3.22}$$

Let us define

$$\alpha = |\mathcal{M}(\tilde{u};\sigma_0^p,\sigma_0^b)(t)|\,, \quad e = \frac{1}{\alpha}\mathcal{M}(\tilde{u};\sigma_0^p,\sigma_0^b)(t)\in\mathbb{T}_d\,, \tag{3.23}$$

then obviously

$$0 < 1-\alpha < \kappa\,, \quad |e| = 1\,. \tag{3.24}$$

The choice of t implies that

$$\begin{aligned} 0 \;&<\; \frac{1}{2}\frac{\mathrm{d}}{\mathrm{dt}}\left(|\mathcal{M}(\tilde{u};\sigma_0^p,\sigma_0^b)(t)|^2\right) = \frac{\alpha}{\Gamma_1}\int_I \gamma(k)\langle\dot{\sigma}_k^b(t),e\rangle\,d\nu(k) \\ &= \frac{\alpha}{\Gamma_1}|\dot{\xi}(t)|\int_I \gamma(k)^2\langle\frac{R(k)}{r}x(t)-\sigma_k^b(t),e\rangle\,d\nu(k)\,, \end{aligned} \tag{3.25}$$

so in particular $|\dot{\xi}(t)|>0$ and therefore

$$\begin{aligned} \int_I \gamma(k)^2\langle\sigma_k^b(t),e\rangle\,d\nu(k) \;&<\; \frac{\Gamma_2}{r}\langle x(t),e\rangle \\ &= \frac{\Gamma_2}{r}\left(\langle\sigma_d(t),e\rangle - \int_I\langle\sigma_k^b(t),e\rangle\,d\nu(k)\right)\,, \end{aligned} \tag{3.26}$$

hence

$$\int_I\left(\gamma(k)^2+\frac{\Gamma_2}{r}\right)\langle\sigma_k^b(t),e\rangle\,d\nu(k) < \frac{\Gamma_2}{r}\|\,\sigma_d\,\|_\infty\,. \tag{3.27}$$

On the other hand, the a priori estimate $|\sigma_k^b(t)|\le R(k)$ shows that

$$0<1-\alpha = \frac{1}{\Gamma_1}\int_I\gamma(k)\langle R(k)e-\sigma_k^b(t),e\rangle\,d\nu(k) < \kappa\,, \tag{3.28}$$

hence the definition of β in (3.13) yields

$$\begin{aligned} &\int_I\left(1+\frac{r}{\Gamma_2}\gamma(k)^2\right)\langle R(k)e-\sigma_k^b(t),e\rangle\,d\nu(k) \\ &\qquad\le \beta\int_I\gamma(k)\langle R(k)e-\sigma_k^b(t),e\rangle\,d\nu(k) \le \beta\Gamma_1\kappa\,, \end{aligned} \tag{3.29}$$

and therefore

$$\begin{aligned} \|\,\sigma_d\,\|_\infty > &\int_I\left(1+\frac{r}{\Gamma_2}\gamma(k)^2\right)R(k)\,d\nu(k) \\ &- \int_I\left(1+\frac{r}{\Gamma_2}\gamma(k)^2\right)\langle R(k)e-\sigma_k^b(t),e\rangle\,d\nu(k) \ge \Gamma_0 + r - \beta\Gamma_1\kappa\,, \end{aligned} \tag{3.30}$$

which contradicts our assumption (3.13). Thus, such a t cannot exist, and the second step is proved. Applying the two steps in an alternate fashion we are able to cover the whole interval $[t_0, t_1]$, thus completing the existence proof. □

Proof of uniqueness and Lipschitz continuous dependence. We combine a Gronwall type argument with the Lipschitz continuity property of the hysteresis operators $\mathcal{P}$ and $\mathcal{S}$. As the arguments are essentially the same as for the single surface case, i.e. the model of Armstrong and Frederick, we can use the results of [2] to a large extent.

Proposition 3.3. *Let two sets of data* $(\theta_1, u_1^0, \sigma_{10}^p, \sigma_{10}^b)$, $(\theta_2, u_2^0, \sigma_{20}^p, \sigma_{20}^b)$ *with* $\theta_i \in \Theta$, $u_i^0 \in X$, $\sigma_{i0}^p \in \mathbb{T}_r^p$ *and* $\sigma_{i0}^b \in \mathcal{T}^b$ *be given, let* (u_1, ξ_1) *and* (u_2, ξ_2) *be corresponding solutions in* $W^{1,1}(t_0, t_1; \mathbb{T}_d)$ *of the Cauchy problem (3.1), (3.2) which satisfy (3.4) and (3.5). Assume that*

$$(3.31) \quad \max_{s\in[t_0,t]} |\mathcal{M}(u_1; \sigma_{10}^p, \sigma_{10}^b)(s) - \mathcal{M}(u_2; \sigma_{20}^p, \sigma_{20}^b)(s)| \\ \leq A\Big(|\sigma_{10}^p - \sigma_{20}^p| + \|\sigma_{10}^b - \sigma_{20}^b\|_{L^1_\nu(I;\mathbb{T}_d)} + |u_1^0 - u_2^0| + \int_{t_0}^t |\dot{u}_1 - \dot{u}_2|\, ds\Big)$$

holds for all $t \in [t_0, t_1]$. *Then there holds*

$$(3.32) \quad \| u_1 - u_2 \|_{1,1} \\ \leq L\left(|u_1^0 - u_2^0| + |\sigma_{10}^p - \sigma_{20}^p| + \|\sigma_{10}^b - \sigma_{20}^b\|_{L^1_\nu(I;\mathbb{T}_d)} + \| \theta_1 - \theta_2 \|_{1,1}\right),$$

where L *depends only upon* A, κ, r *and*

$$(3.33) \qquad c := \max\{\| \theta_1 \|_{1,1}, \| \theta_2 \|_{1,1}\}.$$

Proof. See Theorem 3.3 in [2]. □

The operator $\mathcal{M}$ as defined by (2.43) - (2.48) satisfies

$$(3.34) \quad |\mathcal{M}(u_1; \sigma_{10}^p, \sigma_{10}^b)(t) - \mathcal{M}(u_2; \sigma_{20}^p, \sigma_{20}^b)(t)| \\ \leq \int_I \gamma(k)|\sigma_{10}^b(k) - \sigma_{20}^b(k)|\, d\nu(k) + \left(\frac{2\Gamma_2}{C} + \frac{\Gamma_3}{C^2}\int_{t_0}^t |\dot{\xi}_1(s)|\, ds\right)\int_{t_0}^t |\dot{\xi}_1 - \dot{\xi}_2|\, ds,$$

as a repeated use of the triangle inequality as well as of the inequality $|\exp(-t) - \exp(-s)| \leq |t-s|$, valid for $t, s \geq 0$, shows. It was proved in [2], Theorem A.5, that

$$(3.35) \quad \int_{t_0}^t |\dot{\xi}_1 - \dot{\xi}_2|\, ds \leq |\sigma_{10d}^p - \sigma_{20d}^p| + \int_{t_0}^t |\dot{u}_1 - \dot{u}_2|\, ds + \frac{\sqrt{2}}{r}\int_{t_0}^t |\dot{u}_1||x_1 - x_2|\, ds$$

holds. Moreover, by the standard uniqueness argument for variational inequalities (see also Proposition A.1 in [2]), one has

$$(3.36) \qquad |x_1(t) - x_2(t)| \leq |\sigma_{10d}^p - \sigma_{20d}^p| + \int_{t_0}^t |\dot{u}_1 - \dot{u}_2|\, ds.$$

Putting together the estimates (3.34) - (3.36), one sees that $\mathcal{M}$ satisfies the assumption (3.31) with some constant A which depends only on $\| u_1 \|_{1,1}, \| u_2 \|_{1,1}$ and on the problem data. Therefore the Lipschitz estimate (3.32) holds for the difference $u_1 - u_2$ of the two solutions. It extends to all the unknown functions in the Chaboche model, since they can be expressed in terms of u and ξ as shown at the end of Section 2, both for the stress controlled and the strain controlled case. Thus, the proof of Theorems 2.3 and 2.4 is complete.

References

1. P.J. ARMSTRONG and C.O. FREDERICK, 1966, *A mathematical representation of the multiaxial Bauschinger effect*, C.E.G.B., Report RD/B/N 731.
2. M. BROKATE, P. KREJČÍ, Wellposedness of kinematic hardening models in elastoplasticity, *Math. Model. Numer. Anal.*, to appear.
3. M. BROKATE, J. SPREKELS, 1996, *Hysteresis and phase transitions*, Springer-Verlag, Berlin.
4. J.-L. CHABOCHE, 1989, Constitutive equations for cyclic plasticity and cyclic viscoplasticity, *Int. J. Plasticity*, **5**, pp. 247–302.
5. J.-L. CHABOCHE, 1991, On some modifications of kinematic hardening to improve the description of ratchetting effects, *Int. J. Plasticity*, **7**, pp. 661–678.
6. J.-L. CHABOCHE, 1994, Modeling of ratchetting: evaluation of various approaches, *Eur. J. Mech., A/Solids*, **13**, pp. 501–518.
7. M. KAMLAH, M. KORZEŃ and CH. TSAKMAKIS, Uniaxial ratchetting in rate-independent plasticity laws, *Acta Mechanica*, to appear.
8. M.A. KRASNOSEL'SKII and A.V. POKROVSKII, 1989, *Systems with hysteresis*, Springer-Verlag, Berlin. Russian edition: Nauka, Moscow 1983.
9. P. KREJČÍ, 1996, *Hysteresis, convexity and dissipation in hyperbolic equations*, Gakkotosho, Tokyo.
10. J. LEMAITRE and J.-L. CHABOCHE, 1990, *Mechanics of solid materials*, Cambridge University Press, Cambridge 1990. French edition: Dunod, Paris 1985.
11. G.A. MAUGIN, 1992, *The thermomechanics of plasticity and fracture*, Cambridge University Press, Cambridge 1992.
12. E. MELAN, 1938, Zur Plastizität des räumlichen Kontinuums, *Ingenieur-Archiv*, **9**, 116–126.
13. W. PRAGER, 1949, Recent developments in the mathematical theory of plasticity, *J. Appl. Phys.*, **20**, pp. 235–241.
14. A. VISINTIN, 1994, *Differential models of hysteresis*, Springer-Verlag, Berlin.

Martin Brokate
Mathematisches Seminar
Universität Kiel
D-24098 Kiel, Germany

Pavel Krejčí
Institute of Mathematics
Academy of Sciences
Žitná 25
CZ-11567 Praha, Czech Republic

International Series of Numerical Mathematics
Vol. 126, © 1998 Birkhäuser Verlag, Basel

On the Behaviour of the Value Function of a Mayer Optimal Control Problem along Optimal Trajectories

PIERMARCO CANNARSA AND MARIA ELISABETTA TESSITORE

Dipartimento di Matematica
Università di Roma "Tor Vergata"

ABSTRACT. We consider a Mayer optimal control problem for a system governed by a semilinear evolution equation of parabolic type.

We are interested in the smoothness of the related value function V along an optimal trajectory $x^*(\cdot)$. We obtain an estimate on the superdifferential of V at $(t, x^*(t))$ which states that

$$dim D^+V(t, x^*(t)) \leq 1.$$

This result may also be regarded as a necessary condition for optimality.

1991 *Mathematics Subject Classification.* 49K20, 49L10

Key words and phrases. Mayer problem, value function, semi–concavity, optimality condition.

1. Introduction

In this paper we are concerned with the following optimal control problem:

$$\text{minimize } g(x(T)) \tag{1.1}$$

over all trajectory–control pairs $\{x, \gamma\}$, subject to the semilinear state equation

$$\begin{cases} x'(t) + Ax(t) + f(t, x(t), \gamma(t)) = 0, & t \in [t_0, T] \\ x(t_0) = x_0 \end{cases} \tag{1.2}$$

Here, x_0 belongs to a real Hilbert space X, $t_0 \in [0, T]$ and $-A$ is the infinitesimal generator of an analytic semigroup. For simplicity, we assume that A is self–adjoint.

A control $\gamma^*(\cdot)$ is said to be optimal, if the minimum in (1.1) is attained at γ^*; the corresponding trajectory $x^*(\cdot)$ is said to be an optimal trajectory.

The value function V of problem (1.1)–(1.2) is defined as

$$V(t_0, x_0) = \inf\{g(x(T)) | \{x, \gamma\} \text{subject to } (1.2)\} \tag{1.3}$$

and satisfies the Dynamic Programming (or Hamilton–Jacobi–Bellman) equation

$$\begin{aligned} &-\partial_t V(t, x) + \langle D_x V(t, x), Ax \rangle + H(t, x, D_x V(t, x)) = 0 \\ &\forall t \in (0, T), \ \forall x \in X \end{aligned} \tag{1.4}$$

see [7]. In the above equation, $H : [0, T] \times X \times X \to \mathbb{R}$ is the Hamiltonian defined as

$$H(t, x, p) = \sup_{\gamma \in U} -\langle p, f(t, x, \gamma) \rangle. \tag{1.5}$$

It is well known that V is not differentiable in general. We are interested in studying if it gains regularity along an optimal trajectory. The condition we derive in the sequel can therefore be seen as a necessary condition for optimality.

There are several reasons suggesting that the behaviour of V should be better along optimal trajectories. In finite dimensions this improvement in regularity is known for problems with a "strictly convex structure", such as Calculus of Variations (see [8]), some Minimum Time problems (see [4]), and problems with infinite horizon (see [10]). Indeed, in such examples, the value function V is differentiable along an optimal trajectory.

For Mayer problems, however, the Hamiltonian H is homogenuos of degree one in p, and so it is not strictly convex. Nevertheless, by a careful application of the Dynamic Programming equation (1.4), we deduce that the value function V cannot be too singular along an optimal trajectory $x^*(t), t \in (t_0, T)$, of a sufficiently "smooth" problem. In fact, we prove the estimate

$$dim D^+ V(t, x^*(t)) \leq 1, \quad \forall t \in (t_0, T), \tag{1.6}$$

which bounds the dimension of the superdifferential of V along $x^*(t)$. This result is new also for finite dimensional problems.

Finally, we would like to observe that (1.6) does not hold for an arbitrary Mayer problem, but requires a smoothness assumption on the set $f(t, x^*(t), U)$ of admissible velocities. Indeed, if H vanishes, then equation (1.4) becomes too weak a condition on V to prevent the generation of higher singularities. We discuss this phenomenon in Example 3.5.

2. Preliminaries

Let X be a real Hilbert space and U a complete separable metric space. Fix $T > 0$ and let $(t_0, x_0) \in [0, T] \times X$. Consider the problem of minimizing the functional

$$J(t_0, x_0; \gamma) = g(x(T; t_0, x_0, \gamma)) \tag{2.1}$$

over all measurable functions $\gamma : [0, \infty) \to U$ (usually called controls). Here $g : X \to \mathbb{R}$ is a given continuous function and $x(\cdot; t_0, x_0, \gamma)$ is the mild solution of the semilinear state equation

$$\begin{cases} x'(t) + Ax(t) + f(t, x(t), \gamma(t)) = 0, & t \in [t_0, T] \\ x(t_0) = x_0, \end{cases} \tag{2.2}$$

that is the solution of the integral equation

$$x(t) = e^{-(t-t_0)A} x_0 - \int_{t_0}^{t} e^{-(t-s)A} f(s, x(s), \gamma(s)) ds \tag{2.3}$$

for all $t \in [t_0, T]$.

In the above Mayer optimal control problem we impose the following assumptions on the data:

$$(2.4)\qquad \begin{cases} (i) & A : D(A) \subset X \to X \text{ is self-adjoint and generates} \\ & \text{an analytic semigroup, } \ e^{tA}, t \geq 0; \\ & \\ (ii) & f : [0,T] \times X \times U \to X \text{ is continuous and such that} \\ & |f(t,x,\gamma)| \leq C_0(1+|x|), \\ & |f(t,x,\gamma) - f(s,y,\gamma)| \leq C_0\left[|t-s| + |x-y|\right] \\ & \text{for some } C_0 > 0 \text{ and all } t,s \in [0,T], x,y \in X, \gamma \in U; \\ & \\ (iii) & g \text{ is Lipschitz on all bounded subsets of } X. \end{cases}$$

It is well known that, under the above assumptions, problem (2.3) has a unique mild solution $x(\cdot) \in C([t_0,T];X)$.

Let Ω be an open subset of X and $\varphi : \Omega \to \mathbb{R}$.

Definition 2.1. *For any fixed $x_0 \in \Omega$, the semi-differentials of φ at x_0 are defined as*

$$D^+\varphi(x_0) = \left\{ p \in X \,|\, \limsup_{x \to x_0} \frac{\varphi(x) - \varphi(x_0) - \langle p, x - x_0 \rangle}{|x - x_0|} \leq 0 \right\}$$

$$D^-\varphi(x_0) = \left\{ p \in X \,|\, \liminf_{x \to x_0} \frac{\varphi(x) - \varphi(x_0) - \langle p, x - x_0 \rangle}{|x - x_0|} \geq 0 \right\}$$

and called super and subdifferential of φ at x_0, respectively (see [5]).

The semi-differentials $D^+\varphi(x_0)$ and $D^-\varphi(x_0)$ are both non-empty if and only if φ is Fréchet differentiable at x_0. In this case we have

$$D^+\varphi(x_0) = D^-\varphi(x_0) = \{\nabla\varphi(x_0)\}$$

where $\nabla\varphi$ denotes the gradient of φ.

Definition 2.2. *We denote by $D^*\varphi(x_0)$ the set of all points $p \in X$ for which there exists a sequence $\{x_n\}_{n \in N}$ in X with the following properties*

$$(2.5)\qquad \begin{cases} (i) & x_n \text{ converges to } x_0 \text{ as } n \to \infty \\ (ii) & \varphi \text{ is Fréchet differentiable at } x_n, \forall n \in N \\ (iii) & \nabla\varphi(x_n) \text{ weakly converges to } p \text{ as } n \to \infty \end{cases}$$

If φ is Lipschitz in a neighborhood of x_0, then φ is Fréchet diffentiable on a dense subset of Ω. Consequently, $D^*\varphi(x_0) \neq \phi$.

Let now Ω be convex and set $B_r(x_0) = \{x \in X \mid |x - x_0| < r\}$.

Definition 2.3. *We say that φ is semi-concave if there exists a function*

$$\omega : [0,+\infty) \times [0,+\infty) \to [0,+\infty)$$

satisfying

$$\begin{cases} (i) & \omega(r,s) \leq \omega(R,S), \quad \forall 0 \leq r \leq R, \quad \forall 0 \leq s \leq S \\ (ii) & \lim_{s \downarrow 0} \omega(r,s) = 0, \qquad \forall r > 0 \end{cases}$$

and such that

$$\lambda\varphi(x) + (1-\lambda)\varphi(y) - \varphi(\lambda x + (1-\lambda)y) \leq \lambda(1-\lambda)|x-y|\omega(r,|x-y|) \tag{2.6}$$

for every $r > 0, \lambda \in [0,1]$ *and* $x, y \in \Omega \cap B_r(0)$.

The superdifferential of a semi–concave function has several useful properties, some of which are recalled in the following

Proposition 2.4. *If* φ *is semi–concave in* $B_r(x_0)$ *for some* $r > 0$*, then*

$$D^+\varphi(x_0) = \overline{co}D^*\varphi(x_0) \tag{2.7}$$

where $\overline{co}$ *denotes the closed convex hull. In particular* $D^+\varphi(x_0) \neq \phi$.

Remark 2.5. For a semi–concave function the semidifferential of interest is the superdifferential, since either the subdifferential is empty or it coincides with the superdifferential and the function is differentiable.

We define the value function of problem (2.1)–(2.2) as

$$V(t_0, x_0) = \inf\{g(x(T; t_0, x_0, \gamma)) |\ \gamma : [t_0, T] \to U \text{ is measurable }\}. \tag{2.8}$$

The result below will be applied in the next section. We denote by $D_x^+V(t,x)$ the superdifferential of $V(t,\cdot)$ at x. The proof of the Proposition below is given in [3].

Proposition 2.6. *Assume* (2.4) *and let* $(t_0, x_0) \in [0,T) \times X$*. Then, for all* $\alpha \in [0,1)$,

$$D_x^+V(t_0,x_0) \subset D(A^\alpha) \ \ \& \ \ D_x^-V(t_0,x_0) \subset D(A^\alpha).$$

Under additional assumptions on the data, the value function V is semi–concave in (t,x) on $[0,T) \times X$. More precisely, the following result is obtained in [3].

Theorem 2.7. *Assume* (2.4)*, and suppose that there exists* $\alpha \in (0,1]$ *such that, for all* $R > 0$ *and for some constant* $C_R > 0$,

$$\begin{cases} (i) & f(\cdot,\cdot,\gamma) \text{ is differentiable and} \\ & \left\| \frac{\partial f}{\partial(t,x)}(t,x,\gamma) - \frac{\partial f}{\partial(t,x)}(s,y,\gamma) \right\| \leq C_R(|x-y| + |t-s|)^\alpha \\ & \text{for all } s,t \in [0,T], x,y \in B_R(0), \gamma \in U; \\ (ii) & g(x) + g(y) - 2g(\frac{x+y}{2}) \leq C_R|x-y|^{1+\alpha}, \quad \forall x,y \in B_R(0). \end{cases} \tag{2.9}$$

Then for any $r > 0$ *there exists* $C_r > 0$ *such that*

$$V(t_1,x_1) + V(t_0,x_0) - 2V\left(\frac{t_1+t_0}{2}, \frac{x_1+x_0}{2}\right) \leq qC_r(|t_1-t_0| + |x_1-x_0|)^{1+\alpha} \tag{2.10}$$

for all $t_1, t_0 \in [0, T - \frac{1}{r}]$ *and all* $x_1, x_0 \in B_r(0)$.

We conclude this section with few remarks about the Dynamic Programming equation (1.4) of a Mayer optimal control problem. As we recalled above, (1.4) is satisfied in a suitable generalized sense, as V is in general not differentiable and the coefficient of the linear term $\langle D_xV(t,x), Ax\rangle$ is defined only for $x \in D(A)$. However, we know from Proposition 2.6 that $D_x^+V(t,x) \subset D(A^\theta)$ for all $\theta \in [0,1)$. Moreover, every trajectory of (2.2) enters the fractional domain $D(A^{1-\theta})$ as soon as $t > t_0$. Therefore, all the terms in the following equality are well defined.

Theorem 2.8. *Assume* (2.4), (2.9). *Let* $x^*(\cdot)$ *be an optimal trajectory of problem* (2.2)–(2.1) *and* $\theta \in (0,1)$. *Then,*

$$-p_t + \left\langle A^\theta p_x, A^{1-\theta} x^*(t) \right\rangle + H(t, x^*(t), p_x) = 0 \tag{2.11}$$

for all $t \in (t_0, T)$ *and all* $(p_t, p_x) \in D^+V(t, x^*(t))$.

The above result is all we need to know about equation (1.4) in the sequel. For its proof, see [3], Theorem 5.2.

3. An Optimality Condition

We now show that the results of the previous section, toghether with some hypotheses on the control set and on the dynamics f in the state equation (2.2), can be used to study the structure of the singular set of the value function V associated with a Mayer optimal control problem.

In addition to the Dynamic Programming equality (2.11), the proof of our main result is based on some notions of convex analysis. These results are adapted from [4], where they were obtained in a finite dimensional set–up.

We recall that the support function to a convex set $K \subset X$ is given by

$$\sigma_K(p) = \sup_{k \in K} \langle k, p \rangle.$$

If $\overline{k} \in K$, the normal cone to K at $\overline{k}$ is the set

$$N_K(\overline{k}) = \{p \in X : \langle p, \overline{k} - k \rangle \geq 0 \ \forall k \in K\}.$$

Moreover, given a convex set K, we consider the smallest affine set that contains K. This set is called the affine hull of K and is denoted by aff(K). The relative interior of a convex set K is defined as the interior which results when K is regarded as a subset of its affine hull aff(K).

Lemma 3.1. *Let* $K_1, K_2 \subset X$ *be bounded closed convex sets and suppose that* K_2 *is not a singleton. Then the following two properties are equivalent:*

(3.1)

(*i*) *there exists* $\overline{k} \in K_1$ *such that* $K_2 \subset N_{K_1}(\overline{k})$;

(*ii*) $\sigma_{K_1}(\lambda p_0 + (1-\lambda)p_1) = \lambda \sigma_{K_1}(p_0) + (1-\lambda)\sigma_{K_1}(p_1)$ $\forall p_0, p_1 \in K_2, \ \forall \lambda \in [0,1]$.

Proof. From the definition of normal cone

$$p \in N_{K_1}(\overline{k}) \iff \sigma_{K_1}(p) = \langle p, \overline{k} \rangle.$$

This yields that (i) implies (ii).

Conversely, let us suppose that (ii) holds. Let $\overline{p}$ be a point in the relative interior of K_2. Since K_1 is a weakly compact set and the map $k \to \langle k, \overline{p} \rangle$ is weakly continuous, there exists $\overline{k} \in K_1$ such that $\sigma_{K_1}(\overline{p}) = \langle \overline{p}, \overline{k} \rangle$.

By definition,

$$\sigma_{K_1}(p) \geq \langle p, \overline{k} \rangle, \ \forall p \in K_2. \tag{3.2}$$

Let us suppose that there exists $p_0 \in K_2$ such that $\sigma_{K_1}(p_0) > \langle p_0, \overline{k}\rangle$. Since $\overline{p}$ is in the relative interior of K_2, there exist $p_1 \in K_2, p_1 \neq \overline{p}$, and $\lambda \in (0,1)$ such that $\overline{p} = \lambda p_0 + (1-\lambda)p_1$. Then

$$\sigma_{K_1}(p_1) = (1-\lambda)^{-1}(\sigma_{K_1}(\overline{p}) - \lambda\sigma_{K_1}(p_0)) < (1-\lambda)^{-1}\langle \overline{p} - \lambda p_0, \overline{k}\rangle = \langle p_1, \overline{k}\rangle,$$

in contraddiction with (3.2). It follows that (3.2) holds as an equality for every $p \in K_2$. Therefore $K_2 \subset N_{K_1}(\overline{k})$. ■

We denote the segment from a point x to a point y as

$$[x,y] = \{\lambda x + (1-\lambda)y|\ \lambda \in [0,1]\}.$$

Theorem 3.2. *Assume* (2.4), (2.9) *and let* $x^*(\cdot)$ *be an optimal trajectory of problem* (2.2)–(2.1) *such that*

$$f(t, x^*(t), U) \text{ is a closed convex set with boundary of class } C^1 \tag{3.3}$$

for any $t \in (t_0, T)$. *Then,*

$$\dim D^+V(t, x^*(t)) \leq 1\,, \quad \forall t \in (t_0, T). \tag{3.4}$$

Moreover, for any $t \in (t_0,T)$ *there exists* $p(t) \in \mathbb{R} \times X$ *and* $\mu(t) \in [0,1]$ *such that*

$$D^+V(t, x^*(t)) = [\mu(t)p(t), p(t)].$$

Proof. Fix $t \in (t_0, T)$. Then either $dimD^+V(t,x^*(t)) = 0$ or $dimD^+V(t,x^*(t)) > 0$. Clearly if the first case holds there is nothing to prove. Let us consider the second case. Assume that $dimD^+V(t,x^*(t)) \geq 1$ and take two elements (p_t, p_x) and $(p_t, p_x^{'})$ in $D^+V(t,x^*(t))$. Consider $p_x^\lambda = \lambda p_x + (1-\lambda)p_x^{'}$, $\lambda \in [0,1]$. Then, since $D^+V(t,x^*(t))$ is a convex set, $(p_t, p_x^\lambda) \in D^+V(t,x^*(t))$. Recalling that

$$H(t, x^*(t), p_x) = \sigma_{f(t,x^*(t),U)}(p_x),$$

we evaluate equation (2.11) at (p_t, p_x^λ) to obtain

$$\sigma_{f(t,x^*(t),U)}(\lambda p_x + (1-\lambda)p_x^{'}) =$$

$$p_t - \left\langle A^\theta(\lambda p_x + (1-\lambda)p_x^{'}), A^{1-\theta}x^*(t)\right\rangle = \lambda\sigma_{f(t,x^*(t),U)}(p_x) + (1-\lambda)\sigma_{f(t,x^*(t),U)}(p_x^{'}).$$

Therefore, on the sections of $D^+V(t,x^*(t))$ at the level p_t, $\sigma_{f(t,x^*(t),U)}(\cdot)$ is a linear function. By Lemma 3.1 and assumption (3.3), we derive that $p_x^{'} = \rho p_x$ for some $\rho \geq 0$. We now evaluate equation (2.11) at (p_t, p_x) and $(p_t, p_x^{'})$ to obtain

$$\begin{aligned} \sigma_{f(t,x^*(t),U)}(p_x) + \left\langle A^\theta p_x, A^{1-\theta}x^*(t)\right\rangle &= p_t \\ \text{and}\quad \rho\sigma_{f(t,x^*(t),U)}(p_x) + \rho\left\langle A^\theta p_x, A^{1-\theta}x^*(t)\right\rangle &= p_t. \end{aligned}$$

Hence,

$$\sigma_{f(t,x^*(t),U)}(p_x) + \left\langle A^\theta p_x, A^{1-\theta}x^*(t)\right\rangle = \rho[\sigma_{f(t,x^*(t),U)}(p_x) + \left\langle A^\theta p_x, A^{1-\theta}x^*(t)\right\rangle],$$

which yields $\rho = 1$. Therefore, for any p_t there exists at most one p_x such that $(p_t, p_x) \in D^+V(t,x^*(t))$.

Let $p^0(t), p^1(t) \in D^+V(t, x^*(t))$ be such that

$$\begin{aligned} p_t^0 &= \min\{p_t | \exists p_x : (p_x, p_t) \in D^+V(t, x^*(t))\} \\ \text{and} \quad p_t^1 &= \max\{p_t | \exists p_x : (p_x, p_t) \in D^+V(t, x^*(t))\}. \end{aligned}$$

By the convexity of the set $D^+V(t, x^*(t))$, we derive that there are no points p_x such that $(p_t, p_x) \in D^+V(t, x^*(t))$ with $p_t < p_t^0$ or $p_t > p_t^1$. On the other hand, for any p_t such that $p_t^0 \le p_t \le p_t^1$, there exists at least a p_x so that (p_t, p_x) belongs to $D^+V(t, x^*(t))$. Moreover p_x is unique since we proved before that on any section of $D^+V(t, x^*(t))$ at the level p_t there is only one p_x, hence $dim D^+V(t, x^*(t)) \le 1$. This yields $D^+V(t, x^*(t)) = [p^0(t), p^1(t)]$.

We now prove the last statement of the theorem. Plugging $p^0 = p^0(t), p^1 = p^1(t)$ and $p^\lambda = \lambda p^0 + (1-\lambda)p^1$, $\lambda \in [0, 1]$ in equation (2.11) we obtain

$$-p_t^0 + \left\langle A^\theta p_x^0, A^{1-\theta} x^*(t) \right\rangle + \sigma_{f(t,x^*(t),U)}(p_x^0) = 0,$$

$$-p_t^1 + \left\langle A^\theta p_x^1, A^{1-\theta} x^*(t) \right\rangle + \sigma_{f(t,x^*(t),U)}(p_x^1) = 0,$$

$$-p_t^\lambda + \left\langle A^\theta p_x^\lambda, A^{1-\theta} x^*(t) \right\rangle + \sigma_{f(t,x^*(t),U)}(p_x^\lambda) = 0.$$

From the above equations we derive

$$\sigma_{f(t,x^*(t),U)}(p_x^\lambda) = \lambda \sigma_{f(t,x^*(t),U)}(p_x^0) + (1-\lambda)\sigma_{f(t,x^*(t),U)}(p_x^1).$$

Therefore, $\sigma_{f(t,x^*(t),U)}(\cdot)$ is linear when it is restricted to the ortogonal projection of $D^+V(t, x^*(t))$ on X, denoted by $\Pi_X(D^+V(t, x^*(t)))$. Hence

$$\Pi_X(D^+V(t, x^*(t))) = [\mu(t)p_x^1, p_x^1],$$

for some $\mu(t) \in [0, 1]$, which yields $p_x^0 = \mu(t)p_x^1$. Again from equation (2.11) we get $p_t^0 = \mu(t)p_t^1$ and the proof is concluded. ∎

Remark 3.3. The above Theorem holds for any real Hilbert space X such that $dim\ X \ge 1$. However, if $dim\ X = 1$, then assumption (3.3) is meaningless. In this case we only require that $f(t, x^*(t), U) \neq \{0\}$ for all $t \in (t_0, T)$, in order to assure that $\sigma_{f(t,x^*(t),U)} \neq 0$.

Remark 3.4. The last statement of the above Theorem shows that $D^+V(t, x^*(t))$ must be a radial segment, contained in a half–line starting from the origin. This fact is a very strict requirement and suggests that, even for Mayer problem, the value function might be differentiable along optimal trajectories provided that $f(t, x, U)$ is sufficiently smooth. This question is, however, still an open problem.

The above argument heavily relies on assumption (3.3), that allows to use the Hamilton–Jacobi–Bellman equation (2.11) to study D^+V. However, such a condition is also necessary for an estimate like (2.11) to hold true, as we now show with the following example.

Example 3.5. We consider the state equation in $X = \mathbb{R}^2$

$$\begin{cases} \overline{x}'(s) = \gamma_1(s)\overline{x}(s), & s \in [t, T], \quad \overline{x}(t) = x \\ \overline{y}'(s) = \gamma_2(s)\overline{y}(s), & s \in [t, T], \quad \overline{y}(t) = y. \end{cases}$$

with $U = [0,1] \times [0,1]$. Notice that assumption (3.3) is not satisfied on $xy = 0$. The value function V is defined as

$$V(t,x,y) = \inf\{\overline{x}(T) + \overline{y}(T) | \ (\gamma_1, \gamma_2) : [t,T] \to U \text{ measurable}\}.$$

Then is not difficult to see that

$$V(t,x,y) = \begin{cases} x+y & x,y \geq 0 \\ e^{T-t}x + y & x < 0, \ y \geq 0 \\ x + e^{T-t}y & x \geq 0, \ y < 0 \\ e^{T-t}(x+y) & x,y < 0 \end{cases}$$

We note that the (unique) optimal trajectory at $(0,0)$ is given by $\overline{x}(t) = 0, \ \overline{y}(t) = 0, \ t \in [0,T]$. Moreover, by Proposition 2.4, we have

$$D^+V(t,0,0) = \text{ co } \{(1,1,0), (e^{T-t},1,0), (1,e^{T-t},0), (e^{T-t},e^{T-t},0)\},$$

and so $\dim D^+V(t,0,0) = 2$ for all $t \in [0,T)$.

References

1. P. Cannarsa, *Regularity properties of solutions to Hamilton-Jacobi equations in infinite dimensions and nonlinear optimal control*, Differential and Integral Equations 2, (1989), pp. 479–493.
2. P. Cannarsa and H. Frankowska, *Value function and optimality conditions for semilinear control problems*, Applied Math. Optim., 26 (1992), pp. 139–169.
3. P. Cannarsa and H. Frankowska *Value function and optimality conditions for semilinear control problems.II:parabolic case*, Applied Math. Optim., 33, (1996), pp. 1–33.
4. P. Cannarsa and C. Sinestrari, *Convexity properties of the minimum time function*, Calc. Var. 3, (1995), pp. 273–298.
5. M. G. Crandall, L. C. Evans and P. L. Lions, *Some properties of viscosity solutions of Hamilton–Jacobi equations*, Trans. Amer. Soc., 282 (1984), pp. 487–502.
6. M. G. Crandall and P. L. Lions, *Viscosity solutions of Hamilton–Jacobi equations*, Trans. Amer. Soc., 277 (1983), pp. 1–42.
7. M. G. Crandall and P. L. Lions, *Hamilton–Jacobi equations in infinite dimensions.* Part IV: *Hamiltonians with unbounded linear terms*, J. Funct. Anal. 90 (1990), pp. 237–283.
8. W.H. Fleming, *The Cauchy problem for a nonlinear first order partial differential equation*, J. Diff. Eqs. 5 (1969), pp. 515–530.
9. A. Pazy, *Semigroups of linear operators and applications to partial differential equations*, Springer Verlag, New York–Heidelberg–Berlin, 1983.
10. M.E. Tessitore, *"Optimality conditions for infinite horizon control problems"*, Bollettino dell'Unione Matematica Italiana 7, 9-B, (1995), pp. 785–814.

Piermarco Cannarsa
Dipartimento di Matematica
Università di Roma "Tor Vergata"
Via della Ricerca Scientifica
I-00133, Roma, Italy

Maria Elisabetta Tessitore
Dipartimento di Matematica
Università di Roma "Tor Vergata"
Via della Ricerca Scientifica
I-00133, Roma, Italy

International Series of Numerical Mathematics
Vol. 126, © 1998 Birkhäuser Verlag, Basel

Optimal Control Problem Governed by Semilinear Elliptic Equations with Integral Control Constraints and Pointwise State Constraints

E. CASAS, J.-P. RAYMOND, AND H. ZIDANI

Departamento de Matemática Aplicada y Ciencias
de la Computatión
E.T.S.I. Industriales y de Telecomunicatión
Universidad de Cantabria

UMR CNRS MIP, UFR MIG
Université Paul Sabatier

ABSTRACT. We consider optimal control problems governed by semilinear elliptic equations with pointwise state constraints. The set of control constraints is a subset of a L^p-Lebesgue space with $1 \leq p < \infty$, it is described by pointwise and integral constraints. We prove a Pontryagin principle in integral form.

1991 *Mathematics Subject Classification.* 49K20

Key words and phrases. Optimal control, Pontryagin's principle, unbounded controls, state constraints, integral control constraints.

1. Introduction

During the ten last years, many papers have been devoted to Pontryagin principles for control problems governed by partial differential equations (see [18], [6]). In the presence of state constraints, Pontryagin's principles are often proved thanks to the Ekeland's variational principle (applied to problems in which state constraints are penalized), coupled with methods of spike perturbations. To apply the Ekeland's variational principle, the space of controls V_{ad} endowed with the so-called Ekeland distance, must be complete. This assumption of completeness is in general not satisfied if V_{ad} is a subset of a L^p-Lebesgue space with $1 \leq p < \infty$ (see [14]). This is the reason why most applications of that method deal with bounded controls.
Extensions to problems with unbounded controls are considered in [14], [12], [19]. The idea is to use perturbations whose difference with an optimal solution is bounded in a L^p-space with $1 \leq p < \infty$. Extensions of [14], [12] are given in [11]. The method in [19] has been improved in [20] by considering perturbations whose difference with an optimal solution is bounded in L^∞.
In this paper we want to show that the method developed by Li-Yong [17], [18] and by Casas [6] (first introduced by Li-Yao [16]), and extended to unbounded controls in [19], [20], can also be extended to problems with integral control constraints (see the beginning of Section 3.2). Let us explain what is new. In [11], [12], [20], the sets of admissible controls are patch complete in the sense introduced by H. O. Fattorini [12]

(that is, a perturbation of an admissible control, over a subset of small measure, by an other admissible control, is still admissible). Notice that constraints of the form $\|v\|_{L^p} \leq C$ do not correspond to a patch complete subset of L^p if $1 \leq p < \infty$. Here the novelty is that we still obtain a Pontryagin's principle, even in cases when the set of admissible controls is not patch complete. When the set of admissible controls is not patch complete, we cannot use, as in [20], perturbations whose difference with an optimal solution is bounded. This is the reason of another difficulty because there is a gap between convergence properties implied by the Ekelend variational principle (Lemma 3.1) and convergence properties necessary to prove some continuity properties of the cost functional (see [6], Theorem 5.1 and [20], Lemma 3.1). Contrary to [6], [20], here we only prove a lower semicontinuity property (Proposition 3.1, ii) under some convexity condition of the cost functional with respect to the control variable (observe that this convexity condition is not needed in [6], [20]).
For simplicity, we present this extension for problems governed by semilinear elliptic equations, but the same kind of results can be obtained for problems governed by semilinear parabolic equations as those considered in [6], [13], [19], [20].

2. Setting of the Problem

Let Ω be an open bounded subset of $\mathbb{R}^N$ ($N \geq 2$) with a Lipschitz boundary Γ, q, r and $\bar{r}$ denote positive numbers satisfying

$$q > \frac{N}{2}, \quad r > \bar{r} > N - 1$$

We consider a second order differential operator defined by:

$$Ay = -\sum_{i,j=1}^{N} D_i(a_{ij}(x) D_j y),$$

(D_i denotes the partial derivative with respect to x_i) with coefficients a_{ij} belonging to $L^\infty(\Omega)$ and satisfying for some $m_0 > 0$

$$\sum_{i,j=1}^{N} a_{ij}(x)\xi_i\xi_j \geq m_0|\xi|^2 \quad \text{for all } \xi \in \mathbb{R}^N \text{ and a.e. } x \in \Omega.$$

We consider the following boundary value problem:

$$Ay + \Phi(x, y) = 0 \text{ in } \Omega, \qquad \frac{\partial y}{\partial n_A} + \Psi(s, y, v) = 0 \text{ on } \Gamma. \tag{2.1}$$

The function Φ (resp. Ψ) is a Carathéodory function from $\Omega \times \mathbb{R}$ (resp. from $\Gamma \times \mathbb{R}^2$) into $\mathbb{R}$. For almost every $x \in \Omega$, (resp. almost every $s \in \Gamma$ and every $v \in \mathbb{R}$), $\Phi(x, \cdot)$ (resp. $\Psi(s, \cdot, v)$) is of class C^1 and we have the following estimates

$$|\Phi(x, 0)| \leq M_1(x), \quad 0 \leq a_0 \leq \Phi'_y(x, y) \leq M_1(x)\eta(|y|),$$

$$|\Psi(s, 0, v)| \leq M_2(s) + m_1|v|, \quad 0 \leq b_0 \leq \Psi'_y(s, y, v) \leq (M_2(s) + m_1|v|)\eta(|y|),$$

where $M_1 \in L^q(\Omega)$, $M_2 \in L^r(\Gamma)$, m_1 is a positive constant, and η is a nondecreasing function from $\mathbb{R}^+$ into $\mathbb{R}^+$ (we have denoted by Φ'_y the partial derivative of Φ with respect to y, we adopt in all the sequel the same kind of notation for other functions).

In addition, we assume that the pair (a_0, b_0) satisfies the ellipticity condition (E_m) stated below.

We say that a pair of nonnegative functions $(a, b) \in L^q(\Omega) \times L^r(\Gamma)$ satisfies the ellipticity condition (E_m) if:

$$(E_m) \quad \inf\{\frac{1}{\|y\|_{H^1(\Omega)}}[\int_\Omega (a_{ij}D_jyD_iy + ay^2)\,dx + \int_\Gamma by^2\,ds] \mid y \in H^1(\Omega)\} \geq m > 0.$$

Remark 2.1. Since a and b are nonnegative functions, if Ω is connected, the ellipticity condition (E_m) is satisfied for some $m > 0$ if and only if (a, b) is not identically zero.

Remark 2.2. For every $v \in L^r(\Gamma)$, equation (2.1) admits a unique weak solution in $H^1(\Omega) \cap C(\bar{\Omega})$ (see Section 2).

We consider the following optimal control problem:

$$(P) \quad \inf\{J(y, v) \mid (y, v) \in (H^1(\Omega) \cap C(\bar{\Omega})) \times V,\ (y, v) \text{ satisfies (2.1)-(2.2)}\},$$

where

$$J(y, v) = \int_\Omega F(x, y(x))dx + \int_\Gamma G(s, y(s), v(s))ds,$$

$$(2.2) \qquad f(x, y(x)) \leq 0 \qquad \text{for every } x \in \bar{\Omega},$$

$$V = \{v \in L^r(\Gamma) \mid v(s) \in K(s),\ \int_\Gamma g(s, v(s))ds = 0,\ \int_\Gamma h(s, v(s))\,ds \leq 0\},$$

where K is a measurable multimapping from Γ with closed and nonempty values into $\mathcal{P}(\mathbb{R})$.

We suppose in the sequel that F (resp. G) is a Carathéodory function from $\Omega \times \mathbb{R}$ (resp. from $\Gamma \times \mathbb{R}^2$) into $\mathbb{R}$. For almost every $x \in \Omega$, (resp. almost every $s \in \Gamma$ and every $v \in \mathbb{R}$), $F(x, \cdot)$ (resp. $G(s, \cdot, v)$) is of class C^1 and we have

$$|F(x, 0)| \leq M_3(x), \quad |F'_y(x, y)| \leq M_3(x)\eta(|y|),$$

$$0 \leq G(s, 0, v) \leq M_4(s) + m_1|v|^r, \quad |G'_y(s, y, v)| \leq (M_4(s) + m_1|v|^{\bar{r}})\eta(|y|),$$

where $M_3 \in L^1(\Omega)$, $M_4 \in L^1(\Gamma)$, m_1 and η are the same as before. For almost every $s \in \Gamma$ and every $y \in \mathbb{R}$, $G(s, y, \cdot)$ is convex on $\mathbb{R}$. The functions g and h are Carathéodory functions from $\Gamma \times \mathbb{R}$ into $\mathbb{R}$ satisfying

$$|g(s, v)| \leq M_4(s) + m_1|v|^{\bar{r}}, \quad -M_4(s) - m_1|v|^{\bar{r}} \leq h(s, v) \leq M_4(s) + m_1|v|^r,$$

m_1 and M_4 are the same as before. We also suppose that the function $f : \bar{\Omega} \times \mathbb{R} \longrightarrow \mathbb{R}$ is continuous and that for every $x \in \bar{\Omega}$, $f(x, \cdot)$ is of class C^1.

The main result is the following Pontryagin principle.

Theorem 2.1. *If $(\bar{y}, \bar{v})$ is a solution of (P), there then exist $(\bar{\nu}, \bar{\mu}) \in \mathbb{R}^+ \times \mathcal{M}(\bar{\Omega})$ and $\bar{p}$ belonging to $W^{1,\tau}(\Omega)$ for all $1 \leq \tau < N/(N-1)$ such that*

$$(2.3) \qquad (\bar{\nu}, \bar{\mu}) \neq 0, \quad \bar{\mu} \geq 0, \quad \langle \bar{\mu}, f(\cdot, \bar{y}(\cdot)) \rangle_{\mathcal{M}(\bar{\Omega}) \times C(\bar{\Omega})} = 0,$$

$$
(2.4) \qquad \begin{cases} A^*\bar p + \Phi'_y(x,\bar y)\bar p = \bar\nu F'_y(x,\bar y) + \bar\mu_\Omega f'_y(x,\bar y) & \text{in } \Omega, \\ \frac{\partial \bar p}{\partial n_{A^*}} + \Psi'_y(s,\bar y,\bar v)\bar p = \bar\nu G'_y(s,\bar y,\bar v) + \bar\mu_\Gamma f'_y(s,\bar y) & \text{in } \Gamma, \end{cases}
$$

$$
(2.5) \qquad \int_\Gamma \mathcal{H}(s,\bar y(s),\bar v(s),\bar p(s),\bar\nu)ds \le \int_\Gamma \mathcal{H}(s,\bar y(s),v(s),\bar p(s),\bar\nu)ds
$$

for every $v \in V$, *where* $\mathcal{H}(s,y,v,p,\nu) = \nu G(s,y,v) + p\Psi(s,y,v)$, $A^*p = -\sum_{i,j=1}^N D_j(a_{ij}(x)D_ip)$, $\bar\mu_\Omega$ *is the restriction of* $\bar\mu$ *to* Ω *and* $\bar\mu_\Gamma$ *is the restriction of* $\bar\mu$ *to* Γ.

3. Technical results

3.1. State equation. Adjoint equation.

Theorem 3.1. *For every* $v \in L^{\bar r}(\Gamma)$, *there exists a unique weak solution* $y_v \in H^1(\Omega) \cap L^\infty(\Omega)$ *of (2.1). This solution belongs to* $C(\bar\Omega)$ *and we have*

$$
\|y_v\|_{H^1(\Omega)} + \|y_v\|_{C(\bar\Omega)} \le C_4(1 + \|v\|_{L^{\bar r}(\Gamma)}),
$$

where C_4 *depends on* $\bar r$ *but does not depend on* v. *Moreover, the mapping* $v \longmapsto y_v$ *is continuous from* $L^{\bar r}(\Gamma)$ *into* $C(\bar\Omega)$.

Proof. See [1].

Let $(a,b) \in L^q(\Omega) \times L^r(\Gamma)$ be a pair of nonnegative functions satisfying (E_m). We consider the following boundary value problem

$$
(3.1) \qquad A^*p + ap = \mu_\Omega \text{ in } \Omega, \qquad \frac{\partial p}{\partial n_{A^*}} + bp = \mu_\Gamma \text{ in } \Gamma,
$$

where $\mu = \mu_\Omega + \mu_\Gamma$ is a Radon measure on $\bar\Omega$. We shall say that $p \in W^{1,1}(\Omega)$ is solution of (3.1) if $(ap, bp) \in L^1(\Omega) \times L^1(\Gamma)$, and if :

$$
\int_\Omega (\sum_{i,j} a_{ij} D_i p D_j \varphi + ap\varphi)\, dx + \int_\Gamma b\varphi p\, ds = \langle \mu, \varphi \rangle_{\mathcal{M}(\bar\Omega)\times C(\bar\Omega)}
$$

for every $\varphi \in W^{1,\infty}(\Omega)$. Following [1], we have

Theorem 3.2. *For every pair of nonnegative functions* $(a,b) \in L^q(\Omega) \times L^r(\Gamma)$ *satisfying* (E_m) *and every* $\mu \in \mathcal{M}(\bar\Omega)$ *there exists a unique solution* $p \in W^{1,1}(\Omega)$ *of (3.1) satisfying*

$$
\int_\Omega p(Ay + ay)\, dx + \int_\Gamma p(\frac{\partial y}{\partial n_A} + by)\, ds = \langle \mu, y \rangle_{\mathcal{M}(\bar\Omega)\times C(\bar\Omega)}
$$

for every $y \in \{y \in H^1(\Omega) \mid Ay \in L^q(\Omega),\ \frac{\partial y}{\partial n_A} \in L^r(\Gamma)\}$. *Moreover, this solution belongs to* $W^{1,\tau}(\Omega)$ *for every* $1 \le \tau < N/(N-1)$, *and there exists a positive constant* $C_5 = C_5(\tau)$, *not depending on* a, b *and* μ, *such that*

$$
\|p\|_{W^{1,\tau}(\Omega)} \le C_5 \|\mu\|_{\mathcal{M}(\bar\Omega)}.
$$

Proof. The proof is given in [1], Theorem 4.2. It is well known that (3.1) can admit more than one solution [21]. However, there is a unique one satisfying the Green formula stated in Theorem 3.2. □

We also need in Section 3.3 the following regularity results.

Theorem 3.3. *For every pair of nonnegative functions $(a,b) \in L^q(\Omega) \times L^r(\Gamma)$ satisfying (E_m) and every $\phi \in L^q(\Omega)$, $\psi \in L^r(\Gamma)$, there exists a unique weak solution $y \in H^1(\Omega)$ of the equation:*

$$Ay + ay = \phi \quad \text{in } \Omega, \qquad \frac{\partial y}{\partial n_A} + by = \psi \quad \text{on } \Gamma,$$

this solution belongs to $C^\alpha(\bar{\Omega})$ for some $0 < \alpha < 1$ and we have the following estimates

$$\|y\|_{C(\bar{\Omega})} \leq C_6(\|\phi\|_{L^q(\Omega)} + \|\psi\|_{L^r(\Gamma)}),$$

$$\|y\|_{C^\alpha(\bar{\Omega})} \leq C_7(\|\phi\|_{L^q(\Omega)} + \|\psi\|_{L^r(\Gamma)})(1 + \|a\|_{L^q(\Omega)} + \|b\|_{L^r(\Gamma)}),$$

where C_6 and C_7 are independent of ϕ, ψ, a and b.

Proof. The first part is proved in [1]. Thanks to the first estimate, the second one can be proved with regularity results in [15].

3.2. Metric space of controls. In methods developed in [4], [6], [18] to prove a Pontryagin principle the set of admissible controls V_{ad} is endowed with the so-called Ekeland distance d_E. To apply the Ekeland variational principle, the space (V_{ad}, d_E) must be complete and the mapping $v \longmapsto y_v$ must be continuous from this metric space to $C(\bar{\Omega})$. Approximate optimal solutions are then characterized by an approximate Pontryagin's principle thanks to some method of perturbation. When the control set is defined by

$$V_{ad} = \{v \in L^\infty(\Gamma) \mid v(s) \in K \text{ for a.e. } s \in \Gamma\}$$

(where K is a compact subset in $\mathbb{R}$), the above conditions are satisfied and the method of spike perturbations can be used to recover a Pontryagin's principle [4]. In our case, (V, d_E) is not complete, the mapping $v \longmapsto y_v$ is not continuous from (V, d_E) into $C(\bar{\Omega})$, and the method of spike perturbations cannot be used because a spike perturbation of an admissible control is not necessarily admissible.
An other kind of perturbation, first introduced by Li-Yao [16], has been developed by Li-Yong [17], [18], and by Casas [6]. This kind of perturbation, called diffuse perturbation, has been adapted in [20] to treat problems with unbounded controls. The idea is to use diffuse perturbations whose difference with an optimal solution is bounded in $L^\infty(\Gamma)$. Here, by using (as in [19]) diffuse perturbations whose difference with an optimal solution is bounded in $L^r(\Gamma)$, we can consider problems with integral control constraints.

We define a new metric space in the following way. Let $\bar{v}$ be in V ($\bar{v}$ will be an optimal boundary control that we want to characterize). For $0 < M < \infty$, let us set:

$$V(M) = \{v \in V \mid \|v - \bar{v}\|_{L^r(\Gamma)} \leq M\},$$

and define:

$$d_E(v_1, v_2) = \mathcal{L}^{N-1}(\{s \in \Gamma \mid v_1(s) \neq v_2(s)\}).$$

Lemma 3.1. *Let $M > 0$ and $\{(v_n)_n, v\} \subset V(M)$. If $(v_n)_n$ tends to v in $(V(M), d_E)$, then $(v_n)_n$ tends to v in $L^{\bar{r}}(\Gamma)$.*

Proof. The proof is immediate if we remark that, since $1 \leq \bar{r} < r$, we have

$$\int_\Gamma |v - v_n|^{\bar{r}}\, ds \leq \|v - v_n\|_{L^r(\Gamma)}^{\bar{r}} (d_E(v_n, v))^{\frac{r-\bar{r}}{r}} \leq (2M)^{\bar{r}} (d_E(v_n, v))^{\frac{r-\bar{r}}{r}}. \qquad \square$$

Proposition 3.1. *For every $M > 0$, we have :*
i) $(V(M), d_E)$ is a complete metric space,
ii) the mapping which associates y_v with v is continuous from $(V(M), d_E)$ into $C(\bar{\Omega})$,
ii) the mapping which associates $J(y_v, v)$ with v is lower semicontinuous from $(V(M), d_E)$ into $\mathbb{R}$.

Proof.

i) Let $(v_n)_n$ be a Cauchy sequence in $(V(M), d_E)$. Following [9], we can prove that $(v_n)_n$ converges for the distance d_E to some measurable function v such that $v(s) \in K(s)$ for a.e. $s \in \Gamma$. As in the proof of Lemma 3.1, we can prove that $(v_n)_n$ is a Cauchy sequence in $L^{\bar{r}}(\Gamma)$. Moreover, $(v_n)_n$ is bounded in $L^r(\Gamma)$. Therefore $(v_n)_n$ converges to v strongly in $L^{\bar{r}}(\Gamma)$ and weakly in $L^r(\Gamma)$. On the other hand, by using Fatou's Lemma (applied to the sequence of functions $h(\cdot, v_n(\cdot)) + M_4(\cdot) + m_1|v_n(\cdot)|^{\bar{r}}$), it yields :

$$\int_\Gamma h(s, v)\, ds \leq \liminf_{n\to\infty} \int_\Gamma h(s, v_n)\, ds \leq 0,$$

and we also have

$$\|v - \bar{v}\|_{L^r(\Gamma)} \leq \liminf_{n\to\infty} \|v_n - \bar{v}\|_{L^r(\Gamma)} \leq M.$$

Moreover, we have $\lim_n \int_\Gamma g(s, v_n)\, ds = 0$ and (because $(v_n)_n$ converges to v in $L^{\bar{r}}(\Gamma)$) there exists a subsequence, still indexed by n, such that $\lim_n \int_\Gamma g(s, v_n)\, ds = \int_\Gamma g(s, v)\, ds$. Therefore, $\int_\Gamma g(s, v)\, ds = 0$ and v belongs to $V(M)$.

ii) This assertion follows from Lemma 3.1 and from the continuity result of Theorem 3.1.

iii) We consider $\{(v_n)_n, v\} \subset V(M)$, such that $(v_n)_n$ converges to v for the metrics d_E. Recall (see (i)) that $(v_n)_n$ also converges to v for the weak topology of $L^r(\Gamma)$. We complete the proof thanks to the assumptions on F, G and thanks to the continuity results stated in (ii). (In particular, we use the convexity of $G(s, y, \cdot)$ and estimates on G to prove the lower semicontinuity of $v \longmapsto J(y_v, v)$.) $\square$

3.3. Diffuse perturbations.

Lemma 3.2. *Let v_1, v_2 and v_3 be in V and let y_1 be in $C(\bar{\Omega})$. For every $0 < \rho < 1$, there exists a sequence of measurable subsets $(E_\rho^n)_n$ in Γ such that*

$$\mathcal{L}^{N-1}(E_\rho^n) = \rho \mathcal{L}^{N-1}(\Gamma), \tag{3.2}$$

$$\int_{\Gamma\setminus E_\rho^n} h(s, v_1)\, ds + \int_{E_\rho^n} h(s, v_2)\, ds = (1-\rho)\int_\Gamma h(s, v_1)\, ds + \rho \int_\Gamma h(s, v_2)\, ds, \tag{3.3}$$

$$\int_{\Gamma\setminus E_\rho^n} g(s, v_1)\, ds + \int_{E_\rho^n} g(s, v_2)\, ds = (1-\rho)\int_\Gamma g(s, v_1)\, ds + \rho \int_\Gamma g(s, v_2)\, ds, \tag{3.4}$$

$$(3.5)\quad \int_{\Gamma\setminus E_\rho^n} |v_1 - v_3|^r\, ds + \int_{E_\rho^n} |v_2 - v_3|^r\, ds = (1-\rho)\int_\Gamma |v_1 - v_3|^r\, ds + \rho\int_\Gamma |v_2 - v_3|^r\, ds,$$

$$(3.6)\quad \int_{E_\rho^n} (G(s,y_1,v_2) - G(s,y_1,v_1))\, ds = \rho\int_\Gamma (G(s,y_1,v_2) - G(s,y_1,v_1))\, ds,$$

$$(3.7)\quad \frac{1}{\rho}\chi_{E_\rho^n} \rightharpoonup 1 \quad \textit{weakly-star in } L^\infty(\Gamma) \quad \textit{when } n \textit{ tends to infinity,}$$

where $\chi_{E_\rho^n}$ *is the characteristic function of* E_ρ^n.

Proof. This lemma is an easy consequence of the Lyapunov convexity Theorem (see [20]).

Theorem 3.4. *Let* v_1, v_2 *and* v_3 *be in* V. *For every* $0 < \rho < 1$, *there exists a measurable subset* $E_\rho \subset \Gamma$ *such that*

$$(3.8)\quad \mathcal{L}^{N-1}(E_\rho) = \rho\mathcal{L}^{N-1}(\Gamma),$$

$$(3.9)\quad \int_{\Gamma\setminus E_\rho} h(s,v_1)\, ds + \int_{E_\rho} h(s,v_2)\, ds = (1-\rho)\int_\Gamma h(s,v_1)ds + \rho\int_\Gamma h(s,v_2)ds,$$

$$(3.10)\quad \int_{\Gamma\setminus E_\rho} g(s,v_1)\, ds + \int_{E_\rho} g(s,v_2)\, ds = (1-\rho)\int_\Gamma g(s,v_1)\, ds + \rho\int_\Gamma g(s,v_2)\, ds,$$

$$(3.11)\quad \int_{\Gamma\setminus E_\rho} |v_1 - v_3|^r\, ds + \int_{E_\rho} |v_2 - v_3|^r\, ds = (1-\rho)\int_\Gamma |v_1 - v_3|^r\, ds + \rho\int_\Gamma |v_2 - v_3|^r\, ds,$$

$$(3.12)\quad \int_{E_\rho} (G(s,y_1,v_2) - G(s,y_1,v_1))\, ds = \rho\int_\Gamma (G(s,y_1,v_2) - G(s,y_1,v_1))\, ds,$$

$$(3.13)\quad y_\rho = y_1 + \rho z + r_\rho, \ \lim_{\rho\to 0}\frac{1}{\rho}\|r_\rho\|_{C(\bar\Omega)} = 0,$$

$$(3.14)\quad J(y_\rho, v_\rho) = J(y_1, v_1) + \rho\Delta J + o(\rho),$$

where

$$v_\rho(s) = \begin{cases} v_1(s) & \textit{if } s \in \Gamma\setminus E_\rho, \\ v_2(s) & \textit{if } s \in E_\rho, \end{cases}$$

y_ρ *and* y_1 *are the weak solutions of (2.1) corresponding respectively to* v_ρ *and to* v_1, z *is the weak solution in* $H^1(\Omega)$ *of*

$$Az + \Phi'_y(x,y_1)z = 0 \quad \textit{in } \Omega, \qquad \frac{\partial z}{\partial n_A} + \Psi'_y(s,y_1,v_1)z + \Psi(s,y_1,v_2) - \Psi(s,y_1,v_1) = 0 \quad \textit{in } \Gamma$$

and

$$\Delta J = \int_\Gamma (G'_y(s,y_1,v_1)z + G(s,y_1,v_2) - G(s,y_1,v_1))\, ds.$$

Proof. Let $(E^n_\rho)_n$ be the sequence of measurable subsets defined in Lemma 4.1. We set

$$v^n_\rho(s) = \begin{cases} v_1(s) & \text{if } s \in \Gamma \setminus E^n_\rho, \\ v_2(s) & \text{if } s \in E^n_\rho. \end{cases}$$

Let y^n_ρ be the solution of (2.1) corresponding to v^n_ρ and let z be the function defined in the statement of Theorem 3.4. It is clear that $\zeta^n_\rho = (y^n_\rho - y_1)/\rho - z$ is the weak solution in $H^1(\Omega)$ of

$$A\zeta + a^n_\rho \zeta = \phi^n_\rho \text{ in } \Omega, \qquad \frac{\partial \zeta}{\partial n_A} + b^n_\rho \zeta = \psi^n_\rho + \varphi^n_\rho \text{ on } \Gamma,$$

where $a^n_\rho = \int_0^1 \Phi'_y(x, y_1 + t(y^n_\rho - y_1))dt$, $b^n_\rho = \int_0^1 \Psi'_y(s, y_1 + t(y^n_\rho - y_1), v^n_\rho)dt$, $\phi^n_\rho = (\Phi'_y(x, y_1) - a^n_\rho)z$, $\psi^n_\rho = (\Psi'_y(s, y_1, v_1) - b^n_\rho)z$, $\varphi^n_\rho = (1 - \frac{1}{\rho}\chi_{E^n_\rho})(\Psi(s, y_1, v_2) - \Psi(s, y_1, v_1))$ and $\chi_{E^n_\rho}$ is the characteristic function of E^n_ρ. We denote by ζ^{n1}_ρ the solution in $H^1(\Omega)$ of

$$A\zeta + a^n_\rho \zeta = \phi^n_\rho \text{ in } \Omega, \qquad \frac{\partial \zeta}{\partial n_A} + b^n_\rho \zeta = \psi^n_\rho \text{ on } \Gamma,$$

by ζ^{n2}_ρ the solution in $H^1(\Omega)$ of

$$A\zeta + a^n_\rho \zeta = 0 \text{ in } \Omega, \qquad \frac{\partial \zeta}{\partial n_A} + b^n_\rho \zeta = \varphi^n_\rho \text{ on } \Gamma,$$

and by ξ^n_ρ the solution in $H^1(\Omega)$ of

$$A\xi + a\xi = 0 \text{ in } \Omega, \qquad \frac{\partial \xi}{\partial n_A} + b\xi = \varphi^n_\rho \text{ on } \Gamma,$$

where $a = \Phi'_y(x, y_1)$, $b = \Phi'_y(s, y_1, v_1)$. We also have

$$A(\xi^n_\rho - \zeta^{n2}_\rho) + a^n_\rho(\xi^n_\rho - \zeta^{n2}_\rho) = (a^n_\rho - a)\xi^n_\rho \text{ in } \Omega, \qquad \frac{\partial(\xi^n_\rho - \zeta^{n2}_\rho)}{\partial n_A} + b^n_\rho(\xi^n_\rho - \zeta^{n2}_\rho) = (b^n_\rho - b)\xi^n_\rho \text{ on } \Gamma.$$

Notice that (a, b) and (a^n_ρ, b^n_ρ) satisfy the ellipticity condition (E_m). Thanks to Theorem 3.3, we have

$$\|\zeta^{n1}_\rho\|_{C(\bar{\Omega})} \leq C_6(\|\phi^n_\rho\|_{L^q(\Omega)} + \|\psi^n_\rho\|_{L^r(\Gamma)}), \tag{3.15}$$

$$\|\zeta^{n2}_\rho - \xi^n_\rho\|_{C(\bar{\Omega})} \leq C_6(\|a^n_\rho - a\|_{L^q(\Omega)} + \|b^n_\rho - b\|_{L^r(\Gamma)})\|\xi^n_\rho\|_{C(\bar{\Omega})}. \tag{3.16}$$

The operator $\mathcal{T}$ which associates ξ, the solution in $H^1(\Omega)$ of

$$A\xi + a\xi = 0, \qquad \frac{\partial \xi}{\partial n_A} + b\xi = \varphi,$$

with φ, is continuous from $L^r(\Gamma)$ into $C^\alpha(\bar{\Omega})$ (see Theorem 3.3). Since the embedding from $C^\alpha(\bar{\Omega})$ into $C(\bar{\Omega})$ is compact, the operator $\mathcal{T}$ is compact from $L^r(\Gamma)$ into $C(\bar{\Omega})$. Because of (3.7), for every $0 < \rho < 1$, the sequence $(\varphi^n_\rho)_n$ converges to zero for the weak topology in $L^r(\Gamma)$, therefore the sequence $(\xi^n_\rho)_n$ converges to zero in $C(\bar{\Omega})$. There then exists an integer depending on ρ, denoted by $n(\rho)$, such that

$$\|\xi^{n(\rho)}_\rho\|_{C(\bar{\Omega})} \leq \rho.$$

Since we have
$$\int_\Gamma |v_\rho^{n(\rho)} - v_1|^r\, ds = \int_{E_\rho^{n(\rho)}} |v_2 - v_1|^r\, ds,$$
$(v_\rho^{n(\rho)})_\rho$ converges to v_1 in $L^r(\Gamma)$ and $(y_\rho^{n(\rho)})_\rho$ converges to y_1 in $C(\bar\Omega)$. From assumptions on Φ and Ψ, we deduce that, $(\phi_\rho^{n(\rho)})_\rho$, $(a_\rho^{n(\rho)} - a)_\rho$ both converge to zero in $L^q(\Omega)$ and $(\psi_\rho^{n(\rho)})_\rho$, $(b_\rho^{n(\rho)} - b)_\rho$ both converge to zero in $L^r(\Gamma)$ when ρ tends to zero. Thus, thanks to (3.15) and to (3.16), we get
$$\lim_{\rho\to 0} \|\zeta_\rho^{n(\rho)}\|_{C(\bar\Omega)} \le \lim_{\rho\to 0} \|\zeta_\rho^{n(\rho)1}\|_{C(\bar\Omega)} + \lim_{\rho\to 0} \|\xi_\rho^{n(\rho)} - \zeta_\rho^{n(\rho)2}\|_{C(\bar\Omega)} + \lim_{\rho\to 0} \|\xi_\rho^{n(\rho)}\|_{C(\bar\Omega)} = 0.$$
Now we set $E_\rho = E_\rho^{n(\rho)}$, $\frac{1}{\rho} r_\rho = \zeta_\rho^{n(\rho)}$ and $v_\rho = v_\rho^{n(\rho)}$, conditions (3.8) to (3.13) are clearly satisfied. Moreover, taking into account (3.12), (3.13) and the definition of v_ρ we easily verify (3.14). □

Remark. Thanks to (3.9) and to (3.10), we conclude that, even if V is not patch complete (in the sense given in Introduction), for every $v_1, v_2 \in V$, we can construct a diffuse perturbation $(v_\rho)_\rho$, of v_1 by v_2, such that v_ρ still belongs to V. By setting $v_3 = \bar v$ in (3.11), we can see that for $M > 0$, if $v_1, v_2 \in V(M)$ then v_ρ belongs to $V(M)$, for every $0 < \rho < 1$.

4. Proof of the Pontryagin principle

4.1. Penalized problem. Let $|\cdot|_{C(\bar\Omega)}$ be a norm on $C(\bar\Omega)$, equivalent to the usual norm $\|\cdot\|_{C(\bar\Omega)}$, such that $(C(\bar\Omega), |\cdot|_{C(\bar\Omega)})$ be strictly convex and $\mathcal{M}(\bar\Omega)$, endowed with the dual norm of $|\cdot|_{C(\bar\Omega)}$ (denoted by $|\cdot|_{\mathcal{M}(\bar\Omega)}$), be also strictly convex (see [8], Corollary 2 p. 148, or Corollary 2 p. 167). We have

$$(4.1)\quad \limsup_{\substack{\rho\searrow 0,\\ \varphi'\to\varphi}} \frac{|(\varphi' + \rho z)^+|_{C(\bar\Omega)} - |(\varphi')^+|_{C(\bar\Omega)}}{\rho} = \max\{\langle \xi, z\rangle_{\mathcal{M}(\bar\Omega)\times C(\bar\Omega)} \mid \xi \in \partial|(\cdot)^+|_{C(\bar\Omega)}(\varphi)\}$$

for every $\varphi, z \in C(\bar\Omega)$, where $\partial|(\cdot)^+|_{C(\bar\Omega)}$ is the subdifferential, of $|(\cdot)^+|_{C(\bar\Omega)}$, in the sense of convex analysis [7] and $(\cdot)^+ = \max(\cdot, 0)$. Therefore, for a given $\varphi \in C(\bar\Omega)$ we have

$$(4.2)\quad \langle \xi, z-\varphi\rangle_{\mathcal{M}(\bar\Omega)\times C(\bar\Omega)} + |\varphi^+|_{C(\bar\Omega)} \le |z^+|_{C(\bar\Omega)}\ \forall\, \xi \in \partial|(\cdot)^+|_{C(\bar\Omega)}(\varphi),\ \forall\, z \in C(\bar\Omega),$$
$$|\xi|_{\mathcal{M}(\bar\Omega)} \le 1 \quad \text{for every } \xi \in \partial|(\cdot)^+|_{C(\bar\Omega)}(\varphi).$$
Moreover it is proved in ([17], Lemma 3.4) that, since $\partial|(\cdot)^+|_{C(\bar\Omega)}(\varphi)$ is convex in $\mathcal{M}(\bar\Omega)$ and $(\mathcal{M}(\bar\Omega), |\cdot|_{\mathcal{M}(\bar\Omega)})$ is strictly convex, then if $\varphi^+ \neq 0$, $\partial|(\cdot)^+|_{C(\bar\Omega)}(\varphi)$ is a singleton and $|(\cdot)^+|_{C(\bar\Omega)}$ is Gâteaux-differentiable at φ.

Let $(\bar y, \bar v)$ be a solution of (P). We consider the penalized functional
$$J_n(y, v) = \{[(J(y, v) - J(\bar y, \bar v) + \frac{1}{n^2})^+]^2 + |(f(\cdot, y(\cdot)))^+|^2_{C(\bar\Omega)}\}^{\frac{1}{2}}$$
and, for every $n > 0$, we define the penalized problem
$$(P_n)\qquad \inf\{J_n(y, v) \mid (y, v) \in C(\bar\Omega) \times V(M_n),\ (y, v) \text{ satisfies (2.1)}\},$$

where $M_n = n^{(\frac{1}{2\bar{r}}-\frac{1}{2r})}$. It is easy to see that for every $n > 0$, $(\bar{y}, \bar{v})$ is $\frac{1}{n^2}$-solution of (P_n). Thanks to the Ekeland's principle, there exists $v_n \in V(M_n)$ such that

$$(4.3) \qquad d_E(v_n, \bar{v}) \leq \frac{1}{n}, \quad J_n(y_n, v_n) \leq J_n(y_v, v) + \frac{1}{n} d_E(v_n, v)$$

for every $v \in V(M_n)$ (y_n and y_v are the solutions of (2.1) corresponding respectively to v_n and to v).

4.2. Proof of Theorem 2.1.

1. Approximate optimality conditions for the boundary control v_n satisfying (4.3).

Let v_0 be in V. For n_0 large enough, $v_0 \in V(M_n)$ for every $n \geq n_0$. Applying Theorem 3.4, we deduce the existence of measurable subsets E_n^ρ, such that $\mathcal{L}^{N-1}(E_n^\rho) = \rho \mathcal{L}^{N-1}(\Gamma)$,

$$(4.4) \quad \int_{\Gamma \setminus E_n^\rho} h(s, v_n)\, ds + \int_{E_n^\rho} h(s, v_0)\, ds = (1-\rho) \int_\Gamma h(s, v_n) ds + \rho \int_\Gamma h(s, v_0) ds,$$

$$(4.5) \quad \int_{\Gamma \setminus E_n^\rho} g(s, v_n)\, ds + \int_{E_n^\rho} g(s, v_0)\, ds = (1-\rho) \int_\Gamma g(s, v_n)\, ds + \rho \int_\Gamma g(s, v_0)\, ds,$$

$$(4.6) \quad \int_{\Gamma \setminus E_n^\rho} |v_n - \bar{v}|^r\, ds + \int_{E_n^\rho} |v_0 - \bar{v}|^r\, ds = (1-\rho) \int_\Gamma |v_n - \bar{v}|^r\, ds + \rho \int_\Gamma |v_0 - \bar{v}|^r\, ds,$$

$$(4.7) \qquad y_n^\rho = y_n + \rho z_n + r_n^\rho, \quad \lim_{\rho \to 0} \frac{1}{\rho} ||r_n^\rho||_{C(\bar{\Omega})} = 0,$$

$$(4.8) \qquad J(y_n^\rho, v_n^\rho) = J(y_n, v_n) + \rho \Delta J_n + o(\rho),$$

where v_n^ρ is defined by

$$(4.9) \qquad v_n^\rho(s) = \begin{cases} v_n(s) & \text{on } \Gamma \setminus E_n^\rho, \\ v_0(s) & \text{on } E_n^\rho, \end{cases}$$

y_n^ρ and y_n are the state variables corresponding respectively to v_n^ρ and to v_n, z_n is the weak solution of

$$A z_n + \Phi'_y(x, y_n) z_n = 0 \text{ in } \Omega, \quad \frac{\partial z_n}{\partial n_A} + \Psi'_y(s, y_n, v_n) z_n = \Psi(s, y_n, v_n) - \Psi(s, y_n, v_0) \text{ on } \Gamma,$$

and

$$\Delta J_n = \int_\Omega F'_y(x, y_n(x)) z_n(x)\, dx + \int_\Gamma G'_y(s, y_n(s), v_n(s)) z_n(s)\, ds + \int_\Gamma [G(s, y_n(s), v_0(s)) - G(s, y_n(s), v_n(s))]\, ds.$$

On the other hand, thanks to (4.4), (4.5) and (4.6), for every $n \geq n_0$ and every $0 < \rho < 1$, v_n^ρ belongs to $V(M_n)$. If we set $v = v_n^\rho$ in (4.3), it yields

$$(4.10) \qquad \lim_{\rho \to 0} \frac{J_n(y_n, v_n) - J_n(y_n^\rho, v_n^\rho)}{\rho} \leq \frac{1}{n} \mathcal{L}^{N-1}(\Gamma).$$

Taking (4.1), (4.8) and the definition of J_n into account, we get

$$-\nu_n \Delta J_n - \langle \mu_n, f_y'(y_n) z_n \rangle_{\mathcal{M}(\bar{\Omega})\times C(\bar{\Omega})} \leq \frac{1}{n}\mathcal{L}^{N-1}(\Gamma) \tag{4.11}$$

where

$$\nu_n = \frac{(J(y_n, v_n) - J(\bar{y}, \bar{v}) + \frac{1}{n^2})^+}{J_n(y_n, v_n)}, \tag{4.12}$$

$$\mu_n = \begin{cases} \frac{|(f(y_n))^+|_{C(\bar{\Omega})} \nabla |(f(y_n))^+|_{C(\bar{\Omega})}}{J_n(y_n, v_n)} & \text{if } (f(y_n))^+ \neq 0, \\ 0 & \text{if not.} \end{cases} \tag{4.13}$$

For every $n \geq n_0$, we consider p_n the unique weak solution of

$$\begin{cases} A^* p_n + \Phi_y'(x, y_n) p_n = \nu_n F_y'(x, y_n) + \mu_{n\Omega} f_y'(\cdot, y_n) & \text{in } \Omega, \\ \\ \frac{\partial p_n}{\partial n_{A^*}} + \Psi_y'(s, y_n, v_n) p_n = \nu_n G_y'(s, y_n, v_n) + \mu_{n\Gamma} f_y'(\cdot, y_n) & \text{on } \Gamma, \end{cases} \tag{4.14}$$

which satisfies the Green formula :

$$\begin{aligned} &\int_\Omega p_n (Ay + \Phi_y'(x, y_n) y)\, dx + \int_\Gamma p_n (\frac{\partial y}{\partial n_A} + \Psi_y'(s, y_n, v_n) y)\, ds \\ &= \int_\Omega \nu_n F_y'(x, y_n) y\, dx + \int_\Gamma \nu_n G_y'(s, y_n, v_n) y\, ds + \langle \mu_n, f_y'(\cdot, y_n) y \rangle_{\mathcal{M}(\bar{\Omega})\times C(\bar{\Omega})}, \end{aligned} \tag{4.15}$$

for every $y \in \{y \in H^1(\Omega) \mid Ay \in L^q(\Omega),\ \frac{\partial y}{\partial n_A} \in L^r(\Gamma)\}$ ($\mu_{n\Omega}$ is the restriction of μ_n to Ω and $\mu_{n\Gamma}$ is the restriction of μ_n to Γ). With this Green formula, with (4.11) and the definition of ΔJ_n, we get

$$\begin{aligned} &\int_\Gamma [\nu_n G(s, y_n, v_n) - p_n \Psi(s, y_n, v_n)]\, ds \\ &\qquad \leq \int_\Gamma [\nu_n G(s, y_n, v_0) - p_n \Psi(s, y_n, v_0)]\, ds + \frac{1}{n}\mathcal{L}^{N-1}(\Gamma) \end{aligned} \tag{4.16}$$

for every $n \geq n_0$.

2. Convergence of sequences $(\nu_n)_n$, $(\mu_n)_n$ and $(p_n)_n$.

We remark that

$$|\mu_n|^2_{\mathcal{M}(\bar{\Omega})} + \nu_n^2 = 1.$$

The sequences $(\nu_n)_n$ and $(\mu_n)_n$ are respectively bounded in $\mathbb{R}$ and in $\mathcal{M}(\bar{\Omega})$, there then exist $\bar{\nu} \in R^+$, $\bar{\mu} \in \mathcal{M}(\bar{\Omega})$ and a subsequence, still denoted by $(\nu_n, \mu_n)_n$, such that

$$\nu_n \longrightarrow \bar{\nu}, \quad \mu_n \rightharpoonup \bar{\mu} \ \text{ weakly star in } \mathcal{M}(\bar{\Omega}). \tag{4.17}$$

Let $1 < \tau < N/(N-1)$ for which the following embeddings are continuous

$$W^{1,\tau}(\Omega) \hookrightarrow L^{q'}(\Omega), \quad W^{1-\frac{1}{\tau},\tau}(\Gamma) \hookrightarrow L^{\bar{r}'}(\Gamma).$$

From Theorem 3.2, it yields

$$\|p_n\|_{W^{1,\tau}(\Omega)} \leq C_5\{\nu_n \|F_y'(\cdot, y_n)\|_{L^1(\Omega)} + \nu_n \|G_y'(\cdot, y_n, v_n)\|_{L^1(\Gamma)} + |\mu_n|_{\mathcal{M}(\bar{\Omega})} |f_y'(\cdot, y_n)|_{C(\bar{\Omega})}\}.$$

Since the sequences $(\nu_n)_n, (\mu_n)_n, (y_n)_n$ and $(v_n)_n$ are bounded respectively in $\mathbb{R}$, $\mathcal{M}(\bar\Omega)$, $C(\bar\Omega)$ and in $L^{\bar r}(\Gamma)$, the sequence $(p_n)_n$ is bounded in $W^{1,\tau}(\Omega)$. There then exist $\bar p \in W^{1,\tau}(\Omega)$ and a subsequence, still denoted by $(p_n)_n$, such that $(p_n)_n$ weakly converges to $\bar p$ in $W^{1,\tau}(\Omega)$. Therefore, $(p_n)_n$ weakly converges to $\bar p$ in $L^{q'}(\Omega)$ and the sequence of traces $(p_n|_\Gamma)_n$ weakly converges to the trace $\bar p|_\Gamma$ in $L^{\bar r'}(\Gamma)$. Let us prove that $\bar p$ is a weak solution of equation (2.4).
Let φ be in $W^{1,\infty}(\Omega)$, for every $n \ge n_0$, we have :

$$
\begin{aligned}
(4.18)\quad & \int_\Omega \{\sum_{i,j=1}^N a_{ij} D_i p_n D_j\varphi + \Phi'_y(x,y_n)p_n\varphi\}\,dx + \int_\Gamma \Psi'_y(s,y_n,v_n)p_n\varphi\,ds \\
&= \int_\Omega \nu_n F'_y(x,y_n)\varphi\,dx + \int_\Gamma \nu_n G'_y(s,y_n,v_n)\varphi\,ds + \langle \mu_n, f'_y(\cdot,y_n)\varphi\rangle_{\mathcal{M}(\bar\Omega)\times C(\bar\Omega)},
\end{aligned}
$$

and the Green formula (4.15). Moreover, since $\int_\Gamma |v_n - \bar v|^{\bar r}\,ds \le (d_E(v_n,\bar v))^{\frac{r-\bar r}{r}} M_n^{\bar r} \le (\frac{1}{n})^{\frac{r-\bar r}{2r}}$, $(v_n)_n$ converges to $\bar v$ in $L^{\bar r}(\Gamma)$ and $(y_n)_n$ converges to $\bar y$ in $C(\bar\Omega)$. Thanks to assumptions on Φ, Ψ, F, G, f, we have

$$
\lim_n \|\Phi'_y(\cdot,y_n) - \Phi'_y(\cdot,\bar y)\|_{L^q(\Omega)} = 0, \quad \lim_n \|F'_y(\cdot,y_n) - F'_y(\cdot,\bar y)\|_{L^1(\Omega)} = 0,
$$
$$
\lim_n \|\Psi'_y(\cdot,y_n,v_n) - \Psi'_y(\cdot,\bar y,\bar v)\|_{L^{\bar r}(\Gamma)} = 0, \quad \lim_n \|G'_y(\cdot,y_n,v_n) - G'_y(\cdot,\bar y,\bar v)\|_{L^1(\Gamma)} = 0,
$$
$$
\lim_n |f'_y(\cdot,y_n) - f'_y(\cdot,\bar y)|_{C(\bar\Omega)} = 0.
$$

Thus, by passing to the limit in (4.18) and in (4.15), we prove that $\bar p$ is the unique weak solution of the equation (2.4) which satisfies the Green formula :

$$
\begin{aligned}
(4.19)\quad & \int_\Omega \bar p(Ay + \Phi'_y(x,\bar y)y)\,dx + \int_\Gamma \bar p(\frac{\partial y}{\partial n_A} + \Psi'_y(s,\bar y,\bar v)y)\,ds \\
&= \int_\Omega \bar\nu F'_y(x,\bar y)y\,dx + \int_\Gamma \bar\nu G'_y(s,\bar y,\bar v)y\,ds + \langle \bar\mu, f'_y(\cdot,\bar y)y\rangle_{\mathcal{M}(\bar\Omega)\times C(\bar\Omega)},
\end{aligned}
$$

for every $y \in \{y \in H^1(\Omega) \mid Ay \in L^q(\Omega),\ \frac{\partial y}{\partial n_A} \in L^r(\Gamma)\}$. Because of uniqueness of the weak solution of (2.4) satisfying (4.19), we can deduce by classical arguments that $\bar p$ is independent of τ and that all the sequence $(p_n)_n$ weakly converges to $\bar p$ in $W^{1,\tau}(\Omega)$ for every $1 \le \tau < N/(N-1)$.

3. Pontryagin's principle.

Notice that $(v_n)_n$ tends to $\bar v$ in $L^{\bar r}(\Gamma)$. By letting n tend to infinity in (4.16), with Fatou's Lemma (applied to the sequence of functions $\nu_n G(\cdot,0,v_n(\cdot))$) and the convergence results stated in step 2, we obtain

$$
(4.20)\qquad \int_\Gamma \mathcal{H}(s,\bar y(s),\bar v(s),\bar p(s),\bar\nu)\,ds \le \int_\Gamma \mathcal{H}(s,\bar y(s),v_0(s),\bar p(s),\bar\nu)\,ds,
$$

for every $v_0 \in V$.
On the other hand, from the definition of μ_n and from (4.2), we deduce

$$
(4.21)\qquad \langle \mu_n, z - f(\cdot,y_n)\rangle_{\mathcal{M}(\bar\Omega)\times C(\bar\Omega)} \le 0 \quad \text{for every } z \in \{z \in C(\bar\Omega) \mid z \le 0\}.
$$

By passing to the limit in this expression, we obtain

$$
(4.22)\qquad \langle \bar\mu, z - f(\cdot,y)\rangle_{\mathcal{M}(\bar\Omega)\times C(\bar\Omega)} \le 0 \qquad \text{for every } z \in \{z \in C(\bar\Omega) \mid z \le 0\},
$$

which is equivalent to $\bar{\mu} \geq 0$ and $\langle \bar{\mu}, f(\cdot, y) \rangle_{\mathcal{M}(\bar{\Omega}) \times C(\bar{\Omega})} = 0$. To prove that $(\bar{\nu}, \bar{\mu})$ is nonzero, we recall that $\nu_n^2 + |\mu_n|^2_{\mathcal{M}(\bar{\Omega})} = 1$. If $\bar{\nu} > 0$, the proof is complete. If $\bar{\nu} = 0$, we can prove that $|\bar{\mu}|_{\mathcal{M}(\bar{\Omega})} > 0$ by using $\lim_n |\mu_n|_{\mathcal{M}(\bar{\Omega})} = 1$. Indeed there exists a ball $B(\bar{z}; 2\epsilon) \subset \{z \in C(\bar{\Omega}) \mid z \leq 0\}$ centered at $\bar{z}$ and with radius $2\epsilon > 0$. We can choose $z_n \in B(0; 2\epsilon)$ such that $\langle \mu_n, z_n \rangle_{\mathcal{M}(\bar{\Omega}) \times C(\bar{\Omega})} = \epsilon |\mu_n|_{\mathcal{M}(\bar{\Omega})}$. Since $\bar{z} + z_n \in \{z \in C(\bar{\Omega}) \mid z \leq 0\}$, from (4.21), we get

$$\langle \mu_n, \bar{z} + z_n - f(\cdot, y_n) \rangle_{\mathcal{M}(\bar{\Omega}) \times C(\bar{\Omega})} \leq 0.$$

By passing to the limit, we get

$$\epsilon + < \bar{\mu}, \bar{z} - f(\cdot, \bar{y}) \rangle_{\mathcal{M}(\bar{\Omega}) \times C(\bar{\Omega})} \leq 0,$$

thus $\bar{\mu} \neq 0$.

References

1. J. J. Alibert, J. P. Raymond, Optimal Control Problems Governed by Semilinear Elliptic Equations with Pointwise State Constraints, preprint 1994.
2. V. Barbu, “Analysis and Control of Nonlinear Infinite Dimensional Systems”, Academic Press, New York, 1993.
3. J. F. Bonnans, E. Casas, A Boundary Pontryagin's principle for the optimal control of state constrained elliptic systems, Intern. Series of Num. Math., Vol. 107, Birkhäuser Verlag, Basel, 1992, p. 241–249.
4. J. F. Bonnans, E. Casas, An Extension of Pontryagin's Principle for State Constrained Optimal Control of Semilinear Elliptic Equations and Variational Inequalities, SIAM J. Cont. and Optim., Vol. 33 (1995).
5. E. Casas, Boundary Control of Semilinear Elliptic Equations with Pointwise State Constraints, SIAM J. Control Optim. Vol. 31 (1993), 993–1006.
6. E. Casas, Pontryagin's Principle for State-Constrained Boundary Control Problems of Semilinear Parabolic Equations, preprint, 1995.
7. F. H. Clarke, “Optimization and Nonsmooth Analysis”, John Wiley, New York, 1983.
8. J. Diestel, “Geometry of Banach Spaces – Selected Topics”, Lecture Notes in Mathematics No 485, Springer-Verlag/Berlin/Heidelberg/New York, 1975.
9. I. Ekeland, On the Variational Principle, J. Math. Anal. Appl., Vol. 47 (1974), p. 324–353.
10. I. Ekeland, Nonconvex Minimization Problems, Bull. Amer. Math. Soc. N.S., Vol. 1 (1979), p. 443–474.
11. H. O. Fattorini, Nonlinear Infinite Dimensional Optimal Control Problems with State Constraints and Unbounded Control Sets, to appear.
12. H. O. Fattorini, Optimal Control Problems with State Constraints for Semilinear Distributed Parameter Systems, J. O. T. A., Vol. 88 (1996), 25–59.
13. H. O. Fattorini, Optimal Control Problems for Distributed Parameter Systems in Banach Spaces, Appl. Math. Optim., Vol. 28 (1993), p. 225–257.
14. H. O. Fattorini, S. Sritharan, Necessary and Sufficient Conditions for Optimal Controls in Viscous Flow Problems, Proceedings of the Royal Society of Edinburgh, 124A (1994), p. 211–251.
15. D. Kinderlehrer, G. Stampacchia, “An Introduction to Variational Inequalities”, Academic Press, 1980.
16. X. J. Li, Y. Yao, Maximum Principle of Distributed Parameter Systems with Time Lags, Proceedings on the Conference on Control Theory of Distributed Parameter Systems and Applications, F. Kappel and K. Kunisch Eds., Springer-Verlag, New York, 1985, p. 410–427.
17. X. J. Li, J. Yong, Necessary Conditions for Optimal Control of Distributed Parameter Systems, SIAM J. Control and Optim., Vol. 29 (1991), p. 895–908.
18. X. J. Li, J. Yong, “Optimal Control Theory for Infinite Dimensional Systems”, Birkhäuser, Boston/Basel/Berlin, 1995.

19. J. P. Raymond, Pontryagin's Principle for State-Constrained Control Problems with Unbounded Controls, Proceedings of the 2nd Catalan Days on Applied Mathematics, Presses Universitaires de Perpignan, M. Sofonea and J. N. Corvellec Eds., 1995, p. 227–237.
20. J. P. Raymond, H. Zidani, Pontryagin's Principles for State-Constrained Control Problems Governed by Parabolic Equations with Unbounded Controls, preprint 1996.
21. J. Serrin, Pathological Solutions of Elliptic Differential Equations, Ann. Scuola Norm. Sup. Pisa, Vol. 18 (1964), 385–387.

E. Casas
Departamento de Matemática Aplicada
y Ciencias de la Computatión
E.T.S.I. Industriales y de Telecomunicatión
Universidad de Cantabria
E-39071 Santander, Spain

J.-P. Raymond
Université Paul Sabatier
UMR CNRS MIP
UFR MIG
F-31062 Toulouse Cedex 4, France

H. Zidani
Université Paul Sabatier
UMR CNRS MIP
UFR MIG
F-31062 Toulouse Cedex 4, France
e-mail:zidani@mip.ups-tlse.fr

International Series of Numerical Mathematics
Vol. 126, © 1998 Birkhäuser Verlag, Basel

Designing for Optimal Energy Absorption II, The Damped Wave Equation

STEVEN J. COX

Department of Computational and Applied Mathematics
Rice University

ABSTRACT. We consider the wave equation in a bounded domain with zero Dirichlet data and damping proportional to velocity and pose the problem of minimizing, with respect to damping, the maximum, over all initial data of unit energy, of the infinite time integral of the instantaneous energy. We show the minimum to exist over those dampings that uniformly avoid zero and infinity. We provide an exact minimum over the class of constant dampings and proceed to show it to be a critical point over the class of bounded dampings.

1991 *Mathematics Subject Classification.* 35B20, 35L05, 47A55

Key words and phrases. Eigenvalue, damping, Lyapunov, optimal design, perturbation.

1. Introduction

In [1] we minimized two standard merit functions, decay rate and greatest total energy, over a class of finite dimensional damped linear systems. Analogous results for the decay rate in the context of the damped wave equation can be found in [5], [4], and [2]. In this note we minimize, over a, the greatest total energy associated with the damped wave equation,

$$u_{tt}(x,t) - \Delta u(x,t) + 2a(x)u_t(x,t) = 0, \quad u(\cdot,t) \in H_0^1(\Omega) \tag{1.1}$$

on the open bounded connected set $\Omega \subset \mathbb{R}^d$. The greatest total energy is simply the maximum, over all initial data of unit energy, of the infinite time integral of the instantaneous energy. In order to make these notions precise let us take $U(t) = [u(t)\ u_t(t)]$ and interpret (1.1) as $U_t = A(a)U$ where

$$A(a) = \begin{pmatrix} 0 & I \\ \Delta & -2a \end{pmatrix}, \qquad D(A) = (H^2(\Omega) \cap H_0^1(\Omega)) \times H_0^1(\Omega),$$

is densely defined in the Hilbert space $X = H_0^1(\Omega) \times L^2(\Omega)$ with inner product

$$\langle [f,g],[u,v] \rangle_X = \int_\Omega \nabla f \cdot \nabla \overline{u} + g\overline{v}\,dx.$$

This $A(a)$ is the infinitesimal generator of a semigroup $T(t;a)$ and $U(t) = T(t;a)V$ solves the Cauchy problem $U_t = A(a)U$, $U(0) = V$. The associated instantaneous and total energies are, respectively

$$E(t) = \|T(t;a)V\|_X^2 \quad \text{and} \quad \int_0^\infty \|T(t;a)V\|_X^2\,dt.$$

This work was supported by NSF Grant DMS–9258312.

The greatest total energy, $\mathcal{T}_1(a)$, is now the supremum of the total energy over all initial data, V, of unit energy. That is,

$$\mathcal{T}_1(a) \equiv \sup_{\|V\|_X=1} \int_0^\infty \|T(t;a)V\|_X^2\,dt. \tag{1.2}$$

In §2 we show that $\mathcal{T}_1(a)$ is finite over the class of bounded nonnegative a and in fact that it attains its minimum there. In §3 we compute $\mathcal{T}_1(a)$, by hand, for constant a and so readily identify the best constant damping. In §4 we show this best constant to be a critical point of $\mathcal{T}_1$ in the class of bounded a.

2. Existence of an Optimal Design

We denote by $\{\lambda_n\}_{n=1}^\infty$ the increasing sequence of eigenvalues of $-\Delta$ on $H_0^1(\Omega)$ and by $\{q_n\}_{n=1}^\infty$ the corresponding orthonormal base of eigenfunctions.

We choose δ small and positive and define

$$ad \equiv \{a \in L^\infty(\Omega) : \delta\sqrt{\lambda_1}/2 \le a(x) \le \sqrt{\lambda_1}/(2\delta)\}$$

From the literature we may conclude, for such a, that $\mathcal{T}_1(a)$ is finite and that the total energy may be represented as a quadratic form. More precisely,

Theorem 2.1. *If $a \in ad$ then*

$$\|T(t;a)\|_{L(X)}^2 \le 4e^{-\delta\sqrt{\lambda_1}t}, \quad t \ge 0, \tag{2.1}$$

and there exists an Hermitian positive semidefinite endomorphism $B(a)$ on X for which

$$2\langle B(a)A(a)V, V\rangle_X = -\|V\|_X^2, \qquad \forall V \in D(A), \tag{2.2}$$

and

$$\langle B(a)V, V\rangle_X = \int_0^\infty \|T(t;a)V\|_X^2\,dt, \qquad \forall V \in X. \tag{2.3}$$

Proof. The estimate in (2.1) follows directly from Theorem 1 of Rauch [8]. Datko [6] has shown that (2.1) is a necessary and sufficient condition for the existence of the stated B. ∎

As ad is compact with respect to the weak* topology we need only show $a \mapsto \mathcal{T}_1(a)$ to be weak* lower semicontinuous. Let us denote weak* convergence by $\stackrel{*}{\rightharpoonup}$, weak convergence by $\rightharpoonup$, and strong convergence by $\to$.

Theorem 2.2. *If $\{a_n\} \subset ad$ and $a_n \stackrel{*}{\rightharpoonup} a$ in $L^\infty(\Omega)$ then $T(t;a_n)V \to T(t;a)V$ in X for each $V \in X$.*

Proof. Recalling Kato [7, IX.2.16] we note that it suffices to show that $A^{-1}(a_n)V \to A^{-1}(a)V$. Set $V = [f,g]$ and define $[y_n, z_n] = A^{-1}(a_n)[f,g]$ and $[y,z] = A^{-1}(a)[f,g]$ so

$$[y_n, z_n] = [\Delta^{-1}(2a_n f + g), f] \quad \text{and} \quad [y,z] = [\Delta^{-1}(2af+g), f].$$

As a result,

$$\|A^{-1}(a_n)V - A^{-1}(a)V\|_X = 2\|\nabla\Delta^{-1}(a_n - a)f\|_2, \tag{2.4}$$

where $\|\cdot\|_2$ denotes the $L^2(\Omega)$ norm. If $w_n \equiv \Delta^{-1}(a-a_n)f$ then w_n satisfies

$$-\Delta w_n = (a_n - a)f, \quad w_n \in H_0^1(\Omega).$$

Upon integrating each side against w_n we find

$$\|w_n\|_2^2 \le \frac{1}{\lambda_1}\|\nabla w_n\|_2^2 \le \frac{1}{\delta\sqrt{\lambda_1}}\|f\|_2\|w_n\|_2. \tag{2.5}$$

As a result

$$\|w_n\|_2 \le \frac{1}{\delta\sqrt{\lambda_1}}\|f\|_2 \quad \text{and} \quad \|\nabla w_n\|_2 \le \frac{1}{\delta}\|f\|_2.$$

From these bounds it follows that $w_n \rightharpoonup w$ in $H_0^1(\Omega)$. Hence, passing to the limit in the weak form

$$\int_\Omega \nabla w_n \cdot \nabla \phi \, dx = \int_\Omega (a_n - a) f \phi \, dx \qquad \forall\, \phi \in H_0^1(\Omega)$$

we find

$$\int_\Omega \nabla w \cdot \nabla \phi \, dx = 0 \qquad \forall\, \phi \in H_0^1(\Omega),$$

i.e., w is identically zero in Ω. As $w_n \to 0$ in $L^2(\Omega)$ it now follows from (2.5) that $w_n \to 0$ in $H_0^1(\Omega)$. Recalling (2.4), we have shown that $A^{-1}(a_n)V \to A^{-1}(a)V$ in X. ∎

Theorem 2.3. *If $a_n \overset{*}{\rightharpoonup} a$ in $L^\infty(\Omega)$ then $\mathcal{T}_1(a) \le \liminf_{n\to\infty} \mathcal{T}_1(a_n)$.*

Proof. It follows from the previous theorem that $\|T(t;a_n)V\|_X^2 \to \|T(t;a)V\|_X^2$ for each $t \ge 0$. In addition it follows from (2.1) that $t \mapsto \|T(t;a_n)V\|_X^2$ is uniformly dominated by an integrable function. The Lebesgue dominated convergence theorem now yields

$$\int_0^\infty \|T(t;a_n)V\|_X^2 \, dt \to \int_0^\infty \|T(t;a)V\|_X^2 \, dt,$$

or, in the language of (2.2), $\langle B(a_n)V, V\rangle_X \to \langle B(a)V, V\rangle_X$. By the nature of the supremum in (1.2) it follows that to each $\varepsilon > 0$ there corresponds a unit vector V_ε for which

$$\mathcal{T}_1(a) - \varepsilon \le \langle B(a)V_\varepsilon, V_\varepsilon\rangle_X.$$

In addition, on taking the limit inferior of each side of $\langle B(a_n)V_\varepsilon, V_\varepsilon\rangle_X \le \mathcal{T}_1(a_n)$ we find

$$\langle B(a)V_\varepsilon, V_\varepsilon\rangle_X \le \liminf_{n\to\infty} \mathcal{T}_1(a_n),$$

and hence

$$\mathcal{T}_1(a) - \varepsilon \le \liminf_{n\to\infty} \mathcal{T}_1(a_n).$$

As ε is arbitrary our claim has been established. ∎

Corollary 2.4. *$a \mapsto \mathcal{T}_1(a)$ attains its minimum on ad.*

As a candidate for the global minimizer we now carry out the exact minimization of $\mathcal{T}_1$ over constant a.

3. The Case of Constant Damping

In a manner analogous to the case of friction damping of finite dimensional systems, see [1, §2], we obtain an explicit representation of $B(a)$.

Theorem 3.1. *If a is constant then*

$$B(a) = \begin{pmatrix} \frac{1}{2a} - a\Delta^{-1} & -\frac{1}{2}\Delta^{-1} \\ \frac{1}{2}I & \frac{1}{2a} \end{pmatrix}.$$

Proof. It is a simple matter to check that this operator is an Hermitian, positive semidefinite endomorphism on X satisfying (2.1). There are a number of means by which this $B(a)$ may be derived. One way is to note that for smooth solutions the instantaneous energy obeys

$$E(t) = -\frac{d}{dt}\left\{\frac{1}{2a}E(t) + \int_\Omega (uu_t + au^2)\,dx\right\} = -\frac{d}{dt}\langle B(a)U(t), U(t)\rangle_X,$$

and hence that the total energy is simply $\langle B(a)V, V\rangle_X$. A second approach is to note that $B(a)$ is (formally) a solution to the associated Liapunov equation $A^*(a)B(a) + B(a)A(a) = -I$. ∎

The eigenvalues of $B(a)$ are

$$\mathcal{T}_{\pm n}(a) = \frac{1}{2a} + \frac{a \pm \sqrt{a^2+\lambda_n}}{2\lambda_n}, \qquad n = 1, 2, \ldots \tag{3.1}$$

and its (unit) eigenvectors are

$$V_{\pm n} = \mu_{\pm n}[1 \quad c_{\pm n}]q_n \tag{3.2}$$

where

$$c_{\pm n} = -a \pm \sqrt{a^2+\lambda_n} \quad \text{and} \quad \mu_{\pm n} = (\lambda_n + c_{\pm n}^2)^{-1/2} \tag{3.3}$$

The greatest of the $\mathcal{T}_{\pm n}(a)$ is $\mathcal{T}_1(a)$ and, as λ_1 is simple, so too is $\mathcal{T}_1(a)$. With this explicit expression in hand we may easily establish (compare [1, Theorem 2.4])

Corollary 3.2. *The greatest total energy, $\mathcal{T}_1 : \mathbb{R}_+ \to \mathbb{R}_+$, is strictly convex and attains its global minimum at*

$$\breve{a} = \sqrt{\lambda_1}\sqrt{\frac{\sqrt{5}-1}{2}}.$$

Its minimum value is

$$\mathcal{T}_1(\breve{a}) = \frac{1}{\sqrt{\lambda_1}}\frac{\sqrt{5}+2}{2}\sqrt{\frac{\sqrt{5}-1}{2}}.$$

We record for future use the fact that

$$c_1^2 = 2\breve{a}^2 + \lambda_1 \tag{3.4}$$

follows on substitution of $\breve{a}$ into (3.3)

4. The Perturbed Operator

We now show that $\breve{a}$ is in fact a critical point for $\mathcal{T}_1 : ad \to \mathbb{R}_+$. In particular, we shall show that, for each $b \in L^\infty(\Omega)$,

$$\mathcal{T}_1(\breve{a} + \kappa b) = \mathcal{T}_1(\breve{a}) + O(\kappa^2), \quad \text{as } \kappa \to 0.$$

In fact we compute the gradient, $\partial\mathcal{T}_1$, at an arbitrary admissible constant, a. To begin, we fix $b \in L^\infty(\Omega)$ and consider $A(\kappa) \equiv A_0 + \kappa A_1$ where

$$A_0 = \begin{pmatrix} 0 & I \\ \Delta & -2a \end{pmatrix} \quad \text{and} \quad A_1 = \begin{pmatrix} 0 & 0 \\ 0 & -2b \end{pmatrix}.$$

We denote by $T(t;\kappa)$ the semigroup generated by $A(\kappa)$ and recall, see, e.g., Kato [7, IX.2.1], that, for fixed t, $\kappa \mapsto T(t;\kappa)$ is entire and

$$T(t;\kappa) = \sum_{n=0}^{\infty} \kappa^n T_n(t)$$

where $T_0(t)$ is the semigroup generated by A_0 and

$$T_{n+1}(t) = \int_0^t T_0(t-s) A_1 T_n(s)\, ds, \quad n > 1.$$

If κ is sufficiently small then $a + \kappa b$ will lie in ad and so, by Theorem 2.1, there exists a $B(a+\kappa b)$. We now express this as a power series in κ by following the construction in Datko [6]. In particular, we define,

$$B(t;\kappa) \equiv \int_0^t T^*(s;\kappa) T(s;\kappa)\, ds = \sum_{n=0}^{\infty} \kappa^n B_n(t),$$

where

$$B_0(t) = \int_0^t T_0^*(s) T_0(s)\, ds \quad \text{and} \quad B_n(t) = \int_0^t \{T_0^*(s) T_n(s) + T_n^*(s) T_0(s)\}\, ds.$$

As $t \to \infty$ each $B_n(t)$ converges in the strong operator topology to an operator B_n. It follows that

$$B(a + \kappa b) = \sum_{n=0}^{\infty} \kappa^n B_n$$

is, in the language of Kato [7, §VII.7], a selfadjoint bounded–holomorphic family of operators. As such we may avail ourselves of the perturbation series of [7, §II.2.2]. In particular, as $\mathcal{T}_1(a)$ is simple, the greatest eigenvalue, $\mathcal{T}_1(a+\kappa b)$, of $B(a+\kappa b)$ satisfies

$$\mathcal{T}_1(a+\kappa b) = \mathcal{T}_1(a) + \sum_{n=1}^{\infty} \kappa^n \mathcal{T}_1^{(n)},$$

where $\mathcal{T}_1^{(n)}$ may be expressed in terms of the eigenvectors of $B(a)$. More precisely,

$$\mathcal{T}_1^{(1)} = \langle B_1 V_1, V_1 \rangle_X \quad \text{and} \quad \mathcal{T}_1^{(2)} = \langle B_2 V_1, V_1 \rangle_X + \sum_{n \neq 1} \frac{\langle B_1 V_1, V_n \rangle_X^2}{\mathcal{T}_1 - \mathcal{T}_n}$$

where $\mathcal{T}_n$ and V_n are the eigenvalues and eigenvectors of $B(a)$, see (3.1)–(3.3). Accordingly we begin the evaluation of

$$\langle B_1V_1, V_1\rangle_X = \int_0^\infty \langle\{T_0^*(t)T_1(t) + T_1^*(t)T_0(t)\}V_1, V_1\rangle_X \, dt$$
$$= 2\int_0^\infty \langle T_1(t)V_1, T_0(t)V_1\rangle_X \, dt.$$

It will be convenient to suppose that $a^2 < \lambda_1$ and to label the frequencies

$$\omega_j = \sqrt{\lambda_j - a^2}.$$

As V_1 is a constant multiple of q_1 and A_0 is a constant coefficient operator it follows that $T_0(t)V_1$ is simply

$$T_0(t)V_1 = \mu_1 e^{-at}\begin{pmatrix} \cos(\omega_1 t) + (c_1 + a)\sin(\omega_1 t)/\omega_1 \\ c_1\cos(\omega_1 t) - (ac_1 + \lambda_1)\sin(\omega_1 t)/\omega_1 \end{pmatrix} q_1$$

For ease of reference let us express this as

$$T_0(t)V_1 = \begin{pmatrix} y_1(t) \\ y_1'(t) \end{pmatrix} q_1.$$

Regarding $T_1(t)V_1$ we must first compute $T_0(t-s)A_1T_0(s)V_1$. This is the solution, at time $t-s$, to $W' = A_0W$ subject to

$$W(0) = A_1T_0(s)V_1 = -2bq_1y_1'(s)\begin{pmatrix} 0 \\ 1 \end{pmatrix}.$$

We find, via separation of variables, that

$$W(t-s) = -2e^{-a(t-s)}y_1'(s)\sum_{j=1}^\infty \begin{pmatrix} \sin(\omega_j(t-s))/\omega_j \\ \cos(\omega_j(t-s)) - a\sin(\omega_j(t-s))/\omega_j \end{pmatrix} q_j\langle bq_1, q_j\rangle_2.$$

Hence,

$$T_1(t)V_1 = \int_0^t W(t-s)\, ds = -2\sum_{j=1}^\infty \begin{pmatrix} \alpha_j(t) \\ \alpha_j'(t) \end{pmatrix} q_j\langle bq_1, q_j\rangle_2$$

where

$$\alpha_j(t) = \int_0^t e^{-a(t-s)}y_1'(s)\sin(\omega_j(t-s))/\omega_j \, ds.$$

It follows that

$$\langle T_1(t)V_1, T_0(t)V_1\rangle_X = -2\langle bq_1, q_1\rangle_2\xi_1(t)$$

where

$$\xi_1(t) = \lambda_1\alpha_1(t)y_1(t) + \alpha_1'(t)y_1'(t).$$

It is important to note that ξ_1 is simply a product of sums of terms involving sine, cosine, and the exponential and so may be integrated by hand. Its length (greater than one page) however precludes us from effectively presenting it here and compels us to adopt a symbolic means (Maple) in the final evaluation of

$$\begin{aligned}\mathcal{T}_1^{(1)} &= \langle B_1 V_1, V_1\rangle_X = 2\int_0^\infty \langle T_1(t)V_1, T_0(t)V_1\rangle_X\,dt = -4\langle bq_1, q_1\rangle_2 \int_0^\infty \xi_1(t)\,dt \\ &= \langle bq_1, q_1\rangle \frac{2a^2 - c_1^2 - \lambda_1}{2a^2(\lambda_1 + c_1^2)}.\end{aligned}$$

If $b \equiv 1$ this indeed agrees with what one finds on differentiating the $\mathcal{T}_1$ offered in (3.1). In addition, from (3.4) it follow that $\mathcal{T}_1^{(1)} = 0$ when evaluated at the best constant, $\check{a}$.

As to whether or not $\check{a}$ is a local minimizer we note that components for the calculation of $\mathcal{T}_1^{(2)}$ are all here. The complexity of the terms however has rendered this calculation a formidable exercise.

Even with $\mathcal{T}_1^{(2)}$ in hand one has only a local result. A numerical study, analogous to [2], is currently underway.

Finally, we note that $V \mapsto \langle B(a)V, V\rangle_X$ is a Liapunov function, see [3], with which one may study the stability of the trivial solution of (1.1) with a right hand side depending on u, u_t, and ∇u.

References

1. Cox, S.J., *Designing for optimal energy absorption I, lumped parameter systems*, ASME J. on Vibration and Acoustics, to appear.
2. Cox, S.J., *Designing for optimal energy absorption III, numerical minimization of the spectral abscissa*, Structural Optimization 13(1), pp. 17–22, 1997.
3. Cox, S.J., and Moro, J., *A Liapunov function for systems whose linear part is almost classically damped*, ASME J. of Applied Mechanics, to appear.
4. Cox, S.J., and Overton, M.L., *Perturbing the Critically Damped Wave Equation*, SIAM J. Appl. Math. 56(5), pp. 1353–1362, 1996.
5. Cox, S.J., and Zuazua, E., *The Rate at Which Energy Decays in a Damped String*, Commun. in Partial Differential Equations, Vol. 19, Nos. 1& 2, pp. 213–243, 1994.
6. Datko, R., *Extending a theorem of A.M. Liapunov to Hilbert space*, J. Math. Anal. & Appl. 32, pp. 610–616, 1970.
7. Kato, T., *Perturbation Theory for Linear Operators*, 2nd ed., Springer-Verlag, New York, 1976.
8. Rauch, J., *Qualitative behavior of dissipative wave equations on bounded domains*, Arch. Rat. Mech. Anal., pp. 77–85, 1977.

Steven J. Cox
Department of Computational
and Applied Mathematics
Rice University
PO Box 1892
Houston, TX 77251, USA

International Series of Numerical Mathematics
Vol. 126, © 1998 Birkhäuser Verlag, Basel

On the Approximate Controllability for Higher Order Parabolic Nonlinear Equations of Cahn-Hilliard Type

J.I. DÍAZ AND A.M. RAMOS

Dpto. Matemática Aplicada
Universidad Complutense de Madrid

Dpto. Informática y Automática
Universidad Complutense de Madrid

ABSTRACT. We prove the approximate controllability property for some higher order parabolic nonlinear equations of Cahn-Hilliard type when the nonlinearity is of sublinear type at infinity. We also give a counterexample showing that this property may fail when the nonlinearity is of superlinear type.

1991 *Mathematics Subject Classification.* 93B05, 93C20, 35K55

Key words and phrases. Approximate controllability, higher order nonlinear parabolic boundary value problems, Cahn-Hilliard type equations.

1. Introduction

Let Ω be a bounded open subset of $\mathbb{R}^N$ of class C^{2m}, $T > 0$, ω a nonempty open subset of Ω, f a continuous real function and $k \in \mathbb{N}$ such that $0 \leq 2k \leq m$. The main goal of this work is the study of the approximate controllability of the following semilinear equation with Dirichlet boundary conditions:

$$\text{(1.1)} \qquad \begin{cases} y_t + (-\Delta)^m y + (-\Delta)^k f(y) = h + v\chi_\omega & \text{in } Q := \Omega \times (0,T), \\ \dfrac{\partial^j y}{\partial \nu^j} = 0 \quad , \quad j = 0, 1, \ldots, m-1 & \text{on } \Sigma := \partial\Omega \times (0,T), \\ y(0) = y_0 & \text{in } \Omega, \end{cases}$$

where v is a suitable output control, χ_ω is the characteristic function of ω, ν is the unit outward normal vector, $h \in L^2(0,T : H^{-m}(\Omega))$ and $y_0 \in L^2(\Omega)$. Due to the term χ_ω the controls are assumed supported on the set $\mathcal{O} := \omega \times (0,T)$. Problems as (1.1), sometimes known as Cahn-Hilliard problems, appear, with $m = 2$, in the study of phase separation in cooling binary solutions and in other contexts generating spatial pattern formation (see [6], [8] and the references cited therein).

We recall that problem (1.1) satisfies the approximate controllability property, at time T with states space X and controls space Y, if the set

$$\{\, y(T, \cdot : v) : \ v \in Y, \ y \text{ solution of } (1.1)\}$$

is dense in X.

The main goal of this paper is to extend the approximate controllability results on second order problems, $m = 1$ and $k = 0$ (see e.g. [9], [10] and [7]) to the case of higher order equations for which the maximum principle does not hold, in general.

Our first result gives a positive answer when f is assumed to be sublinear at the infinity:

Theorem 1.1. *Assume that f satisfies the following conditions: there exist some positive constants c_1 and c_2 such that*

$$|f(s)| \leq c_1 + c_2|s| \quad \textit{for all } s \in \mathbb{R} \tag{1.2}$$

and

$$\textit{there exists } f'(s_0) \textit{ for some } s_0 \in \mathbb{R}. \tag{1.3}$$

Then problem (1.1) satisfies the approximate controllability property at time T with states space $X = L^2(\Omega)$ and controls space $Y = L^2(\mathcal{O})$.

In contrast to the above result, we shall prove that when f is superlinear the approximate controllability property does not hold in general, as explained in Section 4. Therefore if, for instance, $f(s) = |s|^{p-1}s$ Theorem 1.1 gives a positive approximate controllability result for $0 < p \leq 1$. The results of section 6 provide a negative approximate controllability answer when $1 < p < \infty$. The similar alternative was obtained in Díaz-Ramos [7] for second order parabolic semilinear problems.

We remark that the existence of solutions in the class

$$y \in L^2(0,T; H_0^m(\Omega)) \cap C([0,T]; L^2(\Omega)),\ f(y) \in L^2(Q),\ \Delta^k f(y) \in L^2(0,T; H^{-m}(\Omega)),$$

is also obtained as a by-product of Theorem 1.1 for a suitable subclass of controls. The uniqueness of solutions can be easily proved if, for instance, f is nondecreasing or Lipschitz continuous. Those uniqueness results are not needed in our arguments.

2. Approximate controllability for an associated linear problem

In order to prove Theorem 1.1 we follows the same scheme of proof than in [9], [10] and [7]. We define the function

$$g(s) = \frac{f(s) - f(s_0)}{s - s_0}.$$

From assumptions (1.2) and (1.3) we have that $g \in L^\infty(\mathbb{R}) \cap \mathcal{C}(\mathbb{R})$. The conclusion will be derived from a fixed point argument. As $f(s) = f(s_0) + g(s)s - g(s)s_0$, we shall start by considering the approximate controllability for a linear problem obtained by replacing the term $f(y)$ by

$$g(z)y + f(s_0) - g(z)s_0,$$

where z is an arbitrary function in $L^2(Q)$. Notice that when $z = y$ this expression coincides with $f(y)$ and that if we denote $g(z(t,x)) := a(t,x)$ and

$$h(a) := -(-\Delta)^k f(s_0) + (-\Delta)^k (a(t,x)s_0), \tag{2.1}$$

then $a \in L^\infty(Q)$ and $h(a) \in L^\infty(0,T;H^{-2k}(\Omega))$. More in general, given $a \in L^\infty(Q)$ and $h(a)$ defined by (2.1), we consider the approximate controllability property corresponding to the linear problem

$$
(2.2)\quad \begin{cases} y_t+(-\Delta)^m y+(-\Delta)^k(a(t,x)y)=h+h(a)+u\chi_\omega & \text{in } Q:=\Omega\times(0,T),\\ \dfrac{\partial^j y}{\partial\nu^j}=0 \quad , \quad j=0,1,\ldots,m-1 & \text{on } \Sigma:=\partial\Omega\times(0,T),\\ y(0)=y_0 & \text{in } \Omega. \end{cases}
$$

Before stating an approximate controllability result for this problem, following Lions [14] and Fabre-Puel-Zuazua [9], [10], we consider $\varepsilon > 0$ and $y_d \in L^2(\Omega)$ and we introduce the functional $J = J(\cdot; a, y_d) : L^2(\Omega) \to \mathbb{R}$ defined by

$$
(2.3)\quad J(\varphi_0;a,y_d)=J(\varphi^0)=\frac{1}{2}\left(\int_{\mathcal{O}}|\varphi(t,x)|dxdt\right)^2+\varepsilon\parallel\varphi^0\parallel_{L^2(\Omega)}-\int_\Omega y_d\,\varphi^0,dx
$$

where $\varphi(t,x)$ is the solution of the backward problem

$$
(2.4)\quad \begin{cases} -\varphi_t+(-\Delta)^m\varphi+a(t,x)\Delta^k\varphi=0 & \text{in } Q:=\Omega\times(0,T),\\ \dfrac{\partial^j\varphi}{\partial\nu^j}=0 \quad , \quad j=0,1,\ldots,m-1 & \text{on } \Sigma:=\partial\Omega\times(0,T),\\ \varphi(T)=\varphi^0 & \text{in } \Omega. \end{cases}
$$

To study the above backward problem we introduce the space

$$
W:=\{y\in L^2(0,T;H_0^m(\Omega)),\ y_t\in L^2(0,T;H^{-m}(\Omega))\}.
$$

The following result will be used later

Proposition 2.1. *Given $h \in L^2(0,T;H^{-m}(\Omega))$ and $y_0 \in L^2(\Omega)$, there exists a unique function $y \in W$ satisfying*

$$
(2.5)\quad \begin{cases} y_t+(-\Delta)^m y+a(t,x)\Delta^k y=h & \text{in } Q,\\ \dfrac{\partial^j y}{\partial\nu^j}=0 \quad , \quad j=0,1,\ldots,m-1 & \text{on } \Sigma,\\ y(0)=y_0 & \text{in } \Omega. \end{cases}
$$

Furthermore, we have the estimate

$$
(2.6)\quad \|y\|_{L^2(0,T;H_0^m(\Omega))}+\|y_t\|_{L^2(0,T;H^{-m}(\Omega))}\le C\left(\|h\|_{L^2(0,T;H^{-m}(\Omega))}+\|y_0\|_{L^2(\Omega)}\right),
$$

where the constant C depends only on $M :=\parallel a \parallel_{L^\infty(Q)}$ (provided that Ω, T and m are kept fixed). Moreover, if $h \in L^2(Q)$, the solution y also satisfies that

$$
(2.7)\quad y\in L^2(\delta,T;H^{2m}(\Omega)) \quad \text{and} \quad y_t\in L^2((\delta,T)\times\Omega) \quad \text{for all } \delta\in(0,T).
$$

Proof. For all $n \in \mathbb{N}$ we define y^{n+1} as the solution of the following iterative problem

$$
\begin{cases} y_t^{n+1}+(-\Delta)^m y^{n+1}=h-a(t,x)\Delta^k y^n & \text{in } Q,\\ \dfrac{\partial^j y^{n+1}}{\partial\nu^j}=0 \quad , \quad j=0,1,\ldots,m-1 & \text{on } \Sigma,\\ y^{n+1}(0)=y_0 & \text{in } \Omega, \end{cases}
$$

where $y^0(t) := 0$ for all $t \in [0,T]$. The existence of a solution $y^n \in W$ can be found, for instance, in Theorem 3.4.1 of Lions-Magenes [15]. Thus, for all $n \in \mathbb{N}\backslash\{0,1\}$, $y^{n+1} - y^n$ satisfies

$$(2.8) \quad \begin{cases} (y^{n+1} - y^n)_t + (-\Delta)^m (y^{n+1} - y^n) = -a(t,x)\Delta^k (y^n - y^{n-1}) & \text{in } Q, \\ \dfrac{\partial^j (y^{n+1} - y^n)}{\partial \nu^j} = 0 \quad , \quad j = 0,1,\ldots,m-1 & \text{on } \Sigma, \\ (y^{n+1} - y^n)(0) = 0 & \text{in } \Omega \end{cases}$$

and therefore

$$y^{n+1} - y^n \in H^{1,2m}(Q) := H^1(0,T;L^2(\Omega)) \cap L^2(0,T;H^{2m}(\Omega))$$

and

$$\| y^{n+1} - y^n \|_{H^{1,2m}(Q)} \leq c_1 \| a\Delta^k (y^n - y^{n-1}) \|_{L^2(Q)}$$

(see, for instance, Theorem 4.6.1 of Lions-Magenes [16]). Then, since

$$H^{1,2m}(Q) \subset \mathcal{C}([0,T];H^m(\Omega))$$

with continuous embedding (see, for instance, Theorems 1.3.1 and 1.9.6 of Lions-Magenes [15]), there exists $c_2 = c_2(T)$ such that

$$\| y^{n+1} - y^n \|_{\mathcal{C}([0,T];H_0^m(\Omega))} \leq c_2 \| a\Delta^k (y^n - y^{n-1}) \|_{L^2(Q)} .$$

Further, it is clear that we can choose $C_2 = C_2(T)$ such that for all $t \in [0,T]$

$$\| y^{n+1} - y^n \|_{\mathcal{C}([0,t];H_0^m(\Omega))} \leq C_2 \| a\Delta^k (y^n - y^{n-1}) \|_{L^2((0,t)\times\Omega)} .$$

Hence,

$$\| (y^{n+1} - y^n)(t) \|^2_{H_0^m(\Omega)} \leq (C_2 M)^2 \int_0^t \| \Delta^k (y^n - y^{n-1})(\tau) \|^2_{L^2(\Omega)} \, d\tau, \quad \text{for all } t \in [0,T]$$

and therefore, by using the Poincaré inequality, there exists a constant K, independent of M, such that

$$\| (y^{n+1} - y^n)(t) \|^2_{H_0^m(\Omega)} \leq (KM)^2 \int_0^t \| (y^n - y^{n-1})(\tau) \|^2_{H_0^m(\Omega)} \, d\tau, \quad \text{for all } t \in [0,T].$$

Then, for every $t \in [0,T]$ we deduce that

$$\begin{aligned} \| (y^{n+1} - y^n)(t) \|^2_{H_0^m(\Omega)} &\leq (K^2M^2)^{n-1} \int_0^t \int_0^{\tau_1} \cdots \int_0^{\tau_{n-1}} \| (y^2 - y^1)(\tau_n) \|^2_{H_0^m(\Omega)} \, d\tau_n \ldots d\tau_1 \\ &\leq (K^2M^2)^{n-1} \int_0^t \int_0^{\tau_1} \cdots \int_0^{\tau_{n-1}} \| y^2 - y^1 \|^2_{\mathcal{C}([0,T];H_0^m(\Omega))} \, d\tau_n \ldots d\tau_1 \\ &\leq (K^2M^2)^{n-1} \frac{t^{n-1}}{(n-1)!} \| y^2 - y^1 \|^2_{\mathcal{C}([0,T];H_0^m(\Omega))} \\ &\leq \frac{(K^2M^2T)^{n-1}}{(n-1)!} \| y^2 - y^1 \|^2_{\mathcal{C}([0,T];H_0^m(\Omega))}, \end{aligned}$$

which implies that

$$\| y^{n+1} - y^n \|_{\mathcal{C}([0,T];H_0^m(\Omega))} \to 0 \quad \text{as } n \to \infty$$

and therefore, by (2.8), we deduce that

$$\| (y^{n+1} - y^n)_t \|_{L^2(0,T;H^{-m}(\Omega))} \to 0 \quad \text{as } n \to \infty.$$

Then, there exists $y \in W$ such that

$$y_n \to y \quad \text{in } W \quad \text{as } n \to \infty.$$

In order to prove that y satisfies (2.5) we point out that

$$\Delta^m y^n \to \Delta^m y \quad \text{in } L^2(0,T;H^{-m}(\Omega)) \quad \text{as } n \to \infty,$$

$$\Delta^k y^n \to \Delta^k y \quad \text{in } L^2(\Omega) \quad \text{as } n \to \infty,$$

and

$$y_t^n \to y_t \quad \text{in } L^2(0,T;H^{-m}(\Omega)) \quad \text{as } n \to \infty.$$

this implies (passing to the limit) that y is the solution of (2.5). In order to prove (2.6), we "multiply" in (2.5) by y. Then it is easy to see that

$$\text{(2.9)} \quad \|y\|_{L^2(0,T;H_0^m(\Omega))} + \|y_t\|_{L^2(0,T;H^{-m}(\Omega))} \leq C\left(\|h\|_{L^2(0,T;H^{-m}(\Omega))} + \|y_0\|_{L^2(\Omega)} + \|y\|_{L^2(Q)}\right).$$

Furthermore,

$$\| y(t) \|_{L^2(\Omega)}^2 \leq \left(\| y(0) \|_{L^2(\Omega)}^2 + c_2 \| h \|_{L^2(0,T;H^{-m}(\Omega))}^2\right) + c_3 \int_0^t \| y(s) \|_{L^2(\Omega)}^2 \, ds.$$

Then, applying Gronwall's inequality (see, for instance, Lemma 4 of Haraux [11]), we deduce that

$$\| y(t) \|_{L^2(\Omega)}^2 \leq \left(\| y(0) \|_{L^2(\Omega)}^2 + c_2 \| h \|_{L^2(0,T;H^{-m}(\Omega))}^2\right) e^{c_3 t} \qquad \forall\, t \in [0,T].$$

From here, we obtain that

$$\| y \|_{L^2(Q)} \leq c_4 \left(\|h\|_{L^2(0,T;H^{-m}(\Omega))} + \|y_0\|_{L^2(\Omega)}\right)$$

which implies, together with (2.9), inequality (2.6). Now, thanks to (2.6) and the linearity of Problem (2.5), we deduce the uniqueness of solution.

Finally, if $h \in L^2(Q)$, since $y(\delta) \in H_0^m(\Omega)$ for all $\delta \in (0,T)$, taking $y(\delta)$ as initial datum and applying Theorem 4.6.1 of [16], we get (2.7). ■

As usual in Controllability Theory we shall use a *unique continuation* property for solutions of the *dual problem* (in our case Problem (2.4)).

Lemma 2.1. *Let ω be a nonempty open subset of Ω. Assume that*

$$\varphi \in L^2(0,T;H_0^m(\Omega)) \,\cap\, C([0,T];L^2(\Omega))$$

satisfies (2.4) and that $\varphi \equiv 0$ in $\mathcal{O} = \omega \times (0,T)$. Then $\varphi \equiv 0$ in Q.

Proof. From Proposition 2.1 (applied with backward time) we deduce that $\varphi \in L^2(0,T-\delta;H^{2m}(\Omega))$ for all $\delta \in (0,T)$. Then Lemma 2.1 follows from Theorem 3.2 of Saut-Scheurer [17]. ■

The following two results are easy adaptations (by using Lemma 2.1) of the similar ones given in [9], [10] for second order parabolic problems.

Proposition 2.2. *The functional $J(\cdot; a, y_d)$ is continuous, strictly convex on $L^2(\Omega)$ and verifies*

$$\liminf_{\|\varphi^0\|_{L^2(\Omega)}\to\infty} \frac{J(\varphi^0; a, y_d)}{\| \varphi^0 \|_{L^2(\Omega)}} \geq \varepsilon. \tag{2.10}$$

Further $J(\cdot; a, y_d)$ attains its minimum at a unique point $\widehat{\varphi}^0$ in $L^2(\Omega)$ and

$$\widehat{\varphi}^0 = 0 \quad \Leftrightarrow \quad \| y_d \|_{L^2(\Omega)} \leq \varepsilon. \tag{2.11}$$

Proposition 2.3. *Let $\mathcal{M}$ be the mapping*

$$\begin{array}{rccc} \mathcal{M}: & L^\infty(Q) \times L^2(\Omega) & \rightarrow & L^2(\Omega) \\ & (a(t,x), y_d) & \longrightarrow & \widehat{\varphi}^0. \end{array}$$

If B is a bounded subset of $L^\infty(Q)$ and K is a compact subset of $L^2(\Omega)$, then $\mathcal{M}(B \times K)$ is a bounded subset of $L^2(\Omega)$.

In order to characterize the duality of problem (2.4), we recall that given a convex and proper function $V : X \rightarrow \mathbb{R} \cup \{+\infty\}$ on the Banach space X, it is said that a element p_0 of V' belongs to the set $\partial V(x_0)$ (subdifferential of V at $x_0 \in X$) if

$$V(x_0) - V(x) \leq (p_0, x_0 - x) \quad \forall\, x \in X.$$

It is well known that that if V is Gateaux differentiable its differential coincides with its subdifferential and that x_0 minimizes V over X (or over a convex subset of X) if and only if $0 \in \partial V(x_0)$. Finally, if V is a lower semicontinuous function, then $p_0 \in \partial V(x_0)$ if and only if

$$(p_0, x) \leq \lim_{h\to 0^+} \frac{V(x_0 + hx) - V(x_0)}{h} (< +\infty) \quad \forall\, x \in X.$$

(See, for instance, Aubin-Ekeland [3]). Coming back to the functional J we have:

Lemma 2.2. *For every $\varphi^0 \in L^2(\Omega)$ ($\varphi^0 \neq 0$), if φ is the solution of (2.4) satisfying $\varphi(T) = \varphi^0$, we have that*

$$\partial J(\varphi^0; a, y_d) = \{\xi \in L^2(\Omega),\ \exists\, v \in sgn(\varphi)\chi_{\mathcal{O}} \ \textit{satisfying}$$

$$\begin{aligned} \int_\Omega \xi(x)\theta^0(x)dx &= \left(\int_{\mathcal{O}} |\varphi(t,x)| d\Sigma\right)\left(\int_{\mathcal{O}} v(t,x)\theta(t,x)d\Sigma\right) \\ &\quad + \varepsilon \int_\Omega \frac{\varphi^0(x)}{\| \varphi^0 \|_{L^2(\Omega)}} \theta^0(x)dx - \int_\Omega y_d(x)\theta^0(x)dx \ \forall \theta^0 \in L^2(\Omega)\}, \end{aligned}$$

where θ is the solution of (2.4) satisfying $\theta(T) = \theta^0$.

Proof. It is an easy modification of Proposition 2.4 of [10].

Let us prove the approximate controllability property for an special version of the linear problem given in (2.2).

Theorem 2.1. *If $\| y_d \|_{L^2(\Omega)} > \varepsilon$ and $\widehat{\varphi}$ is the solution of (2.4) corresponding to $\widehat{\varphi}(T) = \widehat{\varphi}^0$, with $\widehat{\varphi}^0$ minimum of $J(\cdot; a, y_d)$. Then there exists $v \in sgn(\widehat{\varphi})\chi_{\mathcal{O}}$ such that the solution of*

$$(2.12) \qquad \begin{cases} y_t + (-\Delta)^m y + (-\Delta^k)(a(t,x)y) = \| \widehat{\varphi} \|_{L^1(\mathcal{O})} v\chi_{\mathcal{O}} & \text{in } Q, \\ \dfrac{\partial^j y}{\partial \nu^j} = 0 \quad (j = 0 \ldots (m-1)) & \text{on } \Sigma, \\ y(0) = 0 & \text{in } \Omega, \end{cases}$$

satisfies

$$y(T) = y_d - \varepsilon \frac{\widehat{\varphi}^0}{\| \widehat{\varphi}^0 \|_{L^2(\Omega)}},$$

and then $\| y(T) - y_d \|_{L^2(\Omega)} = \varepsilon$.

Remark 2.1. In the case $\| y_d \|_{L^2(\Omega)} \leq \varepsilon$, if we use the null control, we obtain $y = 0$ and therefore $\| y(T) - y_d \|_{L^2(\Omega)} \leq \varepsilon$.

First of all we prove the existence and uniqueness to problem given by (2.2).

Proposition 2.4. *Assumed $y_0 \in L^2(\Omega)$, $h \in L^2(0,T; H^{-m}(\Omega))$ and $a(t,x) \in L^\infty(Q)$, there exists a unique function $y \in W$ satisfying*

$$(2.13) \qquad \begin{cases} y_t + (-\Delta)^m y + \Delta^k (a(t,x)y) = h & \text{in } Q, \\ \dfrac{\partial^j y}{\partial \nu^j} = 0 \quad , \quad j = 0, 1, \ldots, m-1 & \text{on } \Sigma, \\ y(0) = y_0 & \text{in } \Omega. \end{cases}$$

Moreover, we have the estimate

$$(2.14) \quad \|y\|_{L^2(0,T;H_0^m(\Omega))} + \|y_t\|_{L^2(0,T;H^{-m}(\Omega)} \leq C \left(\|h\|_{L^2(0,T;H^{-m}(\Omega))} + \|y_0\|_{L^2(\Omega)} \right),$$

where the constant C depends only on M (provided that Ω, T and m are kept fixed).

Proof. For all $n \in \mathbb{N}$ we define again y^{n+1} as the solution of the iterative problem

$$\begin{cases} y_t^{n+1} + (-\Delta)^m y^{n+1} = h - \Delta^k (a(t,x) y^n) & \text{in } Q, \\ \dfrac{\partial^j y^{n+1}}{\partial \nu^j} = 0 \quad , \quad j = 0, 1, \ldots, m-1 & \text{on } \Sigma, \\ y^{n+1}(0) = y_0 & \text{in } \Omega, \end{cases}$$

where $y^0(t) := 0$ for all $t \in [0,T]$. The existence of a solution $y^n \in W$ can be found, for instance, in Theorem 3.4.1 of Lions-Magenes [15]. Thus, for all $n \in \mathbb{N} \backslash \{0,1\}$, $y^{n+1} - y^n$ is solution of

$$(2.15) \quad \begin{cases} (y^{n+1} - y^n)_t + (-\Delta)^m (y^{n+1} - y^n) = -\Delta^k [a(t,x)(y^n - y^{n-1})] & \text{in } Q, \\ \dfrac{\partial^j (y^{n+1} - y^n)}{\partial \nu^j} = 0 \quad , \quad j = 0, 1, \ldots, m-1 & \text{on } \Sigma, \\ (y^{n+1} - y^n)(0) = 0 & \text{in } \Omega \end{cases}$$

and therefore (see again Theorem 3.4.1 of Lions-Magenes [15]) $y^{n+1} - y^n \in W$ and

$$(2.16) \qquad \| y^{n+1} - y^n \|_W \leq c_1 \| a(y^n - y^{n-1}) \|_{L^2(Q)} \cdot$$

Then, since $W \subset \mathcal{C}([0,T];L^2(\Omega))$ with continuous embedding (see, for instance, [12] or [15]), we have that

$$\| y^{n+1} - y^n \|_{\mathcal{C}([0,T];L^2(\Omega))} \leq c_2 \| a(y^n - y^{n-1}) \|_{L^2(Q)} .$$

Further, as in the proof of Proposition 2.1, we can choose $C_2 = C_2(T)$ such that

$$\| y^{n+1} - y^n \|_{\mathcal{C}([0,t];L^2(\Omega))} \leq C_2 \| a(y^n - y^{n-1}) \|_{L^2((0,t)\times\Omega)}, \quad \text{for all } t \in [0,T].$$

Hence,

$$\| (y^{n+1} - y^n)(t) \|^2_{L^2(\Omega)} \leq (C_2 M)^2 \int_0^t \| (y^n - y^{n-1})(\tau) \|^2_{L^2(\Omega)} \, d\tau, \quad \text{for all } t \in [0,T]$$

Then, for every $t \in [0,T]$ we deduce that

$$\begin{aligned} \| (y^{n+1} - y^n)(t) \|^2_{L^2(\Omega)} &\leq (C_2^2 M^2)^{n-1} \int_0^t \int_0^{\tau_1} \cdots \int_0^{\tau_{n-1}} \| (y^2 - y^1)(\tau_n) \|^2_{L^2(\Omega)} \, d\tau_n \ldots d\tau_1 \\ &\leq (C_2^2 M^2)^{n-1} \int_0^t \int_0^{\tau_1} \cdots \int_0^{\tau_{n-1}} \| y^2 - y^1 \|^2_{\mathcal{C}([0,T];L^2(\Omega))} \, d\tau_n \ldots d\tau_1 \\ &\leq (C_2^2 M^2)^{n-1} \frac{t^{n-1}}{(n-1)!} \| y^2 - y^1 \|^2_{\mathcal{C}([0,T];L^2(\Omega))} \\ &\leq \frac{(C_2^2 M^2 T)^{n-1}}{(n-1)!} \| y^2 - y^1 \|^2_{\mathcal{C}([0,T];L^2(\Omega))}, \end{aligned}$$

which implies that

$$\| y^{n+1} - y^n \|_{\mathcal{C}([0,T];L^2(\Omega))} \to 0 \quad \text{as } n \to \infty$$

and therefore, by (2.16), we deduce that

$$\| y^{n+1} - y^n \|_W \to 0 \quad \text{as } n \to \infty.$$

Then, there exists $y \in W$ such that

$$y_n \to y \quad \text{in } W \text{ as } n \to \infty.$$

The end of the proof is similar to the end of the proof of Proposition 2.1. ■

Proof of Theorem 2.1. Using the subdifferentiability of $J(.;a,y_d)$ at $\widehat{\varphi}^0$ ($\neq 0$ by (2.11)), we know that

$$0 \in \partial J(\widehat{\varphi}^0),$$

which is equivalent, from Lemma 2.2, to the existence of $v \in sgn(\widehat{\varphi})\chi_{\mathcal{O}}$, such that

$$(2.17) \quad \begin{aligned} - \| \widehat{\varphi} \|_{L^1(\mathcal{O})} \left(\int_{\mathcal{O}} v(x,t)\theta(x,t)dxdt \right) &= \frac{\varepsilon}{\| \widehat{\varphi}^0 \|_{L^2(\Omega)}} \int_\Omega \widehat{\varphi}^0(x)\theta^0(x)dx \\ &\quad - \int_\Omega y_d(x)\theta^0(x)dx. \end{aligned}$$

On the other hand, as $y \in W$, if we "multiply" by θ in (2.12) we obtain, by (2.4), that

$$(2.18) \qquad \int_\Omega y(T,x)\theta^0(x)dxdt = \| \widehat{\varphi} \|_{L^1(\mathcal{O})} \left(\int_{\mathcal{O}} v(x,t)\theta(x,t)dxdt \right)$$

Then, from (2.17) and (2.18), we obtain

$$\int_\Omega y(T,x)\theta^0(x)dxdt = \int_\Omega (y_d(x) - \varepsilon \frac{\widehat{\varphi}^0(x)}{\| \widehat{\varphi}^0 \|_{L^2(\Omega)}})\theta^0(x)dxdt \quad \forall\, \theta^0 \in L^2(\Omega)$$

and we conclude that $y(T) = y_d - \varepsilon \dfrac{\widehat{\varphi}^0}{\| \widehat{\varphi}^0 \|_{L^2(\Omega)}}$. ∎

Now we are ready to prove a linear version of Theorem 1.1 for problem (2.2)

Corollary 2.1. *Let* $\| y_d \|_{L^2(\Omega)} > \varepsilon$ *and* $\widehat{\varphi}$ *the solution of (2.4) corresponding to* $\widehat{\varphi}(T) = \widehat{\varphi}^0$, *with* $\widehat{\varphi}^0$ *minimum of* $J(\cdot; a, y_d - y(T; a, 0))$, *where in general* $y(t; a, u)$ *denotes the solution of (2.2) corresponding to the control* u. *Then there exists* $v \in sgn(\widehat{\varphi})\chi_{\mathcal{O}}$ *such that the solution of*

$$\begin{cases} y_t + (-\Delta)^m y + (-\Delta^k)(a(t,x)y) = h + h(a) + \| \widehat{\varphi} \|_{L^1(\mathcal{O})} v\chi_{\mathcal{O}} & \text{in } Q, \\ \dfrac{\partial^j y}{\partial \nu^j} = 0 \;\; (j = 0 \ldots (m-1)) & \text{on } \Sigma, \\ y(0) = y_0 & \text{in } \Omega, \end{cases}$$

satisfies

$$\| y(T) - y_d \|_{L^2(\Omega)} \leq \varepsilon.$$

Proof. We put $y = L + Y$, where $L = L(a)$ satisfies

$$\begin{cases} L_t + (-\Delta)^m L + (-\Delta^k)(a(t,x)L) = h + h(a) & \text{in } Q, \\ \dfrac{\partial^j L}{\partial \nu^j} = 0 \;\; (j = 0 \ldots (m-1)) & \text{on } \Sigma, \\ L(0) = y_0 & \text{in } \Omega \end{cases} \tag{2.19}$$

and $Y = Y(a)$ is taken associated to the approximate controllability problem

$$\begin{cases} Y_t + (-\Delta)^m Y + (-\Delta^k)(a(t,x)Y) = u(a)\chi_{\mathcal{O}} & \text{in } Q, \\ \dfrac{\partial^j Y}{\partial \nu^j} = 0 \;\; (j = 0 \ldots (m-1)) & \text{on } \Sigma, \\ Y(0) = 0 & \text{in } \Omega, \end{cases}$$

with desired state $y_d - L(T)$, i.e. such that $\| Y(T) - (y_d - L(T)) \| \leq \varepsilon$. Notice that the existence of such a control $u(a)$ is consequence of Theorem 2.1. In particular, if $\| y_d - L(T) \| \leq \varepsilon$, we can take $u(a) \equiv 0$ and if $\| y_d - L(T) \| > \varepsilon$, then we take $u(a) = \| \widehat{\varphi}(a) \|_{L^1(\Omega)} v(a)$, where $v(a) \in sgn(\widehat{\varphi}(a))\chi_{\mathcal{O}}$ and $\widehat{\varphi}(a)$ is the solution of (2.4) with initial value $\mathcal{M}(\ (a(x,t), y_d - L(T))\ )$ defined in Proposition 2.3. It is obvious that such function y and such control $u(a)$ lead to the conclusion. ∎

3. Controllability for the nonlinear problem

As mentioned before, we shall use a fixed point argument to prove Theorem 1.1. In fact we shall deal with multivalued operators. Let us recall a well-known result: the Kakutani's fixed point Theorem. The usual continuity assumption in other fixed pont theorems is replaced here by the following notion:

Definition 3.1. *Let X, Y two Banach spaces and, $\Lambda : X \to \mathcal{P}(Y)$ a multivalued function. We say that Λ is* <u>*upper hemicontinuous*</u> *at $x_0 \in X$, if for every $p \in Y'$, the function*

$$x \to \sigma(\Lambda(x), p) = \sup_{y \in \Lambda(x)} < p, y >_{Y' \times Y}$$

is upper semicontinuous at x_0. We say that the multivalued function is upper hemicontinuous on a subset K of X, if it satisfies this properties for every point of K.

Theorem 3.1. ***(Kakutani's fixed point Theorem).*** *Let $K \subset X$ be a convex and compact subset and $\Lambda : K \to K$ an upper hemicontinuous application with convex, closed and nonempty values. Then, there exists a fixed point x_0, of Λ.*

For a proof see, for instance, Aubin [2].

Proof of Theorem 1.1. We fix $y_d \in L^2(\Omega)$ and $\varepsilon > 0$. By using Corollary 2.1, for each $z \in L^2(Q)$ and $\varepsilon > 0$ it is possible to find two functions $\varphi(z) \in L^1(Q)$ and $v(z) \in sgn(\varphi(z))\chi_{\mathcal{O}}$ such that the solution $y = y^z$ of

$$\text{(3.1)} \qquad \begin{cases} y_t + (-\Delta)^m y + (-\Delta)^k (g(z)y) = h + h(g(z)) + u\chi_{\mathcal{O}} & \text{in } Q, \\ \dfrac{\partial^j y}{\partial \nu^j} = 0 \ , \ j = 0, 1, \dots m-1 & \text{on } \Sigma, \\ y(0) = y_0 & \text{in } \Omega, \end{cases}$$

(where $u = u(z) = |\varphi(z)|_{L^1(\mathcal{O})} v(z))$ satisfies

$$\text{(3.2)} \qquad |y(T) - y_d|_{L^2(\Omega)} \leq \varepsilon.$$

Here $\varphi(z)$ is the solution of (2.4) with initial value $M(\ (g(z), y_d - L(z;T))\ )$ (see Proposition 2.3) and $a(t,x) = g(z)$, where is $L(z;T)$ the solution of (2.19), with $a = g(z)$, at time T .

Lemma 3.1. *The set*

$$\{y_d - L(z;T),\ \ z \in L^2(Q)\},$$

is relatively compact in $L^2(\Omega)$.

Proof of Lemma 3.1. Applying Proposition 2.4 it is easy to see that the set of solutions $L(z)$ of

$$\text{(3.3)} \qquad \begin{cases} L_t + (-\Delta)^m L + (-\Delta)^k (g(z)y) = h + h(g(z)) & \text{in } Q, \\ \dfrac{\partial^j L}{\partial \nu^j} = 0 \ , \ j = 0, 1, \dots m-1 & \text{on } \Sigma, \\ L(0) = y_0 & \text{in } \Omega, \end{cases}$$

satisfy

$$\text{(3.4)} \qquad \| L(z) \|_W \leq K(1 + \| y_0 \|_{L^2(\Omega)} + \| h \|_{L^2(0,T;H^{-m}(\Omega))}) \ \ \forall\ z \in L^2(Q)$$

with $K > 0$ independent of z. Recall that $\| g(z) \|_{L^\infty(Q)} \leq M$ with M independent of z. Now, let $L(z_n)$ be a sequence of solutions (3.3) with $z_n \in L^2(Q)$. We must prove that there exists a subsequence (that we rewrite as $L(z_n)$), such that

$$\| L(z_n;T) - L(z_{n+1};T) \|_{L^2(\Omega)} \to 0 \quad \text{as } n \to \infty.$$

By a compactness result due to Aubin [1], we know that

$$W \subset L^2(0,T;H^{m-1}(\Omega)) \quad \text{with compact embedding.}$$

Therefore, by (3.4), we can suppose that

$$\| L(z_n) - L(z_{n+1}) \|_{L^2(0,T;H^{m-1}(\Omega))} \to 0 \quad \text{as } n \to \infty.$$

Further, it is easy to prove that $L(z_n) - L(z_{n+1})$ satisfies

$$\| L(z_n;T) - L(z_{n+1};T) \|^2_{L^2(\Omega)}$$

$$\leq - \int_0^T \langle D^k \left(g(z_n)L(z_n) - g(z_{n+1})L(z_{n+1})\right), D^k \left(L(z_n) - L(z_{n+1})\right)\rangle_{H^{-k}(\Omega)\times H_0^k(\Omega)} dt$$

$$+ \int_0^T \langle D^k \left(g(z_n)s_0 - g(z_{n+1})s_0\right), D^k \left(L(z_n) - L(z_{n+1})\right)\rangle_{H^{-k}(\Omega)\times H_0^k(\Omega)} dt.$$

Then, by (3.4), since $k \leq m-1$ (notice that $k = 0$ if $m = 1$),

$$\| L(z_n;T) - L(z_{n+1};T) \|^2_{L^2(\Omega)} \leq \widetilde{K} \| L(z_n) - L(z_{n+1}) \|^2_{L^2(0,T;H^{m-1}(\Omega))} \to 0 \quad \text{as } n \to \infty$$

and the proof ends. ■

Completion of Proof of Theorem 1.1. From Lemma 3.1, we obtain that $y_d - L(z;T)$ belongs to a compact set for all $z \in L^2(Q)$ and so, by using Propositions 2.3 and 2.1, we obtain that

$$\{\| \varphi(z) \|_{L^1(\mathcal{O})} v(z),\ z \in L^2(Q)\} \quad \text{is bounded in } L^\infty(Q) \tag{3.5}$$

Thus

$$K_1 = \sup_{z\in L^2(Q)} \| \varphi(z) \|_{L^1(\mathcal{O})} < \infty. \tag{3.6}$$

Obviously, $u = u(z)$ satisfies

$$\| u \|_{L^2(Q)} \leq K_2. \tag{3.7}$$

Therefore, if we define the operator

$$\Lambda : L^2(Q) \to \mathcal{P}(L^2(Q))$$

by

$$\Lambda(z) = \{y \text{ satisfies (3.1), (3.2) for some } u \text{ satisfying (3.7) }\},$$

we have seen that for each $z \in L^2(Q)$, $\Lambda(z) \neq \emptyset$. In order to apply Kakutani's fixed point theorem, we have to check that the next properties hold:

(i) There exists a compact subset U of $L^2(Q)$, such that for every $z \in L^2(Q)$, $\Lambda(z) \subset U$.

(ii) For every $z \in L^2(Q)$, $\Lambda(z)$ is a convex, compact and nonempty subset of $L^2(Q)$.

(iii) Λ is upper hemicontinuous.

The proof of these properties is as follows:

(i) From Proposition 2.4 we know that, there exists a bounded subset U of W such that for every $z \in L^2(Q)$, $\Lambda(z) \subset U$. Now, to see that we can choose U compact we shall prove that the set

$$\mathcal{Y} = \{y \text{ satisfying (3.1) for some } z \in L^2(Q) \text{ and } u \text{ satisfying (3.7)}\}$$

is a relatively compact subset of $L^2(Q)$. But this is easy to prove by using that

$$W \subset L^2(Q) \text{ with compact embedding} \tag{3.8}$$

(see Lions [12] or Simon [18]).

(ii) We have already seen that for every $z \in L^2(Q)$, $\Lambda(z)$ is a nonempty subset of $L^2(Q)$. Further $\Lambda(z)$ is obviously convex, because $B(y_d, \varepsilon)$ and $\{u \in L^2(Q) : \text{satisfying (3.7)}\}$ are convex sets. Then, we have to see that $\Lambda(z)$ is a compact subset of $L^2(Q)$. In (i) we have proved that $\Lambda(z) \subset U$ with U compact. Let $(y^n)_n$ be a sequence of elements of $\Lambda(z)$ which converges in $L^2(Q)$ to $y \in U$. We have to prove that $y \in \Lambda(z)$. We know that there exist $u^n \in L^2(Q)$ satisfying (3.7) such that

$$\begin{cases} y^n_t + (-\Delta)^m y^n + (-\Delta)^k (g(z) y^n) = h + h(g(z)) + u^n \chi_{\mathcal{O}} & \text{in } Q, \\ \dfrac{\partial^j y^n}{\partial \nu^j} = 0 \ , \ j = 0, 1, \ldots, m-1 & \text{on } \Sigma, \\ y^n(0) = y_0 & \text{in } \Omega, \\ |y^n(T) - y_d|_2 \leq \varepsilon. \end{cases} \tag{3.9}$$

Now, by using that the controls u^n are uniformly bounded, we deduce that $u^n \to u$ in the weak topology of $L^2(Q)$ and u satisfies (3.7) (see Proposition III.5 of Brezis [5]). Then, using (3.9) and Proposition 2.4 we can see that $(y^n)_n$ converges to y in the weak topology of W (and so, by (3.8), strongly in $L^2(Q)$). Therefore, passing to the limit in (3.9) we obtain

$$\begin{cases} y_t + (-\Delta)^m y + (-\Delta)^k (g(z) y) = h + h(g(z)) + u \chi_{\mathcal{O}} & \text{in } Q, \\ \dfrac{\partial^j y}{\partial \nu^j} = 0 \ , \ j = 0, 1, \ldots, m-1 & \text{on } \Sigma, \\ y(0) = y_0 & \text{in } \Omega. \end{cases}$$

Further, $v^n = y - y^n$ is solution of

$$\begin{cases} v^n_t + (-\Delta)^m v^n + (-\Delta)^k (g(z) v^n) = (u - u^n) \chi_{\mathcal{O}} & \text{in } Q, \\ \dfrac{\partial^j v^n}{\partial \nu^j} = 0 \ , \ j = 0, 1, ..., m-1 & \text{on } \Sigma, \\ v^n(0) = 0 & \text{in } \Omega \end{cases} \tag{3.10}$$

and satisfies $v^n \in W$ (see Proposition 2.4). Further, if we "multiply" in (3.10) by v^n and integrate, we obtain that

$$\| v^n(T) \|^2_{L^2(\Omega)} \leq k \int_Q (u - u^n) \chi_{\mathcal{O}} v^n dx dt \to 0 \quad \text{as } n \to \infty.$$

Thus $y^n(T)$ converges to $y(T)$ in the strong topology of $L^2(\Omega)$ and $\| y(T) - y_d \|_2 \leq \varepsilon$. This prove that $y \in \Lambda(z)$ and concludes the proof of (ii).

(iii) We must prove that for every $z_0 \in L^2(Q)$

$$\limsup_{z_n \xrightarrow{L^2(Q)} z_0} \sigma(\Lambda(z_n), k) \leq \sigma(\Lambda(z_0), k), \ \forall\, k \in L^2(Q).$$

We have seen in (ii) that $\Lambda(z)$ is a compact set, which implies that for every $n \in \mathbb{N}$ there exists $y^n \in \Lambda(z_n)$ such that

$$\sigma(\Lambda(z_n), k) = \int_Q k(x,t) y^n(x,t)\,dxdt.$$

Now, by (i), $(y^n)_n \subset U$ (compact set of $L^2(Q)$). Then, there exists $y \in L^2(Q)$ such that (after extracting a subsequence) $y^n \to y$ in $L^2(Q)$. We shall prove that $y \in \Lambda(z_0)$. We know that there exist $u^n \in L^2(Q)$ satisfying (3.7) such that

$$(3.11) \qquad \begin{cases} y_t^n + (-\Delta)^m y^n + (-\Delta)^k (g(z_n) y^n) = h + h(z_n) + u^n \chi_{\mathcal{O}} & \text{in } Q, \\ \dfrac{\partial^j y^n}{\partial \nu^j} = 0 \ , \ j = 0, 1, \ldots, m-1 & \text{on } \Sigma, \\ y^n(0) = y_0 & \text{in } \Omega, \\ |y^n(T) - y_d|_2 \leq \varepsilon. \end{cases}$$

Then there exists $u \in L^2(Q)$ satisfying (3.7) such that $u^n \to u$ in the weak topology of $L^2(\mathcal{O})$. On the other hand, by using the smoothing effect of the parabolic linear equation (in a similar way to the proof of (ii)) and that $g \in L^\infty(\mathbb{R}) \cap \mathcal{C}(\mathbb{R})$, we deduce that y satisfies (3.1) and (3.2) with $z = z_0$ for some $u \in L^2(Q)$ satisfying (3.7), which implies that $y \in \Lambda(z_0)$. Then, for every $k \in L^2(Q)$,

$$\sigma(\Lambda(z_n), k) = \int_Q k(x,t) y^n(x,t)\,dxdt \to \int_Q k(x,t) y(x,t)\,dxdt$$

$$\leq \sup_{\overline{y} \in \Lambda(z_0)} \int_Q k(x,t) \overline{y}(x,t)\,dxdt = \sigma(\Lambda(z_0), k),$$

which proves that Λ is upper hemicontinuous and conclude the proof of (iii).

Finally, if we restrict Λ to $K = conv(U)$ (the convex envelope of U), which is a compact set of $L^2(Q)$, it satisfies the assumptions of Kakutani's fixed point theorem. Then, Λ has a fixed point $y \in K$. Further, by construction, there exists a control $u \in L^2(Q)$ satisfying (3.7) such that

$$(3.12) \qquad \begin{cases} y_t + (-\Delta)^m y + (-\Delta)^k (f(y)) = h + u \chi_{\mathcal{O}} & \text{in } Q, \\ \dfrac{\partial^j y}{\partial \nu^j} = 0 \ , \ j = 0, 1, \ldots m-1 & \text{on } \Sigma, \\ y(0) = y_0 & \text{in } \Omega, \\ |y(T) - y_d|_2 \leq \varepsilon. \end{cases}$$

Therefore, y is the solution that we were looking for. ∎

Remark 3.1. Several generalizations seem possible. For instance, the equation of (1.1) could be replaced by other ones with a more general nonlinearity

$$y_t + (-\Delta)^m y + \sum_{i=0}^{k} (-\Delta)^i f_i(y) = h + v \chi_\omega$$

or a more general lower order differential operator

$$y_t + (-\Delta)^m y + L(f(y)) = h + v\chi_\omega,$$

with L suitable linear partial differential operator of degree lower than $2m$. The key point in those generalizations is that the unique continuation result of Lemma 2.1, for the associated dual problem, remains true thanks to Theorem 3.2 of Saut-Scheurer [17] and the rest of arguments of the proof of Theorem 1.1 apply.

4. Non-controllability for superlinear problems

In this section we assume $k = 0$. We shall prove a result of non-controllability for a superlinear nonlinear term with $\overline{\omega} \subset \Omega$.

Theorem 4.1. *Let $p > 1$ and let $y(t;u) = y \in L^2(0,T;H^m(\Omega)) \cap \mathcal{C}([0,T];L^2(\Omega))$ a function satisfying*

$$\begin{cases} y_t + (-\Delta)^m y + |y|^{p-1}y = u\chi_\omega & \text{in } Q, \\ y(0) = y_0 & \text{in } \Omega, \end{cases}$$

associated to any "natural" boundary condition and with control $u \in L^2(Q)$. Then we can choose $y_d \in L^2(\Omega)$ and $\varepsilon > 0$ such that

$$\| y(T;u) - y_d \|_{L^2(\Omega)} > \varepsilon \quad \text{for any} \quad u \in L^2(Q). \tag{4.1}$$

In order to prove Theorem 4.1 we introduce, previously, some auxiliar functions. Given $R > 0$ we define, on $\mathbb{R}^N$, the functions

$$\xi_R(x) = (R^2 - |x|^2)/R \ \text{ if } \ |x| < R, \quad \xi_R(x) = 0 \ \text{ if } |x| \geq R$$

and

$$d_R(x) = R - |x| \ \text{ if } \ |x| < R, \quad d_R(x) = 0 \ \text{ if } |x| \geq R. \tag{4.2}$$

It is clear that

$$d_R(x) \leq \xi_R(x) \leq 2d_R(x) \tag{4.3}$$

for all $x \in \mathbb{R}^N$.

The following result was proved in Bernis [4].

Proposition 4.1. *Let $s \geq 2m$ and $R > 0$. Then, for each $\varepsilon > 0$ there exist a constant C depending only on N, m, s and ε (thus independent of R) such that the following inequality holds for all $y \in H^m_{loc}(\mathbb{R}^N)$:*

$$((-\Delta)^m y, \xi_R^s y)_{H^{-m}_{loc}(\mathbb{R}^N)\times H^m_c(\mathbb{R}^N)} \geq (1-\varepsilon)\int_{\mathbb{R}^N} \xi_R^s |D^m y|^2 dx - C\int_{\mathbb{R}^N} \xi_R^{s-2m} y^2 dx.$$

Remark 4.1. *Since $s \geq 2m$, $\xi_R^s \in W_c^{2m,\infty}(\mathbb{R}^N)$. Hence $\xi_R^s \in \mathcal{C}_c^m(\mathbb{R}^N)$ (see e.g. Corollary IX.13 of [5]) and $\xi_R^s u \in H_c^m(\mathbb{R}^N)$ (see e.g. Note IX.4 of [5]).*

Corollary 4.1. *Let $s \geq 2m$ and $R > 0$ such that $\overline{B_R} \subset \Omega$. Then, for each $\varepsilon > 0$ there exist a constant C depending only on N, m, s and ε (thus independent of R) such that the following inequality holds for all $y \in H^m(\Omega)$:*

$$((-\Delta)^m y, \xi_R^s y)_{H^{-m}(\Omega)\times H_0^m(\Omega)} \geq (1-\varepsilon)\int_\Omega \xi_R^s |D^m y|^2 dx - C\int_\Omega \xi_R^{s-2m} y^2 dx.$$

Proof. Let $\overline{y} \in H^m(\Omega)$ such that $\overline{y} = y$ in Ω (such $\overline{y}$ exists by standar results: see, e.g., Chapter IX of Brezis [5]). Then, by Proposition 4.1, the inequality holds for $\overline{y}$, but as $\overline{B_R} \subset \Omega$ we obtain the result. ∎

Theorem 4.2. *Let $p > 1$, $r = p+1$, $y_0 \in L^2(\Omega)$ and $u \in L^{r'}(Q)$. Then any solution $y \in L^r(Q) \cap L^2(0,T;H^m(\Omega))$ of*

$$\text{(4.4)} \qquad \begin{cases} y_t + (-\Delta)^m y + |y|^{p-1}y = u & \text{in } \mathcal{D}'(Q), \\ y(0) = y_0 & \text{on } \Omega, \end{cases}$$

with any "natural" boundary condition, satisfies the local estimate

$$\sup_{0<t<T}\int_{B_R} y(x,t)^2 dx + \int_{B_R\times(0,T)} (|D^m y|^2 + |y|^r)dxdt$$
$$\leq K\left(1 + \int_{B_{R_1}\times(0,T)} |u|^{r'} dxdt + \int_{B_{R_1}} y_0^2 dx\right)$$

if $\overline{B_{R_1}} \subset \Omega$ and $0 < R \leq R_1$. Moreover, the constant K depends only on N, m, p, R, R_1 and T.

Proof of Theorem 4.2. We take $X_r = L^r(Q) \cap L^2(0,T;H_0^m(\Omega))$. Then the equation of (4.4) is satisfied in $X_r' = L^{r'}(Q) + L^2(0,T;H^{-m}(\Omega))$. Then, if $s \geq 2m$, we can multiply (4.4) by $\xi_R^s y$ with the duality product $(\cdot,\cdot)_{X_r'\times X_r}$ and we obtain

$$\frac{1}{2}\int_{B_R} \xi_R^s y(x,T)^2 dx + ((-\Delta)^m y, \xi_R^s y)_{L^2(0,T;H^{-m}(\Omega))\times L^2(0,T;H_0^m(\Omega))}$$
$$+ (|y|^{p-1}y, \xi_R^s y)_{L^{r'}(Q)\times L^r(Q)}$$
$$= \frac{1}{2}\int_{B_R} \xi_R^s y_0(x)^2 dx + (u, \xi_R^s y)_{L^{r'}(Q)\times L^r(Q)}.$$

Now, from Corollary 4.1 it follows that

$$\text{(4.5)} \qquad \begin{aligned} &\frac{1}{2}\int_{B_R} \xi_R^s y(x,T)^2 dx + \int_{B_R\times(0,T)} \xi_R^s(|D^m y|^2 + |y|^r)dxdt \\ &\leq C\int_{B_R} \xi_R^s y_0(x)^2 dx + C\int_{B_R\times(0,T)} \xi_R^{s-2m} y^2 dxdt + C\int_{B_R\times(0,T)} \xi_R^s u y dxdt. \end{aligned}$$

By (4.2) and (4.3) we can replace in (4.5) $\xi_R(x)$ by $R-|x|$ (modifying the constants). Further, writing $s-2m = 2s/r + (s(r-2)/r) - 2m$, we can apply Hölder's or Young's inequality with exponents $q = r/2$ and $q' = r/r-2$ and we obtain

$$\int_{B_R\times(0,T)} (R-|x|)^{s-2m} y^2 dxdt$$
$$\leq \varepsilon\int_{B_R\times(0,T)} (R-|x|)^s |y|^r dxdt + K(\varepsilon,q)\int_{B_R\times(0,T)} (R-|x|)^{s-\gamma} dxdt$$

with

$$K(\varepsilon, q) = \frac{1}{q'(q\varepsilon)^{q'/q}} \quad \text{and} \quad \gamma = \frac{2mr}{r-2} \quad .$$

Hence, if we choose $s > \gamma - 1$, the last integral is finite and equal to $\widetilde{C}R^{s+N-\gamma}$. On the other hand, we can apply again Young's inequality and we have

$$\int_{B_R\times(0,T)} (R-|x|)^s uy dx dt$$
$$\leq \varepsilon \int_{B_R\times(0,T)} (R-|x|)^s |y|^r dx dt + k(\varepsilon, r) \int_{B_R\times(0,T)} (R-|x|)^s |u|^{r'} dx dt.$$

Thus, by changing the constants, we deduce that

$$\frac{1}{2}\int_{B_R} (R-|x|)^s y(x,T)^2 dx + \int_{B_R\times(0,T)} (R-|x|)^s (|D^m y|^2 + |y|^r) dx dt$$
$$\leq C\left(\int_{B_R} (R-|x|)^s y_0(x)^2 dx + R^{s+N-\gamma} + \int_{B_R\times(0,T)} (R-|x|)^s |u|^{r'} dx dt\right).$$

Finally, by replacing R by R_1 and by taking into account that $R_1 - |x| \geq R_1 - R$ and $R_1 - |x| \leq R_1$ if $|x| \leq R$ we deduce the result with

$$K = \max\left\{C(\frac{R_1}{R_1 - R})^s, \; \frac{CR_1^{s+N-\gamma}}{(R_1-R)^s}\right\}. \qquad \blacksquare$$

Proof of Theorem 4.1. It is a trivial consequence of Theorem 4.2 since, if R_1 satisfies $\overline{B_{R_1}} \subset \Omega\backslash\omega$, then

$$\| y(u;T) \|^2_{L^2(\Omega)} \leq K(1 + \| y_0 \|^2_{L^2(\Omega)}) \quad \forall u \in L^{r'}(Q).$$

Therefore, taking y_d with $\| y_d \|_{L^2(\Omega)}$ large enough, we obtain (4.1) for $\varepsilon > 0$ small enough. $\blacksquare$

Acknowledgements. We thank F. Bernis for some useful conversations.

References

1. **Aubin, J.P.**: Un théorème de compacité. *C. R. Acad. Sci.*, Paris, Serie I, T. 256, pp. 5042–5044, (1963).
2. **Aubin, J.P.**: *L'analyse non linéaire et ses motivations économiques.* Masson. (1984).
3. **Aubin, J.P. and Ekeland, I.**: *Applied nonlinear Analysis.* Wiley-Interscience Publication, New York, (1984).
4. **Bernis, F.**: Elliptic and Parabolic Semilinear Problems without Conditions at infinity. *Arch. Rat. Mech. Anal.*, Vol. 106, N. 3, pp. 217–241, (1989).
5. **Brézis, H.**:*Analyse Fonctionnelle: Théorie et applications.* Masson, Paris, (1987).
6. **Cahn, J.W. and Hilliard, J.E.**: Free energy of a nonuniform system. I. Interfacial free energy. *J. Chem. Phys.* N. 28, pp. 258–267, (1958).
7. **Díaz, J.I. and Ramos, A.M.**: Positive and negative approximate controllability results for semilinear parabolic equations. *Revista de la Real Academia de Ciencias Exactas, Físicas y Naturales*, Madrid, LXXXIX, 1º–2º, pp. 11–30 (1995).
8. **Elliot, C.M. and Songmu, Z.**: On the Cahn-Hilliard Equation. *Arch. Rat. Mech. Anal.* N. 96, pp. 339–357, (1986).
9. **Fabre, C., Puel, J.P. and Zuazua, E.**: Contrôlabilité approchée de l'équation de la chaleur semi-linéaire. *C. R. Acad. Sci. Paris*, t. 315, Série I, pp. 807–812, (1992).

10. **Fabre, C., Puel, J.P. and Zuazua, E.**: Approximate controllability of the semilinear heat equation, *Proceedings of the Royal Society of Edinburgh*, 125A, pp. 31–61, (1995).
11. **Haraux, A.**: *Nonlinear Evolution Equations*. Lecture Notes in Mathematics. Springer-Verlag, Heidelberg, (1981).
12. **Lions, J.L.**: *Contrôle optimal de systemes gouvernés par des equations aux derivées partielles*. Dunod, Paris, (1968).
13. **Lions, J.L.**: *Quelques méthodes de résolution des problèms aux limites non linéares*. Dunod, Paris, (1969).
14. **Lions, J.L.**: Remarques sur la contrôlabilité approchée. In Proceedings of *Jornadas Hispano-Francesas sobre Control de Sistemas Distribuidos*, Universidad de Malaga, pp. 77–88, (1990).
15. **Lions, J.L. and Magenes, E.**: *Problèmes aux limites non homogènes et applications*, Vol. 1, Dunod, Paris, (1968).
16. **Lions, J.L. and Magenes, E.**: *Problèmes aux limites non homogènes et applications*, Vol. 2, Dunod, Paris, (1968).
17. **Saut, J.C. and Scheurer, B.**: Unique continuation for some evolution equations. *J. Differential Equations*, Vol. 66, N. 1, pp. 118–139, (1987).
18. **Simon, J.**: Compact Sets in the Space $L^p(0,T;B)$. *Annali di Matematica Pura ed Applicata*. Serie 4, N. 146, pp.65–96, (1987).

J.I. Díaz
Dpto. Matemática Aplicada
Universidad Complutense de Madrid
Avda. Complutense s/n
E-28040 Madrid, Spain

A.M. Ramos
Dpto. Informática y Automática
Universidad Complutense de Madrid
Avda. Complutense s/n
E-28040 Madrid, Spain

International Series of Numerical Mathematics
Vol. 126, © 1998 Birkhäuser Verlag, Basel

Control Problems for Parabolic Equations with State Constraints and Unbounded Control Sets

H.O. FATTORINI

Department of Mathematics
University of California

ABSTRACT. Using nonlinear programming theory we derive a version of Pontryagin's maximum principle for abstract parabolic equations that includes state constraints and allows unbounded control sets. The results are shown to apply to parabolic distributed parameter systems and to the Navier-Stokes equations.

1991 *Mathematics Subject Classification.* 93E20, 93E25

Key words and phrases. Distributed parameter systems, optimal controls, unbounded controls, state constraints.

1. Introduction

Consider the reaction–diffusion distributed parameter system

$$\frac{\partial y(t,x)}{\partial t} = \Delta y(t,x) + \phi(y(t,x), \nabla y(t,x)) + u(t,x) \quad (t \geq 0, x \in \Omega) \tag{1.1}$$

in a domain $\Omega \subseteq \mathbb{R}^m$ with boundary Γ, a boundary condition β acting on Γ. Control problems for (1.1) may include constraints on the state $y(t,x)$ and the gradient, either pointwise

$$y(t,x) \in M_s \subseteq \mathbb{R}, \quad \nabla y(t,x) \in M_g \subseteq \mathbb{R}^m \quad (0 \leq t \leq \bar{t}) \tag{1.2}$$

or of integral type. Target conditions can be also pointwise,

$$y(\bar{t},x) \in Y_s \subseteq \mathbb{R}, \quad \nabla y(\bar{t},x) \in Y_g \subseteq \mathbb{R}^m \tag{1.3}$$

or of integral type, the control interval $[0,\bar{t}\,]$ fixed or variable. Control constraints may include

$$u(t,x) \in U \subseteq \mathbb{R} \quad (0 \leq t \leq \bar{t}) \tag{1.4}$$

and (when U is unbounded) summability conditions in the cylinder $(0,\bar{t}) \times \Omega$.

Interest in optimal control problems with unbounded controls is more than academic. When the control set is unbounded, Pontryagin's maximum principle not only gives optimal controls as solutions of independent maximization problems for each time t but includes the statement that the maximum is finite – sometimes a very potent pronouncement. (For ingenious ways of putting this to use in finite dimensional systems see [8]). As an infinite dimensional possibility of obvious interest we mention that of setting up solutions of the Navier-Stokes equations as solutions of minimization problems involving purely differential equations (i.e. relegating nonlocal operators to the cost functional). This can be done in several ways, and in all these problems the controls are naturally unbounded.

Except for particular cases (such as the linear-quadratic problem) interest in Pontryagin's principle for unbounded control sets seems of recent date. Raymond and Zidani treat in [9] and [10] boundary control problems for parabolic equations. Unbounded control sets were considered in [6] for distributed parameter systems with smooth nonlinearities in reflexive spaces. These conditions fit systems described by nonlinear wave equations but would force unreasonable assumptions on a system like (1.1) (for instance, they would not cover such nonlinearities as $\phi(y) = -y^3$ or $\phi(y, \nabla y) = (y \cdot \nabla)y$. We indicate in this paper how to handle abstract parabolic systems, with applications to reaction-diffusion equations and the Navier-Stokes equations.

2. Abstract differential equations

We study (1.1) via the abstract model

$$y'(t) = Ay(t) + f(t, y(t)) + Bu(t), \qquad y(0) = \zeta, \tag{2.1}$$

The operator A in (1.1) generates a bounded analytic semigroup $S(t)$ in a reflexive separable Banach space E (for a nonreflexive setup see §5) and $0 \in \rho(A)$. The *control space* for (2.1) is $F = X^*$ (X a separable Banach space) and $B : X^* \to E$ is a bounded operator with $B^* : E^* \to X$.

The assumptions on A allow construction of the fractional powers $(-A)^\alpha$; $(-A)^\alpha$ is bounded for $\alpha < 0$. For any α, $(-A)^\alpha S(t)$ is a bounded operator, continuous in (E, E) for $t > 0$ ($(E, F) = \{$linear bounded operators from a Banach space E into the Banach space F equipped with the operator norm$\}$) and

$$\|(-A)^\alpha S(t)\| \le C_\alpha t^\alpha e^{-ct} \quad (t > 0,\ 0 \le \alpha < 1). \tag{2.2}$$

If $\alpha \ge 0$ we set $E_\alpha = D((-A)^\alpha)$ equipped with the norm $\|y\|_{E_\alpha} = \|(-A)^\alpha y\|$. Invertibility of $(-A)^\alpha$ implies that E_α is a Banach space. For $\alpha < 0$, E_α is the completion of E with respect to the norm $\|\cdot\|_{E_\alpha}$. Since E is reflexive, A^* is the infinitesimal generator of the strongly continuous semigroup $S^*(t) = S(t)^*$ and we define the spaces $(E^*)_\alpha$ using the fractional powers $(-A^*)^\alpha = ((-A)^\alpha)^*$ in the same way the E_α are defined from the fractional powers $(-A)^\alpha$. We have

$$(-A)^{-\gamma} E_\alpha = E_{\alpha+\gamma}, \quad (-A^*)^{-\gamma}(E^*)_\alpha = (E^*)_{\alpha+\gamma} \tag{2.3}$$

for $\gamma > 0$ and $-\infty < \alpha < \infty$ [5].

We say that the function $f : [0, T] \times E_\alpha \to E_{-\rho}$ ($\alpha, \rho \ge 0$, $\alpha + \rho < 1$) satisfies **Hypothesis** $D_{\alpha,\rho}$ if the Fréchet derivative $\partial_y f(t, y) \in (E_\alpha, E_{-\rho})$ exists and

(*i*) $f(t, y)$ is continuous in $y \in E_\alpha$ for t fixed and strongly measurable in t for y fixed,

(*ii*) $\partial_y f(t, y)\zeta$ is continuous in $y \in E_\alpha$ for t and $\zeta \in E_\alpha$ fixed and strongly measurable in t for y, ζ fixed,

(iii) For every $c > 0$ there exist constants $K(c)$, $L(c)$ such that

$$\|f(t,y)\|_{E_{-\rho}} \le K(c),\ \|\partial_y f(t,y)\|_{(E_\alpha,E_{-\rho})} \le L(c)\ (0 \le t \le T, \|y\|_{E_\alpha} \le c). \tag{2.4}$$

Controls $u(\cdot)$ in (2.1) are elements of $L^p_w(0,T;X^*)$ with $p > 1/(1-\alpha)$, where $L^p_w(0,T;X^*)$ is the space of all $X-$weakly measurable X^*-valued functions $u(\cdot)$, equipped with the usual L^p norm; this space is the dual of $L^q(0,T;X)$ $(1/p+1/q=1)$ and does not in general coincide with $L^p(0,T;X^*)$ (except when X^* is separable). The *control set* U is an arbitrary subset of X^*, and the *admissible control space* $C_{ad}(0,T;U)$ consists of all $u(\cdot) \in L^p_w(0,T;X^*)$ such that $u(t) \in U$ a.e.

Solutions of (2.1) are $y(t) = (-A)^{-\alpha}\eta(t)$, where $\eta(t)$ solves the integral equation

$$\begin{aligned}\eta(t) = (-A)^\alpha S(t)\zeta &+ \int_0^t (-A)^{\alpha+\rho} S(t-\tau)(-A)^{-\rho} f(\tau,(-A)^{-\alpha}\eta(\tau))d\tau \\ &+ \int_0^t (-A)^\alpha S(t-\tau)Bu(\tau)d\tau.\end{aligned} \tag{2.5}$$

The assumptions guarantee local existence (see [5] for details). Note that if $z \in E^*$ we have $\langle z, Bg(t)\rangle = \langle B^*z, g(t)\rangle$, so that $Bu(\cdot)$ is E^*-weakly measurable; since E is separable, $Bu(\cdot)$ is strongly measurable.

We consider optimal control problems for (2.1) in a fixed or variable interval $0 \le t \le \bar{t}$. State and target conditions are given by

$$y(t) \in M \quad (0 \le t \le \bar{t}), \qquad y(\bar{t},u) \in Y, \tag{2.6}$$

with the *state constraint set* $M \subseteq E_\alpha$ and the *target set* $Y \subseteq E_\alpha$ closed in E_α. The *cost functional* is

$$y_0(t,u) = \int_0^t f_0(\tau, y(\tau), u(\tau))d\tau \tag{2.7}$$

where $f_0 : [0,T] \times E_\alpha \times U \to \mathbb{R}$ satisfies **Hypothesis** D^0_α; this means the Fréchet derivative $\partial_y f_0(t,y,u) \in (E_\alpha)^*$ exists and

(i_0) For every t, u fixed $f_0(t,y,u)$ and $\partial_y f_0(t,y,u)$ are continuous in $y \in E_\alpha$,

(ii_0) For every $u(\cdot) \in C_{ad}(0,T;U)$ and $y \in E_\alpha$ fixed $t \to f_0(t,y,u(t))$ and $t \to \partial_y f_0(t,y,u(t))$ are strongly measurable in their home spaces,

(iii_0) For every $u(\cdot) \in C_{ad}(0,T;U)$ there exist constants $K_0(c)$, $L_0(c)$ such that

$$\begin{aligned}|f_0(t,y,u(t))| \le K_0(c)\|u(t)\|^p,\quad &\|\partial_y f_0(t,y,u(t))\|_{E_{-\alpha}} \le L_0(c)\|u(t)\|^p \\ &(0 \le t \le T, \|y\|_{E_\alpha} \le c).\end{aligned} \tag{2.8}$$

3. The minimum principle

For definiteness, we limit ourselves to the model (2.1) under Hypothesis D^0_α. The admissible control space $C_{ad}(0,T;U)$ is equipped with the distance

$$d(u(\cdot),v(\cdot)) = |\{t \in [0,T]; u(t) \ne v(t)\}|, \tag{3.1}$$

$(|\cdot| =$ Lebesgue measure). In general, $C_{ad}(0,T;U)$ is not complete under d if the control set U is unbounded; for instance, consider U a cone $\ne \{0\}$ in X^*, $u \in U$,

$u_n(t) = t^{-1}\chi_n(t)u$, where $\chi_n(\cdot)$ is the characteristic function of $[1/n, T]$. This is the key difficulty precluding direct application of the methods in [5].

The dual space $(E_\alpha)^*$ is given by $(E_\alpha)^* = (E^*)_{-\alpha}$ with pairing

$$\langle z, y\rangle_{(E^*)_{-\alpha}\times E_\alpha} = \langle (-A^*)^{-\alpha}z, (-A)^\alpha y\rangle_{E^*\times E}. \tag{3.2}$$

The proof follows from (2.3); see [5, Lemma 4.4] for details of a more general result (actually needed in §5).

Given any Banach space E, the space $\Sigma(0,T;E^*)$ consists of all countable additive bounded E^*–valued measures $\mu(ds)$ defined in the field generated by the closed sets of $[0,T]$; this space is the dual of $C(0,T;E)$, the duality given by $\langle \mu, y\rangle_c = \int_0^T \langle \mu(ds), y(s)\rangle$. Since $(E_\alpha)^* = (E^*)_{-\alpha}$ we have $\Sigma(0,T;(E^*)_{-\alpha}) = \Sigma(0,T;(E_\alpha)^*) = C(0,T;E_\alpha)^*$. See [1] for further details on vector–valued measures.

Let F be a Banach space. We call a sequence $\{Q_n\}$, $Q_n \in F$ *precompact* if every sequence $\{q_n\}$, $q_n \in Q_n$ has a convergent subsequence. A closed set $Z \subseteq F$ is *T–full at $\bar{x} \in Z$* if, for every sequence $\{x^n\} \subseteq Z$ such that $x^n \to \bar{x}$ there exists $\rho > 0$ and a precompact sequence $\{Q_n\}, Q_n \subseteq F$ such that the sets $T_Z(x^n) \cap B(0,\rho) + Q_n$ contain a common ball $B(0,\varepsilon)$ for n_0 large enough ($T_Z(x)$ is the Clarke tangent cone to Z at x). If the condition above is satisfied with $Q_n = \{0\}$, Z is *strongly T–full* at $\bar{x}$. Finally, Z is *T–full* (resp. *strongly T–full*) if it is T–full (resp. strongly T– full) at every $\bar{x} \in Z$. Examples of strongly T–full sets are closed convex sets with nonempty interior; for other examples see [6].

Given a sequence $\{Z_n\}$ of subsets of F, $\liminf_{n\to\infty} Z_n$ is the set of all $z = \lim_{n\to\infty} z_n$ with $z_n \in Z_n$. If $Z \subseteq F$, $Z^- \subseteq F^*$ is the *polar cone* of all $z^* \in F^*$ with $\langle z^*, z\rangle \le 0$ $(z \in Z)$. Finally, we define

$$\mathbf{M}(\bar{t}) = \{y(\cdot) \in C(0,\bar{t};E_\alpha);\ y(t) \in M\ (0 \le t \le \bar{t})\}.$$

and assume that $\mathbf{M}(\bar{t})$ is strongly T–full in $C(0,\bar{t};E_\alpha)$ and that Y is strongly T–full in E_α.

Theorem 3.1 below for (2.1) gives necessary conditions for optimality of a control $\bar{u}(\cdot)$. We assume that the set of Lebesgue points of all functions $f_0(\cdot, y(\cdot,\bar{u}), v)$ $(v \in U)$ has full measure in $[0,T]$.

Theorem 3.1. *Let $\bar{u}(\cdot) \in C_{ad}(0,\bar{t};U)$ be an optimal control in $0 \le t \le \bar{t}$. Then there exists a double sequence $\{(\tilde{y}_m^n(\cdot), \tilde{y}_m^n)\} \subseteq \mathbf{M}(\bar{t}) \times Y \subseteq C(0,\bar{t};E_\alpha) \times E_\alpha$ such that $(\tilde{y}_m^n(\cdot), \tilde{y}_m^n) \to (y(\cdot,\bar{u}), y(\bar{t},\bar{u}))$ as $n \to \infty$ for $m = 1, 2, \dots$ and a multiplier $(z_0, \mu, z) \in \mathbb{R} \times \Sigma(0,\bar{t};(E_\alpha)^*) \times (E_\alpha)^*$, $(z_0,\mu,z) \ne 0$, satisfying*

$$z_0 \ge 0,\ \mu \in \Big(\liminf_{m\to\infty}\ \liminf_{n\to\infty} T_{\mathbf{M}(\bar{t})}(\tilde{y}_m^n(\cdot))\Big)^-,\ z \in \Big(\liminf_{m\to\infty}\ \liminf_{n\to\infty} T_Y(\tilde{y}_m^n)\Big)^-, \tag{3.3}$$

and such that

$$z_0\{f_0(s,y(s,\bar{u}),v) - f_0(s,y(s,\bar{u}),\bar{u}(s))\} + \langle B^*\bar{z}(s), v - \bar{u}(s)\rangle \ge 0\ (v \in U) \tag{3.4}$$

a.e. in $0 \le t \le \bar{t}$, where $\bar{z}(s)$ is the solution of

$$\begin{aligned} d\bar{z}(s) = &- \{A^* + \partial_y f(s,y(s,\bar{u}))^*\}\bar{z}(s)ds \\ &- z_0\partial_y f_0(s,y(s,\bar{u}),\bar{u}(s))ds - \mu(ds), \qquad z(\bar{t}) = z. \end{aligned} \tag{3.5}$$

By definition, $\bar{z}(s) = (-A^*)^{-\rho}\bar{v}(s)$, where

$$\bar{v}(s) = R_{\alpha,0}(\bar{t}, s)^*((-A)^{-\alpha})^* z + \int_s^t R_{\alpha,0}(\sigma, s)^*((-A)^{-\alpha})^*\mu(d\sigma) = \bar{v}_h(s) + \bar{v}_i(s), \tag{3.6}$$

and the operator $R_{\alpha,0}(t,s)$ is defined by the integral equation

$$R_{\alpha,0}(t,s)\zeta = (-A)^\alpha S(t-s)\zeta + \int_s^t (-A)^\alpha S(t-\tau)\partial_y f(\tau, y(\tau,\bar{u}))(-A)^{-\alpha}R_{\alpha,0}(\tau,s)\zeta d\tau. \tag{3.7}$$

(the operator $(-A)^{-\alpha}R_{\alpha,0}(t,s)$ is the *propagator* or *solution operator* of the variational equation $\xi'(t) = \{A + \partial_y f(t, y(t,\bar{u}))\}\xi(t))$. Finally, $\bar{v}_i(t)$ in (3.6) is understood as follows: for each s, $\bar{v}_i(s)$ is the only element of E^* satisfying

$$\langle y, \bar{v}_i(s)\rangle = \int_s^{\bar{t}} \langle R_{\alpha,0}(\sigma,s)y, ((-A)^{-\alpha})^*\mu(d\sigma)\rangle \quad (y \in E). \tag{3.8}$$

Using the integral equation (3.7) we obtain

$$|R_{\alpha,0}(t,s)\|_{(E,E)} \le C(t-s)^{-\alpha} \quad (0 \le s < t \le \bar{t}) \tag{3.9}$$

so that

$$\|\bar{v}_i(s)\|_E \le C\omega_\alpha(s) = C\int_s^{\bar{t}} (\bar{t}-\sigma)^{-\alpha}\|\mu\|(d\sigma) \quad (0 \le s < t \le \bar{t}). \tag{3.10}$$

Integrating in $0 \le s \le \bar{t}$ and using Tonelli's theorem we deduce that $\omega_\alpha(\cdot) \in L^1(0,\bar{t})$. It follows from its definition that $\bar{v}(\cdot)$ is E–weakly measurable, thus is strongly measurable by separability.

Here is the strategy for the proof. Let

$$C_{ad}(0,T;U,\bar{u})_m = \{u(\cdot) \in C_{ad}(0,T;U); \|u(\sigma) - \bar{u}(\sigma)\| \le m\} \tag{3.11}$$

for $m = 1, 2, \ldots$. This subspace is complete in the distance (3.1); moreover it is *patch complete* in the sense of [4], [5]: if $u(\cdot), v(\cdot) \in C_{ad}(0,T;U,\bar{u})_m$ and $e \subseteq [0,T]$ is measurable, the control equal to $u(t)$ in e and to $v(t)$ outside of e belongs to $C_{ad}(0,T;U,\bar{u})_m$. Finally, it follows from (2.4) and (2.8) that

$$\|f(t,y)\| + \|Bu(t)\| \le K(c) + C(\|\bar{u}(t)\| + m), \quad \|\partial_y f(t,y)\| \le L(c) \tag{3.12}$$

$$|f_0(t,y,u(t))| \le K_0(c)\|\bar{u}(t) + m\|^p, \ |\partial_y f_0(t,y,u(t))| \le L_0(c)\|\bar{u}(t) + m\|^p \tag{3.13}$$

all bounds valid for $0 \le t \le T$, $\|y\|_{E_\alpha} \le c$ independently of $u(\cdot) \in C_{ad}(0,T;U,\bar{u})_m$. Taking into account that if $\bar{u}(\cdot)$ is optimal in $C_{ad}(0,\bar{t};U)$ it is optimal in any subspace that hosts it, in particular in $C_{ad}(0,\bar{t};U,\bar{u})_m$, we can apply [5, Theorem 6.2] and deduce that for each m there exists a sequence $\{(\tilde{y}^n_m(\cdot), \tilde{y}^n_m)\} \subseteq \mathbf{M}(t) \times Y$ with $(\tilde{y}^n_m(\cdot), \tilde{y}^n_m) \to (y(\cdot,\bar{u}), y(\bar{t},\bar{u}))$ as $n \to \infty$ and a multiplier $(z_{0m}, \mu_m, z_m) \in \mathbb{R} \times \Sigma(0,\bar{t};(E_\alpha)^*) \times (E_\alpha)^*$, $(z_{0m}, \mu_m, z_m) \ne 0$ with

$$z_{0m} \ge 0, \quad \mu_m \in \left(\liminf_{n\to\infty} T_{\mathbf{M}(\bar{t})}(\tilde{y}^n_m(\cdot))\right)^-, \quad z_m \in \left(\liminf_{n\to\infty} T_Y(\tilde{y}^n_m)\right)^-, \tag{3.14}$$

and such that

$$
\begin{aligned}
(3.15)\qquad & z_{0m}\int_0^{\bar t}\Big\{f_0(\sigma,y(\sigma,\bar u),v(\sigma)) - f_0(\sigma,y(\sigma,\bar u),\bar u(s))\Big\}d\sigma \\
&\qquad + \int_0^{\bar t}\Big\langle B^*\bar z_m(\sigma), v(\sigma)-\bar u(\sigma)\Big\rangle d\sigma \ge 0
\end{aligned}
$$

for every $v(\cdot)\in C_{ad}(0,\bar t;U;\bar u)_m$, [†] where $\bar z_m(s)$ is the solution of

$$
\begin{aligned}
(3.16)\qquad d\bar z_m(s) &= -\{A^* + \partial_y f(s,y(s,\bar u))^*\}\bar z_m(s)ds \\
&\quad - z_0\partial_y f_0(s,y(s,\bar u),\bar u(s))ds - \mu_m(ds), \quad z(\bar t) = z_m.
\end{aligned}
$$

It is then clear that, in order to prove Theorem 3.1 it is enough to show: (a) if necessary passing to a subsequence

$$
(3.17)\qquad (z_{0m},\mu_m,z_m)\to(z_0,\mu,z)\neq 0,
$$

$(\mathbb{R}\times C(0,\bar t;E_\alpha)\times E_\alpha)$–weakly in $\mathbb{R}\times\Sigma(0,\bar t;(E_\alpha)^*)\times(E_\alpha)^*$, and (b) If $\bar z(s)$ (resp. $\bar z_m(s)$) is the solution of (3.5) (resp. (3.16)) then

$$
(3.18)\qquad \bar z_m(\cdot)\to\bar z(\cdot)
$$

$L^\infty(0,\bar t;E)$–weakly in $L^1(0,\bar t;E^*)$. In fact, if both (3.17) and (3.18) are satisfied, we take limits in (3.15) and obtain

$$
\begin{aligned}
(3.19)\qquad & z_0\int_0^{\bar t}\Big\{f_0(\sigma,y(\sigma,\bar u),v(\sigma)) - f_0(\sigma,y(\sigma,\bar u),\bar u(s))\Big\}d\sigma \\
&\qquad + \int_0^{\bar t}\Big\langle B^*\bar z(\sigma), v(\sigma)-\bar u(\sigma)\Big\rangle d\sigma \ge 0
\end{aligned}
$$

for all $v(\cdot)\in C_{ad}(0,\bar t;U)$ with $v(\sigma)-\bar u(\sigma)$ bounded, and here is how we obtain the pointwise version (3.4) of the maximum principle. Let $m=1,2,\dots$, and define $d_m=\{\sigma\in[0,\bar t\,]; m<\|\bar u(\sigma)\|\le m+1\}$, χ_m the characteristic function of d_m, $\bar u_m(\sigma)=\chi_m(\sigma)\bar u(\sigma)$, e_m the set of all left Lebesgue points of both functions $\chi_m(\sigma)$ and $\bar u_m(\sigma)$ in $[0,\bar t\,]$; e_m has full measure in $[0,\bar t\,]$, so that $e_m\cap d_m$ has full measure in d_m. If $s\in e_m\cap d_m$ then there exists a set $e_m(h)\subseteq[s-h,s]\cap d_m$ such that

$$
(3.20)\qquad \frac{1}{h}|e_m(h)|\to 1,\qquad \frac{1}{h}\int_{e_m(h)}\|\bar u(\sigma)-\bar u(s)\|d\sigma\to 0\quad\text{as } h\to 0+.
$$

The set $\cup(e_m\cap d_m)$ is total in $[0,\bar t\,]$, thus so is its intersection e with the set of left Lebesgue points of the following functions: (i) $\bar z(\cdot)$, (ii) $f_0(\cdot,y(\cdot,\bar u),\bar u(\cdot))$, (iii) all functions $f_0(\cdot,y(\cdot,\bar u),v)$. Take $s\in e$ and $v\in U$, find m so that $s\in e_m\cap d_m$ and stick in (3.19) the function $v(\sigma)=v\chi_m(\sigma)+(1-\chi_m(\sigma))\bar u(\sigma)$. Then $v(\sigma)-\bar u(\sigma)=v\chi_m(\sigma)-\chi_m(\sigma)\bar u(\sigma)$ is bounded. We take limits using (3.20) and the diverse assumptions on Lebesgue points, and obtain (3.4).

[†]Strictly speaking, Theorem 6.2 is proved in [5] under the assumption that the bounds in (3.12) and (3.13) are uniform, that is, that the functions on the left are bounded by constants independent of the controls. The proof extends almost without changes to this more general situation due to the fact that (3.12) and (3.13) are independent of the particular control $u(\cdot)\in C_{ad}(0,T;U,\bar u)_m$. Boundedness of $v(\sigma)-\bar u(\sigma)$ makes it possible the interchange in the order of integration at the end of Theorem 6.2.

Back to (3.17) and (3.18). The required convergence in (3.17) can be attained invoking Alaoglu's Theorem. To show that the limit is nonzero note that, since $(z_{0m}, \mu_m, z_m) \neq 0$ we may assume that

$$|z_{0m}|^2 + \|\mu_m\|^2_{\Sigma(0,\bar{t};(E_\alpha)^*)} + \|z_m\|^2_{(E_\alpha)^*} = 1. \tag{3.21}$$

If $z_{0m} \to z_0 \neq 0$ there is nothing to prove. If $z_0 = 0$ we may erase z_{0m} from the left side of (3.21) replacing " $=$ " by " $\to$ ". We obtain from (3.14) that

$$\langle \mu_m, -y_m(\cdot)\rangle_{\alpha,c} + \langle z_m, -y_m\rangle_\alpha \geq 0 \tag{3.22}$$

for $(y_m(\cdot), y_m) \in \Delta_m = \mathcal{M}_m \times \mathcal{Y}_m$, where

$$\mathcal{M}_m = \liminf_{n\to\infty} T_{\mathbf{M}(\bar{t})}(\tilde{y}^n_m(\cdot)) \subseteq C(0,\bar{t};E_\alpha), \quad \mathcal{Y}_m = \liminf_{n\to\infty} T_Y(\tilde{y}^n_m) \subseteq E_\alpha, \tag{3.23}$$

where $\langle\,\cdot\,,\,\cdot\,\rangle_\alpha$ indicates the duality of E_α and $(E^*)_{-\alpha}$ and $\langle\,\cdot\,,\,\cdot\,\rangle_{c,\alpha}$ the duality of $C(0,\bar{t};E_\alpha)$ and $\Sigma(0,\bar{t};(E^*)_{-\alpha})$. The assumptions on $\mathbf{M}(\bar{t})$ and Y and [6, Lemma 3.6] guarantee that the $\mathcal{M}_m \times \mathcal{Y}_m$ contain a common ball in E_α for m large enough, so the fact that $(\mu, z) \neq 0$ follows from the result below ([2, Lemma 2.5]).

Lemma 3.2. *Let F be a Banach space, $\{\Delta_m\}$ a sequence of subsets of F, $\{z_m\}$ a sequence in F^* such that*

$$0 < c \leq \|z_m\|_{F^*} \leq C < \infty, \qquad \langle z_m, y\rangle \geq -\varepsilon_m \to 0 \quad (y \in \Delta_m).$$

Assume there exists a precompact sequence $\{Q_m\}$ in F such that the sets $\overline{conv}(\Delta_m) + Q_m$ contain a common ball. ($\overline{conv}$ = closed convex hull). Then every F−weakly convergent subsequence of $\{z_m\}$ has a nonzero limit.

It only remains to show (3.18). Writing $\bar{v}_m(s) = \bar{v}_{mh}(s) + \bar{v}_{mi}(s)$ as in (3.6), the statement is obvious for the homogeneous parts; in fact,

$$\bar{v}_{mh}(s) \to v_h(s) \quad 0 \leq s \leq \bar{t},$$

$E-$ weakly in E^*, so that (3.18) follows from the dominated convergence theorem. To show (3.18) for $\bar{v}_{mi}(s)$ we use (3.8). Let $\nu_m(d\sigma) = ((-A)^{-\alpha})^*\mu_m(d\sigma)$, $\nu(d\sigma) = ((-A)^{-\alpha})^*\mu(d\sigma)$. Then $\nu_m \to \nu$ $C(0,\bar{t};E)-$weakly in $\Sigma(0,\bar{t};E^*)$. Given $n = 1, 2, \ldots$ define an operator $R_{\alpha,\rho,n}(t,s)$ in the square $[0,\bar{t}\,] \times [0,\bar{t}\,]$ by $R_{\alpha,\rho,n}(t,s) = R_{\alpha,\rho}(t,s)$ in $t - s \geq 1/n$, $R_{\alpha,\rho,n}(t,s) = 0$ in $t \leq s$, and extend $R_{\alpha,\rho,n}(t,s)$ linearly to $0 \leq t - s \leq 1/n$, so that it shares the bound (3.9). Define $\bar{v}^n_i(s)$ with $\nu(ds)$ in the same way as $\bar{v}_i(s)$ but using $R_{\alpha,\rho,n}(t,s)$ instead of $R_{\alpha,\rho}(t,s)$; likewise, define $\bar{v}^n_{mi}(s)$ using $\nu_m(ds)$ and $R_{\alpha,\rho,n}(t,s)$. We have

$$\|\bar{v}_i(s) - \bar{v}^n_i(s)\|_{E^*} \leq C\rho_n(s) = C\int_s^{\min(\bar{t},s+1/n)} (\sigma - s)^{-\alpha}\|\nu\|(d\sigma)\,. \tag{3.24}$$

By Tonelli's theorem,

$$\int_0^{\bar{t}} \rho_n(s)ds \leq \int_0^{\bar{t}} \|\nu\|(d\sigma)\int_{\sigma-1/n}^{\sigma} (\sigma - s)^{-\alpha}ds = \frac{1}{1-\alpha}\left(\frac{1}{n}\right)^{1-\alpha}\int_0^{\bar{t}} \|\nu\|(d\sigma)\,, \tag{3.25}$$

so that $\|\rho_n(\cdot)\|_{L^1(0,\bar{t})} \to 0$ as $n \to \infty$. Since $\{\nu_m(ds)\}$ is bounded in $\Sigma(0,\bar{t};E^*)$ we obtain in the same way that

$$\|\bar{v}_{mi}(s) - \bar{v}^n_{mi}(s)\|_{E^*} \leq C\rho_{mn}(s) \tag{3.26}$$

for each m, with $\|\rho_{mn}(\cdot)\|_{L^1(0,\bar{t})} \to 0$ as $n \to \infty$, uniformly with respect to m. Pick $f(\cdot) \in L^\infty(0,\bar{t};E)$ and write

$$\begin{aligned}(3.27)\quad &\langle v_i(s) - \bar{v}_{mi}(s), f(s)\rangle \\ &= \langle \bar{v}_i(s) - \bar{v}_i^n(s), f(s)\rangle + \langle \bar{v}_i^n(s) - \bar{v}_{mi}^n(s), f(s)\rangle + \langle \bar{v}_{mi}^n(s) - \bar{v}_{mi}(s), f(s)\rangle\ .\end{aligned}$$

We have that $|\langle \bar{v}_i(s) - \bar{v}_i^n(s), f(s)\rangle| \le \|f\|_{L^\infty(0,\bar{t};E)}\,\rho_n(s)$ with a similar estimate for $|\langle \bar{v}_{mi}^n(s) - \bar{v}_{mi}(s), f(s)\rangle|$, thus the first and third terms on the right of (3.27) are disposed of. For the middle term we note that

$$\int_0^{\bar{t}} \Big\langle \bar{v}_i^n(s) - \bar{v}_{mi}^n(s), f(s)\Big\rangle ds = \int_0^{\bar{t}} \left\langle \int_0^{\sigma} R_{\alpha,0,n}(\sigma,s)f(s)ds, (\nu - \nu_m)(d\sigma)\right\rangle ,$$

hence it is enough to show that the function in the left side of the angled brackets inside the integral on the right is continuous. This is obvious and left to the reader.

The first relation (3.3) is plain; the third relation follows taking limits in $\langle z_m, y_m\rangle \le 0$ $(y_m \in T_Y(\tilde{y}_m^n))$, and the second results in the same way.

4. The point target case

There is a version of the maximum principle for (2.1) [5, Theorem 9.1] that covers point targets. It requires that $f : [0,T] \times E_\alpha \to E_\delta$ $(\alpha, \delta > 0)$ and that $(-A)^\delta f(t,y)$ satisfy Hypothesis $D_{\alpha,0}$. The proof requires again $M \subseteq E_\alpha$ closed and $\mathbf{M}(\bar{t})$ $T-$full in $C(0,\bar{t};E_\alpha)$, but we only need Y to be a closed subset of $E_1 = D(A)$, thus, in particular we may handle the point target condition $y(\bar{t},u) = \bar{y} \in D(A)$. The control space in the present application is $C(0,\bar{t};U,\bar{u})_m$, and the multiplier (z_{0m}, μ_m, z_m) belongs to $\mathbb{R}\times\Sigma(0,\bar{t};(E_\alpha)^*)\times(E^*)_{-1}^1$, where $(E^*)_{-1}^1$ is the subspace of $(E^*)_{-1} = (E_1)^*$ determined by the condition

$$(4.1)\qquad \int_0^1 \|B^*S(t)^*z\|_X dt < \infty\ ,$$

and we have

$$(4.2)\qquad z_{0m} \ge 0,\quad \mu_m \in \Big(\liminf_{n\to\infty} T_{\mathbf{M}(\bar{t})}(\tilde{y}_m^n(\cdot))\Big)^-,\quad z_m \in \Big(\liminf_{n\to\infty} T_Y(\tilde{y}_m^n)\Big)^-,$$

where, for each m, $\{(\tilde{y}_m^n(\cdot), \tilde{y}_m^n)\} \subseteq \mathbf{M}(t) \times Y$ with $(\tilde{y}_m^n(\cdot), \tilde{y}_m^n) \to (y(\cdot,\bar{u}), y(\bar{t},\bar{u}))$ as $n \to \infty$. See [6] on extension of the adjoint semigroup $S(t)^*$ to the spaces $(E^*)_{-1}$ and other details.[‡] The costate is defined as in (3.6) but the homogeneous term is $\bar{v}_{mh}(s) = ((-A)^{1-\alpha}R_{\alpha,0}(\bar{t},s;\bar{u}))^*((-A)^{-1})^*z_m$. The minimum principle in integral form is again (3.15),

$$\begin{aligned}(4.3)\quad z_{0m}\int_0^{\bar{t}} &\Big\{f_0(\sigma,y(\sigma,\bar{u}),v(\sigma)) - f_0(\sigma,y(\sigma,\bar{u}),\bar{u}(s))\Big\}d\sigma \\ &+ \int_0^{\bar{t}} \Big\langle B^*\bar{z}_m(\sigma), v(\sigma) - \bar{u}(\sigma)\Big\rangle d\sigma \ge 0\end{aligned}$$

[‡]Again, Theorem 9.1 in [5] is proved assuming that the bounds in (3.12) and (3.13) are uniform and the controls are bounded. The proof works under (3.12) and (3.13) as long as $v(\sigma) - \bar{u}(\sigma)$ is bounded.

for every $v(\cdot) \in C_{ad}(0, \bar{t}; U; \bar{u})_m$ and

$$\int_0^{\bar{t}} S(\bar{t} - \sigma) B v(\sigma) d\sigma \in D(A) . \tag{4.4}$$

(note that, since $y(\bar{t}, \bar{u}) \in Y \subseteq D(A)$, $\bar{u}(\cdot)$ itself satisfies (4.3)). The conditions above do not guarantee that the multiplier is nonzero. A sufficient condition comes below. Given $v(\cdot) \in C_{ad}(0, \bar{t}; U)$, denote by $\xi(t, \bar{u}, v)$ the solution $\xi(t)$ of

$$\xi'(t) = \{A + \partial_y f(t, y(t, \bar{u}))\}\xi(t) + B(v(t) - \bar{u}(t)), \qquad \xi(0) = 0 \tag{4.5}$$

and assume the *reachable space* $R(\bar{t}, U; \bar{u})_m = \{\xi(\bar{t}, \bar{u}, v); v \in C(0, \bar{t}; U, \bar{u})_m\}$ satisfies

$$R(\bar{t}, U; \bar{u})_1 \supseteq B_1(\varepsilon), \tag{4.6}$$

$B_1(\varepsilon)$ a ball of radius $\varepsilon > 0$ in E_1. Then

$$(z_{0m}, \mu_m, z_m) \neq 0. \tag{4.7}$$

When controls $u(\cdot)$ are bounded, a sufficient condition for (4.6) is that U contain a ball of positive radius around some control $v(\cdot)$ satisfying (4.4). In the general case, it is enough to require that the set

$$\left\{ \int_0^{\bar{t}} S(\bar{t} - \sigma) B(v(\sigma) - \bar{u}(\sigma)) d\sigma;\ v(\cdot) \in C_{ad}(0, \bar{t}; U, \bar{u})_1 \right\} \tag{4.8}$$

contain a ball $B_1(\varepsilon) \subseteq E_1$. This does not seem like a good condition to verify since it contains the unknown optimal control, but it can be checked easily in many cases without any information on $\bar{u}$; one example is $X^* = L^\infty(\Omega)$, U defined by the condition $u(x) \geq 0$ a.e. (see Section 6).

We shall take limits in (4.3) in the same way as in (3.15), and we need to show (3.17) and (3.18). Note that, this time, $(z_{0m}, \mu_m, z_m) \in \mathbb{R} \times \Sigma(0, \bar{t}; (E^*)_{-\alpha}) \times (E^*)_{-1}$ and convergence in (3.17) is $(\mathbb{R} \times E_\alpha \times E_1)$−weak convergence.

Inequality (4.3) is equivalent to

$$z_{0m}\xi_0(\bar{t}, \bar{u}, v) + \langle \mu_m, \xi(\cdot, \bar{u}, v) \rangle_{\alpha, c} + \langle z_m, \xi(\bar{t}, \bar{u}, v) \rangle_1 \geq 0, \tag{4.9}$$

for all $v(\cdot) \in C_{ad}(0, \bar{t}; U, \bar{u})_m$ with $\xi(\bar{t}, \bar{u}, v) \in E_1$ (the latter condition equivalent to (4.4)), where

$$\begin{aligned} \xi_0(t, \bar{u}, v) &= \int_0^t \Big\langle \partial_y f_0(\tau, y(\tau, \bar{u}), \bar{u}(\tau)), \xi(\tau, \bar{u}, v) \Big\rangle_\alpha d\tau \\ &\quad + \int_0^t \Big\{ f_0(\tau, y(\tau, \bar{u}), v(\tau)) - f_0(\tau, y(\tau, \bar{u}), \bar{u}(\tau)) \Big\} d\tau; \end{aligned} \tag{4.10}$$

(see [5] for a proof of the equivalence of (4.3) and (4.9)). Since $(z_{0m}, \mu_m, z_m) \neq 0$ we may assume

$$z_{0m}^2 + \|\mu_m\|^2_{\Sigma(0, \bar{t}, (E_\alpha)^*)} + \|z_m\|^2_{(E_1)^*} = 1 \quad (m = 1, 2, \dots). \tag{4.11}$$

Select a subsequence of the sequence $\{(z_{0m}, \mu_m, z_m)\}$ such that the limit (3.17) exists $(\mathbb{R} \times E_\alpha \times E_1)$−weakly. If $z_{0m} \to z_0 \neq 0$, the limit is nonzero and there is nothing to prove. If $z_0 = 0$, we combine (4.2) with (4.9) *keeping* $v(\cdot)$ *in* $C_{ad}(0,\bar t;U,\bar u)_1 \subseteq C_{ad}(0,\bar t;U,\bar u)_m$. The result is

$$\langle \mu_m, \xi(\cdot,\bar u, v) - y_m(\cdot)\rangle_{\alpha,c} + \langle z_m, \xi(\bar t,\bar u,v) - y_m\rangle_1 \ge -\delta_m \to 0 \tag{4.12}$$

for $(y_m(\cdot), y_m) \in \mathcal{M}_m \times \mathcal{Y}_m$, $\mathcal{M}_m$ and $\mathcal{Y}_m$ given by (3.23), the second lim inf (and tangent cones) computed in the norm of E_1. The expression on the right of (4.12) is justified by the fact that $\|v(\sigma) - \bar u(\sigma)\| \le 1$, so that $\xi(t,\bar u,v)$, hence $\xi_0(t,\bar u,v)$, is bounded independently of $v(\cdot)$ (take a look at (4.5) and (4.10)). Accordingly, to apply Lemma 3.2 it is enough to show that the sets Δ_m defined by

$$(\xi(\cdot,\bar u,v) - y_m(\cdot),\ \xi(\bar t,\bar u,v) - y_m) \subseteq C(0,\bar t;E_\alpha) \times E_1 \tag{4.13}$$

(where $v(\cdot) \in C_{ad}(0,\bar t;U,\bar u)_1$ and we may take $y_m = 0$) contain a common ball in $C(0,\bar t;E_\alpha)\times E_1$. The first coordinate is covered by $y_m(\cdot)$ alone on the strength of the assumptions on $\mathbf{M}(\bar t)$ and of the fact that $\xi(\cdot,\bar u,v)$ is bounded in $C(0,\bar t;E_\alpha)$; for the second, we use (4.6). For reference, we state the final result:

Theorem 4.1. *Let* $\bar u(\cdot) \in C_{ad}(0,\bar t;U)$ *be an optimal control. Then there exists a double sequence* $\{(\tilde y_m^n(\cdot), \tilde y_m^n)\} \subseteq \mathbf{M}(t)\times Y \subseteq C(0,\bar t;E_\alpha)\times E_1$ *such that* $(\tilde y_m^n(\cdot),\tilde y_m^n) \to (y(\cdot,\bar u), y(\bar t,\bar u))$ *in* $C(0,\bar t;E_\alpha)\times E_1$ *as* $n\to\infty$ *and a multiplier* $(z_0,\mu,z) \in \mathbb{R}\times \Sigma(0,\bar t;(E^*)_{-\alpha})\times (E^*)_{-1}$, $(z_0,\mu,z)\neq 0$, *satisfying*

$$z_0 \ge 0,\ \mu \in \Big(\liminf_{m\to\infty}\ \liminf_{n\to\infty} T_{\mathbf{M}(\bar t)}(\tilde y_m^n(\cdot))\Big)^-,\ z\in\Big(\liminf_{m\to\infty}\ \liminf_{n\to\infty} T_Y(\tilde y_m^n)\Big)^- \tag{4.14}$$

and

$$z_0\{f_0(s,y(s,\bar u),v) - f_0(s,y(s,\bar u),\bar u(s))\} + \langle B^*\bar z(s), v-\bar u(s)\rangle \ge 0 \ (v\in U) \tag{4.15}$$

a. e. in $0\le t\le \bar t$, *where* $\bar z(s)$ *is the solution of* (3.5).

Remark 4.2. The integral form (3.19) of the maximum principle guarantees that $z \in (E^*)^1_{-1}$; the argument is the same in [5, Theorem 9.1].

5. Nonreflexive spaces

The setup in §4 covers reaction–diffusion equations and the Navier–Stokes equations in L^p spaces $1 < p < \infty$ (see §6); however, there is an advantage in treating parabolic equations in spaces $C(\overline{\Omega})$ of continuous functions (see §6), thus it is convenient to extend the results in §4 to nonreflexive spaces. This can be done with minor changes; we only need to assume that E is separable. The assumptions on the semigroup $S(t)$ and its infinitesimal generator A are the same and the spaces E_α are defined in the same way. On the other hand, A^* may not be a semigroup generator (or even densely defined) thus the spaces $(E^*)_\alpha$ are defined using $((-A)^\alpha)^*$ rather than $(-A^*)^\alpha$. If $D(A^*)$ is not dense in E^*, it is no longer true that $(E_\alpha)^* = (E^*)_{-\alpha}$; however, the dual is algebraically and metrically isomorphic under (3.2) to a larger space $(E_\alpha)^*$ with $(E^*)_{-\alpha} \hookrightarrow (E_\alpha)^* \hookrightarrow (E^*)_{-(\alpha+\varepsilon)}$ for all $\varepsilon > 0$. Some of the functions of the dual E^* are taken over by the *Phillips dual* $E^\odot$ = closure of $D(A^*)$ in E^*; the adjoint semigroup $S(t)^*$, restricted to $E^\odot$ is called $S^\odot(t)$ and is strongly continuous; the space $E^\odot$ is maximal with respect to this property. We name $A^\odot$ the infinitesimal generator of $S^\odot(t)$. The same considerations applied to the semigroup $S^\odot(t)$ produce

$(E^{\odot})^{\odot}$, $(A^{\odot})^{\odot}$ and $(S(t)^{\odot})^{\odot}$. We assume that E is $\odot-$*reflexive* in the sense that $(E^{\odot})^{\odot} = E$; this implies that $(A^{\odot})^{\odot} = A$ and $(S(t)^{\odot})^{\odot} = S(t)$. Hypothesis $D_{\alpha,\rho}$ is formulated in the same way, but in the case $\rho = 0$, $f(t, y)$ is allowed to take values in $(E^{\odot})^* \supseteq E$.

The statement and proof of the minimum principle are the same; the multiplier (z_0, μ, z) is in $\mathbb{R} \times \Sigma(0, \bar{t}; (E_\alpha)^*) \times (E_\alpha)^*$ $(\mathbb{R} \times \Sigma(0, \bar{t}; (E_\alpha)^*) \times (E_1)^*$ for point targets). We still have $((-A)^{-\alpha})^*(E_\alpha)^* = E^*$, so that the costate $\bar{z}(s)$ is defined in the same way as in the reflexive case; due to the smoothing properties of the semigroup, $\bar{z}(s)$ takes values in $E^{\odot}$ rather than E^*.

6. Applications

Theorems 3.1 and 4.1 for reflexive separable E can be applied to the controlled Navier–Stokes equations

$$\mathbf{y}'(t) = A_p\mathbf{y}(t) + P_p\Big((\mathbf{y} \cdot \nabla)\mathbf{y} + I_p\mathbf{u}(t)\Big) . \tag{6.1}$$

The space is $E = X^p(\Omega)^m$ = closure in $L^p(\Omega)^m$ of all divergence-free $m-$vectors $\mathbf{y}(x) = (y_1(x), \ldots, y_m(x))$; P_p is the projection of $L^p(\Omega)^m$ into $X^p(\Omega)^m$ and A_p is the Stokes operator $A_p = P_p\Delta_p$, Δ_p the m-vector Laplacian in $L^p(\Omega)^m$. The control space is a (possibly unbounded) subset of $L^r(0, T; L^r(\Omega)^m)$, $p \le r$ (for $r = \infty$ we take $L^\infty_w(0, T; L^\infty(\Omega)^m)$) and I_p is the embedding operator from $L^r(\Omega)^m$ into $L^p(\Omega)^m$. Hypothesis D^0_α is satisfied for $\alpha < 1$ if $p > m$ and $1/2 + m/2p < \alpha$; in Theorem 3.1 one may take $r \ge p$ arbitrary, in particular $r = \infty$. On the other hand, the assumptions in Theorem 4.1 hold for p large enough if $1/2 < \alpha < 1$, $\delta < \alpha - 1/2$); in this case, $r = p$. The treatment admits state constraints of the form

$$\mathcal{S}\mathbf{y}(t, x) \in \mathcal{M}_S \subseteq \mathbb{R}^k \quad (0 \le t \le \bar{t}, x \in \Omega), \tag{6.2}$$

$(\mathcal{S}\mathbf{y}(x) = \Sigma\Sigma\eta_{jk}(x)\partial_j y_k(x) + \Sigma\eta_j(x)y_j(x)$ a first order differential operator with $k-$vector coefficients in $C(\overline{\Omega})^k)$ and target conditions of the same type. Nonlinearities more general than the one in (6.1) are tractable. See [7] for details.

The main application in nonreflexive spaces is to uniformly elliptic partial differential operators $A(\beta)$ coupled with a boundary condition β in a bounded domain $\Omega \in \mathbb{R}^m$ with boundary Γ. Here, $E = C(\overline{\Omega}) = \{$all continuous functions in $\overline{\Omega}\}$ equipped with the supremum norm; for the Dirichlet boundary condition the space is the subspace $C_0(\overline{\Omega})$ of $C(\overline{\Omega})$ determined by $y(x) = 0$ $(x \in \Gamma)$. Assuming smoothness of the coefficients, the domain and the nonlinearity $f(t, y)(x) = \phi(t, x, y(x), \nabla y(x))$ all results apply with $\alpha > 1/2$, and state and target constraints of the form (6.2) are tractable. See [5] for further details.

References

1. J. DIESTEL AND J. J. UHL, **Vector Measures,** Amer. Math. Soc., Providence (1977).
2. H. O. FATTORINI, A unified theory of necessary conditions for nonlinear nonconvex control systems, *Applied Math & Optimization* **15** (1987) 141–185
3. H. O. FATTORINI, Optimal control problems in Banach spaces, *Appl. Math. & Optimization* **28** (1993) 225–257
4. H. O. FATTORINI, Optimal control problems with state constraints for semilinear distributed parameter systems, *J. Optimization Theory & Applications* **34** (1996) 132–156

5. H. O. FATTORINI, Optimal control problems with state constraints for distributed parameter systems: the parabolic case, to appear.
6. H. O. FATTORINI, Nonlinear infinite dimensional optimal control problems with state constraints and unbounded control sets, to appear
7. H. O. FATTORINI and S. S. SRITHARAN, Optimal control problems with state constraints in fluid mechanics and combustion, to appear.
8. E. J. McSHANE, The calculus of variations from the beginning through optimal control theory, **Optimal Control and Differential Equations** (A. B. Schwartzkopf, W. B. Kelley and S. B. Eliason, eds.) Abridged version in *SIAM J. Control & Optimization* **27** (1989) 446–455
9. J. P. RAYMOND and H. ZIDANI, Pontryagin's principle for state-constrained control problems governed by parabolic equations with unbounded controls, to appear
10. J. P. RAYMOND and H. ZIDANI, Hamiltonian Pontryagin's principles for control problems governed by semilinear parabolic equations, to appear.

H.O. Fattorini
Department of Mathematics
University of California
Los Angeles, California 90024, USA

International Series of Numerical Mathematics
Vol. 126, © 1998 Birkhäuser Verlag, Basel

Remarks on the Controllability of Some Stochastic Partial Differential Equations

E. FERNÁNDEZ-CARA AND J. REAL
Department of Differential Equations
and Numerical Analysis
University of Sevilla

ABSTRACT. In this paper, we analyze the approximate controllability and the exact-to-zero controllability in quadratic mean of systems governed by stochastic partial differential equations of a particular kind. We obtain several results which are similar to those known for similar deterministic systems.

1991 *Mathematics Subject Classification.* 93B05, 93E99, 35B37, 60H15

Key words and phrases. Approximate controllability, null controllability, stochastic linear partial differential equations.

1. Introduction

During the last years, controllability for deterministic distributed parameter systems has been intensively studied. In particular, it is known that the heat equation and the Stokes system with control concentrated in an arbitrary subdomain are approximately controllable (see [11]; see also [5]).

However, the analysis of the controllability for stochastic partial differential equations seems to remain an almost open field of research. The unique works we know on the subject are [3] and [15].

In this paper, we present some approximate and exact-to-zero controllability results for a class of linear stochastic partial differential systems. This includes, as a particular case, a stochastic heat equation of the form

$$\begin{cases} \partial_t y - \Delta y = 1_{\mathcal{O}} v + B(t)\, \dot{w}_t \quad \text{in} \quad Q = D \times (0,T) \quad P-\text{a.s.}, \\ y = 0 \quad \text{on} \quad \Sigma = \partial D \times (0,T) \quad P-\text{a.s.}, \\ y(0) = y_0 \quad \text{in} \quad D \quad P-\text{a.s.}, \end{cases}$$

where the control is v. Here, $\mathcal{O}$ and D are bounded open sets in $\mathbb{R}^N$ with $\mathcal{O} \subset D$, $1_{\mathcal{O}}$ is the characteristic function of the set $\mathcal{O}$ and $\dot{w}_t = \partial_t w_t$ is a Gaussian random field, white noise in time.

Roughly speaking, we are going to prove that, for general y_0, y_d and B, one can obtain final states $y(T)$ arbitrarily close to y_d in quadratic mean by choosing v appropriately (an approximate controllability result). We will also prove that, if B is not random and in some sense small, then one can also choose v such that $y(T) = 0$ (a null-controllability result).

Partially supported by D.G.I.C.Y.T. (Spain), Proyecto PB95–1242.

We will also study questions of the same kind for Stokes and (more generally) quasi-Stokes stochastic systems

$$
(1.1)\qquad \begin{cases} \partial_t y_j - \Delta y_j - \partial_i\left(a_i(x)y_j\right) + c_{ij}(x)y_i + \partial_j \Pi = 1_{\mathcal{O}} v_j + B_j(t)\,\dot{w}_t, \\ \partial_j y_j = 0, \\ y_j = 0 \quad \text{on} \quad \Sigma, \\ y_j(0) = y_{0j} \quad \text{in} \quad D \end{cases}
$$

$(1 \leq j \leq N)$, where the usual summation convention is assumed.

This work is a continuation of [15]. An extended version, where other more general problems will be considered, will appear in the next future.

2. Approximate controllability results

Assume a bounded and connected open set $D \subset \mathbb{R}^N$ with regular boundary ∂D, a nonempty open subset $\mathcal{O} \subset D$, a positive number T and a complete probability space $\{\Omega,\, \mathcal{F},\, P\}$ are given.

We will use the following notation: $H = L^2(D)$, $V = H_0^1(D)$, $|.|$ and $(.,.)$ are resp. the usual norm and scalar product in H. If X is a Banach space and $f \in L^1(\Omega, \mathcal{F}; X)$, we denote by Ef the expectation of f:

$$
Ef = \int_\Omega f(\omega)\, dP(\omega).
$$

Also, assume that a separable Hilbert space K and a Wiener process w_t on $\{\Omega,\mathcal{F},P\}$ with values in K are given. This means that

$$
w_t = \sum_{k=1}^{\infty} \beta_t^k e_k \quad \forall\, t \geq 0,
$$

where the β_t^k are mutually independent real Wiener processes satisfying

$$
(2.1)\qquad E\left(\beta_t^k\right)^2 = \lambda_k t, \qquad \sum_{k=1}^{\infty} \lambda_k < +\infty
$$

and $\{e_k\}$ is an orthonormal basis in K (for the definition of a real Wiener process, see for example [1]). Notice that, in particular, w_t has Hölder-continuous sample paths.

In the sequel, we put $\mathcal{F}_t = \sigma(\, w_s,\ 0 \leq s \leq t\,)$ (the σ-algebra spanned by w_s for $0 \leq s \leq t$). Obviously, $\{\mathcal{F}_t\}$ is an increasing family of sub σ-algebras of $\mathcal{F}$ and, among other things, one has:

$$
(2.2)\qquad \mathcal{F}_t = \sigma\left(\bigcup_{s<t} \mathcal{F}_s\right) \quad \forall\, t > 0.
$$

For any $f \in L^1(\Omega, \mathcal{F}; H)$, we denote by $E[f|\mathcal{F}_t]$ the conditional expectation of f with respect to $\mathcal{F}_t$, i.e. the unique element in $L^1(\Omega, \mathcal{F}_t; H)$ such that

$$
\int_A E[f|\mathcal{F}_t]\, dP = \int_A f\, dP \quad \forall\, A \in \mathcal{F}_t
$$

(cf. [14] for the main properties of the conditional expectation).

Let X be a Banach space. We denote by $I^2(0,T;\,X)$ the space formed by all stochastic processes $\Phi \in L^2(\Omega \times (0,T), dP \otimes dt;\, X)$ which are $\mathcal{F}_t$-adapted a.e. in $(0,T)$, i.e. such that

$$\Phi(t) \text{ is } \mathcal{F}_t\text{-measurable for almost all } t \in (0,T)$$

(in the case $X = \mathcal{L}(K;\, H)$, measurability will mean strong measurability). Then, $I^2(0,T;\,X)$ is a closed subspace of $L^2(\Omega \times (0,T), dP \otimes dt;\, X)$.

Assume that a stochastic process B is given, with

$$B \in I^2(0,T;\, \mathcal{L}(K;\, H)). \tag{2.3}$$

Then the stochastic integral of B with respect to w_t is defined by the formula

$$\int_0^t B(s)\, dw_s = \sum_{k=1}^{\infty} \int_0^t B(s)e_k\, d\beta_s^k \qquad \forall\, t \in [0,T].$$

Here, the convergence of the series is understood in the sense of $L^2(\Omega, \mathcal{F}_t; H)$. The stochastic integrals in the right side are defined by the equalities

$$(\int_0^t B(s)e_k\, d\beta_s^k, h) = \int_0^t (B(s)e_k, h)\, d\beta_s^k \qquad \forall\, h \in H,$$

where the latter are usual Ito stochastic integrals with respect to the real-valued processes β_t^k (see [1]).

Assume we are given an arbitrary but fixed initial state

$$y_0 \in H \tag{2.4}$$

and set $A = \Delta$ (the usual Laplace operator). For each $v \in I^2(0,T;\, H)$, there exists exactly one solution y_v to the problem

$$\begin{cases} y_v \in I^2(0,T;\, V) \cap L^2(\Omega;\, C^0([0,T];\, H)), \\ y_v(t) = y_0 + \int_0^t \{Ay_v(s) + 1_{\mathcal{O}} v(s)\}\, ds + \int_0^t B(s)\, dw_s \quad \forall\, t \in [0,T]. \end{cases} \tag{2.5}$$

In (2.5), the equalities hold $P-$a.s in V'. Let $S(t)$ be the semigroup generated in H by A, with domain $D(A) = \{h \in V;\, Ah \in H\}$. Then

$$\begin{cases} y_v(t) = S(t)y_0 + \int_0^t S(t-s)(1_{\mathcal{O}} v(s))\, ds + \int_0^t S(t-s)B(s)\, dw_s \\ \forall\, t \in [0,T] \end{cases} \tag{2.6}$$

(see [2], [13]). Our first result is the following:

Theorem 2.1. *The linear manifold $Y_T = \{\, y_v(T)\,;\, v \in I^2(0,T;\, H)\,\}$ is dense in the space $L^2(\Omega, \mathcal{F}_T; H)$.*

Proof. Using (2.6), it suffices to check that, if $f \in L^2(\Omega, \mathcal{F}_T; H)$ and

$$E(\int_0^T S(T-s)(1_{\mathcal{O}} v(s))\, ds,\, f) = 0 \quad \forall\, v \in I^2(0,T;\, H), \tag{2.7}$$

then necessarily $f = 0$. Let f be a function in $L^2(\Omega, \mathcal{F}_T; H)$ satisfying (2.7) and assume that $\varphi \in I^2(0,T;H)$ is given by

$$\begin{cases} -\partial_t\varphi - A\varphi = 0 \quad \text{in} \quad Q, \\ \varphi = 0 \quad \text{on} \quad \Sigma, \\ \varphi(T) = f, \end{cases}$$

i.e. $\varphi(t) = S(T-t)f$. It will be sufficient to prove that $E[\varphi(t)|\mathcal{F}_t] = 0$ for all $t \in (0,T)$. Indeed, this and the continuity property (2.2) of the family $\{\mathcal{F}_t\}$ clearly imply $f = E[\varphi(T)|\mathcal{F}_t] = 0$. We know that

$$E\int_0^T (v(s),\, 1_{\mathcal{O}}\varphi(s))\, ds = 0 \qquad \forall\, v \in I^2(0,T;\, H).$$

Consequently, $1_{\mathcal{O}}E[\varphi(t)|\mathcal{F}_t]$ is a stochastic process in $I^2(0,T;\,H)$ such that

$$\begin{cases} E\displaystyle\int_0^T (v(s),\, 1_{\mathcal{O}}E[\varphi(s)|\mathcal{F}_s])\, ds = \int_0^T E\left(E[(v(s),\, 1_{\mathcal{O}}\varphi(s))|\mathcal{F}_s]\right) ds \\ \qquad = \displaystyle\int_0^T E(v(s),\, 1_{\mathcal{O}}\varphi(s))\, ds = 0 \qquad \forall\, v \in I^2(0,T;\, H) \end{cases}$$

and one has

$$1_{\mathcal{O}}E[\varphi(t)|\mathcal{F}_t] = 0. \tag{2.8}$$

For each $t \in (0,T)$, $E[\varphi(t)|\mathcal{F}_t] = S(T-t)E[f|\mathcal{F}_t]$ is real analytic in the variable $x \in D$. Hence, one must necessarily have $E[\varphi(t)|\mathcal{F}_t] = 0$ for all $t \in (0,T)$ and the theorem is proved.

Remark 2.1. We deduce from theorem 2.1 that, for all $y_d \in L^2(\Omega, \mathcal{F}_T; H)$, $\varepsilon > 0$ and $\delta > 0$, a control $v \in I^2(0,T;\,H)$ can be found such that

$$P\{|y_v(T) - y_d| < \varepsilon\} \geq 1 - \delta.$$

The existence of a control $v \in I^2(0,T;\,H)$ such that $P\{|y_v(T) - y_d| < \varepsilon\} = 1$ is an open question.

The assertion in theorem 2.1 remains true for systems governed by more general equations. More precisely, one has:

Theorem 2.2. *Assume that, in* (2.5), $A \in \mathcal{L}(V;V')$ *is an operator of the form*

$$Ay = \partial_i(a_{ij}\partial_j y) + \partial_i(b_i y) + cy,$$

where the coefficients satisfy

$$a_{ij} \in C^1(\overline{D}), \quad b_i, c \in L^\infty(D)$$

and the usual ellipticity condition

$$a_{ij}(x)\lambda_j\lambda_i \geq \alpha|\lambda|^2 \quad \forall\, \lambda \in \mathbb{R}^N,\ \forall\, x \in D, \quad \alpha > 0.$$

Then the corresponding linear manifold $Y_T = \{\, y_v(T)\,;\, v \in I^2(0,T;\,H)\,\}$ *is dense in the space* $L^2(\Omega, \mathcal{F}_T; H)$.

Proof. It is analogous to the proof of theorem 2.1. Thus, let us denote again by $S(t)$ the semigroup generated by A in H. Let A^* and $S^*(t)$ stand for the corresponding adjoint operators. Assume that $f \in L^2(\Omega, \mathcal{F}_T; H)$ and (2.7) is satisfied. By putting $\varphi(t) = S^*(T-t)f$, one finds again (2.8).

Unfortunately, now $E[\varphi(t)|\mathcal{F}_t]$ is not in general real analytic in x for $0 < t < T$. So, we cannot deduce directly from (2.8) that $E[\varphi(t)|\mathcal{F}_t] = 0$ for all t (this would suffice). However, using the fact that the equation $-\partial_t\varphi - A^*\varphi = 0$ has the unique continuation property, this can be arranged. This is shown in the following

Proposition 2.1. *Let the assumptions in theorem 2.2 be satisfied. Assume that* $f \in L^2(\Omega, \mathcal{F}; H)$ *and* $1_{\mathcal{O}} E[S^*(T-t)f|\mathcal{F}_t] = 0$ *for all* $t \in (0,T)$. *Then*

$$E[S^*(T-t)f|\mathcal{F}_t] = 0 \quad a.e.\ in\ D \quad \forall t \in (0,T). \tag{2.9}$$

Proof. Our assumptions on A imply unique continuation for all functions $S^*(T-t)h$ with $h \in H$ (cf. [16]). In particular, if h is such that $1_{\mathcal{O}} S^*(T-t)h = 0$ for all $t \in [\tau_1, \tau_2]$, then $S^*(T-t)h = 0$ a.e. in D for all $t \in [\tau_1, \tau_2]$. Let us fix $\tau \in (0,T)$ and $F \in \mathcal{F}_\tau$ and let us prove that

$$\int_F E[S^*(T-\tau)f|\mathcal{F}_\tau]\, dP = 0. \tag{2.10}$$

Since τ and F are chosen arbitrarily, this will imply (2.9). We observe that, for each $t \in [0,T]$,

$$\int_F S^*(T-t)f\, dP = S^*(T-t)(E(1_F f)).$$

Also, from the properties of conditional expectation, one has

$$\int_F S^*(T-t)f\, dP = \int_F E[S^*(T-t)f|\mathcal{F}_t]\, dP \quad \forall t \in [\tau, T].$$

Hence, $E(1_F f)$ is a function in H such that

$$1_{\mathcal{O}} S^*(T-t)(E(1_F f)) = \int_F 1_{\mathcal{O}} E[S^*(T-t)f|\mathcal{F}_t]\, dP = 0 \quad \forall t \in [\tau, T).$$

From the unique continuation property, (2.10) is obtained.

Remark 2.2. We can extend to the stochastic framework the penalization methods in [7]. More precisely, assume that y_d is given in the space $L^2(\Omega, \mathcal{F}_T; H)$ and set

$$J_k(v) = E\int_0^T |v|^2\, dt + kE|y_v(T) - y_d|^2 \quad \forall v \in I^2(0,T;\, H)$$

for each $k \geq 1$. It is not difficult to prove the existence and uniqueness of a process $\hat{v}_k$ minimizing J_k in $I^2(0,T;\, H)$. Let us set $\hat{y}_k = y_{\hat{v}_k}$. Then one has

$$\lim_{k\to\infty} E|\hat{y}_k(T) - y_d|^2 = 0. \tag{2.11}$$

Indeed, if $\varepsilon > 0$ is given and $E|y_v(T) - y_d|^2 \leq \frac{\varepsilon}{2}$, the following holds:

$$kE|\hat{y}_k(T) - y_d|^2 \leq J_k(\hat{v}_k) \leq J_k(v) \leq k\frac{\varepsilon}{2} + E\int_0^T |v|^2\, dt \quad \forall k \geq 1.$$

From this, (2.11) follows easily.

Remark 2.3. If B is not random and satisfies

$$B \in C^0([0,T];\ \mathcal{L}(K;\ H))$$

and $y_d = 0$, then it follows from the results in [10] that $\hat{v}_k$ is a feedback control. To be more precise, one has

$$\hat{v}_k(t) = -1_{\mathcal{O}} Q_k(t) \hat{y}_k(t),$$

where Q_k is the unique solution in $C^0([0,T];\ \mathcal{L}_s^+(H))$ to the Riccati equation

$$Q_k(t) = kS(T-t)S(T-t) - \int_t^T S(s-t)Q_k(s)1_{\mathcal{O}}Q_k(s)S(s-t)\,ds$$

(here, $\mathcal{L}_s^+(H)$ denotes the set formed by all self-adjoint nonnegative operators in $\mathcal{L}(H)$). Furthermore, in this particular case the optimal cost is given by

$$J_k(\hat{v}_k) = (Q_k(0)y_0, y_0) + \int_0^T \mathit{trace}\ B^*(t)Q_k(t)B(t)W\,dt,$$

with $W \in \mathcal{L}(H)$ being the covariance operator of w_t.

Remark 2.4. Let us finally mention that the duality methods in [7] also work in this context. More specifically, let Z stand for the Hilbert space $L^2(\Omega, \mathcal{F}_T; H)$ (with norm $\|\cdot\|_Z$ and scalar product $(\cdot,\cdot)_Z$). Assume $\varepsilon > 0$, $y_0 \in H$ and $y_d \in Z$ are given. It is then natural to minimize

$$\frac{1}{2} E \iint_{\mathcal{O}\times(0,T)} |v|^2\,dx\,dt \tag{2.12}$$

over the (nonempty) set

$$I^2(0,T;H) \cap \{\, v;\ \|y_v(T) - y_d\|_Z \le \varepsilon \,\}. \tag{2.13}$$

In the conditions of theorem 2.2, there exists exactly one minimizer (up to an additive function vanishing on $\mathcal{O}\times(0,T)$). This is given as follows. Let us introduce the dual functional

$$\begin{cases} J(f) = \frac{1}{2} E \iint_{\mathcal{O}\times(0,T)} |\tilde{\varphi}|^2\,dx\,dt + \varepsilon \|f\|_Z - (\bar{y}_d, f)_Z \\ \forall f \in Z, \end{cases}$$

where $\bar{y}_d = y_d - S(T)y_0$. Here, we have used the notation $\tilde{\varphi}(t) = E[\varphi(t)|\mathcal{F}_t]$, with $\varphi \in I^2(0,T;H)$ being given by

$$\begin{cases} -\partial_t \varphi - A^*\varphi = 0 \quad \text{in} \quad Q, \\ \varphi = 0 \quad \text{on} \quad \Sigma, \\ \varphi(T) = f, \end{cases}$$

It can be proved that $J : Z \mapsto \mathbb{R}$ is strictly convex, continuous and, due to unique continuation, coercive. Consequently, there exists one and only one $\hat{f} \in Z$ satisfying

$$J(\hat{f}) \le J(f) \quad \forall f \in Z. \tag{2.14}$$

Let $\hat{f}$ be the solution to (2.14). Then, if we set $\hat{v}(t) = E[\hat{\varphi}(t)|\mathcal{F}_t]$ for all t, it is not difficult to check that $1_{\mathcal{O}}\hat{v}$ is the control process minimizing (2.12) in the set (2.13).

3. A null-controllability result

In this section, we present a null-controllability result for (2.5). Again, this is the analog of a deterministic result.

Let us fix a positive function $\gamma \in C^\infty(0,T)$ such that $\gamma(t) = t$ near $t = 0$ and $\gamma(t) = T - t$ near $t = T$. It will be assumed that the hypotheses in theorem 2.2 hold, that B is not random and satisfies $B \in C^1([0,T]; \mathcal{L}(K;\, H))$ and, also, that the support of $B(t)$ does not intersect $\mathcal{O}$ for all t.

Theorem 3.1. *There exists a positive function $\rho = \rho(x)$ such that, if*

$$\iint_Q t\left(\gamma(t)^{-1}\|B\|^2_{\mathcal{L}(K;\,H)} + \gamma(t)^3\|\partial_t B\|^2_{\mathcal{L}(K;\,H)}\right) e^{2\frac{\rho(x)}{\gamma(t)}} < +\infty, \tag{3.1}$$

then, for each $y_0 \in H$ there exists $v \in I^2(0,T;H)$ satisfying $y_v(T) = 0$.

Sketch of the Proof. We will adapt the arguments in [8] in the context of (2.5). We will previously rewrite the null-controllability problem as an equivalent problem for which $y_0 = 0$.

(i) Let $\theta = \theta(t)$ be a C^∞ function such that $\theta(t) = 1$ near $t = 0$, $\theta(t) = 0$ near $t = T$ and $0 \le \theta \le 1$. Let us introduce the function ξ, with

$$\xi(t) = S(t)y_0 \quad \forall t.$$

Then, by setting $g = -\theta'(t)\xi$ and $z_v = y_v - \theta(t)\xi(t)$, one sees that $y_v(T) = 0$ if and only if the unique solution to

$$\begin{cases} \partial_t z_v - A z_v = 1_{\mathcal{O}} v + g + B\dot{w}_t & \text{in} \quad Q \quad P-\text{a.s.}, \\ z_v = 0 \quad \text{on} \quad \Sigma \quad P-\text{a.s.}, \\ z_v(0) = 0 \quad \text{in} \quad D \quad P-\text{a.s.} \end{cases} \tag{3.2}$$

satisfies $z_v(T) = 0$.

(ii) Following the ideas of [8], let us introduce (and solve) an auxiliary variational problem.

Let $\mathcal{O}'$ be a nonempty open set satisfying $\mathcal{O}' \subset\subset \mathcal{O}$. We will put $D' = D \setminus \overline{\mathcal{O}'}$, $Q' = D' \times (0,T)$ and $\Sigma' = \partial D' \times (0,T)$. As usual, $\mathcal{O}'_\eta$ will stand for the open η-neighborhood of $\mathcal{O}'$. The usual co-normal derivative operator associated to A and A^* will be denoted by ∂_A. We will need the following

Proposition 3.1. *There exist a positive function $\rho \in C^2(\overline{D})$ and a positive constant C_* such that*

$$\begin{cases} \displaystyle\iint_{Q'} \gamma\left(|\partial_t q|^2 + |D^2 q|^2\right) e^{-2\frac{\rho}{\gamma}} + \iint_{Q'} \gamma^{-1}|\nabla q|^2 e^{-2\frac{\rho}{\gamma}} + \iint_{Q'} \gamma^{-3}|q|^2 e^{-2\frac{\rho}{\gamma}} \\ \qquad \displaystyle\le C_*\left(\iint_{Q'} |\partial_t q + A^* q|^2 e^{-2\frac{\rho}{\gamma}}\right) \end{cases} \tag{3.3}$$

for all functions $q \in C^2(\overline{Q'})$ such that $q = 0$ on Σ' and $\partial_A q = 0$ on $\partial\mathcal{O}' \times (0,T)$.

This Carleman inequality is proved in [9] (see also [6] and [17] for other more general estimates of the same kind). In the sequel, it will be assumed that B satisfies (3.1) with ρ furnished by proposition 3.1. We will prove that, for some v, one has $z_v(T) = 0$.

Let us introduce the linear space

$$\Psi_0 = \{\, q \in C^2(\overline{Q'})\,;\ q = 0 \text{ on } \Sigma',\ \partial_n q = 0 \text{ on } \partial\mathcal{O}' \times (0,T)\,\}.$$

From proposition 3.1, we know that

$$[p,q] = \iint_{Q'} (\partial_t p + A^* p)(\partial_t q + A^* q)\, e^{-2\frac{\rho}{\gamma}}$$

is a scalar product in Ψ_0. Let Ψ be the completion of Ψ_0 for the scalar product $[\cdot,\cdot]$. Then (3.3) is satisfied for all $q \in \Psi$. Let us put $P = L^2(\Omega, \mathcal{F}; \Psi)$, a Hilbert space for the scalar product

$$E[p,q] = E \iint_{Q'} (\partial_t p + A^* p)(\partial_t q + A^* q)\, e^{-2\frac{\rho}{\gamma}}.$$

Let us also put

$$\langle l, q\rangle = -E \iint_{Q'} gq + E \iint_{Q'} (q\partial_t B + \partial_t q B)\, w_t \quad \forall\, q \in P.$$

Then, using (2.1), (3.1) and proposition 3.1, it can be shown that l is a bounded linear form on P. Arguing as in [8], we introduce the following problem:

$$E[p,q] = \langle l, q\rangle \quad \forall\, q \in P, \quad p \in P. \tag{3.4}$$

Obviously, (3.4) possesses exactly one solution p. Let us put $z = e^{-2\frac{\rho}{\gamma}}(\partial_t p + A^* p)$. Among other things, one has $z \in L^2(\Omega, \mathcal{F}; L^2(Q'))$.

(iii) It can be seen that z has sample paths in $C^0([0,T]; H^{-1}(D'))$ and satisfies

$$(z(t), q_0) = \int_0^t \{(z(s), A^* q_0) + (g(s), q_0)\}\, ds + (\int_0^t B(s)\, dw_s, q_0)$$

P-a.s. for all $t \in (0,T)$ whenever q_0 is (for instance) a function in $C_0^\infty(D')$. The stochastic integral arises as a consequence of Ito's formula (here, the fact that B is not random is needed). It is thus clear that $z \in I^2(0,T; L^2(D'))$. Let δ be such that $0 < \delta < \frac{1}{3}$ dist $(\mathcal{O}', \partial\mathcal{O})$ and let $\chi \in C^\infty(D)$ be a cut-off function satisfying

$$0 \le \chi \le 1, \qquad \chi \equiv 0 \quad \text{in } \mathcal{O}'_\delta, \qquad \chi \equiv 1 \quad \text{in } D \setminus \mathcal{O}'_{2\delta}.$$

Then, $\tilde{z} = \chi\, z$ can be extended by zero to the whole domain Q. Its extension, also denoted by $\tilde{z}$, satisfies $\tilde{z} \in I^2(0,T;H)$ and also the following equalities for all $q_0 \in D(A^*)$ and all $t \in (0,T)$:

$$\left\{\begin{aligned} &(\tilde{z}(t), q_0) = \int_0^t \{(\tilde{z}(s), A^* q_0) + (g(s), q_0)\}\, ds + (\int_0^t B(s)\, dw_s, q_0) \\ &\quad + \int_0^t \{(\tilde{b}_i z(s), \partial_i q_0) + (\tilde{c} z(s) + (\chi - 1) g(s), q_0)\}\, ds. \end{aligned}\right.$$

Here, $\tilde{b}_i = 2a_{ij}\,\partial_j \chi$ and $\tilde{c} = a_{ij}\,\partial^2_{ij}\chi - \tilde{b}_i\,\partial_i \chi$. From known results, we are now able to ensure that $\tilde{z} \in I^2(0,T;V)$ (for example, see [2], [13]).

(iv) Let us introduce a second cut-off function $\tilde{\chi} \in C^\infty(D)$, with

$$0 \le \tilde{\chi} \le 1, \qquad \tilde{\chi} \equiv 0 \quad \text{in } \mathcal{O}'_{2\delta}, \qquad \tilde{\chi} \equiv 1 \quad \text{in } D \setminus \mathcal{O}'_{3\delta}.$$

By putting $\bar{z} = \widetilde{\chi}\,\tilde{z}$, we see that, for some $v \in I^2(0,T;H)$,

$$(3.5)\qquad \begin{cases} (\bar{z}(t), q_0) = \int_0^t \{(\bar{z}(s), A^* q_0) + (g(s), q_0)\}\, ds + (\int_0^t B(s)\, dw_s, q_0) \\ \quad + \int_0^t (1_{\mathcal{O}} v(s), q_0)\, ds \quad \forall\, q_0 \in D(A^*), \quad \forall\, t \in (0,T) \quad P-\text{a.s.} \end{cases}$$

From (3.4), (3.5) and the fact that $\bar{z} \in I^2(0,T;V)$, it is easy to deduce that $\bar{z} = z_v$ (the unique solution to (3.2)) and also that $\bar{z}(T) = 0$.

4. The case of the Stokes and quasi-Stokes systems

The assertion in theorem 2.2 also holds in the case of a quasi-Stokes stochastic system of the form (1.1) with bounded coefficients a_i and c_{ij}. To be more precise, let us introduce the space

$$\mathcal{V} = \{\, \varphi \in C_0^\infty(D)^N\,;\ \nabla \cdot \varphi = 0 \quad \text{in } D \,\}$$

and let us denote by V (resp. H) the closure of $\mathcal{V}$ in $H_0^1(D)^N$ (resp. $L^2(D)^N$). In this section, A will stand for the operator in $\mathcal{L}(V;\, V')$ given by

$$\langle Ay, z\rangle = -\int_\Omega \{\nabla y \cdot \nabla z + a_i(x) y_j \partial_i z_j + c_{ij}(x) y_i z_j\}\, dx \quad \forall\, y,\, z \in V,$$

where it is assumed that

$$a_i,\ c_{ij} \in L^\infty(D).$$

Regarded as an unbounded operator on H with domain

$$D(A) = \{\, y \in V\,;\ Ay \in H \,\},$$

A is the generator of a semigroup on H, again denoted by $S(t)$. Assume B and y_0 are given and satisfy (2.3) and (2.4). For each $v \in I^2(0,T;\, L^2(D)^N)$, there exists one and only one solution y_v to the problem

$$\begin{cases} y_v \in I^2(0,T;\, V) \cap L^2(\Omega;\, C^0([0,T];\, H)), \\ y_v(t) = y_0 + \int_0^t \{Ay_v(s) + 1_{\mathcal{O}} v(s)\}\, ds + \int_0^t B(s)\, dw_s \quad \forall\, t \in [0,T]. \end{cases}$$

In fact, if P_H stands for the orthogonal projector from $L^2(D)^N$ onto H (the Leray operator), then y_v is given by the following identities:

$$\begin{cases} y_v(t) = S(t)y_0 + \int_0^t S(t-s)\,[P_H(1_{\mathcal{O}} v(s))]\, ds + \int_0^t S(t-s)B(s)\, dw_s \\ \forall\, t \in [0,T]. \end{cases}$$

Using this and the unique continuation property established in [4], we can argue as in the proofs of theorem 2.2 and proposition 2.1. The conclusion is:

Theorem 4.1. *With the notation used in this section, the linear manifold* $Y_T = \{\, y_v(T);\ v \in I^2(0,T;\, L^2(D)^N) \,\}$ *is dense in the space* $L^2(\Omega, \mathcal{F}_T; H)$.

Remark 4.1. For systems governed by stochastic quasi-Stokes problems of this kind, we can adapt the arguments in Remarks 2.2, 2.3 and 2.4. Thus, similar penalization, feedback and duality results can be obtained.

Remark 4.2. If (1.1) is the stochastic Stokes problem (i.e. $a_i \equiv c_{ij} \equiv 0$), it is possible to prove that, for fixed $1 \le j \le N$, the set

$$\{\, y_v(T)\,;\, v \in I^2(0,T;\, L^2(D)^N),\, v_j = 0\,\}$$

is also dense in $L^2(\Omega, \mathcal{F}_T; H)$. The proof is as in the similar deterministic case (cf. [11]).

Remark 4.3. For the stochastic 3D Stokes problem in a cylindrical domain $D = G \times (0, L)$, one also has approximate controllability in a "generic" sense with respect to G with controls in the set

$$\mathcal{U}_{ad} = \{v \in I^2(0,T;\, L^2(D)^3)\,;\, v_1 = v_2 = 0\,\}.$$

More precisely, using the results in [12] and arguing as above, it can be seen that, for any given bounded domain $\tilde{G} \subset \mathbb{R}^2$ of class C^k (with $k \ge 3$), there exists another domain G arbitrarily close to $\tilde{G}$ in the C^k topology such that the corresponding set $\{y_v(T)\,;\, v \in \mathcal{U}_{ad}\}$ is dense in $L^2(\Omega, \mathcal{F}_T; H)$.

To our knowledge, whether or not null-controllability (i.e. theorem 3.1) holds for systems governed by (1.1) is an open question.

References

1. L. Arnold: *Stochastic differential equations,* Wiley, New York 1974.
2. G. DaPrato, J. Zabczyk: *Stochastic equations in infinite dimensions,* Cambridge University Press, Cambridge 1992.
3. W.L. Chan, C.K. Lau: *Constrained Stochastic Controllability of Infinite-Dimensional Linear Systems,* J. Math. Anal. Appl., 85, 1982, p. 46–78.
4. C. Fabre, G. Lebeau: *Prolongement unique des solutions des équations de Stokes,* to appear.
5. A.V. Fursikov, O.Yu. Imanuvilov: *On approximate controllability of the Stokes system,* Ann. Fac. Sc. Toulouse, Vol II, no. 2, 1993, p. 205–232.
6. A.V. Fursikov, O.Yu. Imanuvilov: *Local exact boundary controllability of the Boussinesq equation,* to appear.
7. R. Glowinski, J.L. Lions: *Exact and approximate controllability for distributed parameter systems,* Acta Numerica 1994, p. 269–378.
8. O.Yu. Imanuvilov: *Thesis,* Moscow 1991 (in russian); see also: *Exact boundary controllability of the parabolic equation,* Russian Math. Surveys 48, No. 3 (1993), p. 211–212.
9. O.Yu. Imanuvilov: *Boundary controllability of parabolic equations,* Russian Acad. Sci. Sb. Math. 186 (1995), No. 6, p. 109–132.
10. A. Ichikawa: *Dynamic programming approach to stochastic evolution equations,* SIAM J. Control Optim., 17, no. 1, 1979, p. 152–173.
11. J.L. Lions: *Remarques sur la contrôlabilité approchée,* in Proc. of "Jornadas Hispano-Francesas sobre Control de Sistemas Distribuidos", Málaga (Spain), october 1990, p. 77–87.
12. J.L. Lions, E. Zuazua: *A generic uniqueness result for Stokes system and its control theoretical consequences,* to appear.
13. E. Pardoux: *Thèse,* Université Paris XI, 1975.

14. E. PARDOUX: *Intégrales stochastiques hilbertiennes,* Cahiers de Mathématiques de la Décision No. 7617, Université Paris IX,1976.
15. J. REAL: *Some results on controllability for stochastic heat and Stokes equations,* C. R. Acad. Sc. Paris, t. 322, p. 1198–1202, 1996.
16. J.C. SAUT, B. SCHEURER: *Unique continuation for some evolution equations,* J. Diff. Equations 66 (1), p. 118–139, 1987.
17. D. TATARU: *A priori estimates of Carleman's type in domains with boundary,* J. Math. Pures Appl., 73, 1994, p. 355–387; see also: *Carleman estimates and unique continuation for solutions to boundary value problems,* preprint.

E. Fernández-Cara
Department of Differential Equations
and Numerical Analysis
University of Sevilla
Tarfia s/n
E-41012 Sevilla, Spain

J. Real
Department of Differential Equations
and Numerical Analysis
University of Sevilla
Tarfia s/n
E-41012 Sevilla, Spain

International Series of Numerical Mathematics
Vol. 126, © 1998 Birkhäuser Verlag, Basel

A Reduced Basis Method for Control Problems Governed by PDEs

K. ITO AND S.S. RAVINDRAN

Center for Research in Scientific Computation
Department of Mathematics
North Carolina State University

ABSTRACT. This article presents a reduced basis method for constructing a reduced order system for control problems governed by nonlinear partial differential equations. The major advantage of the reduced basis method over others based on finite element, finite difference or spectral method is that it may capture the essential property of solutions with very few basis elements. The feasibility of this method is demonstrated for boundary control problems modeled by the incompressible Navier-Stokes and related equations with the boundary temperature control and boundary electromagnetic control in channel flows.

1991 *Mathematics Subject Classification.* 93B40, 49M05, 76D05, 49K20, 65H10, 76W05, 80A20

Key words and phrases. Reduced basis method, finite elements, optimal control, flow control, boundary temperature control, electromagnetic control.

1. Introduction

Real time simulations of control problems that involve partial differential equations as state equations are often formidable problems to solve. Our work was motivated by the recent interest in *optimal flow control* of viscous flows which are control problems involving Navier-Stokes equations as state equations; see [11] for a review. These problems are by far the most challenging control problems in computational engineering and science. The major difficulty is mainly due to the nonlinearity in the state equations and these state equations when discretized can number in millions. Thus the conventional approaches cannot be adequate for solving such large scale control problems

In this article we discuss a reduction type method which may provide an avenue to overcome this difficulty. In this method hereafter called reduced basis method one uses basis functions which are closely related to and generated from the problem that is being solved. This is in contrast to the traditional numerical methods such as finite difference method and finite elements method which uses grid functions and piecewise polynomials, respectively, as basis functions.

There are several approaches available for the selection of basis functions in reduced basis method. One such approach is Taylor approach in which one uses solutions at a reference point in the parameter space along with their derivatives as basis functions.

This work was supported by the Air Force Office of Scientific Research under grants AFOSR F49620-95-1-0437 and AFOSR F49620-95-1-0447. This work was partially supported by the Office of Naval Research Grant N00014-96-1-0265.

Another approach which we call Lagrange approach uses solutions of the problem at various parameter values as basis functions. Finally the Hermite approach is a hybrid of Lagrange and Taylor approaches which uses solutions and their first derivatives of the problem at various parameter values as basis functions. The applications of reduced basis method to structural mechanics problems can be found in [1] and [6]–[7].

Our goal here is to demonstrate the applicability and feasibility of reduced basis method for control problems governed by Navier-Stokes type partial differential equations. We will consider vorticity minimization problems in backward-facing step type channel geometry as a prototype control problem. Two fluid flow situations are considered: An electrically conducting fluid under applied magnetic field and a thermally convective fluid. In the first situation the control is effected by boundary electric potential and in the latter the control is boundary temperature.

Electromagnetic Control. When the fluid is electrically conducting, such as sea water, one can obtain an interesting control mechanism by appropriately placing electrodes and magnets along the boundary of the flow domain such that there is a coupling between magnetic field $\mathbf{B}$ and the current $\mathbf{j}$, see Figure 1.1 for one such setup. This coupling produces a forcing $\mathbf{j} \times \mathbf{B}$ which appears in the Navier-Stokes equations and is known as Lorenz force, see [10]. This forcing can be exploited to control fluid flows. In §5, we will descirbe this control mechanism and in §6 we will use this for a vorticity minimization problem in fluid flows to demonstrate the feasibility of reduced basis method.

FIGURE 1.1. A simple diagram of MHD setting

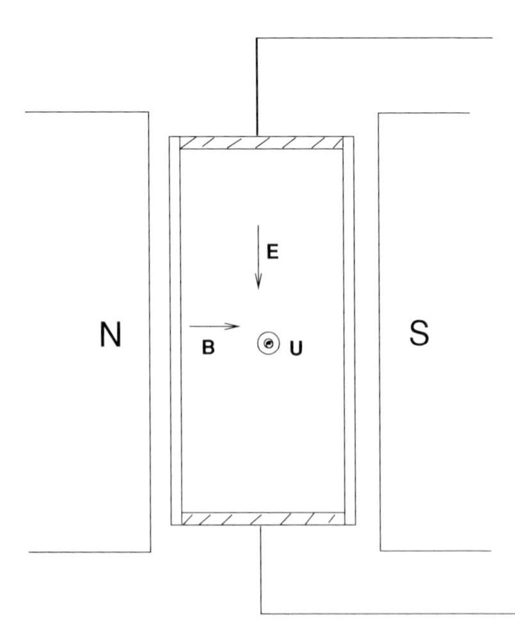

Boundary Temperature Control. By imposing a temperature gradient in the flow by adjusting boundary temperature, one can either enhance or counteract the flow effects. In §5-6, we will use this control mechanism for a vorticity minimization problem in fluid flows to demonstrate the feasibility of reduced basis method.

A Vorticity Minimization Problem. As a prototype problem for vorticity minimization in fluid flows, let us consider the flow through backward-facing step channel at Reynolds number 200. For this Reynolds number range flows can be assumed to be steady and two dimensional. Thus we consider a steady plane flow in a backward-facing step channel as shown in Figure 6.1. The corresponding flow at this Reynolds number is shown in Figure 6.3 which has a corner circulation near the corner. Our objective is to suppress this recirculation by using the control mechanisms discussed previously. We formulate this as an optimal control problem with a cost functional representing vorticity in the flow which is thus minimized subject to the governing equations of the flow under consideration.

We will consider the two control mechanisms described previously in two different flow setting. That is flow is either electrically conducting or thermally convective and we will consider electromagnetic control in the former and boundary temperature control in the latter.

2. Reduced Basis Spaces and Reduced-Order Equation

In order to illustrate the reduced basis method, we assume for ease in exposition that we are dealing with nonlinear dynamics about the equilibrium points. Consider the the parameterized stationary problem

$$\mathcal{E}(y,\mu) = 0 \quad \text{for } \mu \in \Lambda,\ y \in X, \tag{2.1}$$

where μ represents some physical parameter, for example, Reynolds number or viscosity and $\mathcal{E} : X \times \Lambda \to X^*$ is C^2. Equation (2.1) defines a solution function $\mu \in \Lambda \to y(\mu) \in X$. We construct the reduced basis elements by the interpolation of solution function $\mu \to y(\mu)$ as follows.

The Taylor Subspace. In this choice, one uses the Taylor expansion of function $y(\mu)$ at a reference value of μ, say μ^*, and the reduced basis subspace X_R is defined as

$$X_R = \text{span}\{y_j | y_j = \frac{\partial^j y}{\partial \mu^j}|_{\mu=\mu^*}, j = 0,\ldots,M\} \tag{2.2}$$

The jth derivative y_j can be calculated from the equations resulted by successive differentiation of (1.1), i.e.

$$\mathcal{E}_y(y_0,\mu_0)y_j = \mathcal{F}_j(y_0,y_1,\ldots,y_{j-1},\mu^*). \tag{2.3}$$

For example, y_1 satisfies the equation

$$\mathcal{E}_y(y_0,\mu^*)y_1 = -\mathcal{E}_\mu(y_0,\mu^*).$$

We note here that each y_j can be obtained from its predecessors by solving a linear system with the same linear operator $\mathcal{E}_y(y_0,\mu^*)$. However, one cannot continue to use the same basis elements generated at fixed parameter μ^* to compute solutions when the parameter of interest is significantly away from it. In such cases reduced basis elements have to be updated and the solution is sought in the new reduced basis space. Moreover, generating the right hand side of (2.3) could be quite complicated in certain problems. This choice has been used in [7] for structural analysis problems and in [8] for high Reynolds number steady state fluid flow calculations.

The Lagrange Subspace. In this case, the basis elements are solutions of the nonlinear problem under study at various parameter values μ_j. The reduced subspace is given by

$$X_R = \text{span}\{y^j | y^j = y(\mu_j), j = 1, \ldots, M\}. \tag{2.4}$$

The Lagrange basis was used to study structural problems in [1]. A possible advantage in this choice is that updating the basis elements can be done one basis element at a time instead of generating the whole space.

The Hermite Subspace. This is a hybrid of the Lagrange and Taylor approach. The basis elements are solutions and their first derivatives at various parameter values μ_j. The reduced subspace is given by

$$X_R = \text{span}\{y^j = y(\mu_j) \text{ and } \frac{\partial y}{\partial \mu}|_{\mu=\mu_j}, j = 1, \ldots, \widetilde{M}\}. \tag{2.5}$$

Suppose we have a reduced basis space X_R in X. Let $m = dim(X_R)$ and $\{\phi_i\}$ is a basis of X_R. Then we can construct the reduced-order equation by the Galerkin approximation, i.e., for $y_m = \sum_{i=1}^m \alpha_i \phi_i \in X_R$

$$\mathcal{E}_m(y_m, \mu)_i = \langle \mathcal{E}(y_m, \mu), \phi_i \rangle_{X^* \times X} \tag{2.6}$$

for $1 \le i \le m$.

For the evolution equation

$$\frac{d}{dt} y(t) + E(t, y(t)) = 0 \tag{2.7}$$

we, for example, generate reduced basis elements $\{\phi_i\}_{i=1}^m$ in X by the solutions at m different time instants to (2.7). Given the reduced basis space X_R, we define the reduced-order equation for $y_m(t) = \sum_{i=1}^m \alpha_i(t)\phi_i$,

$$\langle \frac{d}{dt} y_m(t) + E(t, y_m(t)), \phi_i \rangle_{X^* \times X} = 0 \tag{2.8}$$

for all $1 \le i \le m$. In [4] this method has been carried out and its feasibility has been demonstrated for channel flow simulations in which reduced order solution $\mathbf{u}^m$ is formed by setting

$$\mathbf{u}^m(t) = \sum_{i=1}^m \alpha_i(t) \boldsymbol{\phi}_i,$$

where $\boldsymbol{\phi}_i = \mathbf{u}_{i+1} - \mathbf{u}_i$, $i = 1, 2, \ldots, m-1$ and $\boldsymbol{\phi}_m = \mathbf{u}_m$. We further take $\alpha_m = 1$ so that the boundary conditions are satisfied. The solution $\mathbf{u}^m$ is computed from

$$(\tfrac{\partial}{\partial t}\mathbf{u}^m, \mathbf{v}^m) + \tfrac{1}{Re} a(\mathbf{u}^m, \mathbf{v}^m) + c(\mathbf{u}^m, \mathbf{u}^m, \mathbf{v}^m) = (\mathbf{f}, \mathbf{v}^m) \quad \text{for all } \mathbf{v}^m \in \mathbf{V}_0^m,$$

where $\mathbf{V}_0^m = \text{span}\{\boldsymbol{\phi}_i : i = 1, \ldots, m-1\}$ is the span of the test functions.

The basis elements were generated by computing the flow from the full model at eleven time instances between 1 and 11. The time step used in the reduced order model was .001.

The dimension of the reduced basis space is very much problem-dependent. The reduced basis elements constructed by the above mentioned approach can be nearly linearly dependent. So we may further reduce the dimension by the conditioning of the mass matrix Q:

$$Q_{i,j} = (\phi_i, \phi_j)_X.$$

Our computational experiments and the computations reported for structural problems in the references mentioned earlier seem to indicate that an accurate approximation can be obtained for large range of parameter values using 5 to 10 basis elements. Therefore, although the resulting reduced order model is dense, they are small compared to the sparse but large system that result from the standard basis functions.

According to our comparison study carried out in [4] for driven cavity incompressible Navier-Stokes calculations, the performance of Hermite approach is better than that of Lagrange. The basis elements for the Lagrange approach were selected at Reynolds numbers 100, 300, 500 and 700, and that for the Hermite was selected at 300 and 700. The comparison was carried out by computing the driven cavity flow at Reynolds number 1200. The L_2-norm difference between the full mixed-finite element solution u_f and the reduced basis solution using these two approaches are as follows: $\|\mathbf{u}_l - \mathbf{u}_f\|_2 = 0.0889$ and $\|\mathbf{u}_h - \mathbf{u}_f\|_2 = 0.0766$, where $\mathbf{u}_l$ is the solution obtained using Lagrange approach and $\mathbf{u}_h$ is that obtained using Hermite approach.

3. Error Analysis

In order to justify the reduced basis solution y_m we need to have a post verification criterion. In general we formulate it as an error analysis as follows. Let X and Y be two Banach spaces and Λ be a compact set. Given a C^2 mapping

$$\mathcal{E} : (y, \mu) \in X \times \Lambda \to \mathcal{E}(y, \mu) \in Y,$$

and we consider the equation

$$\mathcal{E}(y, \mu) = 0. \tag{3.1}$$

The family $\{(y(\mu), \mu) : \mu \in \Lambda\}$ is said to be a branch of nonsingular solutions of equation (3.1), i.e.,

$\mu \to y(\mu)$ is a continuous function from Λ into X and $D_y\mathcal{E}(y, \mu)$ is an isomorphism from X onto Y for all $\mu \in \Lambda$.

Let us consider the reduced order problem

$$\mathcal{E}_m(y_m, \mu) = 0. \tag{3.2}$$

defined on the reduced basis space X_m. We assume that $\mathcal{E}_m : X_m \times \Lambda \to Y_m$ is C^2. For the ease of our discussions we assume that $X_m \subset X$ and $Y_m \subset Y$. The norms on X_m and Y_m are induced from X and Y norms, respectively. The problem is to find the solution $y_m \in X_m$ such that (3.2) is satisfied for a given $\mu \in \Lambda$.

We assume that $D_y\mathcal{E}_m(\tilde{y}_m,\mu)$ is an isomorphism from X_m onto Y_m where $\tilde{y}_m$ is a given element in X_m. We introduce the following notations;

$$\epsilon_m(\mu) = \|\mathcal{E}_m(\tilde{y}_m,\mu)\|_{Y_m},$$

$$\gamma_m(\mu) = \|D_y\mathcal{E}_m(\tilde{y}_m,\mu)^{-1}\|_{\mathcal{L}(Y_m,X_m)},$$

$$S_m(u;\alpha) = \{v \in X_m : \|u-v\|_{X_m} \leq \alpha\},$$

$$L_m(\mu;\alpha) = \sup_{v\in S(\tilde{y}_m,\alpha)} \|D_y\mathcal{E}_m(\tilde{y}_m,\mu) - D_y\mathcal{E}_m(v,\mu)\|_{\mathcal{L}(X_m,Y_m)}.$$

We next state a theorem regarding the error estimate which is derived from Theorem IV.3.1 in [2] for the approximation of branches of nonsingular solutions.

Theorem 3.1. *Suppose $D_y\mathcal{E}_m(\tilde{y}_m,\mu)$ is an isomorphism of X_m onto Y_m and*

$$2\gamma_m(\mu)L_m(\mu,2\gamma_m(\mu)\epsilon_m(\mu)) < 1.$$

Then the problem (3.2) has a unique solution $(y_m(\mu),\mu)$ such that:

$$y_m(\mu) \in S(\tilde{y}_m;2\gamma_m(\mu)\epsilon_m(\mu)).$$

In addition, $y_m(\mu)$ is the only solution of (3.2) in the ball $S_m(\tilde{y}_m;\alpha)$ for all $\alpha \geq 2\gamma_m(\mu)\epsilon_m(\mu)$ that satisfy $\gamma_m(\mu)L_m(\mu;\alpha) < 1$ and we have the estimate:

$$\|y_m(\mu)-v_m\|_X \leq [\gamma_m(\mu)/(1-\gamma_m(\mu)L_m(\mu;\alpha)]\|\mathcal{E}_m(v_m,\mu)\|_Y \quad \textit{for all } v_m \in S_m(\tilde{y}_m,\alpha).$$

■

Moreover, we have the following corollary.

Corollary 3.2. *Suppose there exists an element $\tilde{y}_m \in X_m$ such that $D_y\mathcal{E}(\tilde{y}_m,\mu)$ is an isomorphism of X onto Y and*

$$2\gamma(\mu)L(\mu,2\gamma(\mu)\epsilon(\mu)) < 1 \tag{3.3}$$

where

$$\epsilon(\mu) = \|\mathcal{E}(\tilde{y}_m,\mu)\|_Y,$$

$$\gamma(\mu) = \|D_y\mathcal{E}(\tilde{y}_m,\mu)^{-1}\|_{\mathcal{L}(Y,X)},$$

$$S(y;\alpha) = \{v \in X : \|u-v\|_X \leq \alpha\},$$

$$L(\mu;\alpha) = \sup\nolimits_{v\in S(\tilde{y}_m,\alpha)} \|D_y\mathcal{E}(\tilde{y}_m,\mu) - D_y\mathcal{E}(v,\mu)\|_{\mathcal{L}(X,Y)}.$$

Then the problem (3.1) has a solution $(y(\mu),\mu)$ such that:

$$y(\mu) \in S(\tilde{y}_m;2\gamma(\mu)\epsilon(\mu)).$$

In addition, $y(\mu)$ is the only solution of (3.1) in the ball $S(\tilde{y}_m;\alpha)$ for all $\alpha \geq 2\gamma(\mu)\epsilon(\mu)$ that satisfy $\gamma(\mu)L(\mu;\alpha) < 1$ and we have the estimate:

$$\|y(\mu)-v\|_X \leq [\gamma(\mu)/(1-\gamma(\mu)L(\mu;\alpha)]\|\mathcal{E}(v,\mu)\|_Y$$

for all $v \in S(\tilde{y}_m,\alpha)$.

■

We can apply Theorem 3.1 and Corollary 3.2 to obtain the following error estimate.

Theorem 3.3. *(i) Suppose* $y_m(\mu) \in X_m$ *is a solution to* (3.2) *and assume* $\tilde{y}_m = y_m(\mu)$ *satisfies the condition in Corollary* 3.2. *Then we have a solution* $y(\mu) \in S(\tilde{y}_m; 2\gamma(\mu)\epsilon(\mu))$ *to* (3.1) *and the estimate*

$$\|y(\mu) - y_m(\mu)\|_X \leq [\gamma(\mu)/(1 - \gamma(\mu)L(\mu;\alpha)]\|\mathcal{E}(y_m(\mu),\mu)\|_Y \tag{3.4}$$

with $\alpha = \gamma(\mu)\epsilon(\mu)$.

(ii) Suppose there exits an element $\tilde{y}_m \in X_m$ *such that the conditions in Theorem* 3.1 *and Corollary* 3.2 *are satisfied. Furthermore, we assume that* $\alpha_m = 2\gamma_m(\mu)\epsilon_m(\mu)$ *satisfies* $\gamma(\mu)L(\mu;\alpha_m) < 1$. *Then we have* (3.4) *with* $\alpha = \alpha_n$. ■

4. Optimal Control Problems

In this section we discuss the optimal control problem and the application of reduced basis method. Consider the minimization problem

$$\min \quad J(y,u) \quad \text{subject to } \mathcal{E}(y,u) = 0 \tag{4.1}$$

over $u \in U_{ad} \subset U$. Here X and U are Hilbert spaces and $\mathcal{E} : X \times U \to X^*$ is C^2. We assume that $U = R^m$ and u_{ad} is closed and convex. The Lagrange reduced space can be defined by

$$X_R = \text{span}\{y^j \in X | \mathcal{E}(y^j, u^j) = 0, \ j = 1,\dots,M\},$$

where u^j is a sampled point in U_{ad}. In order to obtain a lower-order reduced basis space, if m is large then we may consider the following pre-selecting step:

- Let u^α, $\alpha \in A$, be the points in U_{ad} defined by
$$u^\alpha = \bar{u} + \sum \delta_i \alpha_i e_i$$
where α is the integer-valued vector, and δ_i and e_i are the step size and unit vector in the i-th direction.
- We determine $y^\alpha \in X$ by solving $\mathcal{E}(y, u^\alpha)$ for each $\alpha \in A$.
- We find an index α_0 in A such that $J(y^\alpha, u^\alpha)$ is smallest.
- Then, we select the sampling set u_j by
$$u^1 = u^{\alpha_0}, \quad u^{2i} = u^{\alpha_0} + \delta_i e_i, \quad \text{and} \quad u^{2i+1} = u_{\alpha_0} - \delta_i e_i.$$

The Hermite reduced space can be defined by

$$X_R = \text{span}\{y^j \in X \times U | \mathcal{E}(y^j, u^j) = 0 \text{ and } \frac{\partial y}{\partial u_i}(u^j), \ 1 \leq i \leq m, \ j = 1,\dots,\widetilde{M}\}.$$

Here, $\xi_i^j = \frac{\partial y}{\partial u_i}(u_j)$ can be calculated by solving the sensitivity equation

$$\mathcal{E}_y(y^j, u^j)\xi_i^j = -\mathcal{E}_{u_i}(y_0, u^j). \tag{4.2}$$

Here we can use the pre-selecting step to select u^j as for the Lagrange case.

Suppose we have the reduced basis space X_R. Then we use the Galerkin method to project the equation onto X_R, i.e., $y^m = \sum_{i=1}^m \alpha_i \phi_i \in X_R$ satisfies

$$\mathcal{E}_m(y^m, u)_i = \langle \mathcal{E}(y,u), \phi_i \rangle_{X^* \times X}.$$

Then we consider the reduced-order control problem

$$(4.3) \qquad \min \quad J(y^m,u) \quad \text{subject to} \quad \mathcal{E}_m(y^m,u)=0 \quad \text{and} \quad u\in U_{ad}.$$

It is a finite dimensional constrained minimization problem and can be readily solved by the constrained optimization methods. A necessary optimality condition is given by

$$(4.4) \qquad \begin{cases} (D_y\mathcal{E}_m(y_m,u_m))^t\lambda_m + D_yJ(y_m,u_m)=0 \\ \\ (D_u\mathcal{E}_m(y_m,u_m)(u-u_m),\lambda_m) + D_uJ(y_m,u_m)(u-u_m)\geq 0 \end{cases}$$

for all $u\in U_{ad}$, assuming $D_y\mathcal{E}_m$ at the optimal pair (y_m,u_m) to (4.3) is an isomorphism. Similarly, we have the necessary optimality condition for (4.1): there exists a Lagrange multiplier $\lambda\in X$ such that

$$(4.5) \qquad \begin{cases} (D_y\mathcal{E}(y^*,u^*))^*\lambda + D_yJ(y^*,u^*)=0 \\ \\ \langle D_u\mathcal{E}(y^*,u^*)(u-u^*),\lambda\rangle_{X^*\times X} + D_uJ(y^*,u^*)(u-u^*)\geq 0 \end{cases}$$

for all $u\in U_{ad}$, assuming $D_y\mathcal{E}_m\mathcal{L}(y^*,u^*)$ at the optimal pair (y^*,u^*) to (4.1) is an isomorphism. Suppose u_m and u^* is interior points of U_{ad}. Then we can apply Theorem 3.3 to equation for $(y,\lambda,u)\in X\times X\times U$

$$(4.6) \qquad \begin{cases} \mathcal{E}(y,u)=0 \\ (D_y\mathcal{E}(y,u))^*\lambda + D_yJ(y,u)=0 \\ (D_u\mathcal{E}(y,u))^*\lambda + D_uJ(y,u)=0 \end{cases}$$

In general we have

$$J(y_m,u_m)\leq J(y^m(u),u)=J(y^m(u),u)-J(y(u),u)+J(y(u),u)$$

for $u\in U_{ad}$ and thus setting $u=u^*$, we obtain

$$J(y(u_m),u_m)-J^*\leq J(y(u_m),u_m)-J(y_m,u_m)+J(y^m(u^*),u^*)-J(y(u^*),u^*).$$

Hence, if U_{ad} is compact then

$$J(y(u_m),u_m)-J^*\leq 2M\max_{u\in\mathcal{U}_{ad}}\|y^m(u)-y(u)\|_X$$

for some constant M.

5. The Reduced Basis Method for Flow Control

In this section we discuss vorticity minimization in fluid flows using boundary temperature control and electromagnetic control. We first present the weak variational formulation of the optimal control problems and then discuss their approximations in finite element and reduced basis setting.

Preliminaries. We denote by $L^2(\Omega)$ the collection of Lebesgue square-integrable functions defined on Ω. Let $H^1(\Omega)=\left\{v\in L^2(\Omega):\frac{\partial v}{\partial x_i}\in L^2(\Omega) \text{ for } i=1,2.\right\}$ and $H_0^1(\Omega)=\{v\in H^1(\Omega): v|_\Gamma=0\}$. Vector-valued counterparts of these spaces are denoted by bold-face symbols, for e.g., $\mathbf{H}(\Omega)1=[H^1(\Omega)]^d$ where $d=2$. We denote the norms and inner products for $H^s(\Omega)$ or $\mathbf{H}(\Omega)s$ by $\|\cdot\|_s$ and $(\cdot,\cdot)_s$, respectively. The $L^2(\Omega)$ or $\mathbf{L}^2(\Omega)$ inner product is denoted by $(\cdot,\cdot)$.

5.1. Electrically Conducting Flow Equations and Variational Formulation. In this section we describe the governing equations for a steady electrically conducting flow and their variational formulation. Suppose there is a length scale ℓ, a velocity scale $\mathbf{U}$ and a magnetic-field scale B_0 in the flow, then one can define nondimensional magnetic Reynolds number $R_m = \mu_0 \sigma U \ell$, where μ_0 is the magnetic permiability, Alfven number $Al = B_0^2/\mu_0 \rho U^2$ and Reynolds number $Re = \rho_0 U \ell/\mu$. Next, if we nondimensionalize according to $\mathbf{x} \leftarrow \mathbf{x}/\ell$, $\mathbf{u} \leftarrow \mathbf{u}/\mathbf{U}$, $\mathbf{j} \leftarrow B_0/\ell$, $\mathbf{E} \leftarrow UB_0$ and $p \leftarrow (p - \mathbf{g} \cdot \mathbf{x})/(\rho_0 \mathbf{U}^2)$, we obtain the dimensionless equations for electrically conducting flow:

$$\mathbf{u} \cdot \nabla \mathbf{u} = -\nabla p + Al\,(\mathbf{j} \times \mathbf{B}) + \tfrac{1}{Re}\,\Delta \mathbf{u} \quad \text{and} \quad \nabla \cdot \mathbf{u} = 0 \quad \text{in } \Omega\,,$$

$$\mathbf{j} = R_m\,[-\nabla \phi + (\mathbf{u} \times \mathbf{B})] \quad \text{and} \quad \nabla \cdot \mathbf{j} = 0 \quad \text{in } \Omega\,,$$

$$\nabla \times \mathbf{B} = \mathbf{j} \quad \text{and} \quad \nabla \cdot \mathbf{B} = 0 \quad \text{in } \Omega\,.$$

Here, $\mathbf{u}$ denotes the velocity field, p the pressure field, $\mathbf{j}$ the electric current density, $\mathbf{B}$ the magnetic field and ϕ the electric potential. Moreover, we denote by Ω the flow domain which is bounded in $\mathbb{R}^2$ with boundary Γ.

To simplify the exposition, let us assume that we are dealing with a special case in which the externally applied magnetic field is undisturbed by the flow. That is, we assume that $\mathbf{B}$ is given. Such an assumption can be met in a variety of physical applications, for example in the modeling of electromagnetic pumps and the flow of liquid lithium for fusion reactor cooling blankets, see [10] and [12]. Under this assumption, the term $\mathbf{j} \times \mathbf{B}$ in the Navier-Stokes equations can be written as

$$Al(\mathbf{j} \times \mathbf{B}) = N\,(-\nabla \phi + \mathbf{u} \times \mathbf{B}) \times \mathbf{B}$$

where $N = Al \cdot R_m$ and if we eliminate $\mathbf{j}$ by applying charge conservation condition $\nabla \cdot \mathbf{j} = 0$ to $\mathbf{j} = R_m\,[-\nabla \phi + (\mathbf{u} \times \mathbf{B})]$, we arrive at the following simplified system modeling the flow:

$$(5.1) \qquad \begin{cases} -\frac{1}{Re}\,\Delta \mathbf{u} + \mathbf{u} \cdot \nabla \mathbf{u} + \nabla p + N\,(\nabla \phi - \mathbf{u} \times \mathbf{B}) \times \mathbf{B} = 0 \quad \text{in } \Omega\,, \\[2ex] \nabla \cdot \mathbf{u} = 0 \quad \text{in } \Omega\,, \\[2ex] -\Delta \phi + \nabla \cdot (\mathbf{u} \times \mathbf{B}) = 0 \quad \text{in } \Omega\,. \end{cases}$$

where N is the interaction parameter. The system (5.1) is supplemented with boundary conditions

$$(5.2) \qquad \mathbf{u} = \mathbf{u}_0 \ \text{on } \Gamma, \quad \phi = g \ \text{on } \Gamma_0 \quad \text{and} \quad \mathbf{n} \cdot \mathbf{j} = 0 \ \text{on } \Gamma_1$$

where Γ is the disjoint union $\Gamma = \Gamma_0 \cup \Gamma_1$ and g denotes the control function, namely, electric potential on Γ_0. Such a control can be effected by attaching electric sources with adjustable resistors to the electrode along the boundary. We assume that the flow is two-dimensional and the applied magnetic field $\mathbf{B}$ is perpendicular to the flow plane, i.e., $\mathbf{B} = (0, 0, B_0)^t$, and that the cross product $\mathbf{u} \times \mathbf{B}$ is understood as $(u_1, u_2, 0)^t \times (0, 0, B_0)^t$. Let $\bar{\mathbf{u}} \in \mathbf{H}^1(\Omega)$ and $\bar{\phi} \in H^1(\Omega)$ be such that

$$\bar{\mathbf{u}} = \mathbf{u}_0 \ \text{ on } \Gamma \quad \text{and} \quad \bar{\phi} = g \ \text{ on } \Gamma_0$$

and $V_1 = \{\phi \in H^1(\Omega) : \phi = 0 \text{ on } \Gamma_0\}$. Then, variational formulation of (5.1)–(5.2) is given as follows: seek $\mathbf{u} \in \mathbf{H}_0^1(\Omega) + \bar{\mathbf{u}}$, $p \in L^2(\Omega)$ and $\phi \in H_0^1(\Omega) + \bar{\phi}$ such that

$$(5.3) \qquad \begin{cases} \frac{1}{Re}(\nabla \mathbf{u}, \nabla \mathbf{v}) + b_0(\mathbf{u}, \mathbf{u}, \mathbf{v}) + N(\nabla \phi - \mathbf{u} \times \mathbf{B}, \mathbf{v} \times \mathbf{B}) \\ \qquad -(p, \nabla \cdot \mathbf{v}) = 0 \quad \forall \mathbf{v} \in \mathbf{H}_0^1(\Omega), \\ (\nabla \cdot \mathbf{u}, q) = 0 \quad \forall q \in L^2(\Omega), \\ (\nabla \phi - \mathbf{u} \times \mathbf{B}, \nabla \psi) = 0 \quad \forall \psi \in V_1. \end{cases}$$

Here, the trilinear form $b_0(\cdot,\cdot,\cdot)$ is defined by

$$(5.4) \qquad b_0(\mathbf{u}, \mathbf{v}, \mathbf{w}) = \langle \mathbf{u} \cdot \nabla \mathbf{v}, \mathbf{w} \rangle$$

for $\mathbf{u}, \mathbf{v}, \mathbf{w} \in \mathbf{H}^1(\Omega)$.

5.2. Thermally Convective Flow Equations and Variational Formulation. In this section we describe the governing equations for a steady thermally conducting flow and their variational formulation.

If we assume there is a length scale ℓ, a velocity scale $\mathbf{U}$ and a temperature scale $T_1 - T_0$ in the flow, then one can define nondimensional Prandtl number $Pr = \mu c_p/\kappa$, Grashof number $Gr = \beta \ell^3 \rho_0^2 |\mathbf{g}|(T_1 - T_0)/\mu^2$ and Reynolds number $Re = \rho_0 U \ell/\mu$. Next, if we nondimensionalize according to $\mathbf{x} \leftarrow \mathbf{x}/\ell$, $\mathbf{u} \leftarrow \mathbf{u}/\mathbf{U}$, $T \leftarrow (T - T_0)/(T_1 - T_0)$, and $p \leftarrow (p - \mathbf{g} \cdot \mathbf{x})/(\rho_0 \mathbf{U}^2)$, we obtain the following nondimensional form of Boussinesq equations:

$$(5.5) \qquad \begin{cases} -\frac{1}{Re}\Delta \mathbf{u} + (\mathbf{u} \cdot \nabla)\mathbf{u} + \nabla p + \frac{Gr}{Re^2} T \mathbf{g} = \mathbf{0} \quad \text{in } \Omega, \\ \nabla \cdot \mathbf{u} = 0 \quad \text{in } \Omega, \\ -\frac{1}{RePr}\Delta T + \mathbf{u} \cdot \nabla T = 0 \quad \text{in } \Omega, \end{cases}$$

where $\mathbf{g}$ is a unit vector in the direction of gravitational acceleration.

We consider the boundary condition as follows.

$$(5.6) \qquad \mathbf{u} = \mathbf{u}_0 \text{ on } \Gamma, \quad T = g \text{ on } \Gamma_0 \quad \text{and} \quad \frac{\partial T}{\partial n} = 0 \text{ on } \Gamma_1$$

where g represents the boundary temperature control function. Let $\bar{T} \in H^1(\Omega)$ be such that $\bar{T} = g$ on Γ_0 and $\bar{\mathbf{u}}$ be as defined previously. Then, variational formulation of (5.5)–(5.6) is given by

$$(5.7) \qquad \begin{cases} \frac{1}{Re}(\nabla \mathbf{u}, \nabla \mathbf{v}) + b_0(\mathbf{u}, \mathbf{u}, \mathbf{v}) + (\beta T\, \mathbf{g}, \mathbf{v}) - (p, \nabla \cdot \mathbf{v}) = 0 \quad \forall \mathbf{v} \in \mathbf{H}_0^1(\Omega), \\ (\nabla \cdot \mathbf{u}, q) = 0 \quad \forall q \in L^2(\Omega), \\ b_1(\mathbf{u}, T, \psi) + \kappa(\nabla T, \nabla \psi) \quad \forall \psi \in V_1, \end{cases}$$

for $\mathbf{u} \in \mathbf{H}_0^1(\Omega) + \bar{\mathbf{u}}$, $p \in L^2(\Omega)$ and $T \in H_0^1(\Omega) + \bar{T}$, where $\beta = \frac{Gr}{Re^2}$, $\kappa = \frac{1}{RePr}$ and the trilinear form $b_1(\cdot,\cdot,\cdot)$ is given by

$$b_1(\mathbf{u}, T, \psi) = \langle \mathbf{u} \cdot \nabla T, \psi \rangle.$$

for T, $\psi \in H^1(\Omega)$ and $\mathbf{u} \in \mathbf{H}^1(\Omega)$.

We can establish the existence of solutions to (5.3) and (5.7) using the above properties of the trilinear forms and the Hopf's lemma, see [5].

5.3. Mixed Finite Element Approximation. In order to construct the reduced basis element we use the mixed finite element method [3] to approximate solution to (5.1)–(5.2) and (5.5)–(5.6).

Let us define, using standard finite element notations,

$$X^h = \{v \in C^0(\overline{\Omega}) : v|_K \in P_2(K), \text{ on each element } K\},$$

$$\mathbf{X}^h = \{\mathbf{v} = (v_1, v_2)^T \in \mathbf{C}^0(\overline{\Omega}) : v_i \in X^h,\ i = 1, 2\}$$

and

$$S^h = \{q \in C^0(\overline{\Omega}) : q|_K \in P_1(K), \text{ on each element } K\}.$$

Also we define $\mathbf{X}_0^h = \{v \in \mathbf{X}^h : v = 0 \text{ on } \Gamma\}$ and $X_1^h = X^h \cap V_1$. That is, we choose continuous piecewise quadratic polynomials for both components of the velocity $\mathbf{u}^h$ and electric potential ϕ^h for (5.1) and temperature T^h for (5.5), continuous piecewise linear polynomials for the pressure p^h. On each triangle, the degrees of freedom for quadratic elements are the function values at the vertices and midpoints of each edge; the degrees of freedom for linear elements are the function values at the vertices. Here, the spaces are defined over the same triangulation of the domain $\Omega = \bigcup K$. This selection is known to satisfy the *inf-sup* condition, see [2].

The finite element equation of (5.3) for $\mathbf{u}^h \in \mathbf{X}^h$, $p^h \in S^h$ and $\phi^h \in X^h$ is given by

$$(5.3)^h \qquad \begin{cases} \frac{1}{Re}(\nabla \mathbf{u}^h, \nabla \mathbf{v}^h) + b_0(\mathbf{u}^h, \mathbf{u}^h, \mathbf{v}^h) \\ \qquad + N\,(\nabla\phi^h - \mathbf{u}^h \times \mathbf{B}, \mathbf{v}^h \times \mathbf{B}) - (p^h, \nabla\cdot\mathbf{v}^h) = 0 \quad \forall \mathbf{v}^h \in \mathbf{X}_0^h, \\ (\nabla\cdot\mathbf{u}^h, q^h) = 0 \quad \forall q^h \in S^h, \\ (\nabla\phi^h - \mathbf{u}^h \times \mathbf{B}, \nabla\psi^h) = 0, \quad \forall \psi^h \in X_1^h. \end{cases}$$

where $\mathbf{u}^h|_\Gamma = \mathbf{u}_0^h$ and $\phi^h|_{\Gamma_0} = g^h$ and $\mathbf{u}_0^h$, g^h are the projection of $\mathbf{u}_0$, g onto the finite element spaces, respectively. Similarly, for the Boussinesq equation (5.7) we have

$$(5.7)^h \qquad \begin{cases} \frac{1}{Re}(\nabla \mathbf{u}^h, \nabla \mathbf{v}^h) + b_0(\mathbf{u}^h, \mathbf{u}^h, \mathbf{v}^h) + (\beta T^h \mathbf{g}, \mathbf{v}^h) - (p^h, \nabla\cdot\mathbf{v}^h) = 0\ \forall \mathbf{v}^h \in \mathbf{X}_0^h, \\ (\nabla\cdot\mathbf{u}^h, q^h) = 0 \quad \forall q^h \in S^h, \\ b_1(\mathbf{u}^h, T^h, \psi^h) + \kappa\,(\nabla T^h, \nabla\psi^h) \quad \forall \psi^h \in X_1^h, \end{cases}$$

where $\mathbf{u}^h \in \mathbf{X}^h$, $T^h \in X^h$ satisfy $\mathbf{u}^h|_\Gamma = \mathbf{u}_0^h$ and $T^h|_{\Gamma_0} = g^h$, respectively.

5.4. Boundary Control Problems and Reduced-Order Control Problems. Let U be the control space defined by

$$U = \{g = \sum_{i=1}^{p} g_i\,\chi_i, \quad g_i \in R\}$$

where χ_i is the i-th basis function of U and is the trace of a $H^1(\Omega)$ function onto Γ_0. We consider the minimization of the form

$$(5.8) \qquad \min \quad \|\nabla \times \mathbf{u}\|_\Omega^2 \quad \text{subject to } (5.3) \text{ or } (5.7)$$

where the vorticity $\nabla \times \mathbf{u}$ is defined by $\nabla \times \mathbf{u} = \frac{\partial u_2}{\partial x_1} - \frac{\partial u_2}{\partial x_2}$ and the cost functional defines the total friction forces in Ω.

We define the reduced basis element by the finite element approximation $(5.3)^h$ and $(5.7)^h$, respectively for each control problem. For example, the Lagrange reduced basis element $(\mathbf{u}, p, T)$ given $g^j \in U$ for problem (5.8) subject to (5.7) can be constructed by a solution $(\mathbf{u}^h, p^h, T^h)$ to $(5.7)^h$ given $g^j \in U$. For the case of the boundary control problem, the reduced basis space $X_R \subset \mathbf{X}^h \times X^h$ should consist of the basis element Φ_0^h that corresponds to the reference control $\bar{g} \in U$, the element Φ_j^h that corresponds to the j-th control in the direction of χ_j, and the test functions $\Psi^h \in X_R \cap (\mathbf{X}_0^h \times X_1^h)$. Since $\mathbf{u}^h$ satisfies the pseudo-divergence condition $(\nabla \cdot \mathbf{u}^h, q^h) = 0$ for all $q^h \in S^h$ we have the reduced-order control problem;

$$
(5.9) \qquad \begin{cases} \min \quad \|\nabla \times \mathbf{u}^h\|_\Omega^2 \quad \text{subject to} \\[2ex] \frac{1}{Re}(\nabla \mathbf{u}^h, \nabla \mathbf{v}^h) + b_0(\mathbf{u}^h, \mathbf{u}^h, \mathbf{v}^h) + (\beta T^h\, \mathbf{g}, \mathbf{v}^h) \\[2ex] \qquad + b_1(\mathbf{u}^h, T^h, \psi^h) + \kappa\, (\nabla T^h, \nabla \psi^h) = 0 \end{cases}
$$

for all $\Psi^h = (\mathbf{v}^h, \psi^h) \in X_R \cap (\mathbf{X}_0^h \times X_1^h)$. Here, the element in X_R is represented by

$$
(u^h, T^h) = \Phi_0^h + \sum \alpha_i\, \Psi_i^h + \sum_{j=1}^{p} \frac{(g_j - \bar{g}_j)}{\delta_j}\, \Phi_i^h
$$

where $\Phi_0^h = (u^h, T^h)_0$ is a solution to $(5.7)^h$ corresponding to the reference control, $\Phi_j^h = (u^h, T^h)_j - (u^h, T^h)_0$ for $1 \le j \le p$, with $(u^h, T^h)_j$ being a solution to $(5.7)^h$ corresponding to $\bar{g} + \delta_j\, \chi_j$ and $\{\Psi_i^h\}$ are a basis of the test function space $X_R \cap (\mathbf{X}_0^h \times X_1^h)$.

Similarly, for problem (5.8) subject to (5.3) we have the reduced-order control problem;

$$
(5.10) \qquad \begin{cases} \min \quad \|\nabla \times \mathbf{u}^h\|_\Omega^2 \quad \text{subject to} \\[2ex] \frac{1}{Re}(\nabla \mathbf{u}^h, \nabla \mathbf{v}^h) + b_0(\mathbf{u}^h, \mathbf{u}^h, \mathbf{v}^h) + N\,(\nabla \phi^h - \mathbf{u}^h \times \mathbf{B}, \mathbf{v}^h \times \mathbf{B}) \\[2ex] \qquad + (\nabla \phi^h - \mathbf{u}^h \times \mathbf{B}, \nabla \psi^h) = 0, \end{cases}
$$

for all $\Psi^h = (\mathbf{v}^h, \psi^h) \in X_R \cap (\mathbf{X}_0^h \times X_1^h)$.

6. Computational Results

In this section we will give computational result by implementing the computational procedure for the specific control problem described in §1 using the two proposed control mechanisms. We select the backward facing step channel for our study, a schematic of this geometry and the finite element grid are given in Figure 6.1 and Figure 6.2, respectively. It has been observed in a number of computational and experimental study on this specific channel geometry that a recirculation appears near the corner region for large Reynolds number. Our aim is to remove/suppress the recirculation by means of boundary control.

FIGURE 6.1. Schematic of backward-facing step channel

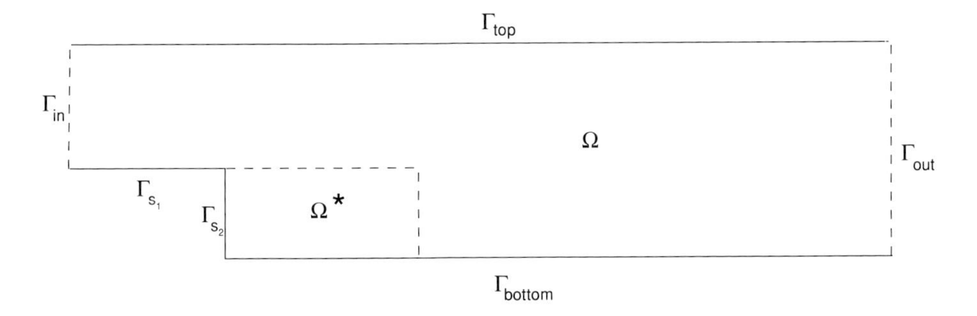

FIGURE 6.2. Finite element grid for the channel geometry

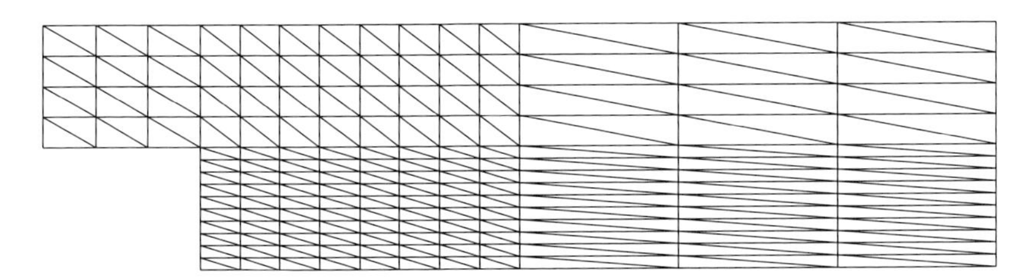

6.1. Boundary Temperature Control. The aim is to shape the flow to a desired configuration which in our study means to remove the recirculation by means of controlling the temperature along the bottom part of the boundary.

We assume that the inflow and outflow are parabolic, i.e. we take the inflow to be $\mathbf{u}_i = 8(y-0.5)(1-y)$ and the outflow to be $\mathbf{u} = \mathbf{u}_o = y(1-y)$. We take the Reynolds number to be 200 and $\frac{Gr}{Re^2}$ to be 1. For the temperature we used the following boundary conditions:

$$\begin{array}{ll} \Gamma_{s_1} \text{ and } \Gamma_{\text{out}} : & \frac{\partial T}{\partial n} = 0 \\ \Gamma_{s_2} \text{ and } \Gamma_{\text{bottom}} : & T = g \\ \Gamma_{\text{in}} \text{ and } \Gamma_{\text{top}} : & T = 1. \end{array}$$

Figure 6.3 qualitatively demonstrate the flow for high Reynolds number. Here our objective is to remove the recirculation that occurs in the corner region Ω^*. Therefore we minimize vorticity in the corner region Ω^* which is chosen to be $\Omega^* = (1,3)\times(0,.5)$. This leads us to a constrained minimization problem of the type (5.8) and this is solved by the reduced basis computational method described in §5.4.

Basis elements are computed with g=1, 0.875, 0.75, 0.625, 0.5, 0.3775, 0.25 and denoted by $(\mathbf{u}_i, T_i)$, $i = 1, \ldots, 7$. The test functions $\{\Psi_1, \ldots, \Psi_5\}$ are chosen so that they have zero boundary conditions. The trial function Φ_1 corresponds to the control force such that $\Phi_1 = 0$ everywhere on the boundary except on the bottom.

Then we set

$$(\mathbf{u}, T) = \Phi_0 + \frac{(g-\bar{g})}{\delta}\Phi_1 + \sum_{i=1}^{5} \alpha_i \Psi_i,$$

FIGURE 6.3. The velocity field for the uncontrolled case with Re=200

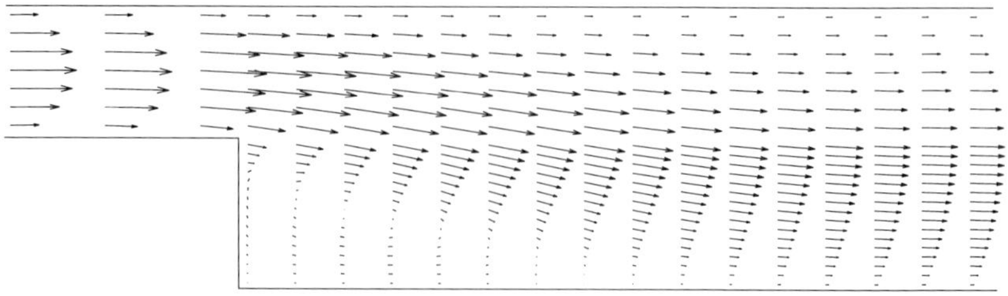

FIGURE 6.4. The velocity field for the controlled case with Re=200: Temperature control

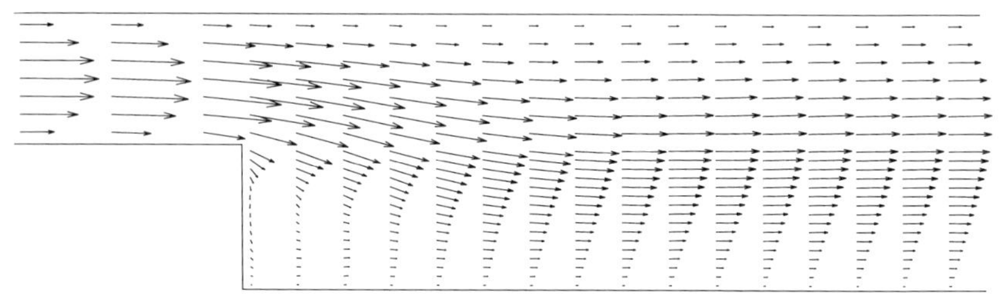

where $\bar{g} = 1$, $\delta = -0.75$ and

$$\begin{cases} \Phi_0 = (\mathbf{u}_1, T_1), \quad \Phi_1 = (\mathbf{u}_7, T_7) - (\mathbf{u}_1, T_1), \\ \Psi_1 = (\mathbf{u}_7, T_7) - 2(\mathbf{u}_6, T_6) + (\mathbf{u}_5, T_5), \quad \Psi_2 = (\mathbf{u}_6, T_6) - 2(\mathbf{u}_5, T_5) + (\mathbf{u}_4, T_4), \\ \Psi_3 = (\mathbf{u}_5, T_5) - 2(\mathbf{u}_4, T_4) + (\mathbf{u}_3, T_3), \quad \Psi_4 = (\mathbf{u}_4, T_4) - 2(\mathbf{u}_3, T_3) + (\mathbf{u}_2, T_2), \\ \Psi_5 = (\mathbf{u}_3, T_3) - 2(\mathbf{u}_2, T_2) + (\mathbf{u}_1, T_1). \end{cases}$$

The constrained minimization problem is solved by employing Newtons method to the necessary optimality condition (4.4). We obtained the boundary temperature control $T = 0.516$ in 7 Newton iterations. The computed control was then used in the full system to simulate the flow. The resulting flow shown in Figure 6.4 shows significant reduction in the size of the recirculation region.

6.2. Electromagnetic Control. In this problem, control is effected through boundary electric potential on the top and bottom boundary of the backward facing step channel. A magnetic field $\mathbf{B} = (0, 0, 1)$ is applied into the fluid. The boundary conditions for the velocity are the same as in the preceding control problem except for the electric potential whose boundary conditions are as follows:

$$\begin{array}{ll} \Gamma_{s2}, \Gamma_{\text{in}} \text{ and } \Gamma_{\text{out}}: & \frac{\partial \phi}{\partial n} = 0 \\ \Gamma_{\text{top}}: & \phi = g_1 \\ \Gamma_{\text{bottom}}: & \phi = g_2 \\ \Gamma_{s2}: & \phi = 1. \end{array}$$

We take the Reynolds number to be $Re = 200$ and the interaction parameter to be $N = 1$. Here also our objective is to suppress the recirculation that occurs in the corner region Ω^* and thereby obtain a relatively smoother flow. Therefore we minimize

FIGURE 6.5. The velocity field for the controlled case with Re=200: Electromagnetic control

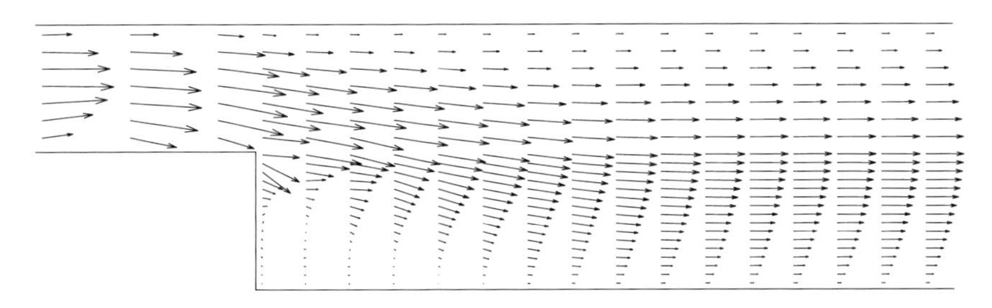

vorticity in the corner region Ω^*. This leads us to a constrained minimization problem of the type (5.8) and we use the reduced basis computational method described in §5.4 to solve it.

The basis elements were computed with

$$(g_1, g_2)=(1,1),\ (1,0.5),(1,0),(0.5,1),(0.5,0.5),\ (0.5,0),(0,1),(0,0.5)$$

and the corresponding elements are denoted by $(\mathbf{u}_i, \phi_i)$, $i = 1, \ldots, 8$. The test functions $\{\Psi_1, \ldots, \Psi_5\}$ are chosen so that they have zero boundary conditions. The trial functions Φ_1 and Φ_2 corresponds to the control force such that $\Phi_1 = 0$ everywhere on the boundary except on the top and $\Phi_2 = 0$ everywhere on the boundary except on the bottom. Then we set

$$(\mathbf{u}, \phi) = \Phi_0 + \frac{(g_1 - \bar{g}_1)}{\delta_1}\Phi_1 + \frac{(g_2 - \bar{g}_2)}{\delta_2}\Phi_2 + \sum_{i=1}^{5} \alpha_i \Psi_i,$$

where $\bar{g}_1 = \bar{g}_2 = 1$, $\delta_1 = \delta_2 = -0.5$ and

$$\begin{cases} \Phi_0=(\mathbf{u}_1,\phi_1), \quad \Phi_1=(\mathbf{u}_3,\phi_3), \quad \Phi_2=(\mathbf{u}_7,\phi_7), \\ \Psi_1=(\mathbf{u}_1,\phi_1)-2(\mathbf{u}_2,\phi_2)-2(\mathbf{u}_4,\phi_4), \quad \Psi_2=(\mathbf{u}_1,\phi_1)-(\mathbf{u}_3,\phi_3)-(\mathbf{u}_7,\phi_7), \\ \Psi_3=(\mathbf{u}_1,\phi_1)-(\mathbf{u}_5,\phi_5)-(\mathbf{u}_2,\phi_2)-(\mathbf{u}_4,\phi_4), \quad \Psi_4=(\mathbf{u}_1,\phi_1)-(\mathbf{u}_8,\phi_8)-(\mathbf{u}_2,\phi_2), \\ \Psi_5=(\mathbf{u}_1,\phi_1)-(\mathbf{u}_6,\phi_6)-(\mathbf{u}_4,\phi_4). \end{cases}$$

We employed the Newtons method to the necessary optimality condition (4.4) and obtained the boundary controls $\phi_{\text{top}} = 1.0423$ and $\phi_{\text{top}} = 1.7735$ respectively, in 5 Newton iterations. The computed control was then used in the full system to simulate the flow. The resulting flow shown in Figure 6.5 shows significant reduction in the size of the recirculation region.

In conclusion, we have demonstrated the feasibility of using reduced basis method in both one parameter and two parameter control setting. Two different control mechanisms have been used in two different fluid flow setting. Our numerical results seem to indicate that the reduced basis method can be successfully used in flow control problems with significant reduction in computational cost compare to the results presented in [3], [5] and [9] for the same problems where computations were performed by directly applying finite element methods to the optimal control problems.

References

1. B.O. Almroth, P. Stern and F.A. Brogan, *Automatic Choice of Global Shape Functions in Structural Analysis*, AIAA Journal, **16**, pp. 525–528, 1978.
2. V. Girault and P. Raviart, *Finite Element Methods for Navier-Stokes Equations*, Springer-Verlag, New York, 1986.
3. L.S. Hou and S.S. Ravindran, *Computations of Boundary Optimal Control Problems for an Electrically Conducting Fluid*, Journal of Computational Physics, **128**, No.2, pp. 319–330, 1996.
4. K. Ito and S.S. Ravindran, *A Reduced Order Method for Simulation and Control of Fluid Flows*, J. Computational Physics (submitted).
5. K. Ito and S.S. Ravindran, *Optimal Control of Thermally Convected Fluid Flows*, SIAM J. Scientific Computing, **19**, No. 5, 1998.
6. A.K. Noor, *Recent Advances in Reduction Methods for Nonlinear Problems*, Computers & Structures, **13**, pp. 31–44, 1981.
7. A.K. Noor, C.M. Anderson and J.M. Peters, *Reduced Basis Technique for Collapse Analysis of Shells*, AIAA Journal, **19**, pp. 393–397, 1981.
8. J.S. Peterson, *The Reduced Basis Method for Incompressible Viscous Flow Calculations*, SIAM J. Scientific Computing, **10**, pp. 777–786, 1989.
9. S.S. Ravindran, *Computations of Optimal Control for Fluid Flows*, in *Optimal Control of Viscous Flows*, S. S. Sritharan, Ed., Frontiers in Applied Mathematics, SIAM, 1997.
10. J.A. Shercliff, *A Textbook of Magnetohydrodynamics*, Pergamon Press, London, 1965.
11. S.S. Sritharan, *Optimal Feedback Control of Hydrodynamics: A Progress Report*, in Flow Control, M.D. Gunzburger, Ed., Springer-Verlag, New York, 1995.
12. J.S. Walker, *Large Interaction Parameter Magnetohydrodynamics and Applications in Fusion Reactor Technology*, in *Fluid Mechanics in Energy Conversion*, J. Buchmaster, Ed., SIAM, Philadelphia, 1980.

K. Ito
Center for Research in Scientific Computation
Department of Mathematics
North Carolina State University
Raleigh, NC 27695-8205, USA
e-mail: kito@eos.ncsu.edu

S.S. Ravindran
Center for Research in Scientific Computation
Department of Mathematics
North Carolina State University
Raleigh, NC 27695-8205, USA
e-mail: ravi@eos.ncsu.edu

International Series of Numerical Mathematics
Vol. 126,

Proximal Penalty Method for Ill-Posed Parabolic Optimal Control Problems*

A. KAPLAN AND R. TICHATSCHKE
Department of Mathematics
Technical University Darmstadt

FB IV (Math.)
University Trier

ABSTRACT. The application of a proximal point approach to ill-posed convex control problems governed by linear parabolic equations is studied. A stable penalty method is constructed by means of multi-step proximal regularization (only w.r.t. the control functions) in the penalized problems. For distributed control problems with state constraints convergence of the approximately determined solutions of the regularized problems to an optimal process is proved.

1991 *Mathematics Subject Classification.* 49J20, 49M37, 65K10, 65M30

Key words and phrases. Prox-regularization, ill-posed parabolic control problems, distributed control, penalty methods.

1. Introduction

In this paper the proximal point approach coupled with the penalty technique is developed for solving ill-posed convex parabolic control problems with state constraints. The investigation is concentrated on problems governed by linear parabolic equations, the objective functional and the sets of admissible controls and states are assumed to be convex.

Usually, convergence of numerical methods for such problems is studied under the additional assumption that the objective functional is strictly (or strongly) convex w.r.t. the control, or that the optimal control possesses the bang-bang property. We refer here to ALT AND MACKENROTH (1989), GLASHOFF AND SACHS (1977), HACKBUSCH AND WILL (1984), KNOWLES (1982), LASIECKA (1980, 1984), MACKENROTH (1982-83, 1987), MALANOWSKI (1981), TROELTZSCH (1987).

The first results, connected with the use of the penalty technique for control problems are obtained by LIONS (1968) and BALAKRISHNAN (1968 A,B), for further applications see LIONS (1985). Penalization of the state equation permits to handle with control and state variables as independent ones. In BERGOUNIOUX (1992, 1994), for convex elliptic and parabolic control problems with state constraints, penalty methods have been used in order to prove the existence of Lagrange multipliers under weak qualification hypothesises.

---

*Supported by the German Research Foundation (DFG).

In all these investigations strong convexity of the objective functional was one of the essential conditions.
The paper presented here deals with convex parabolic control problems without additional assumptions mentioned above. So, the problem may be non-uniquely solvable, moreover, we don't exclude that the set of optimal controls can be unbounded. Using the scheme of multi-step regularization developed in KAPLAN AND TICHATSCHKE (1994) for abstract convex variational problems, a partial proximal regularization (w.r.t. the control only) of the family of penalized problems is performed. This permits to handle with well-posed auxiliary problems and to ensure weak convergence of their approximately determined solutions to an optimal process, as well as convergence of the corresponding values of the objective functional to the optimal value of the original problem.
For convex elliptic control problems an analogous approach has been realized in HETTICH, KAPLAN AND TICHATSCHKE (1994, 1996). In the last two decades proximal point technique is sucessfully developed for solving variational inequalities with monotonous operators, including convex optimization problems and saddle-point problems. ECKSTEIN AND BERTSEKAS (1992) have shown a relationship between the proximal point method and the Douglas-Rachford splitting method, pointing out new application fields, especially in mathematical physics. Nevertheless, besides the papers mentioned here, we don't know publications, where proximal point technique was applied to control problems.

2. Formulation of the control problem

Let $\Omega \subset \mathbb{R}^n$ be a bounded domain with a boundary $\partial\Omega$ of the class C^2, Ω be locally situated on one side of $\partial\Omega$, and

$$Q = \Omega \times]0,T[, \quad \Sigma = \partial\Omega \times]0,T[.$$

In the sequel we use the following notation for functional spaces:
$L_2(0,T;Z)$ – space of functions with range in a Hilbert space Z, square integrable on $(0,T)$,

$$\|v\|_{L_2(0,T;Z)} = \left(\int_0^T \|v(t)\|_Z^2 dt \right)^{1/2};$$

$\|\cdot\|_{0,Q}$ – norm of an element in $L_2(0,T;L_2(\Omega))$;
$C([0,T];Z)$ – space of continuous functions on $[0,T]$ with range in Z,

$$\|v\|_{C([0,T];Z)} = max_{0 \leq t \leq T} \|v(t)\|_Z;$$

$H^s(\Omega)$, $H_0^s(\Omega)$ – standard Sobolev spaces, $L_2(\Omega) = H^0(\Omega)$, $\|\cdot\|_{s,\Omega}$ – norm in $H^s(\Omega)$;
$\|\cdot\|_{0,s,\Omega}$ – norm in $L_2(0,T;H^s(\Omega))$ for $s \geq 1$;
$(\cdot,\cdot)_\Omega$ – inner product in $L_2(\Omega)$;
$X \hookrightarrow H$ – continuous embedding of the space X into H.

We consider the parabolic equation

$$\frac{\partial y}{\partial t}(x,t) + Ay(x,t) = u(x,t) \quad \text{a. e. in } Q, \tag{2.1}$$

$$y(x,0) = 0 \text{ in } \Omega, \tag{2.2}$$

$$y(x,t) = 0 \text{ on } \Sigma, \tag{2.3}$$

where the elliptic operator A is given by

$$Ay = -\sum_{i,j=1}^{n} \frac{\partial}{\partial x_i}\left(a_{ij}(x,t)\frac{\partial y}{\partial x_j}\right) + a_0(x,t)y, \tag{2.4}$$

with $a_{ij} \in C^2(\bar{Q})$, $a_0 \in C^2(\bar{Q})$ such that for all $(x,t) \in \bar{Q}, \xi \in \mathbb{R}^n$ and some $d_0 > 0$

$$a_0(x,t) \geq 0 \text{ and } \sum_{i,j=1}^{n} a_{ij}(x,t)\xi_i\xi_j \geq d_0 \sum_{i=1}^{n} \xi_i^2. \tag{2.5}$$

For each $u \in U \equiv L_2(0,T;L_2(\Omega))$ Problem (2.1)–(2.3) is uniquely solvable, and its solution y_u belongs to

$$W = \{y \in L_2(0,T;H^2(\Omega) \cap H_0^1(\Omega)) : \frac{dy}{dt} \in L_2(0,T;L_2(\Omega)), y(x,0) = 0 \text{ in } \Omega\} \tag{2.6}$$

(see, for instance LIONS AND MAGENES (1968), vol. 1.). The space W endowed by the norm

$$\|y\|_W = \left(\|y\|_{0,2,\Omega}^2 + \left\|\frac{dy}{dt}\right\|_{0,Q}^2\right)^{1/2} \tag{2.7}$$

is a Hilbert space. Moreover, (ibid., Theorem 1.3.1)

$$W \hookrightarrow C([0,T];H_0^1(\Omega)),$$

and the operator $\mathcal{T} : \mathcal{T}u = y_u$ is continuous as a mapping from $L_2(0,T;L_2(\Omega))$ into $C([0,T];H_0^1(\Omega))$ (see LIONS AND MAGENES (1968), vol. 2.).

In order to formulate the control problem we introduce the space

$$Y = \{y \in L_2(0,T;H_0^1(\Omega)) : \frac{dy}{dt} + Ay \in L_2(0,T;L_2(\Omega)),\ y(x,0) = 0 \text{ in } \Omega\} \tag{2.8}$$

which coincides algebraically with W: Indeed, regarding the smoothness of a_{ij}, a_0, the inclusion $W \subset Y$ is obvious, and the inclusion $Y \subset W$ is a consequence of the fact that $y_u \in W$ for each $u \in U$.

Using the inequality

$$\|y\|_W \leq \bar{c}\|y\|_Y \ \forall y \in Y,$$

which follows from the L_p-estimates for the solutions of parabolic equations (see LADYSHENSKAJA, SOLONNIKOV AND URAL'ZEWA (1968), Theorem 4.9.1), one can easily show that the space Y with the norm

$$\|y\|_Y = \left\|\frac{dy}{dt} + Ay\right\|_{0,Q} \tag{2.9}$$

is a Hilbert space, too; moreover,

$$Y \hookrightarrow C([0,T];H_0^1(\Omega)).$$

The approach suggested will be presented with the following model problem

Problem (P) :

$$\text{minimize } J(u) = \|y_u(T) - y_d\|_{0,\Omega}^2 \text{ subject to } u \in U_{ad},\ y_u \in G,$$

where U_{ad} and G are convex and closed sets in the spaces $L_2(0,T;L_2(\Omega))$ and Y, respectively; $y_d \in L_2(\Omega)$ is a given function and it is supposed that $\{u \in U_{ad} : y_u \in G\} \neq \emptyset$.

Due to the continuity of the mapping

$$\mathcal{T} :\ L_2(0,T;L_2(\Omega)) \longrightarrow C([0,T];H_0^1(\Omega))$$

the functional J is continuous on $L_2(0,T;L_2(\Omega))$. Therefore, if U_{ad} is bounded, Problem (P) is solvable, but in general, non-uniquely solvable. If U_{ad} is unbounded, it may happen that the set of optimal controls is empty or unbounded.

We introduce the space

$$\Xi = Y \times L_2(0,T;L_2(\Omega)),$$

endowed by the natural norm: For $z = (y,u)$ with $y \in Y,\ u \in L_2(0,T;L_2(\Omega))$,

$$\|z\|_\Xi = \left(\|y\|_Y^2 + \|u\|_{0,Q}^2\right)^{1/2}. \tag{2.10}$$

3. Regularized penalty method (RP-method)

Method (Multi-step regularization)

Let $\{r_i\}, \{\epsilon_i\}, \{\chi_i\}$, and $\{\delta_i\}$ be positive sequences with

$$lim_{i\to\infty} r_i = 0,\ \ sup_i\ r_i < 1, lim_{i\to\infty} \frac{\epsilon_i}{\chi_i} = 0,\ \ sup_i\ \chi_i \leq 2,$$

and $u^0 \in U_{ad}$.

Step i: Given $u^{i-1} \in U_{ad}$.

a) Set $u^{i,0} := u^{i-1},\ \ s := 1$.

b) Given $u^{i,s-1}$, let

$$(\bar{y}^{i,s}, \bar{u}^{i,s}) = \arg\min\{\Psi_{i,s}(y,u)\ :\ (y,u) \in G \times U_{ad}\} \tag{3.1}$$

with

$$\Psi_{i,s}(y,u) = \|y(T) - y_d\|_{0,\Omega}^2 + \frac{1}{r_i}\left\|\frac{dy}{dt} + Ay - u\right\|_{0,Q}^2 + \frac{\chi_i}{2}\left\|u - u^{i,s-1}\right\|_{0,Q}^2. \tag{3.2}$$

Compute an approximation $(y^{i,s}, u^{i,s}) \in G \times U_{ad}$ of $(\bar{y}^{i,s}, \bar{u}^{i,s})$ such that

$$\left\|(y^{i,s}, u^{i,s}) - (\bar{y}^{i,s}, \bar{u}^{i,s})\right\|_\Xi \leq \frac{\epsilon_i}{\chi_i}. \tag{3.3}$$

c) If $\|u^{i,s} - u^{i,s-1}\|_{0,Q} > \delta_i$, set $s := s+1$ and repeat b).

Otherwise, set $u^i := u^{i,s}, s(i) := s, i := i+1$, and continue with Step $(i+1)$.

Of course, the stopping rule (3.3) is not yet practicable. But, as it will be shown below, the functional $\Psi_{i,s}$ is strongly convex on Ξ. This usually permits to satisfy (3.3) by means of a stopping criterion of an algorithm, minimizing $\Psi_{i,s}$ on $G \times U_{ad}$.

4. Convergence of the RP-method

For shortness, in the sequel we will use the following abbreviations:

$$z = (y, u),\ z^* = (y^*, u^*),\ \bar{z}^{i,s} = (\bar{y}^{i,s}, \bar{u}^{i,s}) \text{ etc.} \tag{4.1}$$

Let

$$|z| = \left(\left\| \frac{dy}{dt} + Ay - u \right\|_{0,Q}^2 + \|u\|_{0,Q}^2 \right)^{1/2}. \tag{4.2}$$

We start with some preliminary statements.

Lemma 4.1. *On the space* Ξ *relation* (4.2) *defines a new norm* $|\cdot|$, *which is equivalent to the norm* $\|\cdot\|_\Xi$*:*

$$\frac{1}{\sqrt{3}} \|z\|_\Xi \le |z| \le \sqrt{3}\|z\|_\Xi. \tag{4.3}$$

Proof. The right-sided inequality in (4.3) is obvious, and

$$\begin{aligned} |z|^2 &= \left\| \frac{dy}{dt} + Ay - u \right\|_{0,Q}^2 + \|u\|_{0,Q}^2 \\ &= \left\| \sqrt{\frac{2}{3}} \left(\frac{dy}{dt} + Ay \right) - \sqrt{\frac{3}{2}} u \right\|_{0,Q}^2 + \frac{1}{3} \left\| \frac{dy}{dt} + Ay \right\|_{0,Q}^2 \\ &\quad + \frac{1}{3}\|u\|_{0,Q}^2 + \frac{1}{6}\|u\|_{0,Q}^2 \ge \frac{1}{3}\|z\|_\Xi^2 \end{aligned}$$

proves the left-sided inequality. □

Lemma 4.2. *The functional* $\Psi_{i,s}$ *is continuous and strongly convex on* Ξ.

Proof. Due to Lemma 4.1, continuity of $\|y(T) - y_d\|_{0,\Omega}$ on Ξ ensures continuity of $\Psi_{i,s}$. Now, let us prove strong convexity. To this end, we rewrite the functional as follows:

$$\begin{aligned} \Psi_{i,s}(y,u) &= \|y(T) - y_d\|_{0,\Omega}^2 + \left(\frac{1}{r_i} - \frac{\chi_i}{2} \right) \left\| \frac{dy}{dt} + Ay - u \right\|_{0,Q}^2 \\ &\quad + \frac{\chi_i}{2} \left\| \frac{dy}{dt} + Ay - u \right\|_{0,Q}^2 + \frac{\chi_i}{2} \left\| u - u^{i,s-1} \right\|_{0,Q}^2 \\ &= \Bigg[\|y(T) - y_d\|_{0,\Omega}^2 + \left(\frac{1}{r_i} - \frac{\chi_i}{2} \right) \left\| \frac{dy}{dt} + Ay - u \right\|_{0,Q}^2 \\ &\quad + \frac{\chi_i}{2} \left\| u^{i,s-1} \right\|_{0,Q}^2 - \chi_i \int_0^T \left(u(t), u^{i,s-1}(t) \right)_\Omega dt \Bigg] + \frac{\chi_i}{2} |(y,u)|^2. \end{aligned} \tag{4.4}$$

Because of $r_i < 1, \chi_i \le 2$ the term in the square brackets is a quadratic functional with a non-negative quadratic term in (y, u), hence, it is a convex functional. Therefore, taking into account Lemma 4.1, $\Psi_{i,s}$ is strongly convex on the space Ξ with the norm $\|\cdot\|_\Xi$ or $|\cdot|$. □

The following result is an analogon of Lemma 22.3 in KAPLAN AND TICHATSCHKE (1994). Let Z be a Hilbert space with an inner product $((\cdot,\cdot))_Z$ and a norm $\|\cdot\|_Z$; Z_1 be a closed subspace of Z and $\mathcal{P}: Z \to Z_1$ be the orthogonal projection operator. We consider the problem

$$\text{minimize } \Phi(z) = a(z,z) - \ell(z) \text{ subject to } z \in K, \tag{4.5}$$

where $a(\cdot,\cdot)$ is a continuous, symmetric and positive semi-definite bilinear form on $Z \times Z$, ℓ is a linear, continuous functional on Z and $K \subset Z$ is a convex, closed set. Further, suppose that $b(\cdot,\cdot)$ is a second symmetric bilinear form on $Z \times Z$ such that

$$0 \le b(z,z) \le a(z,z) \text{ for } z \in Z, \tag{4.6}$$

and, with some $\beta > 0$,

$$b(z,z) + ||\mathcal{P}z||_Z^2 \ge \beta ||z||_Z^2 \text{ for all } z \in Z. \tag{4.7}$$

By

$$|z|_Z^2 = b(z,z) + ||\mathcal{P}z||_Z^2 \tag{4.8}$$

another norm is defined on Z, which is equivalent to $||\cdot||_Z$ according to the obvious relation

$$(M+1)||z||_Z^2 \ge |z|_Z^2 \ge \beta ||z||_Z^2$$

with $M \ge \sup_{z \ne 0} \frac{b(z,z)}{||z||_Z^2}$.

Lemma 4.3. *For each $a^0 \in Z$ and*

$$a^1 = \arg\ \min \left\{\Phi(z) + \frac{\chi}{2}||\mathcal{P}z - \mathcal{P}a^0||_Z^2 \ : z \in K\right\} \tag{4.9}$$

($\chi \in (0,2]$ is kept fixed) the following inequalities are true for all $z \in K$:

$$|a^1 - z|_Z^2 - |a^0 - z|_Z^2 \le -||\mathcal{P}a^1 - \mathcal{P}a^0||_Z^2 + \frac{2}{\chi}[\Phi(z) - \Phi(a^1)] \tag{4.10}$$

and

$$|a^1 - z|_Z \ \le \ |a^0 - z|_Z + \eta(z), \tag{4.11}$$

with

$$\eta(z) = \begin{cases} \left[\frac{2}{\chi}(\Phi(z) - \Phi(a^1))\right]^{1/2} & \text{if } \Phi(z) > \Phi(a^1) \\ 0 & \text{otherwise} \end{cases}.$$

If, moreover, $||\mathcal{P}a^1 - \mathcal{P}a^0||_Z \ge \delta \ge \eta(z)$, then

$$|a^1 - z|_Z \le |a^0 - z|_Z + \frac{\eta^2(z) - \delta^2}{2|a^0 - z|_Z}. \tag{4.12}$$

Now we come back to the control problem.

• **The case of a bounded set U_{ad}**

Assume there exists a point $\tilde{u} \in U_{ad}$ such that

$$y_{\tilde{u}} \equiv \mathcal{T}\tilde{u} \in \ int\ G\ (in\ Y).$$

Lemma 4.4. *Let (y^*, u^*) be an optimal process of Problem (P), $\nu \in (0, \frac{1}{2})$ be an arbitrary number. Suppose that $u^{i,s-1} \in U_{ad}$ is arbitrarily chosen and $(\bar{y}^{i,s}, \bar{u}^{i,s})$ and $(y^{i,s}, u^{i,s})$ are defined by (3.1) and (3.3) with this $u^{i,s-1}$. Then there exist two constants $d(\nu)$ and d_1, independent of i, $s \geq 1$, $u^{i,s-1}$, $\{\epsilon_i\}$, $\{r_i\}$, and $\{\chi_i\}$, such that*

$$J_i(y^*, u^*) - J_i(\bar{y}^{i,s}, \bar{u}^{i,s}) < d(\nu) r_i^{1-2\nu} \tag{4.13}$$

and

$$|(y^{i,s}, u^{i,s}) - (y^*, u^*)| < d_1, \tag{4.14}$$

with

$$J_i(y, u) = \|y(T) - y_d\|_{0,\Omega}^2 + \frac{1}{r_i} \left\| \frac{dy}{dt} + Ay - u \right\|_{0,Q}^2 . \tag{4.15}$$

Underline that $|\cdot|$ is defined by (4.2), and the controlling sequences $\{r_i\}, \{\epsilon_i\}$, and $\{\chi_i\}$ are chosen according to the RP-method.

Proof. The existence of the points $(\bar{y}^{i,s}, \bar{u}^{i,s})$ and $(y^{i,s}, u^{i,s})$ is guaranteed by Lemma 4.2. Now, we introduce the following notation:

$$\hat{y}^{i,s} = \mathcal{T}\bar{u}^{i,s}, \quad \hat{\bar{z}}^{i,s} = \left(\hat{y}^{i,s}, \bar{u}^{i,s}\right), \quad z(u) = (\mathcal{T}u, u),$$

$$\tau_{min} = \min_{w \in \partial G} \|\mathcal{T}\tilde{u} - w\|_Y, \quad \tau_{max} = \max_{u \in U_{ad}} \|z(\tilde{u}) - z(u)\|_\Xi,$$

$$w^{i,s} = \arg\min_{v \in G \times U_{ad}} \left\| \hat{\bar{z}}^{i,s} - v \right\|_\Xi .$$

Note that $\tau_{max} > \tau_{min}$ if $\{\mathcal{T}u : u \in U_{ad}\} \cap \partial G \neq \emptyset$. In case of $\hat{y}^{i,s} \notin G$ we define the points

$$h^{i,s} \in \left\{ z(\tilde{u}) + \lambda \left(\hat{\bar{z}}^{i,s} - z(\tilde{u}) \right) \; : \; \lambda \geq 0 \right\} \cap \{\partial G \times U_{ad}\},$$

and (if $h^{i,s} \neq w^{i,s}$)

$$k^{i,s} \in \left\{ z(\tilde{u}) + \lambda \left(\hat{\bar{z}}^{i,s} - w^{i,s} \right) \; : \; \lambda \geq 0 \right\} \cap \left\{ h^{i,s} + \mu \left(h^{i,s} - w^{i,s} \right) \; \mu \geq 0 \right\}.$$

Obviously, the points $h^{i,s}$ and $k^{i,s}$ are uniquely determined, and

$$\frac{\left\| \hat{\bar{z}}^{i,s} - w^{i,s} \right\|_\Xi}{\|k^{i,s} - z(\tilde{u})\|_\Xi} = \frac{\left\| \hat{\bar{z}}^{i,s} - h^{i,s} \right\|_\Xi}{\|h^{i,s} - z(\tilde{u})\|_\Xi}.$$

Due to the trivial implication $-1 \neq \frac{\alpha}{\beta} = \frac{\gamma}{\delta} \Rightarrow \frac{\alpha}{\alpha+\beta} = \frac{\gamma}{\gamma+\delta}$, we obtain

$$\begin{aligned} \left\| \hat{\bar{z}}^{i,s} - h^{i,s} \right\|_\Xi &= \frac{\left\| \hat{\bar{z}}^{i,s} - z(\tilde{u}) \right\|_\Xi}{\|k^{i,s} - z(\tilde{u})\|_\Xi + \left\| \hat{\bar{z}}^{i,s} - w^{i,s} \right\|_\Xi} \left\| \hat{\bar{z}}^{i,s} - w^{i,s} \right\|_\Xi \\ &< \frac{\tau_{max}}{\tau_{min}} \left\| \hat{\bar{z}}^{i,s} - w^{i,s} \right\|_\Xi . \end{aligned} \tag{4.16}$$

In the standard manner the Gâteaux-differentiability of the functional

$$\frac{1}{r_i} \left\| \frac{dy}{dt} + Ay - u \right\|_{0,Q}^2 + \frac{\chi_i}{2} \|u - u^{i,s-1}\|_{0,Q}^2$$

in the space Ξ with the norm $|\cdot|$ can be established. Regarding the definition of $(\bar{y}^{i,s}, \bar{u}^{i,s})$, we obtain by means of Proposition II.2.2 in EKELAND AND TEMAM (1976) that, for all $(y, u) \in G \times U_{ad}$,

$$
\begin{aligned}
&\|y(T) - y_d\|_{0,\Omega}^2 - \left\|\bar{y}^{i,s}(T) - y_d\right\|_{0,\Omega}^2 \\
&\quad + \frac{2}{r_i} \int_0^T \left(\frac{d\bar{y}^{i,s}(t)}{dt} + A\bar{y}^{i,s}(t) - \bar{u}^{i,s}(t) \, , \right. \\
&\quad \left. \frac{dy(t)}{dt} + Ay(t) - u(t) - \frac{d\bar{y}^{i,s}(t)}{dt} - A\bar{y}^{i,s}(t) + \bar{u}^{i,s}(t) \right)_\Omega dt \\
&\quad + \chi_i \int_0^T \left(\bar{u}^{i,s}(t) - u^{i,s-1}(t) \, , \, u(t) - \bar{u}^{i,s}(t) \right)_\Omega dt \geq 0.
\end{aligned} \tag{4.17}
$$

Setting $y = y^*$, $u = u^*$ in (4.17), in view of $\frac{dy^*}{dt} + Ay^* - u^* = 0$ and the obvious inequality

$$
\begin{aligned}
&\left\|u - u^{i,s-1}\right\|_{0,Q}^2 - \left\|\bar{u}^{i,s} - u^{i,s-1}\right\|_{0,Q}^2 \\
&\qquad \geq 2 \int_0^T \left(\bar{u}^{i,s}(t) - u^{i,s-1}(t) \, , \, u(t) - \bar{u}^{i,s}(t) \right)_\Omega dt,
\end{aligned} \tag{4.18}
$$

one can conclude that

$$
\frac{2}{r_i} \left\| \frac{d\bar{y}^{i,s}}{dt} + A\bar{y}^{i,s} - \bar{u}^{i,s} \right\|_{0,Q}^2 \leq \|y^*(T) - y_d\|_{0,\Omega}^2 + \frac{\chi_i}{2} \left\|u^* - u^{i,s-1}\right\|_{0,Q}^2 . \tag{4.19}
$$

Thus,

$$
\left\|\bar{y}^{i,s}\right\|_Y \leq \left(\frac{r_i}{2}\right)^{1/2} \left(\|y^*(T) - y_d\|_{0,\Omega}^2 + \frac{\chi_i}{2} \left\|u^* - u^{i,s-1}\right\|_{0,Q}^2 \right)^{1/2} + \left\|\bar{u}^{i,s}\right\|_{0,Q} . \tag{4.20}
$$

Now, regarding the boundedness of U_{ad}, $r_i < 1$, $\chi_i \leq 2$, $\lim_{i\to\infty} \frac{\epsilon_i}{\chi_i} = 0$, and condition (3.3), inequality (4.20) yields

$$
\left\|\bar{y}^{i,s}\right\|_Y < c_1, \quad \left\|y^{i,s}\right\|_Y < c_1 \tag{4.21}
$$

(all the constants c_k in this proof don't depend on (i, s)). Estimate (4.14) follows immediately from (4.21) and the equivalence of the norms $\|\cdot\|_\Xi$ and $|\cdot|$.
Due to (4.21) and

$$
\|y\|_W \leq \bar{c} \, \|y\|_Y \quad \text{for all } y \in Y,
$$

one gets

$$
\left\|\bar{y}^{i,s}\right\|_W < c_2, \quad \left\|y^{i,s}\right\|_W < c_2 \text{ with } c_2 = \bar{c}c_1. \tag{4.22}
$$

Inequality (4.19) ensures also that

$$
\left\| \frac{d\bar{y}^{i,s}}{dt} + A\bar{y}^{i,s} - \bar{u}^{i,s} \right\|_{0,Q} < c_3 r_i^{1/2}. \tag{4.23}
$$

Taking into account that $\frac{d\hat{y}^{i,s}}{dt} + A\hat{y}^{i,s} - \bar{u}^{i,s} = 0$, this leads to

$$\left\|\bar{y}^{i,s} - \hat{y}^{i,s}\right\|_Y < c_3 r_i^{1/2} \tag{4.24}$$

and, due to $\|\bar{y}^{i,s} - \hat{y}^{i,s}\|_Y = \left\|\bar{z}^{i,s} - \hat{\bar{z}}^{i,s}\right\|_\Xi$, we obtain

$$\left\|\bar{z}^{i,s} - \hat{\bar{z}}^{i,s}\right\|_\Xi < c_3 r_i^{1/2}. \tag{4.25}$$

Because $\bar{z}^{i,s} \in G \times U_{ad}$, the estimate

$$\left\|w^{i,s} - \hat{\bar{z}}^{i,s}\right\|_\Xi < c_3 r_i^{1/2} \tag{4.26}$$

follows from (4.25) and from the definition of $w^{i,s}$.
Denote by

$$Z_f = \{z = (\mathcal{T}u, u) \ : \ z \in G \times U_{ad}\}$$

the set of feasible processes and

$$\bar{\bar{z}}^{i,s} \equiv \left(\bar{\bar{y}}^{i,s}, \bar{\bar{u}}^{i,s}\right) = \arg \ \min\left\{\|\bar{z}^{i,s} - z\|_\Xi : \ z \in Z_f\right\}.$$

If $\hat{y}^{i,s} \notin G$ and $w^{i,s} \neq h^{i,s}$, then on account of $h^{i,s} \in Z_f$, we obtain from (4.16), (4.25) and (4.26) that

$$\begin{aligned} \left\|\bar{z}^{i,s} - \bar{\bar{z}}^{i,s}\right\|_\Xi &\leq \left\|\bar{z}^{i,s} - h^{i,s}\right\|_\Xi \\ &\leq \left\|\bar{z}^{i,s} - \hat{\bar{z}}^{i,s}\right\|_\Xi + \left\|\hat{\bar{z}}^{i,s} - h^{i,s}\right\|_\Xi < \left(\frac{\tau_{max}}{\tau_{min}} + 1\right) c_3 r_i^{1/2}. \end{aligned} \tag{4.27}$$

If $\hat{y}^{i,s} \notin G$, but $w^{i,s} = h^{i,s}$, estimate (4.27) follows from (4.25), (4.26) and $\tau_{max} > \tau_{min}$.
In case $\hat{y}^{i,s} \in G$, the inequality

$$\left\|\bar{z}^{i,s} - \bar{\bar{z}}^{i,s}\right\|_\Xi < c_3 r_i^{1/2} \tag{4.28}$$

is an immediate consequence of (4.25).
Inserting into relation (4.17) $y = \bar{\bar{y}}^{i,s}, \ u = \bar{\bar{u}}^{i,s}$, one gets

$$\begin{aligned} &\|\bar{y}^{i,s}(T) - y_d\|^2_{0,\Omega} + \frac{2}{r_i}\left\|\frac{d\bar{y}^{i,s}}{dt} + A\bar{y}^{i,s} - \bar{u}^{i,s}\right\|^2_{0,Q} \\ &\quad \leq \left\|\bar{\bar{y}}^{i,s}(T) - y_d\right\|^2_{0,\Omega} + \chi_i \left\|\bar{u}^{i,s} - \bar{\bar{u}}^{i,s}\right\|_{0,Q} \left\|\bar{u}^{i,s} - u^{i,s-1}\right\|_{0,Q}, \end{aligned}$$

and hence,

$$\begin{aligned} \frac{2}{r_i}\left\|\frac{d\bar{y}^{i,s}}{dt} + A\bar{y}^{i,s} - \bar{u}^{i,s}\right\|^2_{0,Q} &\leq \left\|\bar{\bar{y}}^{i,s}(T) - \bar{y}^{i,s}(T)\right\|_{0,\Omega} \left\|\bar{\bar{y}}^{i,s}(T) + \bar{y}^{i,s}(T) - 2y_d\right\|_{0,\Omega} \\ &+ \ \chi_i \left\|\bar{u}^{i,s} - \bar{\bar{u}}^{i,s}\right\|_{0,Q} \left\|\bar{u}^{i,s} - u^{i,s-1}\right\|_{0,Q}. \end{aligned} \tag{4.29}$$

Because of $Y \hookrightarrow C([0,T]; H_0^1(\Omega))$, (4.22), (4.27), $r_i < 1, \ \chi_i \leq 2$ and the boundedness of U_{ad}, inequality (4.29) leads to

$$\left\|\frac{d\bar{y}^{i,s}}{dt} + A\bar{y}^{i,s} - \bar{u}^{i,s}\right\|_{0,Q} \leq c_4 r_i^{3/4}. \tag{4.30}$$

Using (4.30) instead of (4.23), the estimates (4.24)-(4.26) can be improved (w.r.t. the order) and we obtain

$$\left\| \bar{z}^{i,s} - \hat{\bar{z}}^{i,s} \right\|_{\Xi} < c_4 r_i^{3/4}, \tag{4.31}$$

$$\left\| w^{i,s} - \hat{\bar{z}}^{i,s} \right\|_{\Xi} < c_4 r_i^{3/4}. \tag{4.32}$$

Thus, similar to (4.27), (4.28), the inequality

$$\left\| \bar{z}^{i,s} - \bar{\bar{z}}^{i,s} \right\|_{\Xi} \leq \left(\frac{\tau_{max}}{\tau_{min}} + 1 \right) c_4 r_i^{3/4}$$

can be established.
A multiple repetition of this operation (using in each step the current estimates) leads to the conclusion that, with arbitrarily fixed $\nu \in (0, \frac{1}{2})$ and some constant $c(\nu)$, the estimates

$$\left\| \frac{d\bar{y}^{i,s}}{dt} + A\bar{y}^{i,s} - \bar{u}^{i,s} \right\|_{0,Q} \leq c(\nu) r_i^{1-2\nu}, \tag{4.33}$$

$$\left\| \bar{z}^{i,s} - \bar{\bar{z}}^{i,s} \right\|_{\Xi} \leq \left(\frac{\tau_{max}}{\tau_{min}} + 1 \right) c(\nu) r_i^{1-2\nu} \tag{4.34}$$

are valid uniformly w.r.t. (i, s). Now, from the obvious equality

$$\begin{aligned} J_i(z^*) - J_i\left(\bar{z}^{i,s}\right) &= \|y^*(T) - y_d\|_{0,\Omega}^2 - \left\|\bar{\bar{y}}^{i,s}(T) - y_d\right\|_{0,\Omega}^2 + \left\|\bar{\bar{y}}^{i,s}(T) - y_d\right\|_{0,\Omega}^2 \\ &- \left\|\bar{y}^{i,s}(T) - y_d\right\|_{0,\Omega}^2 - \frac{1}{r_i}\left\| \frac{d\bar{y}^{i,s}}{dt} + A\bar{y}^{i,s} - \bar{u}^{i,s} \right\|_{0,Q}^2, \end{aligned}$$

due to $\|y^*(T) - y_d\|_{0,\Omega} \leq \left\|\bar{\bar{y}}^{i,s}(T) - y_d\right\|_{0,\Omega}$, (4.21), (4.34), and the embedding $Y \hookrightarrow C([0,T]; H_0^1(\Omega))$, we get

$$J_i(z^*) - J_i\left(\bar{z}^{i,s}\right) < d(\nu) r_i^{1-2\nu},$$

with $d(\nu)$ independent from (i, s), i.e. estimate (4.13) is true. □

Theorem 4.1. *Assume that U_{ad} is a bounded set and $T\tilde{u} \in intG$ for some $\tilde{u} \in U_{ad}$; that $\nu \in (0, \frac{1}{2})$ is a fixed number and that constants $d(\nu)$, d_1 are defined according to Lemma 4.4. Let the positive sequences $\{r_i\}, \{\epsilon_i\}, \{\chi_i\}$, and $\{\delta_i\}$ in the RP-method satisfy the conditions*

$$\sup_i r_i < 1, \ \sup_i \chi_i \leq 2, \ \sum_{i=1}^{\infty} \frac{r_i^{1/2-\nu}}{\chi_i^{1/2}} < \infty, \ \sum_{i=1}^{\infty} \frac{\epsilon_i}{\chi_i} < \infty \tag{4.35}$$

and

$$\frac{1}{2d_1}\left(2d(\nu)\frac{r_i^{1-2\nu}}{\chi_i} - \left(\delta_i - \frac{\epsilon_i}{\chi_i}\right)^2 \right) + \sqrt{3}\frac{\epsilon_i}{\chi_i} < 0, \ \ \delta_i > \frac{\epsilon_i}{\chi_i}. \tag{4.36}$$

Then, for any starting point $u^0 \in U_{ad}$, the RP-method is well-defined, i.e. $s(i) < \infty$ for each i, and $\{u^{i,s}\}, \{y^{i,s}\}$ converge weakly in $L_2(Q), Y$ to $\bar{u}, \bar{y}$ respectively, where $(\bar{y}, \bar{u})$ is an optimal process for Problem (P).

Proof. Let us assume that $s(i) < \infty$ for $i = 1, \dots, k-1$. Then, starting in step $i = k$ with $u^{k-1} = u^{k-1,s(k-1)}$, due to the definition of $s(i)$, (3.3) and (4.36), we conclude

$$\begin{aligned} \left\|\bar{u}^{k,s} - u^{k,s-1}\right\|_{0,Q} &\geq \left\|u^{k,s} - u^{k,s-1}\right\|_{0,Q} - \left\|u^{k,s} - \bar{u}^{k,s}\right\|_{0,Q} \\ &> \delta_k - \frac{\epsilon_k}{\chi_k} > 0 \text{ for } 1 \leq s < s(k). \end{aligned}$$

Together with inequality (4.13) and

$$2d(\nu)\frac{r_k^{1-2\nu}}{\chi_k} - \left(\delta_k - \frac{\epsilon_k}{\chi_k}\right)^2 < 0$$

(cf. (4.36)), this implies

$$\left\|\bar{u}^{k,s} - u^{k,s-1}\right\|_{0,Q}^2 > \frac{2}{\chi_k}\left[J_k(z^*) - J_k(\bar{z}^{k,s})\right] \text{ for } 1 \leq s < s(k). \tag{4.37}$$

Let $z^{1,0} = (\mathcal{T}u^0, u^0)$. Applying (4.13) and Lemma 4.3 with

$$Z = \Xi, \quad Z_1 = \{z = (y, u) \in \Xi : \ y = 0\}, \quad \Phi = J_k,$$

$$a(z, \hat{z}) = (y(T), \hat{y}(T))_\Omega + \frac{1}{r_k}\int_0^T\left(\frac{dy(t)}{dt} + Ay(t) - u(t), \frac{d\hat{y}(t)}{dt} + A\hat{y}(t) - \hat{u}(t)\right)_\Omega dt,$$

$$b(z, \hat{z}) = \int_0^T \left(\frac{dy(t)}{dt} + Ay(t) - u(t)\ , \ \frac{d\hat{y}(t)}{dt} + A\hat{y}(t) - \hat{u}(t)\right)_\Omega dt,$$

$$\ell(z) = 2\,(y(T), y_d)_\Omega, \quad K = G \times U_{ad}, \quad a^0 = z^{k,s-1}, \quad \chi = \chi_k, \quad \text{and} \quad \delta = \delta_k - \frac{\epsilon_k}{\chi_k},$$

we obtain from (4.12) and (4.14)

$$\begin{aligned} &\left|\bar{z}^{k,s} - z^*\right| \\ &\quad < \left|z^{k,s-1} - z^*\right| + \frac{1}{2d_1}\left(2d(\nu)\frac{r_k^{1-2\nu}}{\chi_k} - \left(\delta_k - \frac{\epsilon_k}{\chi_k}\right)^2\right) \text{ for } 1 \leq s < s(k). \end{aligned} \tag{4.38}$$

Using (3.3), (4.3) and (4.36), inequality (4.38) yields

$$\begin{aligned} &\left|z^{k,s} - z^*\right| - \left|z^{k,s-1} - z^*\right| \\ &\quad < \frac{1}{2d_1}\left(2d(\nu)\frac{r_k^{1-2\nu}}{\chi_k} - \left(\delta_k - \frac{\epsilon_k}{\chi_k}\right)^2\right) + \sqrt{3}\frac{\epsilon_k}{\chi_k} < 0. \end{aligned} \tag{4.39}$$

Inequality (4.39) proves that $s(k) < \infty$, because the middle term in (4.39) is independend from s.
Now, for $s = s(k)$, the use of Lemma 4.3 with the same data as above, leads to

$$\left|\bar{z}^{k,s(k)} - z^*\right| < \left|z^{k,s(k)-1} - z^*\right| + \sqrt{2d(\nu)\frac{r_k^{1-2\nu}}{\chi_k}},$$

hence,

$$\left|z^{k,s(k)} - z^*\right| < \left|z^{k,s(k)-1} - z^*\right| + \sqrt{2d(\nu)\frac{r_k^{1-2\nu}}{\chi_k}} + \sqrt{3}\frac{\epsilon_k}{\chi_k}. \tag{4.40}$$

Taking into account that the finiteness of $s(1)$ can be proved quite analogously, we infer that

$$s(i) < \infty \text{ for each } i,$$

and the inequalities (4.39) and (4.40) are valid for each k.
In view of (4.39), (4.40) and (4.35), Lemma 2.2.2 from POLYAK (1987) ensures the convergence of the sequence $\{|z^{i,s} - z^*|\}$, and with regard to (3.3), (4.3) and the last inequality in (4.35), the sequence $\{|\bar{z}^{i,s} - z^*|\}$ converges to the same limit.
Suppose that $\{\bar{z}^{i_k,s_k}\}$, with $s_k > 0$ for each k, converges weakly to $\bar{z} = (\bar{y}, \bar{u}) \in \Xi$. Due to (4.34), (4.35), $\left\{\bar{\bar{z}}^{i_k,s_k}\right\}$ converges weakly to $\bar{z}$, too. Observing the convexity and the closedness of Z_f and that $\left\{\bar{\bar{z}}^{i_k,s_k}\right\} \subset Z_f$, we conclude that $\bar{z} \in Z_f$.
But Lemma 4.3 yields also

$$\left|\bar{z}^{i,s} - z^*\right|^2 - \left|z^{i,s-1} - z^*\right|^2 \leq \frac{2}{\chi_i}\left[J_i(z^*) - J_i\left(\bar{z}^{i,s}\right)\right],$$

and by definition of J_i (cf (4.15))

$$J_i(z^*) = \|y^*(T) - y_d\|_{0,\Omega}^2 = J(u^*), \quad J_i(\bar{z}^{i,s}) \geq \left\|\bar{y}^{i,s}(T) - y_d\right\|_{0,\Omega}^2,$$

hence,

$$\left|\bar{z}^{i,s} - z^*\right|^2 - \left|z^{i,s-1} - z^*\right|^2 \leq \frac{2}{\chi_i}\left[J(u^*) - \left\|\bar{y}^{i,s}(T) - y_d\right\|_{0,\Omega}^2\right].$$

Due to the convexity and the continuity (in Y), the functional $\|y(T) - y_d\|_{0,\Omega}^2$ is weakly lower semi-continuous. Taking limit in the last inequality w.r.t. the subsequence $\{\bar{z}^{i_k,s_k}\}$, we obtain

$$J(u^*) \geq \|\bar{y}(T) - y_d\|_{0,\Omega}^2,$$

hence, $\bar{z}$ is an optimal process. Finally, Lemma 4.1 in OPIAL (1967) ensures weak convergence of both $\{\bar{z}^{i,s}\}$ and $\{z^{i,s}\}$ to $\bar{z} \in \Xi$. □

• **The case of unbounded set U_{ad}**

Now, we formulate convergence results for the case of an unbounded set U_{ad} and $G = Y$. Hereby solvability of Problem (P) is assumed. Let us choose

$$c_0 \geq \|y^*(T) - y_d\|_{0,\Omega}, \quad \rho_{i,s-1} > 0.$$

Let

$$u^{i,s-1} \in \{u \in U_{ad} \;:\; \|u - u^*\|_{0,Q} < \rho_{i,s-1}\} \tag{4.41}$$

be arbitrarily fixed, where $z^* = (y^*, u^*)$ is an optimal process. Suppose that the points $(\bar{y}^{i,s}, \bar{u}^{i,s})$ and $(y^{i,s}, u^{i,s})$ are defined by (3.1) and (3.3) with this $u^{i,s-1}$.

Lemma 4.5. *Let the sequences $\{r_i\}, \{\epsilon_i\}, \{\chi_i\}$ be chosen according to the RP-method and $\nu \in \left(0, \frac{1}{2}\right)$ be an arbitrary number. Then there exists a constant $\bar{d}(\nu)$, independent from i, $s \geq 1$, $\{r_i\}, \{\epsilon_i\}, \{\chi_i\}$ and $u^{i,s-1}$ in (4.41), such that*

$$J_i(y^*, u^*) - J_i\left(\bar{y}^{i,s}, \bar{u}^{i,s}\right) < \bar{d}(\nu)\,(c_0 + \rho_{i,s-1})^2\, r_i^{1-2\nu}. \tag{4.42}$$

Theorem 4.2. *Let $u^0 \in U_{ad}$, $z^{1,0} = (\mathcal{T}u^0, u^0)$. With $\rho_1 > \|z^{1,0} - z^*\|$ the sequence $\{\rho_i\}$ let be defined recursively by*

$$\rho_{i+1} = \rho_i + \sqrt{\frac{2\bar{d}(\nu)}{\chi_i}}\,(c_0 + \rho_i)\, r_i^{1/2-\nu} + \sqrt{3}\frac{\epsilon_i}{\chi_i}. \tag{4.43}$$

Moreover, assume that the controlling parameters in the RP-method satisfy the conditions (4.35) and, for each i, let

$$\frac{1}{2\rho_i}\left(2\bar{d}(\nu)\,(c_0 + \rho_i)^2\,\frac{r_i^{1-2\nu}}{\chi_i} - \left(\delta_i - \frac{\epsilon_i}{\chi_i}\right)^2\right) + \sqrt{3}\frac{\epsilon_i}{\chi_i} < 0, \quad \delta_i > \frac{\epsilon_i}{\chi_i}. \tag{4.44}$$

Then, applying the RP-method to Problem (P) with $G = Y$, one gets $s(i) < \infty$ for each i, and $\{u^{i,s}\}, \{y^{i,s}\}$ converge weakly in $L_2(Q), Y$ to $\bar{u}, \bar{y}$ respectively, where $(\bar{y}, \bar{u})$ is an optimal process for Problem (P).

Note that, due to (4.35) and Lemma 2.2.2 in Polyak (1987), the sequence $\{\rho_i\}$ is convergent.
The conditions of the Theorems 4.1 and 4.2 permit a slow change of the controlling parameters ϵ_i, r_i, and χ_i : For instance, it is possible to take

$$0 < \underline{\chi} \leq \chi_i \leq 2 \text{ and } r_i = q_1^i;\ \epsilon_i = q_2^i \text{ with arbitrary } q_1, q_2 \in (0,1),$$

and then to choose δ_i according to (4.36) or (4.43), (4.44). However, the calculation of $d(\nu)$ or $\bar{d}(\nu)$ may be difficult.
There are no principal difficulties to extend this consideration to other objective functions of the form $J(u) = \|Cy_u - y_d\|^2_{\mathcal{H}}$, where $\mathcal{H}$ is a Hilbert space (on Ω, Q or Σ), $C \in \mathcal{L}(Y, \mathcal{H})$ and $y_d \in \mathcal{H}$.

References

1. Alt, W. and U. Mackenroth (1989), *Convergence of finite element approximations to state constrained convex parabolic boundary control problems.* SIAM J. Contr. Opt. 27, 718–736.
2. Balakrishnan, A. V. (1968 a), *A new computing technique in system identification.* J. Comput. and System Sci. 2, 102–116.
3. Balakrishnan, A.V. (1968 b), *On a new computing technique in optimal control.* SIAM J. Control 6, 149–173.
4. Bergounioux, M. (1992), *A penalization method for optimal control of elliptic problems with state constraints.* SIAM J. Contr. Opt. 30, 305–323.

5. BERGOUNIOUX, M. (1994), *Optimal control of parabolic problems with state constraints: A penalization method for optimality conditions.* Appl. Math. Opt. 29, 285–307.
6. ECKSTEIN, J. AND D. P. BERTSEKAS (1992), *On the Douglas-Rachford splitting method and the proximal point algorithm for maximal monotone operators.* Math. Programming 55, 293–318.
7. EKELAND, I. AND R. TEMAM (1976), *Convex Analysis and Variational Problems.* North-Holland, Amsterdam and American Elsevier, New York.
8. GLASHOFF, K. AND E. SACHS (1977), *On theoretical and numerical aspects of the bang-bang principle.* Num. Math. 29, 93–114.
9. HACKBUSCH, W. AND TH. WILL (1984) *A numerical method for a parabolic bang-bang problem. In: Optimal Control of Partial Differential Equations (K. H. Hoffmann, W. Krabs, eds.)* ISNM 68, Birkhäuser, Basel.
10. HETTICH, R., KAPLAN, A. AND R. TICHATSCHKE (1994), *Regularized penalty methods for optimal control of elliptic problems (ill-posed case).* In: Schwerpunktprogramm der Deutschen Forschungsgemeinschaft: Anwendungsbezogene Optimierung und Steuerung, Report No. 522, Humboldt-Universität Berlin, 34 pp.
11. HETTICH, R., KAPLAN, A. AND R. TICHATSCHKE (1996), *Regularized penalty methods for ill-posed optimal control problems with elliptic equations, Part I : Distributed control with bounded control sets and state constraints, Part II : Distributed and boundary control with unbounded control set and state constraints.* Control and Cybernetics 1,1997.
12. KAPLAN, A. AND R. TICHATSCHKE (1994), *Stable Methods for Ill-Posed Variational Problems – Prox-Regularization of Elliptic Variational Inequalities and Semi-Infinite Problems.* Akademie Verlag, Berlin.
13. KNOWLES, G. (1982), *Finite element approximation of parabolic time optimal control problems.* SIAM J. Contr. Opt. 20, 414–427.
14. LADYSHENSKAJA, O., SOLONNIKOV, V. AND N. URAL'ZEWA (1968), *Linear and Quasilinear Equations of Parabolic Type.* Transl. Amer. Math. Monographs, vol. 23, American Mathematical Society, Providence, Rhode Island.
15. LASIECKA, I. (1980), *Boundary control of parabolic systems, finite element approximation.* Appl. Math. Opt. 6, 31–62.
16. LASIECKA, I. (1984), *Ritz-Galerkin approximation of the time-optimal boundary control problem for parabolic systems with Dirichlet boundary conditions.* SIAM J. Contr. Opt. 22, 477–500.
17. LIONS, J.L. (1968), *Contrôle Optimal des Systèmes Gouvernés par des Équations aux Dérivées Partielles.* Dunod, Gauthier-Villars, Paris.
18. LIONS, J.L. (1985), *Control of Distributed Singular Systems.* Dunod, Paris.
19. LIONS, J.L. AND E. MAGENES (1968), *Problèmes aux limites non homogenes et applications, t. 1–2.* Dunod, Paris.
20. MACKENROTH, U. (1982–83), *Some remarks on the numerical solution of bang-bang type optimal control problems.* Num. Funct. Anal. Opt. 5, 457–484.
21. MACKENROTH, U. (1987), *Numerical solution of some parabolic boundary control problems by finite elements.* Lect. Notes Contr. Inf. Sci. 97, 325–335.
22. MALANOWSKI, K. (1981), *Convergence of approximations vs. regularity of solutions for convex, control-constrained optimal control problems.* Appl. Math. Opt. 8, 69–95.
23. OPIAL, Z. (1967), *Weak convergence of the successive approximaions for nonexpansive mappings in Banach spaces.* Bull. Amer. Math. Soc. 73, 591–597.
24. POLYAK, B.T. (1987), *Introduction to Optimization.* Optimization Software, Inc. Publ. Division, New York.
25. TRÖLTZSCH, F. (1987), *Semidiscrete finite element approximation of parabolic boundary control problems – convergence of switching points.* ISNM 78, Birkhäuser, Basel, 219–232.

A. Kaplan
Technical University Darmstadt
Department of Mathematics
D-64289 Darmstadt, Germany

R. Tichatschke
University Trier
FB IV (Math.)
D-54286 Trier, Germany

International Series of Numerical Mathematics
Vol. 126, © 1998 Birkhäuser Verlag, Basel

On the Control of Coupled Linear Systems

VILMOS KOMORNIK, PAOLA LORETI, AND ENRIQUE ZUAZUA

Institut de Recherche Mathématique Avancée
Université Louis Pasteur et CNRS

Istituto per le Applicazioni del Calcolo "Mauro Picone"
Consiglio Nazionale delle Ricerche

Departamento de Matemática Aplicada
Universidad Complutense

ABSTRACT. J.-L. Lions proved several observability theorems for coupled linear distributed systems provided the coupling parameters are small enough. It remained an open question whether the assumption on the smallness of these parameters is necessary for the validity of the results. Using nonharmonic analysis the first two authors proved recently that in some cases the observability holds in fact for almost all values of the parameters if the underlying domain is an open ball. Combining this method with a compactness-uniqueness method developed by the third author we extend these results for all sufficiently regular bounded domains.

1991 *Mathematics Subject Classification.* 35L05, 35Q72, 93B05, 93B07, 93C20, 93D15

Key words and phrases. observability, controllability, stabilizability by feedback, partial differential equation, wave equation, Petrovsky system.

1. Introduction

Let Ω be a bounded open domain of class C^4 in $\mathbb{R}^n$. We shall denote by ν the outward unit normal vector to its boundary Γ. Given two real numbers a and b, consider the following system:

$$
\begin{cases}
u_1'' - \Delta u_1 + bu_2 = f_1 & \text{in} \quad \mathbb{R}\times\Omega, \\
u_2'' + \Delta^2 u_2 + au_1 = f_2 & \text{in} \quad \mathbb{R}\times\Omega, \\
u_1 = u_2 = \Delta u_2 = 0 & \text{on} \quad \mathbb{R}\times\Gamma, \\
u_1(0) = u_{10} \quad \text{and} \quad u_1'(0) = u_{11} & \text{in} \quad \Omega, \\
u_2(0) = u_{20} \quad \text{and} \quad u_2'(0) = u_{21} & \text{in} \quad \Omega.
\end{cases}
\tag{1.1}
$$

One can prove by standard methods that for any given functions $f_1 \in L^1_{loc}(\mathbb{R}; L^2(\Omega))$, $f_2 \in L^1_{loc}(\mathbb{R}; H^{-1}(\Omega))$, and initial data

$$(u_{10}, u_{11}, u_{20}, u_{21}) \in H^1_0(\Omega) \times L^2(\Omega) \times H^1_0(\Omega) \times H^{-1}(\Omega),$$

The first author is grateful to the organisers of the conference for their invitation and to the INRIA Lorraine (Projet Numath) for supporting his travel expenses.

The third author is grateful to the organisers of the conference for their invitation. He was supported by DGICYT (Spain) (PB93-1203) and the European Union (CHRX-CT94-0471).

this problem has a unique *weak* solution satisfying

$$u_1 \in C(\mathbb{R}; H_0^1(\Omega)) \cap C^1(\mathbb{R}; L^2(\Omega))$$

and

$$u_2 \in C(\mathbb{R}; H_0^1(\Omega)) \cap C^1(\mathbb{R}; H^{-1}(\Omega)).$$

We define the *initial energy* of the solutions by the formula

$$E_0 = \|u_{10}\|^2_{H_0^1(\Omega)} + \|u_{11}\|^2_{L^2(\Omega)} + \|u_{20}\|^2_{H_0^1(\Omega)} + \|u_{21}\|^2_{H^{-1}(\Omega)}.$$

Let us first assume that the system (1.1) is uncoupled, i.e. $a = b = 0$. Then it follows from some results of [11], [12] (p. 44, théorème 4.1, p. 287, théorème 4.1) that for every bounded interval I containing 0 the solutions of (1.1) satisfy the estimates

$$\int_I \int_\Gamma (\partial_\nu u_1)^2 + (\partial_\nu u_2)^2 \, d\Gamma \, dt \le c_1 E_0 + c_1 \|f_1\|^2_{L^2(I;L^2(\Omega))} + c_1 \|f_2\|^2_{L^2(I;H^{-1}(\Omega))} \tag{1.2}$$

with a constant c_1. Here and in the sequel all constants are assumed to be independent of the particular choice of the initial data.

Now fix two open subsets Γ_1, Γ_2 of Γ and a positive number T_0 such that for any bounded intervals I_1, I_2 of length $|I_1| > T_0$ and $|I_2| > 0$, in case $a = b = 0$ and $f_1 = f_2 = 0$ the solutions of (1.1) also satisfy the inverse inequalities

$$E_0 \le c_2 \int_{I_1} \int_{\Gamma_1} (\partial_\nu u_1)^2 \, d\Gamma \, dt + c_2 \int_{I_2} \int_{\Gamma_2} (\partial_\nu u_2)^2 \, d\Gamma \, dt \tag{1.3}$$

with a suitable constant c_2.

The purpose of this paper is to show that then analogous estimates hold for the coupled system, too, at least for almost all choices of the coupling parameters a and b.

Remarks. According to earlier results of Lions, Zuazua and Komornik [11], [12] (p. 55, théorème 5.1, p. 296, théorème 4.3), [13] (pp. 474–478), [5] (p. 82, theorem 6.11) on the wave equation and on Petrovsky systems, the estimates (1.3) are satisfied for example if there exists an open ball $B(x_1; R_1)$ containing Ω and a point $x_2 \in \mathbb{R}^n$ such that

$$(x - x_1) \cdot \nu(x) \le 0 \quad \text{on} \quad \Gamma - \Gamma_1,$$
$$(x - x_2) \cdot \nu(x) \le 0 \quad \text{on} \quad \Gamma - \Gamma_2,$$

and if the lengths of I_1, I_2 satisfy $|I_1| > 2R_1$ and $|I_2| > 0$. Let us emphasize the fact that I_2 can be arbitrarily small: this is due to the infinite propagation speed for the Petrovsky system.

If Ω is of class C^∞, then much weaker sufficient conditions were obtained by Bardos, Lebeau and Rauch [1], [2], [10]: every ray of geometric optics in Ω meets $\Gamma_1 \times I_1$ at some nondiffractive point, and there exists a bounded interval I_2', possibly longer than I_2, such that every ray of geometric optics in Ω meets $\Gamma_2 \times I_2'$ at some nondiffractive point. Note again that there is no assumption on the length of I_2: it can be arbitrarily small.

See also Burq [3] for an extension of the results of [1] and [2] to the case where Ω is only of class C^3.

Our main result is the

Theorem 1.1. *(a) Given two real numbers a, b arbitrarily and a bounded interval I containing 0, in case $f_1 = f_2 = 0$ the solutions of (1.1) satisfy the estimates*

$$\int_I \int_\Gamma (\partial_\nu u_1)^2 + (\partial_\nu u_2)^2 \, d\Gamma \, dt \leq c_3 E_0 \tag{1.4}$$

with some constant c_3.

(b) For almost all choices of the pair $(a, b) \in \mathbb{R}^2$, if I_1, I_2 are two bounded intervals satisfying $|I_1| > 2R_1$ and $|I_2| > 0$, then in case $f_1 = f_2 = 0$ the solutions of (1.1) satisfy the estimates

$$E_0 \leq c_4 \int_{I_1} \int_{\Gamma_1} (\partial_\nu u_1)^2 \, d\Gamma \, dt + c_4 \int_{I_2} \int_{\Gamma_2} (\partial_\nu u_2)^2 \, d\Gamma \, dt \tag{1.5}$$

with a suitable constant c_4.

Remark. In the special case where Ω is a ball, theorem 1.1 was proved earlier in [7], [8] by a direct computation (leading to explicit constants). The proof given below is different: it is based on a compactness–uniqueness method introduced in [13]. While it is indirect and so the constants are not explicit, it works for all sufficiently regular bounded domains.

Applying the duality method from [4] or the Hilbert Uniqueness Method from [11], [12] one can deduce from theorem 1.1 an exact boundary controllability result concerning the system

$$\begin{cases} y_1'' - \Delta y_1 + a y_2 = 0 \quad \text{in} \quad (0,T) \times \Omega, \\ y_2'' + \Delta^2 y_2 + b y_1 = 0 \quad \text{in} \quad (0,T) \times \Omega, \\ y_1 = v_1 \quad \text{on} \quad (0,T) \times \Gamma, \\ y_2 = 0 \text{ and } \Delta y_2 = v_2 \quad \text{on} \quad (0,T) \times \Gamma, \\ y_1(0) = y_{10} \quad \text{and} \quad y_1'(0) = y_{11} \quad \text{in} \quad \Omega, \\ y_2(0) = y_{20} \quad \text{and} \quad y_2'(0) = y_{21} \quad \text{in} \quad \Omega. \end{cases} \tag{1.6}$$

Let us introduce for brevity the Hilbert space

$$H = L^2(\Omega) \times H^{-1}(\Omega) \times H_0^1(\Omega) \times H^{-1}(\Omega).$$

Theorem 1.2. *Fix positive numbers $0 \leq T_1 < T_2 \leq T$ such that $T > T_0$, and let Γ_1, Γ_2 be as above. For almost all pairs $(a, b) \in \mathbb{R}^2$ the problem (1.6) is exactly controllable in the following sense: Given*

$$(y_{10}, y_{11}, y_{20}, y_{21}) \in H \quad \textit{and} \quad (z_{10}, z_{11}, z_{20}, z_{21}) \in H$$

arbitrarily, there exist control functions $v_1, v_2 \in L^2(0, T; L^2(\Gamma))$ such that v_1 vanishes outside of $(0, T) \times \Gamma_1$, v_2 vanishes outside of $(T_1, T_2) \times \Gamma_2$, and the solution of (1.6) satisfies

$$y_1(T) = z_{10}, \quad y_1'(T) = z_{11} \quad \textit{and} \quad y_2(T) = z_{20}, \quad y_2'(T) = z_{21} \quad \textit{in } \Omega\,.$$

Next, applying another general method developed in [6], one can deduce from theorem 1.1 a uniform boundary stabilization result concerning the system

$$
\begin{cases}
y_1'' - \Delta y_1 + a y_2 = 0 \quad \text{in} \quad (0,\infty)\times\Omega,\\
y_2'' + \Delta^2 y_2 + b y_1 = 0 \quad \text{in} \quad (0,\infty)\times\Omega,\\
y_1 = v_1 \quad \text{on} \quad (0,\infty)\times\Gamma,\\
y_2 = 0 \text{ and } \Delta y_2 = v_2 \quad \text{on} \quad (0,\infty)\times\Gamma,\\
y_1(0) = y_{10} \quad \text{and} \quad y_1'(0) = y_{11} \quad \text{in} \quad \Omega,\\
y_2(0) = y_{20} \quad \text{and} \quad y_2'(0) = y_{21} \quad \text{in} \quad \Omega.
\end{cases} \tag{1.7}
$$

Theorem 1.3. *Fix an (arbitrarily large) positive number* ω, *and let* Γ_1, Γ_2 *be as above. For almost all pairs* $(a,b)\in\mathbb{R}^2$ *there exist two bounded linear maps*

$$P : H \to L^2(\Omega) \quad \textit{and} \quad Q : H \to H_0^1(\Omega)$$

and a positive constant M *such that, setting*

$$v_1 = \partial_\nu P(y_1, y_1', y_2, y_2') \quad \textit{and} \quad v_2 = \partial_\nu Q(y_1, y_1', y_2, y_2')$$

the problem (1.7) *is well-posed for all*

$$(y_{10}, y_{11}, y_{20}, y_{21}) \in H,$$

and its solutions satisfy the estimates

$$\|(y_1, y_1', y_2, y_2')(t)\|_H \le M e^{-\omega t}\|(y_{10}, y_{11}, y_{20}, y_{21})\|_H$$

for all $t \ge 0$.

Since both theorems 1.2 and 1.3 can be obtained from theorem 1.1 in a standard way, we shall only prove theorem 1.1 below.

2. Proof of Theorem 1.1

First we prove the inequality (1.4). Applying (1.2) with $f_1 = -bu_2$ and $f_2 = -au_1$ we obtain the estimate

$$\|\partial_\nu u_1\|^2_{L^2(I;L^2(\Gamma))} + \|\partial_\nu u_2\|^2_{L^2(I;L^2(\Gamma))} \le c\Big(E_0 + \|u_2\|^2_{L^1(I;L^2(\Omega))} + \|u_1\|^2_{L^1(I;H^{-1}(\Omega))}\Big).$$

By the well-posedness of the problem (1.1) we have

$$\|u_1\|^2_{L^1(I;H_0^1(\Omega))} + \|u_2\|^2_{L^1(I;H_0^1(\Omega))} \le cE_0$$

and therefore

$$\|\partial_\nu u_1\|^2_{L^2(I;L^2(\Gamma))} + \|\partial_\nu u_2\|^2_{L^2(I;L^2(\Gamma))} \le cE_0,$$

as stated.

Turning to the proof of the inequality (1.5), let us write the solution u of (1.1) as $u = v + w$ where v and w are solutions of the following two problems:

$$\begin{cases} v_1'' - \Delta v_1 = -bu_2 \quad \text{in} \quad \mathbb{R}\times\Omega, \\ v_2'' + \Delta^2 v_2 = -au_1 \quad \text{in} \quad \mathbb{R}\times\Omega, \\ v_1 = v_2 = \Delta v_2 = 0 \quad \text{on} \quad \mathbb{R}\times\Gamma, \\ v_1(0) = v_1'(0) = v_2(0) = v_2'(0) = 0 \quad \text{in} \quad \Omega, \end{cases} \tag{2.1}$$

and

$$\begin{cases} w_1'' - \Delta w_1 = 0 \quad \text{in} \quad \mathbb{R}\times\Omega, \\ w_2'' + \Delta^2 w_2 = 0 \quad \text{in} \quad \mathbb{R}\times\Omega, \\ w_1 = w_2 = \Delta w_2 = 0 \quad \text{on} \quad \mathbb{R}\times\Gamma, \\ w_1(0) = u_{10} \quad \text{and} \quad w_1'(0) = u_{11} \quad \text{in} \quad \Omega, \\ w_2(0) = u_{20} \quad \text{and} \quad w_2'(0) = u_{21} \quad \text{in} \quad \Omega. \end{cases} \tag{2.2}$$

Applying the estimates (1.3) to the uncoupled system (2.2) we obtain the inequality

$$E_0 \le c\|\partial_\nu w_1\|^2_{L^2(I_1;L^2(\Gamma_1))} + c\|\partial_\nu w_2\|^2_{L^2(I_2;L^2(\Gamma_2))},$$

whence

$$E_0 \le c\|\partial_\nu u_1\|^2_{L^2(I_1;L^2(\Gamma_1))} + c\|\partial_\nu u_2\|^2_{L^2(I_2;L^2(\Gamma_2))} + cR \tag{2.3}$$

with

$$R = \|\partial_\nu v_1\|^2_{L^2(I_1;L^2(\Gamma_1))} + \|\partial_\nu v_2\|^2_{L^2(I_2;L^2(\Gamma_2))}.$$

Next applying the estimates (1.2) to the system (2.1) with a bounded interval I containing 0 , I_1 and I_2, we obtain the inequality

$$R \le c\|u_2\|^2_{L^1(I;L^2(\Omega))} + c\|u_1\|^2_{L^1(I;H^{-1}(\Omega))}.$$

Using the well-posedness of problem (1.1) and the compactness of the embeddings $H_0^1(\Omega) \subset L^2(\Omega)$ and $H_0^1(\Omega) \subset H^{-1}(\Omega)$, this implies that R is compact with respect to E_0.

Due to this compactness we can apply a method developed in [13] for the proof of the inverse inequality (1.5). Using (2.3) first we reduce our problem to the following uniqueness property: if a solution w of (2.2) satisfies $\partial_\nu w_1 = 0$ on $\Gamma_1 \times I_1$ and $\partial_\nu w_2 = 0$ on $\Gamma_2 \times I_2$, then in fact all initial data vanish and therefore w_1, w_2 vanish identically. Next, proceeding always as in [13], we reduce this to the following simpler uniqueness property: if for some functions $w_1, w_2 \in H_0^1(\Omega)$ and for some complex number λ we have

$$\begin{cases} -\Delta w_1 + bw_2 = \lambda w_1 \quad \text{in} \quad \Omega, \\ \Delta^2 w_2 + aw_1 = \lambda w_2 \quad \text{in} \quad \Omega, \\ w_1 = w_2 = \Delta w_2 = 0 \quad \text{on} \quad \Gamma \end{cases} \tag{2.4}$$

and if

$$\begin{cases} \partial_\nu w_1 = 0 \quad \text{on} \quad \Gamma_1, \\ \partial_\nu w_2 = 0 \quad \text{on} \quad \Gamma_2, \end{cases} \tag{2.5}$$

then in fact w_1 and w_2 vanish identically in Ω.

Now we shall prove that this last uniqueness property holds for almost all choices of the pair (a, b). Let us fix an orthonormal basis $z_1, z_2, \dots$ in $L^2(\Omega)$, consisting of eigenfunctions of $-\Delta$ in $H_0^1(\Omega)$:

$$\begin{gathered} -\Delta z_n = \gamma_n z_n \quad \text{in} \quad \Omega, \\ z_n = 0 \quad \text{on} \quad \Gamma, \\ 0 < \gamma_1 \le \gamma_2 \le \cdots, \\ \gamma_n \to \infty. \end{gathered}$$

For every pair of integers (k, l) with $\gamma_k \neq \gamma_l$, set

$$\lambda_{k,l} = \frac{\gamma_k^3 - \gamma_l^3}{\gamma_k + \gamma_k^2 - \gamma_l - \gamma_l^2}$$

and consider the countable set

$$C := \{-(\gamma_k^2 - \gamma_k)^2 \ : \ k = 1, 2, \dots\} \cup \{(\gamma_k - \lambda_{k,l})(\gamma_k^2 - \lambda_{k,l}) \ : \ \gamma_k \neq \gamma_l\}.$$

Since the set $\{(a, b) \in \mathbb{R}^2 \ : \ ab \in C\}$ is the countable union of real analytic curves in $\mathbb{R}^2$, we have $ab \notin C$ for almost all pairs $(a, b) \in \mathbb{R}^2$. Henceforth we assume that $ab \notin C$ and we shall prove the above mentioned uniqueness property.

First we show that every eigenvector $W = (w_1, w_2)$ of (2.4) has the form $W = \beta z$ for some $\beta \in \mathbb{C}^2$ and for some nonzero eigenfunction of $-\Delta$ in $H_0^1(\Omega)$:

$$\begin{cases} -\Delta z = \gamma z \quad \text{in} \quad \Omega, \\ z = 0 \quad \text{on} \quad \Gamma. \end{cases} \tag{2.6}$$

For this first we seek eigenvectors of the form $W = \beta z_k$ for every fixed $k \geq 1$. Substituting into (2.4) we obtain for $\beta = (\beta_1, \beta_2)$ the linear system

$$\begin{cases} (\gamma_k - \lambda)\beta_1 + b\beta_2 = 0, \\ a\beta_1 + (\gamma_k^2 - \lambda)\beta_2 = 0. \end{cases}$$

Since by our assumption $ab \neq -(\gamma_k^2 - \gamma_k)^2$ its determinant has two different roots λ, it follows that the problem (2.4) has two linearly independent eigenvectors of the form $W_{2k-1} = \beta_{2k-1} z_k$ and $W_{2k} = \beta_{2k} z_k$.

Denoting by Z the linear hull of $z_1, z_2, \dots$, it follows that $W_1, W_2, \dots$ span $Z \times Z$, which is dense in $L^2(\Omega) \times L^2(\Omega)$. We have thus found a complete linearly independent sequence of eigenvectors of the problem (2.4). It can be shown that this sequence is in fact a Riesz basis, see [9].

Let us denote by $\lambda_1, \lambda_2, \dots$ the eigenvalues associated with $W_1, W_2, \dots$. If they are pairwise distinct, then the problem (2.4) has no other eigenvectors than the multiples of the vectors W_k. If some of these eigenvalues coincide, then all linear combinations of the corresponding W_ks are also eigenvectors of (2.4) with the same eigenvalue. Now it follows from our assumptions on ab that $\lambda_k = \lambda_l$ implies $\gamma_k = \gamma_l$, and therefore every eigenvector $W = (w_1, w_2)$ of (2.4) has the form $W = \beta z$ for some $\beta \in \mathbb{C}^2$ and for some nonzero function satisfying (2.6). Indeed, if $\gamma_k \neq \gamma_l$ but $\lambda_k = \lambda_l =: \lambda$, then λ solves both characteristic equations

$$(\gamma_k - \lambda)(\gamma_k^2 - \lambda) - ab = 0 \quad \text{and} \quad (\gamma_l - \lambda)(\gamma_l^2 - \lambda) - ab = 0.$$

Hence an easy computation gives that $\lambda = \lambda_{k,l}$, and therefore $ab \in C$, contrary to our assumptions.

We have thus shown that under the assumption $ab \notin C$ every eigenvector $W = (w_1, w_2)$ of (2.4) has the form $(\beta_1 z, \beta_2 z)$ for some complex numbers β_1, β_2 and for some nonzero function satisfying (2.6).

Now using (2.5) and applying Carleman's uniqueness theorem we conclude that $\beta_1 z = \beta_2 z = 0$ in Ω, i.e. $W = 0$ in Ω, as stated.

Remark. Note that the proof of the theorem gives more than stated: the set of exceptional matrixes (α_{ij}) is not only of measure zero, but a union of countably many real analytic curves.

References

1. C. Bardos, G. Lebeau and J. Rauch, *Contrôle et stabilisation dans les problèmes hyperboliques*, Appendice II in [12], pp. 492–537.
2. C. Bardos, G. Lebeau and J. Rauch, *Sharp sufficient conditions for the observation, control and stabilization of waves from the boundary*, SIAM J. Control Opt. 30 (1992), 1024–1065.
3. N. Burq, Contrôle de l'équation des ondes dans des domaines peu réguliers, preprint 1094, Centre de Mathématiques, École Polytechnique, Paris, 1995.
4. S. Dolecki and D. L. Russell, *A general theory of observation and control*, SIAM J. Control Opt. 15 (1977), pp. 185–220.
5. V. Komornik, *Exact Controllability and Stabilization. The Multiplier Method.* Masson, Paris and John Wiley & Sons, Chicester, 1994.
6. V. Komornik, *Rapid boundary stabilization of linear distributed systems*, SIAM J. Control Opt. 35 (1997), to appear.
7. V. Komornik and P. Loreti, *Observabilité frontière de systèmes couplés par analyse non harmonique vectorielle*, C. R. Acad. Sci. Paris Sér I Math. 324 (1997), to appear.
8. V. Komornik and P. Loreti, *Ingham type theorems for vector-valued functions and observability of coupled linear systems*, submitted.
9. V. Komornik and P. Loreti, *Observability of compactly perturbed systems and applications*, in preparation.
10. G. Lebeau, *Contrôle de l'équation de Schrödinger*, J. Math. Pures Appl. 71 (1992), 267–291.
11. J.-L. Lions, *Exact controllability, stabilizability, and perturbations for distributed systems*, Siam Rev. 30 (1988), pp. 1–68.
12. J.-L. Lions, *Contrôlabilité exacte et stabilisation de systèmes distribués*, Vol. 1–2, Masson, Paris, 1988.
13. E. Zuazua, *Contrôlabilité exacte en un temps arbitrairement petit de quelques modèles de plaques*, Appendix 1 in [12].

Vilmos Komornik
Institut de Recherche
Mathématique Avancée
Université Louis Pasteur et CNRS
7, rue René Descartes
F-67084 Strasbourg Cédex , France
e-mail: komornik@math.u-strasbg.fr

Paola Loreti
Istituto per le Applicazioni
del Calcolo "Mauro Picone"
Consiglio Nazionale delle Ricerche
Viale del Policlinico, 137
I-00161 Roma , Italy
e-mail:loreti@vaxiac.iac.rm.cnr.it

Enrique Zuazua
Departamento de Matemática Aplicada
Universidad Complutense
E-28040 Madrid, Spain
e-mail:zuazua@sunma4.mat.ucm.es

International Series of Numerical Mathematics
Vol. 126, © 1998 Birkhäuser Verlag, Basel

On Dynamic Domain Decomposition of Controlled Networks of Elastic Strings and Joint-Masses

G. LEUGERING
Institut für Mathematik
Universität Bayreuth

ABSTRACT. We consider a planar graph representative of the reference configuration of a network of elastic prestretched strings coupled at the vertices of that graph. Some or all of the vertices may carry a point mass, and at those nodes dry friction on the plane may occur. We briefly describe the model and some results on well-posedness and control of such systems obtained in the literature. We then introduce a dynamic domain decomposition based on a Steklov-Poincaré-type operator. The analysis is given for the time-domain and the frequency-domain. Optimal control and problems of exact controllability are formulated and investigated in terms of the decoupling procedure.

1991 *Mathematics Subject Classification.* 93C20, 93C80, 93B05, 65N55

Key words and phrases. Dynamic domain decomposition, strings, networks, joint masses, dry friction, Steklov-Poincaré-operators for networks, differential-delay systems, optimal control, controllability.

1. Introduction

We consider a network of dynamic elastic strings as in [10], [9], [13]. Let $G = (V, E)$ be a planar connected graph with vertices $V, \sharp V = n_v$ and edges $E, \sharp E = n_e$. Each edge is representative of a (possibly prestrechted) linear string in its reference configuration. We label nodes v_J by capital letters and edges by lower case letters. At a typical node v_J we have $d(v_J)$ incident edges, the indices of which we label $i \in \mathcal{E}_J$. To each $i \in \mathcal{E}_J$ we assign $\epsilon_{iJ} := -1$ if the edge i starts at v_J and $\epsilon_{iJ} := 1$ else. We denote by x_{iJ} the number l_i if $\epsilon_{iJ} = 1$, 0 if $\epsilon_{iJ} = -1$. We introduce $\overset{\circ}{V} := \{v \in V | d(v) > 1\}$, $\partial V = V \setminus \overset{\circ}{V}$ as multiple nodes and simple nodes, respectively. Let

$$r_i(x,t) := u_i(x,t)e_i + w_i(x,t)e_i^{\perp}$$

denote the deformation of the i-th string, where $u_i, w_i, e_i, e_i^{\perp}$ denote the longitudinal, vertical displacement, and the unit vectors along the undeflected centerline and its orthogonal complement, respectively.

We signify nodes where Dirichlet conditions hold by V_D; for simplicity of notation we impose $V_D \subset \partial V$. Correspondingly, Neumann-nodes are denoted by $V_N \subset \partial V$. Let K_i denote the local stiffness matrix. Then we have the systems of equations governing the motion of the entire network.

$$\ddot{r}_i = K_i r_i'', \ i = 1 : n_e, x_i \in (0, l_i), \ t \in (0, T) \tag{1.1}$$

$$r_d(x_{dD}) = 0, \ v_D \in V_D, \ d \in \mathcal{E}_D, \ t \in (0, T) \tag{1.2}$$

$$r_i(x_{iJ}) = r_j(x_{jJ}),\ \forall i,j \in \mathcal{E}_J, v_J \in \overset{\circ}{V} \tag{1.3}$$

$$\sum_{i\in\mathcal{E}_J} \epsilon_{iJ} K_i r_i'(x_{iJ}) = -m_J \ddot{r}_i(x_{iJ}) + f_J,\ v_J \in \overset{\circ}{V} \cup V_N,\ t \in (0,T) \tag{1.4}$$

$$r_i(x_i,0) = r_{i0}(x_i),\ \dot{r}_i(x_i,0) = r_{i1}(x_i),\ x_i \in (0,l_i),\ i = 1 : n_e\ . \tag{1.5}$$

While (1.1) is the obvious wave equation for the i-th string with homogeneous Dirichlet conditions (1.2), equation (1.3) gives the continuity of displacements accross the joint v_J. Equation (1.4) is the balance of forces at the joint v_J including the possibility of an additional mass m_J and an external force f_J there. Position (1.5) finally represents the initial conditions.

System (1.1)–(1.5) has been shown to be wellposed in

$$H := \prod_{i=1}^{n_e} L_2(0,l_i),\ V := \{r \in \prod_{i=1}^{n_e} H^1(0,l_i) | r \ \text{ satisfies (1.2), (1.3)}\}$$

with the typical setup $V \subset\subset H \subset\subset V^\star$, and $m_J = 0\ \forall_J$. We note that all spaces L_2, H^1 etc. appearing in this paper are to be considered as spaces of functions into the plane.

Theorem 1.1. [9] *Let $m_J = 0\ \forall\, J$, then*

$$\forall\,(r_0,r_1) \in V \times H,\ f \in \prod_{J=1}^{n_v} L_2(0,T)\,\exists | r :$$
$$r \in C(0,T,V) \cap C^1(0,T,H) \cap C^2(0,T,V^\star)\ ,$$

r satisfying (1.1)–(1.5) in a natural weak sense.

Remark 1.2. *More can be said in terms of regularity, when masses at joints are present. See Section 3 below. In case of nonhomogeneous Dirichlet-data, wellposedness for the system with masses $m_J \neq 0$ has been obtained in [15].*

Remark 1.3. *Problems of exact/approximate controllability have been discussed in detail in [9] where also 3-d-networks have been considered. See also [9], [10].*

Remark 1.4. *The balance of forces at multiple nodes (1.4) can be extended in various ways, in order to account for nonlinear phenomena as e.g. dry friction on the plane or elastic and rigid obstacles, rigid bars between joints ect.. We do not have sufficient space to discuss these models here in detail. To illustrate the possibility of treating such phenomena in the context of the subsequent domain decomposition, we provide a model of dry friction.*

$$\sum_{i\in\mathcal{E}_J} \epsilon_{iJ} K_i r_i'(x_{iJ}) + m_J \ddot{r}_i(x_{iJ}) \in -\mu_J \partial |\dot{r}_i(x_{iJ})|,\ v_J \in \overset{\circ}{V} \cup V_N,\ t \in (0,T). \tag{1.6}$$

In a more refined model we have

$$\sum_{i\in\mathcal{E}_J} \epsilon_{iJ} K_i r_i'(x_{iJ}) =: F_J = -m_J \ddot{r}_i(x_{iJ}) - \mu_J^s \rho_J, \tag{1.7}$$

where

$$\rho_J(t) = \frac{\dot{r}_i(x_{iJ})}{|\dot{r}_i(x_{iJ})|},\quad if\quad \dot{r}_i(x_{iJ}) \neq 0$$

$$\rho_J(t) = \xi \frac{F_J(t)}{|F_J(t)|}, \quad \xi \in [-\frac{\mu_J^a}{\mu_J^s}, \frac{\mu_J^a}{\mu_J^s}], \quad if\ \dot{r}_i(x_{iJ}) = 0, \quad F_J(t) \neq 0$$

$$\rho_J(t) = 0 \quad else,$$

and μ_J^a, μ_J^s are adhesive and shear frictional moduli. See Panagiotopoulos [12] for dry friction models and results on well-posedness for single-element systems. The subsequent analysis will show that it is possible to reduce the discontinuous part of the PDE-problem above to a finite dimensional one.

2. A dynamic Steklov-Poincaré-Operator for networks

We first consider the local Dirichlet problems for each individual edge along e_i

$$\ddot{\varphi}_i = K_i \varphi_i'', \qquad (0, l_i) \times (0, T) \tag{2.1}$$

$$\varphi_i(x_{iJ}) = \lambda_J + r_{i0}(x_{iJ}),\ \varphi_i(l_i - x_{iJ}) = \lambda_M + r_{i0}(l_i - x_{iJ}),\ (0, T) \tag{2.2}$$

$$\varphi_i(x, 0) = r_{i0}(x), \dot{\varphi}_i(x, 0) = r_{i1}(x),\ (0, l_i)\ . \tag{2.3}$$

Obviously, φ_i can be written as a superposition

$$\varphi_i(x, t, r_{i0}, r_{i1}, \lambda_J, \lambda_M) = \psi_i(x, t, 0, 0, \lambda_J, \lambda_M) + \eta_i(x, t, r_{i0}, r_{i1}, 0, 0)\ . \tag{2.4}$$

Lemma 2.1. *Let $(r_0, r_1) \in V \times H$, $\lambda_J, \lambda_M \in H^1_{0-}(0, T)$ (i.e. $\lambda_J, \lambda_M \in \{H^1(0, T)|\ \lambda(0) = 0$). Then for each i there exists a unique weak solution φ_i to (2.1)-(2.4) satisfying*

$$\varphi_i \in C(0, T, H^1(0, l_i)) \cap C^1(0, T, L^2(0, l_i))\ .$$

Proof. The proof is standard. Nevertheless, we need a precise representation anyway. In particular, we will use $\psi_i(x, t, 0, 0, \lambda_J, \lambda_M)$ below. $\eta_i(x, t, r_{i0}, r_{i1}, 0, 0)$ can be constructed, say, by a d'Alembert-Ansatz. We focus on ψ_i, and use a Fourier-Ansatz.

$$K_i = k_i \left((1 - \frac{1}{s_i}) I + \frac{1}{s_i} e_i e_i^T \right)\ ,$$

where $s_i > 1$ represents the amount of stretching. We have $\psi_i =: \psi_i^l e_i + \psi_i^v e_i^\perp$, hence (2.1)–(2.4) decouples into

$$\begin{aligned}
&(2.5) & \ddot{\psi}_i^l &= p_i^2 (\psi_i^l)'', & \ddot{\psi}_i^v &= q_i^2 (\psi_i^v)'' \\
&(2.6) & \psi_i^l(x_{iJ}) &= \lambda_J^l, & \psi_i^v(x_{iJ}) &= \lambda_J^v \\
&(2.7) & \psi_i^l(l_i - x_{iJ}) &= \lambda_M^l, & \psi_i^v(l_i - x_{iJ}) &= \lambda_M^v\ ,
\end{aligned}$$

plus zero initial conditions, where $p_i^2 = k_i$, $q_i^2 = k_i(1 - 1/s_i)$. Hence, solving the in-plane problem (2.1)–(2.4) comes down to solving scalar problems (2.5)–(2.7). We express ψ_i^l, ψ_i^v as follows

$$
\begin{aligned}
\psi_i^l &= z_i^l + y_i^l,\ \psi_i^v = z_i^v + y_i^v \\
z_i^l(x,t) &= \frac{1}{l_i}(l_i - x_{iJ} + \epsilon_{iJ}x)\lambda_J^l(t) + \frac{1}{l_i}(x_{iJ} - \epsilon_{iJ}x)\lambda_M^l(t) \\
z_i^v(x,t) &= \frac{1}{l_i}(l_i - x_{iJ} + \epsilon_{iJ}x)\lambda_J^v(t) + \frac{1}{l_i}(x_{iJ} - \epsilon_{iJ}x)\lambda_M^v(t)\ . \\
\ddot{y}_i^l &= p_i^2(y_i^l)'' - \ddot{z}_i^l\ , \qquad \ddot{y}_i^v = q_i^2(y_i^v)'' - \ddot{z}_i^v \qquad (2.8)\\
y_i^l(x_{iJ}) &= 0 = y_i^l(l_i - x_{iJ})\ , \qquad y_i^v(x_{iJ}) = 0 = y_i^v(l_i - x_{iJ}) \qquad (2.9)\\
& + \text{ zero initial conditions.}
\end{aligned}
$$

By standard calculations we obtain, with $\mu_{ij}^l := \frac{p_i \pi j}{l_i}$

$$
\begin{aligned}
(2.10)\ y_i^l(x,t) = -\sum_{j=1}^{\infty} \frac{l_i}{p_i \pi j} \int_0^t \sin(\mu_{ij}^l)^{1/2}(t-s) \Big\{ -\left(\frac{2}{l_i}\right)^{1/2} \epsilon_{iJ} \frac{l_i}{j\pi} \cos\left(\frac{j\pi}{l_i} x_{iJ}\right) \ddot{\lambda}_J^l(s) \\
+ \left(\frac{2}{l_i}\right)^{1/2} \epsilon_{iJ} \frac{l_i}{j\pi} \cos\left(\frac{j\pi}{l_i}(l_i - x_{iJ})\right) \ddot{\lambda}_M^l(s) \Big\} ds \cdot \left(\frac{2}{l_i}\right)^{1/2} \sin \frac{j\pi}{l_i} x\ .
\end{aligned}
$$

If we integrate (2.10) by parts and approximate $H_{0-}^1(0,T)$ by $H_{0-}^2(0,T)$, the asymptotic behaviour of the Fourier coefficients implies that ψ_i^l have the properties required. The same obviously holds for ψ_i^v. □

Lemma 2.2. *Let φ_i $i = 1, \dots, n_e$ be as in Lemma 2.1. Then $\varphi_i(x,t) = r_i(x,t)$, where r_i $i = 1, \dots, n_e$ solve (1.1)–(1.5) if and only if*

$$
\sum_{i \in \mathcal{E}_J} \epsilon_{iJ} K_i \varphi_i'(x_{iJ}) + m_J \ddot{\lambda}_J = f_J \qquad \forall\ J \in \overset{\circ}{V} \cup V_N\ . \tag{2.11}
$$

Proof. If $\varphi_i = r_i$ then (2.11) is obvious. Conversely, since $\ddot{\lambda}_J(t) = \ddot{\varphi}_i(x_{iJ},t)$, condition (2.11) is just (1.4) with φ_i. By construction

$$
\varphi_i(x_{iJ}) = r_{i0}(x_{iJ}) + \lambda_J = r_{k0}(x_{kJ}) + \lambda_J = \varphi_k(x_{kJ})\ ,
$$

since $r_0 \in V$. For Dirichlet nodes v_D we have $\lambda_D \equiv 0$. The argument is applied to strong solutions and then to mild solutions. □

We return to (2.10) and, consequently, require $\lambda \in H_{0-}^2$ for the time being. We obtain

$$
\begin{aligned}
(y_i^l)'(x_{iJ},t) &= \epsilon_{iJ} \frac{2}{l_i} \sum_{j=1}^{\infty} \frac{l_i}{p_i \pi j} \int_0^t \sin(\mu_{ij}^l)^{1/2}(t-s) \left\{ \ddot{\lambda}_J^l(s) - (-1)^j \ddot{\lambda}_M^l(s) \right\} ds \\
&= \epsilon_{iJ} \frac{2}{l_i p_i} \sum_{j=1}^{\infty} \frac{l_i}{\pi j} \int_0^t \sin(\mu_{ij}^l)^{1/2}(s) \left\{ \ddot{\lambda}_J^l(t-s) - (-1)^j \ddot{\lambda}_M^l(t-s) \right\} ds \\
&=: \epsilon_{iJ} \frac{2}{l_i p_i} \sum_{k=0}^{\infty} \frac{l_i}{\pi j} \sum_{k=0}^{m_t} \int_{k l_i/p_i}^{(k+1)l_i/p_i} G_{ij}(s) ds\ ,
\end{aligned}
$$

where $m_t = \arg\min_k\{(k+1)l_i/p_i \geq t\}$ and the integrand is put to zero for $s \in [t,(m_t+1)l_i/p_i]$. Hence,

$$(y_i^l)'(x_{iJ},t) = \epsilon_{iJ}\frac{2}{l_ip_i}\sum_{j=1}^{\infty}\frac{l_i}{\pi j}\sum_{k=0}^{m_t}\int_0^{l_i/p_i} G_{ij}(\tau + kl_i/p_i)d\tau \ . \tag{2.12}$$

Now,

$$\sin\left(\frac{p_i\pi}{l_i}\left(\tau + k\frac{l_i}{p_i}\right)\right) = \sin\left(\frac{p_i\pi j}{l_i}\tau + k\pi j\right) = (-1)^{kj}\sin\left(\frac{p_i\pi j}{l_i}\tau\right) \ ,$$

and hence (2.12) gives

$$\begin{aligned}(y_i^l)'(x_{iJ},t) &= \epsilon_{iJ}\frac{2}{p_i}\sum_{j=1}^{\infty}\frac{1}{\pi j}\sum_{k=0}^{m_t}(-1)^{kj}\int_0^{l_i/p_i}\sin\left(\frac{p_i\pi j}{l_i}\tau\right)\cdot \\ &\quad \cdot\left\{\ddot\lambda_J^l(t-\tau-kl_i/p_i) - (-1)^j\ddot\lambda_M^l(t-\tau-kl_i/p_i)\right\}d\tau \ .\end{aligned} \tag{2.13}$$

Furthermore

$$\begin{aligned}\frac{p_i}{l_i}\int_0^{l_i/p_i} t\sin\left(\frac{p_i\pi j}{l_i}t\right)dt &= -\frac{1}{\pi j}(-1)^j\frac{l_i}{p_i} \ , \\ \frac{p_i}{l_i}\int_0^{l_i/p_i}\left(1-\frac{p_i}{l_i}t\right)\cdot\sin\left(\frac{p_i\pi j}{l_i}t\right)dt &= \frac{1}{\pi j} \ ,\end{aligned}$$

and hence (2.13) can be written as

$$\begin{aligned}(y_i^l)'(x_{iJ},t) &= \frac{1}{p_i}\epsilon_{iJ}\sum_{\substack{k=0\\ k \text{ even}}}^{m_t}\int_0^{l_i/p_i}\sum_{j=1}^{\infty}\left\{\left(\frac{2p_i}{l_i}\right)^{1/2}\cdot\right. \\ &\int_0^{l_i/p_i}\left(1-\frac{p_i}{l_i}t\right)\sin\left(\frac{p_i\pi j}{l_i}t\right)dt\left(\frac{2p_i}{l_i}\right)^{1/2}\sin\frac{p_i\pi j}{l_i}\tau\cdot \\ &\quad\cdot\ddot\lambda_J^l(t-\tau-k\,l_i/p_i) + \\ &\left(\frac{2p_i}{l_i}\right)^{1/2}\int_0^{l_i/p_i}\frac{p_i}{l_i}t\sin\left(\frac{2p_i\pi j}{l_i}t\right)dt\left(\frac{2p_i}{l_i}\right)^{1/2}\sin\left(\frac{p_i\pi j}{l_i}\tau\right)\cdot \\ &\quad\left.\cdot\ddot\lambda_M^l(t-\tau-k\,l_i/p_i)\right\}d\tau \\ &+\frac{1}{p_i}\epsilon_{iJ}\sum_{\substack{k=1\\ k \text{ odd}}}^{m_t}\int_0^{l_i/p_i}\sum_{j=1}^{\infty}\left\{-\left(\frac{2p_i}{l_i}\right)^{1/2}\int_0^{l_i/p_i}\frac{p_i}{l_i}t\sin\left(\frac{p_i\pi j}{l_i}t\right)dt\right. \\ &\quad\cdot\left(\frac{2p_i}{l_i}\right)^{1/2}\sin\left(\frac{p_i\pi j}{l_i}\tau\right)\cdot\ddot\lambda_J^l(t-\tau-k\,l_i/p_i) - \\ &\left(\frac{2p_i}{l_i}\right)^{1/2}\int_0^{l_i/p_i}\left(1-\frac{p_i}{l_i}t\right)\sin\left(\frac{p_i\pi j}{l_i}t\right)dt\left(\frac{2p_i}{l_i}\right)^{1/2}\sin\left(\frac{p_i\pi j}{l_i}\tau\right)\cdot \\ &\quad\left.\cdot\ddot\lambda_M^l(t-\tau-k\,l_i/p_i)\right\}d\tau\end{aligned}$$

$$
\begin{aligned}
= \; & \frac{1}{p_i}\epsilon_{iJ} \sum_{k=0}^{\lceil m_t/2 \rceil} \Big\{ \dot\lambda^l_J(t-2(k+1)l_i/p_i) + \dot\lambda^l_J(t-2k\,l_i/p_i) \\
& \qquad + p_i/l_i(\lambda^l_J(t-2(k+1)l_i/p_i) - \lambda^l_J(t-2k\,l_i/p_i)) \Big\} \\
& -\frac{1}{p_i}\epsilon_{iJ} \sum_{k=0}^{\lceil (m_t-1)/2 \rceil} \Big\{ \dot\lambda^l_M(t-(2k+3)l_i/p_i + \dot\lambda^l_M(t-(2k+1)l_i/p_i) \\
& \qquad + p_i/l_i(\lambda^l_M(t-(2k+3)l_i/p_i) - \lambda^l_M(t-(2k+1)l_i/p_i)) \Big\}
\end{aligned}
$$

$$
\begin{aligned}
(y^l_i)'(x_{iJ},t) \; = \; & \frac{1}{p_i}\epsilon_{iJ}\, \{ \dot\lambda^l_J(t) + 2 \sum_{k=0}^{\lceil (m_t-2)/2 \rceil} \dot\lambda^l_J(t-2(k+1)l_i/p_i \} - \frac{1}{l_i}\epsilon_{iJ}\lambda^l_J(t) \\
& -\frac{1}{p_i}\epsilon_{iJ} \left\{ 2 \sum_{k=0}^{\lceil (m_t-1)/2 \rceil} \dot\lambda^l_M(t-(2k+1)l_i/p_i) \right\} + \frac{1}{l_i}\epsilon_{iJ}\lambda^l_M(t)\ .
\end{aligned}
$$

Now, $(z^l_i)'(x_{iJ},t) = \epsilon_{iJ}\frac{1}{l_i}\lambda^l_J(t) - \frac{\epsilon_{iJ}}{l_i}\lambda^l_M(t)$, and hence

$$
\begin{aligned}
(\psi^l_i)'(x_{iJ},t) = & \epsilon_{iJ}\frac{1}{p_i}\, \{ \dot\lambda^l_J(t) + 2\sum_{k=0}^{m_e} \dot\lambda^l_J(t-2(k+1)l_i/p_i) \\
& - 2\sum_{k=0}^{m_0} \dot\lambda^l_M(t-(2k+1)l_i/p_i) \}\ ,
\end{aligned} \tag{2.14}
$$

with m_e, m_0, denoting the even and odd boundary for the index k obtained from the previous formula.

A similar expression holds for the vertical component with l replaced with v. With the definitions $P_i\lambda := p_i\lambda^l e_i + q_i\lambda^v e_i^\perp$, $(D_i^\tau\lambda)(t) := p_i\lambda^l(t-\tau l_i/p_i)e_i + q_i\lambda^v(t-\tau l_i/q_i)e_i^\perp$ we can write (1.4) with φ_i replacing r_i as

$$
\begin{aligned}
& m_J\ddot\lambda_J(t) + \left(\sum_{i\in\mathcal{E}_J} P_i\right)\dot\lambda_J(t) + 2\sum_{k=0}^{m_e}\sum_{i\in\mathcal{E}_J}(D_i^{2(k+1)}\dot\lambda_J)(t) \\
& -2\sum_{k=0}^{m_0}\sum_{i\in\mathcal{E}_J}(D_i^{2k+1}\dot\lambda_{iM})(t) = f_J(t) - \sum_{i\in\mathcal{E}_J}\epsilon_{iJ}K_i\eta_i'(x_{iJ},t)\ .
\end{aligned}
$$

Here v_{iM} is adjacent to v_J and belongs to the incident edge i. Note that (2.15) also applies to a simple Neumann-node v_J where $d(v_J) = 1$. We may define $\Lambda := (\lambda_J, \lambda_N, \dots)^T$ and

$$
\begin{aligned}
& S(\Lambda) := \left(\sum_{i\in\mathcal{E}_J} \epsilon_{iJ}K_i\psi_i'(x_{iJ},t,0,0,\lambda_J,\lambda_{iM}) + m_J\ddot\lambda_J(t) \right)_{J=1:N}\ , \\
& S : \prod_J H^1_{0-}(0,T) \longrightarrow \prod_J L^2(0,T)\ , \\
& D(S) = \{\Lambda \in \prod_J H^1_{0-}(0,T) | \lambda_J \in H^2_{0-}(0,T) \text{ for } m_J \neq 0\}\ .
\end{aligned} \tag{2.15}
$$

Then S can be interpreted as a *dynamic Steklov-Poincaré-Operator* for the network. Indeed, for time independent problems, or those obtained by some integral transformation, S is identical with a Steklov-Poincaré operator. However, even in the static case, such operators, in the case of networks of any dimension, do not seem to have received attention yet. The case of coupled membranes is the subject of current research. See Lagnese [8], where a domain decomposition is used implicitly. Its importance is due to the following theorem

Theorem 2.3. *Given $(r_0, r_1) \in V \times H$, $f \in \prod_J L^2(0,T)$. If Λ solves $S(\Lambda) = F$ with*

$$F = \left(f_J - \sum_{i \in \mathcal{E}_J} \epsilon_{iJ} K_i \eta_i'(x_{iJ}, t, r_{i0}, r_{i1}, 0, 0) \right)_{J=1:N},$$

then the corresponding local solutions φ_i of Lemma 2.1 constitute a solution of (1.1)–(1.5) according to Lemma 2.2.

The entire calculus leading to (2.15) was done to give an explicit representation of S defined by (2.15).

Therefore, in order to solve problem (1.1)–(1.5), we have to solve the differential-delay system of second order in time (2.15) together with trivial initial histories $\lambda_J(-s) = 0 \ \forall \ s \geq 0 \ \forall J \in V$. Once that is done *step by step*, according to the individual delays $l_i/p_i, l_i/q_i$, the *solutions φ_i can be computed in parallel.* Also, each individual problem can be solved in parallel using waveform relaxation techniques, see Burrage [1].

Remark 2.4. *If dry friction occurs at a joint v_J as explained in Remark 1.4, one has to either add a maximal monotone operator to S, or use the notation of variational inequalities in order to extend existence and uniqueness to the system with dry friction. In particular, using the method of steps, one obtains a sequence of ordinary differential equations with discontinuous forcing term. The latter systems can then be treated step-by-step in the framework of DAE's (differential-algebraic equations) of index 2. We have implemented this strategy using RADAU5 of Hairer and Wanner (see [4] as a base reference) and projection methods. We have applied this also to beam networks with rotational dry friction between "pin-joints". We refer also to the work of Glowinski et.al. [3] (for matrix problems), where the time is first discretized in order to obtain an "elliptic" problem together with a projection accounting for the dry friction. To apply their approach to elastic networks, using the domain decomposition above, is the subject of current research.*

3. Wellposedness and regularity of solutions

It is well-known that the solution $\eta_i(x, t, r_{i0}, r_{i1}, 0, 0)$ of (2.1)–(2.3) has the regularity

$$\eta_i \in C(0, T, H^1(0, l_i)) \cap C^1(0, T, L^2(0, l_i)) .$$

for $r_{0i} \in H^1(0, l_i)$, $r_{1i} \in L^2(0, l_i)$. In fact, one verifies that

$$\eta_i'(x_{iJ}, \cdot, r_{i0}, r_{i1}, 0, 0) \in L^2(0, T) .$$

Lemma 3.1. *Given* $(r_0, r_1) \in V \times H$, $f_J \in L^2(0,T)$ $\forall$ $J \in \overset{\circ}{V} \cup V_N$, *there exists a unique solution of (2.15) with* $\lambda_J(-s) = 0 \; \forall \; s \geq 0$, $J \in V$. *The regularity of* $\lambda_J(t)$ *depends on the presence of a mass* m_J *at node* v_J*:*

$$\begin{array}{lll} i) & \text{if } m_J = 0, & \text{then } \lambda_J \in H^1(0,T), \; \lambda_J(0) = 0 \\ ii) & \text{if } m_J \neq 0, & \text{then } \lambda_J \in H^2(0,T), \; \lambda_J(0) = \dot{\lambda}_J(0) = 0 \; . \end{array}$$

Proof. The right hand side of (2.15) is in $L^2(0,T)$. Let $\alpha_1 = \min\limits_{i=1}^{n_e}(l_i/p_i, \, l_i/q_i)$, then solve (2.15) on $(0, \alpha_1)$ uniquely for λ_J. Note that because of $\lambda_J(-s) = 0 \; \forall \; s \geq 0$, the sums in (2.15) are empty. Now $\lambda_J(t)$ satisfies i), ii) for $m_J = 0$, $m_J \neq 0$, respectively. We do this for all vertices and obtain Λ on $(0, \alpha_1)$. By the same procedure, we obtain the solution Λ on $(0,T)$ if we use the solutions obtained in the previous step to update the history part in (2.15). This is the classical method of steps for differential delay systems, see [1]. □

Corollary 3.2. *It is immediate from Lemma 3.1 that waves passing a node with nonzero mass* m_J *or waves generated at such a node by applying forces* f_J*, have one more degree of regularity in the edge incident at* v_J.

Theorem 3.3. *Let* $(r_0, r_1) \in V \times H$, $f_J \in L^2(0,T)$ $\forall$ $J \in \overset{\circ}{V} \cup V_N$. *Then there exists a unique mild solution* r *of (1.1)–(1.5), satisfying*

$$r \in C(0,T,V) \cap C^1(0,T,H) \; .$$

Proof. The first part is obvious from Lemma 3.1 and Theorem 2.3. The second part follows from Corollary 3.2. □

Remark 3.4. *We illustrate the smoothing property of the joint-masses by an exemplaric situation. There are many interesting features concerning regularity issues when masses are present at joints. We don't have sufficient space to go into details. See, however, Willé and Baker [16], for the propagation of singularities in DDEs and [5], [15] for two-span strings.*
If $r_{i0}, r_{i1} \neq 0$ *on a set* I *of edges and* $r_{i0} = r_{i1} = 0$ *on* $E \backslash I$ *and* $m_J \neq 0 \quad \forall \, v_J \in I$ *such that* $adj(v_J) \cap (E \backslash I) \neq \emptyset$. *Then on* $E \backslash I$ *the restrictions* r_i *of* r *satisfy* $r_i \in C(0,T,H^2(0,l_i)) \cap C^1(0,T,H^1(0,l_i) \cap C^2(0,T,L^2(0,l_i)$.
Obviously, Theorem 3.3 extends Theorem 1.1 to the case with $m_J \neq 0$.

4. A frequency domain representation

We switch to a nodal description of (1.1)–(1.5) as in [11]. The edge connecting the adjacent nodes v_J, v_M is labeled by JM (i.e. $r_i \leftrightarrow r_{JM}$, $\mathcal{E}_J \leftrightarrow \hat{\mathcal{E}}_J = \{M | v_M \leftrightarrow v_J\}$ etc.). Denote $k^1_{JM} := 1/p_{JM}$, $k^2_{JM} := 1/q_{JM}$. $D_{JM} := \operatorname{diag}(p_{JM}, q_{JM})$, $k_{JM} := \operatorname{diag}(k^1_{JM}, k^2_{JM})$, $R_{JM} := \begin{pmatrix} \cos\theta_{JM} & \sin\theta_{JM} \\ -\sin\theta_{JM} & \cos\theta_{JM} \end{pmatrix}$, where $e_{JM} = (\cos\theta_{JM}, \sin\theta_{JM})^T$, θ_{JM} being the angle between e_i and $(1,0)$ in the reference frame. The notation r_{JM} implies that $r_{JM}(0)$ relates to $r_i(x_{iJ})$. For the sake of convenience, we take each edge twice, namely r_{JM}, r_{MJ}, thereby we obtain a multi-digraph. Of course,

$r_{JM}(x) = r_{MJ}(l_{JM} - x)$, implying $u_{JM}(x) = -u_{MJ}(l_{JM} - x)$, $w_{JM}(x) = -w_{MJ}(l_{JM} - x)$. Consider the Laplace-transform of (1.1)–(1.5):

$$\mathcal{L}f(s) := \int_0^\infty e^{-st} f(t)\, dt = \hat{f}(s)$$

$$\begin{aligned}
&(4.1) && s^2 \hat{r}_{JM}(x,s) - s r_{JM}(x,0) - \dot{r}_{JM}(x,0) = K_{JM} \hat{r}''_{JM}(x,s)\\
&(4.2) && \hat{r}_{DM}(0,s) = 0,\ M \in \mathcal{E}_D,\ D \in V_D\\
&(4.3) && \hat{r}_{JM}(0,s) = \hat{r}_{JN}(0,s) = \hat{\lambda}_J(s)\ \forall\ M, N \in \mathcal{E}_J\\
&(4.4) && -\sum_{M \in \mathcal{E}_J} K_{JM} \hat{r}'_{JM}(0,s) + m_J s^2 \hat{\lambda}(s) - s m_J \lambda(0) - m_J \dot{\lambda}(0) = \hat{f}_J(s)\\
&(4.5) && r_{JM}(x,0) = r^0_{JM}(x),\ r^0_{JM}(x,0) = r^1_{JM}(x) \quad J \in \overset{\circ}{V} \cup V_N\ .
\end{aligned}$$

Again $r_{JM} = u_{JM} e_{JM} + w_{JM} e^{\perp}_{JM}$, hence (4.3), (4.4) translate to

$$(4.6) \qquad \begin{pmatrix} \hat{u}_{JM} \\ \hat{w}_{JM} \end{pmatrix}(0,s) = R_{JM-JK} \begin{pmatrix} \hat{u}_{JK} \\ \hat{w}_{JK} \end{pmatrix}(0,s) = R_{JM} \hat{\lambda}_J(s)$$

$$(4.7) \qquad m_J s^2 \hat{\lambda}(s) - \sum_{M \in \mathcal{E}_J} R_{-JM} K_{JM} \begin{pmatrix} \hat{u}_{JK} \\ \hat{w}_{JM} \end{pmatrix}'(0,s) = F_J(s)\ ,$$

with $F_J(s) := \hat{f}_J(s) + s m_J \lambda(0) + m_J \dot{\lambda}(0)$. Put $a_{JM} := (a^1_{JM}, a^2_{JM})^T$, $d_{JM} = (d^1_{JM}, d^2_{JM})^T$, then

$$\begin{aligned}
(4.8)\ \begin{pmatrix} \hat{u}_{JM} \\ \hat{w}_{JM} \end{pmatrix}(x,s) &= e^{s k_{JM} x} \left(a_{JM}(s) + D_{JM} \frac{1}{2s} \int_0^x e^{-s k_{JM} \tau} b_{JM}(s,\tau)\, d\tau \right)\\
&\quad + e^{-s k_{JM} x} \left(d_{JM}(s) - D_{JM} \frac{1}{2s} \int_0^x e^{s k_{JM} \tau} b_{JM}(s,\tau)\, d\tau \right)\\
(4.9)\ \begin{pmatrix} \hat{u}_{JM} \\ \hat{w}_{JM} \end{pmatrix}(0,s) &= a_{JM}(s) + d_{JM}(s),\\
\begin{pmatrix} \hat{u}_{JM} \\ \hat{w}_{JM} \end{pmatrix}'(0,s) &= s k_{JM} (a_{JM}(s) - d_{JM}(s)) = R_{JM} \hat{\lambda}_J(s)\ .
\end{aligned}$$

There are at least two possibilities to proceed further. One is to derive scattering relations based on d_{JM} (departure) and a_{JM} (arrival) as in [11], [10]. Based on this scattering analysis, we have been able to develop nonclassical control strategies, localization of energy-fluxes, spectral properties and much more. Here we focus on the relation between the data (f_J, r_{i0}, r_{i1}) and the λ_J's, as being proposed in Section 2. Therefore, the upcoming *analysis is suitable for parallelization.* We use (4.8) and conclude, with $G_{JM}(l_{JM}, s) := D_{JM} \frac{1}{2s} \int_0^{l_{JM}} e^{-s k_{JM} \tau} b_{JM}(s,\tau)\, d\tau$, after some calculus,

$$(4.10) \qquad \begin{aligned}
a_{JM}(s) &= -(I - e^{-2s k_{JM} l_{JM}})^{-1} \left\{ e^{-s k_{JM} l_{JM}} R_{MJ} \hat{\lambda}_M(s) \right.\\
&\qquad \left. + e^{-2s k_{JM} l_{JM}} R_{JM} \hat{\lambda}_J(s) - \left(I + e^{-s k_{JM} l_{JM}} \right) G(l_{JM}, s) \right\}\\
d_{JM}(s) &= R_{JM} \hat{\lambda}_J(s) - \hat{a}_{JM}(s)\ .
\end{aligned}$$

Equations (4.10) provide an explicit representation of the complex amplitudes of arriving and departing waves in forms of the traces λ_J, λ_M at the nodes v_J, v_M.

$$
\begin{aligned}
& m_J s^2 \hat{\lambda}_J(s) + s \left(\sum_{M \in \mathcal{E}_J} R_{-JM} D_{JM} R_{JM} \right) \hat{\lambda}_J(s) \\
& + 2s \left(\sum_{M \in \mathcal{E}_J} R_{-JM} D_{JM} (I - e^{-2sk_{JM} l_{JM}})^{-1} e^{-2sk_{JM} l_{JM}} R_{JM} \right) \hat{\lambda}_J(s) \\
& + 2s \sum_{M \in \mathcal{E}_J} R_{-JM} D_{JM} (I - e^{-2sk_{JM} l_{JM}})^{-1} e^{-sk_{JM} l_{JM}} R_{MJ} \hat{\lambda}_M(s) \\
& = F_J + 2s \sum_{M \in \mathcal{E}_J} R_{-JM} D_{JM} (I - e^{-2sk_{JM}})^{-1} (I + e^{sk_{JM} l_{JM}}) G_{JM}(l_{JM} s) \ .
\end{aligned}
\tag{4.11}
$$

Equation (4.11) when relabeled with edge-indices is just (2.15) in the frequency domain. Note, however, that we cannot use such transform techniques when e.g. dry friction is present at joints. As a result, (4.11) appears as an extension of the classical Steklov-Poincaré-equation to networks, and (2.15) as its pull-back into the time domain. See Benamou [2] for a 2-d-problem.

5. Control problems

5.1. Minimization of energy in substructures. We go back to (2.15) and integrate with respect to time from 0 to t.

$$
\begin{aligned}
& m_J \dot{\lambda}_J(t) + \left(\sum_{i \in \mathcal{E}_J} P_i \right) \lambda_J(t) + 2 \sum_{k=0}^{m_e} \sum_{i \in \mathcal{E}_J} (D_i^{2(k+1)} \lambda_J)(t) \\
& -2 \sum_{k=0}^{m_0} \sum_{i \in \mathcal{E}_J} (D_i^{2k+1} \lambda_{iM})(t) = \int_0^t f_J(s)\, ds - \sum_{i \in \mathcal{E}_J} \epsilon_{iJ} K_i \int_0^t \eta_i'(x_{iJ}, s)\, ds \ .
\end{aligned}
\tag{5.1}
$$

Note that $\dot{\lambda}_J(0) = 0$ if $m_J \neq 0$ and $\lambda_J(-s) \equiv 0 \quad s \geq 0$. Similar equations hold for each node v_J. Note, however, that the J-th equation is the only one to contain $\lambda_J(t), \dot{\lambda}_J(t)$ at the actual time t. It is readily seen that one might eliminate the λ_J's corresponding to nodes without extra masses m_J from (2.15),((5.1)). As a result, one can reduce the system of equations (5.1) for the indices J to those having $m_J \neq 0$. Having solved that latter system one can then use the recurrence relations for all those variables λ_K with $m_K = 0$. This amounts to saying, that we may w.l.o.g. consider a reduced system with $m_J \neq 0 \,\forall\, J$. That reduced system can be recast, into a standard format as follows.

$$
\begin{aligned}
\dot{\Lambda}(t) &= A_0 \Lambda(t) + \sum_{j=1}^{l} A_j \Lambda(t - h_j) + Bu(t) + g(t) \\
\Lambda(-t) &\equiv 0, \quad t \geq 0
\end{aligned}
\tag{5.2}
$$

with A_i, B obvious from (5.1). In (5.2) the structure of $B\,u(t)$ is again given by

$$
B_0 F(t) + \sum_{j=1}^{l} B_j F(t - h_j) = Bu(t)
$$

where $F_J(t) = \int_0^t f_J(s)\,ds$. As remarked earlier, dry friction at some joints will result in a discontinuous "right-hand-side", of a type mentioned in Remark 2.4. Further, $g(t)$ is given by the local Dirichlet problems with fixed Dirichlet-data and corresponding initial conditions. It is important to note that the number l of delays increases with the process time. Therefore, if we want to solve (2.15)((5.2)) on a time interval $(0,T)$, then we first have to compute l as the largest integer such that $max(2(l+1)l_i/p_i, 2(l+1)l_i/q_i \leq T \quad i = 1,\ldots,n_e$. Then on $(0,T)$ we have a finite delay problem (5.2). We may, thereby, restrict solutions Λ of (5.2) with l fixed, obtained by semi-group theory, to $(0,T)$ in order to solve our problem, while it is not feasible to use these semi-group-solutions beyond T. Keeping this in mind, we can use the powerful theory of functional /retarded differential equations and the corresponding control theory as surveyed for instance in Kappel [7]. On the numerical side, we can either use approximation methods as surveyed als in [7] or the very recent parallel approach outlined in Burrage [1]. We do not have sufficient space to dwell on this any further. The main purpose of this section is to reformlate the problem of minimizing the energy of a specified *subsystem*, in terms of the nodal variables Λ. In the edge i we have the local "total" energy:

$$\begin{aligned} E_i(t) \;&:=\; \frac{1}{2}\int_0^{l_i} \left\{|\dot{r}_i(x,t)|^2 + K_i r_i'(x,t)\cdot r_i'(x,t)\right\} dx \\ &\quad + \frac{1}{2}m_J|\dot{\lambda}_J(t)|^2 + \frac{1}{2}m_M|\dot{\lambda}_M(t)|^2 := E_{i0}(t) + E_{i1}(t) \end{aligned}$$

Using (2.1)–(2.4) and the results of Section 2 we obtain

$$\begin{aligned} \frac{d}{dt}E_{i0}(t) &= \epsilon_{iJ}K_i(\psi_i'(x_{iJ}t) + \eta_i'(x_{iJ},t))\cdot\dot{\lambda}_J(t) + \epsilon_{iM}K_i(\psi_i'(x_{iM},t) + \eta_i'(x_{iM},t))\cdot\dot{\lambda}_M(t) \\ &= \epsilon_{iJ}K_i\psi_i'(x_{iJ},t)\cdot\dot{\lambda}_J(t) + \epsilon_{iM}K_i\psi_i'(x_{iM},t)\cdot\dot{\lambda}_M(t) \\ &\quad + \epsilon_{iJ}K_i\eta_i'(x_{iJ},t)\cdot\dot{\lambda}_J(t) + \epsilon_{iM}K_i\eta_i'(x_{iM},t)\cdot\dot{\lambda}_M(t)\ . \end{aligned}$$

However, by (2.14) we have

$$K_i\psi_i'(x_{iJ},t) \;=\; \epsilon_{iJ}(P_i\dot{\lambda}_J(t) + 2\sum_{k=0}^{m_e}(D_i^{2(k+1)}\dot{\lambda}_J)(t) - 2\sum_{k=0}^{m_0}(D_i^{2k+1}\dot{\lambda}_{iM})(t)),$$

and hence, upon integration, it is seen that $E_i(T)$ is a quadratic functional in $\lambda_J(t), \lambda_M(t)$ and their histories. In particular, set $n = 2\cdot n_v$, $M^2 = \mathbb{R}^n \times L^2(-h,0;\mathbb{R}^n)$ where $0 = h_0 \leq h_1 \leq \ldots \leq h_l =: h$. Then, using the notation from (5.2) we can define the operator $\mathcal{A}$ in M^2 by

$$\begin{aligned} D(\mathcal{A}) \;&=\; \{(\phi^0,\phi^1) \in M^2 | \phi^1 \in H^1(-h,0;\mathbb{R}^n)\phi^0 = \phi^1(0)\} \\ \mathcal{A}\phi \;&=\; \left(L\phi^1, \frac{\partial}{\partial s}\phi^1\right) \\ L\phi^1 \;&=\; \sum_{j=0}^{l} A_j\phi^1(-h_j) \end{aligned}$$

see Kappel [7]. Obviously, with $z(t) := (z^0(t), z^1(t))^T$,

$$\begin{aligned}
\dot{z}^0(t) &= \sum_{j=0}^{l} A_j z^1(t_j - h_j),\ z^0(0) = 0 \\
\dot{z}^1(t,s) &= (z^1)'(t,s),\ z^1(t,0) = z^0(t),\ z^1(0,s) \equiv 0\, s \in [-h,0] \\
\Rightarrow z^1(t,s) &= z^0(t+s) \\
\Rightarrow \dot{z}^0(t) &= \sum_{j=0}^{l} A_j z^0(t - h_j)\ .
\end{aligned}$$

Therefore, our system (5.2) can be written in the standard format

$$\begin{cases} \dot{z}(t) &= \mathcal{A}z(t) + \mathcal{B}u(t) + g(t) \qquad t \in [0,T] \\ z(0) &= z_0 \end{cases}$$

We have thus shown, that solving the problem of minimizing the vibrational energy on a substructure of an elastic network of strings reduces to a *finite-horizon LQR-problem for a system of delay differential equations* of a simple form with zero initial history. As there are no continuous delays involved, one can, in fact, always proceed with a step-by-step procedure involving the solution of ODEs only. It is also possible to account for the unbounded delay case directly. Then one rewrites the system into a system of ordinary Volterra-integrodifferential equations in the Stieltjes-sense with monotone, piecewise constant kernels. This setup is more suitable when discussing the longtime behaviour of the system, e.g. when considering an *infinite horizon LQR-problem.*

As a result, by the decomposition method outlined so far

- we are able to reduce an infinite dimensional – possibly nonlinear – control problem to a finite dimensional one, without any kind of approximation
- we can solve the delay system and its adjoint occuring in the optimality conditions in parallel using methods from [16] to obtain the optimal nodal positions
- we can solve the local PDE's on each edge in parallel for given nodal positions.

The task of putting this program into numerical algorithms is on its way.

5.2. Controlling the energy flux. The flux of engergy in an edge is given by $-K_i r_i'(x,t) \cdot \dot{r}_i(x)$ and is equal to the energy transported across the section at x in the positive e_i-direction, with respect to a time-unit. Hence, at a node v_J, the energy transported into the direction of the incident edge is $\epsilon_{iJ} K_i r_i'(x_{iJ},t) \cdot \dot{\lambda}_J$ $(-K_{JM} r'_{JM}(0,t) \cdot \dot{\lambda}_J(t))$. Hence, in order to maximize (or minimize) the energy flux, say, along a given forest with roots as sources and sinks as leaves, on could take squares of the flux as a cost. The resulting problem would be similar to those of Section 5.1.

Let us instead use the frequency domain approach of Section 4. There, we have shown that

$$d_{JM}(s) = R_{JM}\hat{\lambda}_J(s) - a_{JM}(s),$$

with $a_{JM}(s)$ given by (4.10), is the complex amplitude of the wave running from v_J towards v_M. The correspondance with the time-domain expression of the flux is obvious. Therefore, the problem of controlling the flux of engery comes down to an *equality-constrained LSQR-problem* or a quadratic programming problem. This approach has been discussed in Benamou [2] in the context of minimizing the scattering of waves incident at a surface of an obstacle.

Also, it is possible to consider the problem of exact anihilation of waves – which is an analogue of the well-known *anti-sound problem.*

We remark that, particularily in the context of beams, where dry friction between "pin-joints" becomes relevant, one might consider changing dry frictional parameters as *variable structure controls.* This is an open field of research.

As is amply demonstrated, there is a wealth of interesting optimal control problems which can be reduced in dimension considerably using the proposed domain-decomposition.

5.3. Controllability. Problems of exact controllability and approximate controllability for non collinear 2-d-or 3-d-networks of strings and beams in the absence of masses m_J have been investigated in Lagnese, Leugering, Schmidt [9]. The case of such networks with additional joint-masses m_J remained open. Results in this direction are available only for serial strings with such masses. See Schmidt and Wei [14], Wei [15] and Hansen, Zuazua [5]. As an exemplaric problem in both papers, a two-span string system with point mass in the middle is considered. The comparatively simpler analysis in sections 2, 3 makes it clear that, in that examplaric situation with one extreme point clamped and the other controlled, the waves originating from the latter input source will be smoothened by one degree of regularity while passing the mass in the middle. Therefore, rough data in the first string (with one end clampded) cannot be compensated for by control inputs at the other extreme. Indeed, for Dirichlet control problems in that context, it was shown in [15], [5] that exact controllability holds, if more regularity is required in the span which is not directly connected to a controlled end, but rather reached by passing a mass. It may appear remarkable then, that in the Neumann case (the Dirichlet case can be handled in a similar fashion) a far simpler analysis, when compared with [15], [5], yields the same results, even for non collinear networks! The argument is quite simple: consider a two-span string system with a mass in the middle. For the clamped node there is no component in (2.15) and the corresponding λ, say λ_N, is zero. Let v_J be the mid-node with mass m_J and displacement λ_J, while the controlled node is v_M, where a Neumann control is applied. For the sake of simplicity and bevity we take all constants equal to 1.

Also, for simplicity, let $\phi_J(t)$, $\phi_M(t)$ denote the (sum of) forces at v_J, v_M caused solely by the initial data $\left(\sum_{i\in\mathcal{E}} \epsilon_{iJ} K_i \eta_i'(x_{iJ}, t\right)$ in (2.15); then (2.15) reads like:

$$
\begin{aligned}
&m_J\ddot{\lambda}_J(t) + 2\dot{\lambda}_J(t) + 4\sum_{k=0}^{m_e} \dot{\lambda}_J(t-2(k+1)) - 2\sum_{k=0}^{m_0} \dot{\lambda}_M(t-(2k+1)) = \phi_J(t)\\
&\dot{\lambda}_M(t) + 2\sum_{k=0}^{m_e} \dot{\lambda}_M(t-2(k+1)) - 2\sum_{k=0}^{m_0} \dot{\lambda}_J(t-(2k+1)) = f_M(t) - \phi_M(t) \qquad (5.3)\\
&\lambda_J(0) = \dot{\lambda}_J(0) = \lambda_M(0) = 0\ .
\end{aligned}
$$

Now, given the initial data for $r_1, r_2 (1 \leftrightarrow NJ,\ 2 \leftrightarrow JM)$ and zero boundary condition at v_N, we can compute the solution $r_1(x,1)$, $\dot{r}_1(x,1)$. In the time interval $[1,3]$ we then solve the Dirichlet boundary control problem in the usual way, and obtain a unique boundary control at v_J on $[1,3]$. This is the $\lambda_J(t)$ required on $[1,3]$. Looking closer at (5.3), it appears that $\lambda_J(t)$ given on $[1,3]$ uniquely determines f_M on $[0,2]$. Then, the Dirichlet data $\lambda_J(t)$ on $[1,2]$ and the initial data of r_1, r_2 will give a solution $r_2(x,2), \dot{r}_2(x,2)$. We take those as initial data, and consider the Dirichlet boundary condition given by $\lambda_J(t)$ on $[2,3]$ and 0 on $[3,4]$. On edge #2 we, therefore, have again the problem of exact null controllability on the domain $[0,1] \times [2,4]$. The solution of this controllability problem determines $f_M(t)$ on $[2,4]$.

This principle can be applied also to different physical constants, and using the arguments in [5] on p. 1390, to varying stiffness-problems. Moreover, and more importantly, we can show exact controllability results for tree-like networks with joint-masses, when all leaves are controlled. It is plain that the regularity of initial data has to increase by one degree, each time a mass has to be passed while following a path to a controlled end. The precise regularity statement is, however, a bit involved and admittedly of rather academic interest in real applications. Therefore, if we do not insist on sharp regularity requirements for the initial data, we can state the result in the following

Theorem 5.1. *Let the initial data be sufficiently smooth. Let G be a tree with the root v_{root} clamped. Let all simple nodes (other than the root) be controlled by $L_2(0,T)$-Neumann controls, where $T \geq 2 * dist(v_{root}, G)$. Then the correpsonding network of elastic strings (*1.1-1.5*) with masses at the joints is exactly controllable.*

As was shown in [11], exact controllability of a 4-node star-graph with one node clamped and only one other simple node controlled holds for a massless multiple joint. Now instead we allow for a mass at the coupling node v_J, and consider the two uncontrolled strings connected at v_J through a mass m_J. Assume that one of the strings is clamped at, say v_N, while the other is free at, say v_M. The two strings satisfy, in addition to appropriately regular initial conditions, a continuity condition at v_J, the nodal displacement being λ_J. The resulting subsystem is generically exactly controllable by an $H^2_{0-}(0,T)$-in-span-control (for $H^2(0,l_i) \times H^1(0,l_i)$-initial data). Once that Dirichlet-control is identified with $\lambda_J(t)$ on $(0,T)$, the controllability from the fourth node, say v_C, by a Neumann-control follows as in the case above. Iterating this procedure, we can prove exact controllability of e.g. a serially connected (not necessarily collinear though) string with one extreme clamped, the other extreme controlled in the Neumann-data, and with further uncontrolled strings attached to all interior nodes, provided the boundary data of those attached strings are of Neumann-type. If we do not attach strings to all interior nodes, controllability (given identical elements) will depend on how many nodes (an even or odd number of nodes) are left out in a row. A detailed analysis goes beyond the scope of this paper and would have to be related to Ho's work [6].

References

1. K. Burrage, *Parallel and sequential methods for ordinary differential equations.* Oxford University Press, 1995.
2. J.-D. Benamou, *A domain decomposition method for the optimal control of systems governed by the Helmholtz equation*, In: Mathematical and numerical aspects of wave propagation (G. Cohen Ed.), SIAM 1995, pp. 653–662.
3. R. Glowinski and A.J. Kearsley, *On the simulation and control of some friction constrained motions*, SIAM J. Optimization **5** (1995) pp. 681–694.
4. E. Hairer and G. Wanner, *Solving Ordinary Differential Equations II*, Springer-Verlag 1991.
5. S. Hansen and E. Zuazua, *Exact controllability and stabilization of a vibrating string with an interior point mass*, SIAM J. Control and Optim. **33/5** (1995), pp. 1357–1391.
6. L.F. Ho, *Controllability and stabilizability of coupled strings with controls applied at the coupled points*, SIAM J. Control and Optim. **31/6** (1993), pp. 1416–1437.
7. F. Kappel, *Approximation of LQR-problems for delay systems: a survey*, In: Computation and Control III (K. Bowers and J. Lund, Eds.), Birkhäuser (1991), pp. 187–224.
8. J.E. Lagnese, *Controllability of a system of interconnected membranes*, Discrete and Continuous Dynamical Systems **1/1** (1995), pp. 17–33.
9. J.E. Lagnese, G. Leugering, and E.J.P.G. Schmidt, *Modelling, analysis and control of multi-link flexible structures*, Birkhäuser, 1994.
10. G. Leugering, *Reverberation analysis and control of networks of elastic strings*, In: Control of partial differential equations and applications (E. Casas Ed.), Marcel Dekker (1996), pp. 193–207.
11. G. Leugering, *On active localization of vibrational energy in trusses and frames*, In: Journal of Structural Control (to appear).
12. P.D. Panagiotopoulos, *Inequality Problems in Mechanics and Applications, Convex and Nonconvex Energy Functions*, Birkhäuser 1985.
13. E.J.P.G. Schmidt, *On the modelling and exact controllability of networks of vibrating strings*, SIAM J. of Control and Optimization **30** (1992), pp. 229–245.
14. E.J.P.G. Schmidt and M. Wei, *On the modelling and analysis of networks of vibrating strings and masses*, Report #91-13, Dept. Math. and Stats., Mc Gill University 1991.
15. Ming Wei, *Controllability of networks of strings and masses*, PhD.-thesis Dept. of Math. and Statistics, McGill University, Montreal, 1993.
16. D.R. Willé and C.T.H. Baker, *The tracking of derivative discontinuities in systems of delay differential equations*, Appl. Num. Math. **9** (1992), 209–222.

G. Leugering
Institut für Mathematik
Universität Bayreuth
D-95440 Bayreuth, Germany
e-mail: leugering@uni-bayreuth.de

The system has a dissipative nature. Indeed, multiplying in (1.5) the first equation by $\vec{v}$, the second equation by p, the fifth equation by W' and integrating by parts, we get, formally, that:

$$dE(t)/dt = -\int_{\Gamma_0} (W')^2 \leq 0.$$

The aim of this article is to study the effect of the damping term, which is concentrated in the string equation, on the asymptotic dynamics of the whole system. We shall prove that the dissipation can force the strong stabilization but it cannot ensure an uniform decay rate.

We remark that this result is not surprising in view of the structure of the damping region. Indeed, as Bardos, Lebeau and Rauch prove in [6], in the context of the control and stabilization of the wave equation in bounded domains, if one characteristic ray escapes to the dissipative region we can not expect an uniform decay to hold (see also Ralston [21]). In our case each segment $\{(x,a),\ x\in(0,1)\}$, $0<a<1$, is such a ray and therefore the decay rate may not be uniform.

Nevertheless, in our problem, the lack of uniform decay is fundamentally due to the hybrid structure of the system. Indeed, the nature of the coupling between the acoustic and elastic components of the system (i.e. the boundary conditions on Γ_0) allows to build solutions with arbitrarily slow decay rate and with the energy distributed in all of the domain and not only along some particular ray of geometrical optics as in [21].

B. P. Rao in [22] has shown that, in various one-dimensional hybrid systems, the coupling is such that the damping term is a compact perturbation of the underlying conservative dynamics. This kind of arguments does not apply in our problem, since we are in space dimension two. Actually, in [17], we have proved that, in a similar system, the difference between the semigroup generated by the dissipative system and the one generated by the corresponding conservative system is not compact.

Let us mention that a similar problem, in which Neumann boundary conditions are considered for the string, was studied in detail in [16] and [17]. From the mathematical point of view this case is easier since it allows us to separate the variables and to obtain explicit informations about the eigenvalues and eigenfunctions of the system. In this way we have showed that there exists a sequence of eigenvalues approaching the imaginary axis at high frequencies and that the corresponding eigenfunctions have the property that the energy concentrated in the string vanishes asymptotically. This implies that, although all solutions tend to zero when the time goes to infinity, the decay rate is not uniform.

In [1] and [2] the strong stability of the following system is studied:

$$(1.7)\qquad \begin{cases} \Phi'' - \Delta\Phi = 0 & \text{in} \quad D\times(0,\infty) \\ \Phi = 0 & \text{on} \quad \gamma\setminus\gamma_0\times(0,\infty) \\ \partial\Phi/\partial\nu + \alpha\Phi' = W' & \text{on} \quad \gamma_0\times(0,\infty) \\ W'' + \Delta^2 W + \Delta^2 W' + \Phi' = 0 & \text{on} \quad \gamma_0\times(0,\infty) \\ W = \partial W/\partial\nu = 0 & \text{on} \quad \partial\gamma_0\times(0,\infty) \\ \Phi(0) = \Phi^0, \quad \Phi'(0) = \Phi^1 & \text{in} \quad D \\ W(0) = W^0, \quad W'(0) = W^1 & \text{on} \quad \gamma_0, \end{cases}$$

where D is a bounded open subset of $\mathbb{R}^n$ with Lipshitz boundary γ, γ_0 is a segment of γ and $\alpha \geq 0$.

Observe that, since we are dealing with an irrotational fluid, the velocity $\vec{v}$ and pressure p can be written in terms of a potential Φ: $\vec{v} = \nabla\Phi$ and $p = -\Phi_t$. When doing this, system (1.5) can be rewritten follows:

$$
(1.8) \qquad \begin{cases} \Phi'' - \Delta\Phi = 0 & \text{in} \quad \Omega \times (0,\infty) \\ \partial\Phi/\partial\nu = 0 & \text{on} \quad \Gamma_1 \times (0,\infty) \\ \partial\Phi/\partial\nu = W' & \text{on} \quad \Gamma_0 \times (0,\infty) \\ W'' - W_{xx} + W' + \Phi' = 0 & \text{on} \quad \Gamma_0 \times (0,\infty) \\ W_x(0) = W_x(1) = 0 & \text{for} \quad t \in (0,\infty) \\ \Phi(0) = \Phi^0, \quad \Phi'(0) = \Phi^1 & \text{in} \quad \Omega \\ W(0) = W^0, \quad W'(0) = W^1 & \text{on} \quad \Gamma_0. \end{cases}
$$

Let us point out some of the differences between systems (1.7) and (1.8). First of all observe that the potential Φ is assumed to vanish on the rigid subset $\gamma \setminus \gamma_0$ of the boundary. This simplifies the set of equilibria of the system that, in this case, is reduced to $(\Phi, W) = (0,0)$. However, the condition $\Phi = 0$ on $\gamma \setminus \gamma_0$ does not seem to be realistic. On the other hand the continuity condition on the velocity fields has been modified. Indeed, the condition

$$\partial\Phi/\partial\nu = W',$$

has been replaced by:

$$\partial\Phi/\partial\nu + \alpha\Phi' = W', \quad \alpha \geq 0.$$

These boundary conditions introduce an extra dissipation on the system, since

$$\frac{dE}{dt}(t) = -\int_{\gamma_0} |\nabla W'|^2 - \alpha \int_{\gamma_0} |\Phi'|^2.$$

Moreover, the displacement W is assumed to satisfy a strongly damped plate equation whose principal part $W'' + \Delta^2 W + \Delta^2 W'$ is known to generate an analytic semigroup. In this sense, this problem is different from ours. An analogous model in which the strongly damped plate equation is replaced by $W'' - W_{xx} - W'_{xx} + \Phi' = 0$ and $\alpha = 0$ has been analyzed in [18].

In [1], taking $\alpha > 0$ and γ_0 sufficiently large, the exponential stability result is proved by using multipier techniques.

The rest of the paper is organized as follows.

In Section 2 we present an abstract formulation of the problem and we give a result of existence, uniqueness and stability of solutions. Since we are dealing with a linear system all these results are direct consequences of the classical theory of maximal-monotone operators.

The asymptotic properties of the solutions are studied in Sections 3 and 4.

In Section 3 we prove the convergence of each solution of the system to an equilibrium point uniquely determined by the corresponding initial data. We do this using classical techniques involving La Salle's Invariance Principle and Holmgren's Uniqueness Theorem.

The rate of the convergence to the equilibrium is studied in Section 4. We prove that the decay rate is not uniform. In order to do this we start from the observation that the same property is true for the system with Neumann boundary conditions for the string and next we use the fact that the difference between these two systems is negligible at high frequencies.

2. Existence and uniqueness of solutions

We define the space of finite energy corresponding to (1.5) by:

$$\mathcal{X}_0 = \mathcal{L} \times L^2(\Omega) \times H_0^1(\Gamma_0) \times L^2(\Gamma_0),$$
$$\mathcal{L} = \left\{ \vec{v} \in L^2(\Omega) \times L^2(\Omega) : \ \text{curl } \vec{v} = 0 \right\} =$$
$$= \left\{ \vec{v} = (v_1, v_2) \in L^2(\Omega) \times L^2(\Omega) : \int_\Omega \left(\frac{\partial \varphi}{\partial x} v_2 - \frac{\partial \varphi}{\partial y} v_1 \right) = 0, \, \forall \varphi \in C_c^\infty(\Omega) \right\}.$$

Remark 1. *Observe that $\vec{v} \in \mathcal{L}$ if and only if there exists a function $\Phi \in H^1(\Omega)$ such that $\nabla \Phi = \vec{v}$.*

$\mathcal{X}_0$ with the natural inner product is a Hilbert space.

We define in $\mathcal{X}_0$ the unbounded operator $(\mathcal{D}(A), \mathcal{A})$ in the following way:

$$\mathcal{A}(\vec{v}, p, W, V) = (\nabla p, \text{div } \vec{v}, -V, -W_{xx} + V - p),$$
$$\mathcal{D}(A) = \{ U = (\vec{v}, p, W, V) \in \mathcal{X}_0 : \ \mathcal{A}(U) \in \mathcal{X}_0, \, \vec{v} \cdot \nu = 0 \text{ on } \Gamma_1, \, \vec{v} \cdot \nu = V \text{ on } \Gamma_0 \}.$$

Remark 2. *Let $(\vec{v}, p, W, V) \in \mathcal{D}(A)$. Observe that $\text{div}\vec{v} \in L^2(\Omega)$ and $\vec{v} \in \mathcal{L}$ imply that there exists $\Phi \in H^1(\Omega)$ with $\nabla \Phi = \vec{v}$ such that $\Delta \Phi \in L^2(\Omega)$. Since, in addition, we have $\vec{v} \cdot \nu = 0$ on Γ_1 and $\vec{v} \cdot \nu = V$ on Γ_0 we obtain that*

$$\begin{cases} \Delta \Phi \in L^2(\Omega) \\ \partial \Phi / \partial \nu = 0 \ \textit{on } \Gamma_1, \quad \partial \Phi / \partial \nu = V \in H_0^1(\Gamma_0) \ \textit{on } \Gamma_0. \end{cases}$$

Since Ω is convex it results that $\Phi \in H^2(\Omega)$ (see [8], Theorem 5.1.3.5, p. 263). It follows that $\mathcal{D}(A) \subseteq (H^1(\Omega))^2 \times H^1(\Omega) \times H^2(\Gamma_0) \cap H_0^1(\Gamma_0) \times H_0^1(\Gamma_0)$ and therefore $\mathcal{D}(A)$ is compact in $\mathcal{X}_0$.

We can consider now the following abstract Cauchy formulation of (1.5):

$$\begin{cases} U' + \mathcal{A}U = 0 \\ U(0) = U_0 \\ U(t) = (\vec{v}, p, W, W')(t) \in \mathcal{D}(A). \end{cases} \tag{2.1}$$

First, we have a classical result of existence, uniqueness and stability for the system (2.1). The terminology we use is the same as in [7].

Theorem 2.1. *i) $\mathcal{A}$ is a maximal monotone operator in $\mathcal{X}_0$ generating a strongly continuous semigroup of contractions, $\{S(t)\}_{t \geq 0}$, in $\mathcal{X}_0$.*

ii) Strong solutions: If $U^0 = (\vec{v^0}, p^0, W^0, W^1) \in \mathcal{D}(A)$ then there exists a unique strong solution $S(t)U^0 = U \in \mathcal{C}([0, \infty), \mathcal{D}(\mathcal{A})) \cap \mathcal{C}^1([0, \infty), \mathcal{X}_0)$ of (2.1).

iii) Weak solutions: If $U^0 = (\vec{v^0}, p^0, W^0, W^1) \in \mathcal{X}_0$ then there exists a unique solution $S(t)U^0 = U \in \mathcal{C}([0, \infty), \mathcal{X}_0)$ of (2.1).

For any weak solution, the associated energy (1.6) *satisfies:*

$$\frac{dE}{dt}(t) = -\int_{\Gamma_0} (W')^2. \tag{2.2}$$

Proof. We prove first that the operator $\mathcal{A}$ is maximal monotone in $\mathcal{X}_0$.

Indeed, if $U^0 = (\vec{v^0}, p^0, W^0, W^1) \in \mathcal{D}(A)$ then $\langle \mathcal{A}U^0, U^0 \rangle \leq -\int_{\Gamma_0} (W^1)^2 \leq 0$, which means that $\mathcal{A}$ is monotone.

On the other hand, for all $F = (\vec{f_1}, f_2, f_3, f_4) \in \mathcal{X}_0$ we can find a unique solution $U = (\vec{v}, p, W, V) \in \mathcal{D}(A)$ for the equation $(\mathcal{A} + I)U = F$. This is equivalent to solve the following system:

$$\begin{cases} \nabla p + \vec{v} = \vec{f_1} \\ \operatorname{div} \vec{v} + p = f_2, \ \vec{v} \cdot \nu = 0 \text{ on } \Gamma_1 \text{ and } \vec{v} \cdot \nu = V \text{ on } \Gamma_0 \\ V + W = f_3 \\ -W_{xx} + V - p + V = f_4 \text{ and } W(0) = W(1) = 0. \end{cases} \tag{2.3}$$

First, we consider the variational formulation of (2.3), which consists in finding (p, W) in $H^1(\Omega) \times H_0^1(\Gamma_0)$ such that, for all $(\varphi, u) \in H^1(\Omega) \times H_0^1(\Gamma_0)$:

$$\begin{aligned} &\int_\Omega \nabla p \cdot \nabla \varphi + \int_\Omega p\varphi + \int_{\Gamma_0} W\varphi + \int_{\Gamma_0} W_x u_x - \int_{\Gamma_0} pu + 2\int_{\Gamma_0} Wu \\ &= \int_\Omega \vec{f_1} \cdot \nabla \varphi + \int_\Omega f_2\varphi + \int_{\Gamma_0} f_3\varphi + \int_{\Gamma_0} (f_4 + 2f_3)u. \end{aligned} \tag{2.4}$$

The left side of the equation (2.4) defines a continuous and coercive bilinear form in $(H^1(\Omega) \times H^1(\Gamma_0))^2$ while the right side defines a continuous linear form in $H^1(\Omega) \times H^1(\Gamma_0)$.

Applying Lax-Milgram's Lemma it results that (2.4) has a unique solution (p, W) in $H^1(\Omega) \times H_0^1(\Gamma_0)$. Finally, in view of the classical regularity results for Laplace's operator, this implies that $\mathcal{A} + I$ is maximal.

Since the operator $\mathcal{A}$ is maximal monotone in $\mathcal{X}_0$ we can apply the Hille-Yosida theory (see [7], Theorem 3.1.1, p.37) and obtain the stated results. □

3. Strong stabilization

Concerning the asymptotic behavior of solutions we prove first the following theorem.

Theorem 3.1. *For each initial data* $U^0 = (\vec{v^0}, p^0, W^0, W^1)$ *in* $\mathcal{X}_0$ *the corresponding weak solution of* (2.1) *tends asymptotically towards the equilibrium point* $(\vec{0}, b, ba(x), 0)$ *where* $b = \frac{12}{13}\left(\int_\Omega p^0 + \int_{\Gamma_0} W^0\right)$ *and* $a(x) = \frac{1}{2}(-x^2 + x)$.

Remark 3. *We obtain that the velocities of the fluid and the string go to zero whereas the pressure of the fluid and the position of the string tend to some functions that are uniquely determined by the initial data. Notice that the pressure stabilizes around a suitable constant while the asymptotic deformation of the string is a parabola.*

Proof. The main tools of our analysis are an extension of the well known Invariance Principle of La Salle and Holmgren's Uniqueness Theorem.

Observe first that it is sufficient to consider only initial data $U^0 = (\vec{v^0}, p^0, W^0, W^1)$ in $\mathcal{D}(A)$. A standard density argument and the property of stability (2.2) enable us to complete the proof. In this case Theorem 2.1 gives an unique strong solution $U(t) = (\vec{v}, p, W, W')(t) = S(t)U^0$ for the equation (1.5), with $\{U(t)\}_{t\geq 0}$ bounded in $\mathcal{D}(A)$. Since $\mathcal{D}(A) \subseteq \mathcal{X}_0$ with compact inclusion, we have that $\{U(t)\}_{t\geq 0}$ is relatively compact in $\mathcal{X}_0$.

We now describe the equilibrium points corresponding to our problem. These are elements $Z = (\vec{u}, r, X, Y) \in \mathcal{D}(A)$ with $S(t)Z = Z$ for all $t \geq 0$. It follows that the equilibrium points are characterized by the system:

$$
\left\{
\begin{array}{ll}
\nabla r = 0 & \text{in } \Omega \\
\operatorname{div} \vec{u} = 0 & \text{in } \Omega \\
\vec{u} \cdot \nu = 0 & \text{on } \Gamma_0 \\
-X_{xx} - r = 0 & \text{on } \Gamma_0 \\
X(1) = X(0) = 0. &
\end{array}
\right.
\tag{3.1}
$$

From (3.1) we deduce that the equilibrium points are $(\vec{0}, b, b\,a(x), 0)$, where b is a real constant and $a(x)$ is the solution of the differential equation:

$$
\left\{
\begin{array}{l}
-a_{xx} - 1 = 0, \ x \in (0,1) \\
a(0) = a(1) = 0.
\end{array}
\right.
$$

On the other hand we remark that the energy function defined by (1.6) is a Lyapunov function for the dynamical system defined by $S(t)U^0 = U(t)$ since it satisfies relation (2.2). We prove now that $E(t)$ is a strict Lyapunov function. To do this let $Z^0 = (\vec{u^0}, r^0, X^0, Y^0) \in \mathcal{X}_0$, $Z(t) = (\vec{u}, r, X, Y)(t) = S(t)Z^0$ for all $t > 0$ and suppose that the energy of the solution $Z(t)$ is constant. Hence $Y(t) = 0$, by (2.2).

It follows that $(\vec{u}, r, X, Y)$ satisfies:

$$
\left\{
\begin{array}{lll}
\vec{u}' + \nabla r = 0 & \text{in} & \Omega \times (0,\infty) \\
r' + \operatorname{div} \vec{u} = 0 & \text{in} & \Omega \times (0,\infty) \\
\vec{u} \cdot \nu = 0 & \text{on} & \partial\Omega \times (0,\infty) \\
-X_{xx} - r = 0 & \text{on} & \Gamma_0 \times (0,\infty) \\
X(0,t) = X(1,t) = 0 & \text{for} & t \in (0,\infty).
\end{array}
\right.
\tag{3.2}
$$

Therefore:

$$
\left\{
\begin{array}{ll}
r'' - \Delta r = 0 & \text{in } \Omega \\
\partial r / \partial \nu = 0 & \text{on } \partial\Omega \\
r' = 0 & \text{on } \Gamma_0.
\end{array}
\right.
\tag{3.3}
$$

We can apply now Holmgren's Uniqueness Theorem (see [10], Theorem 8.6.5, p. 309 and [12], Theorem 8.1, p. 88) which implies that $r' = 0$ in $\Omega \times (1,\infty)$ and so $r(t,x,y) = r(x,y)$ in $\Omega \times (1,\infty)$.

From (3.3) we can deduce that $r = b$ in $\Omega \times (1,\infty)$ where b is a real constant.

Moreover, from (3.2), it follows that $\vec{u} = 0$ in $\Omega \times (1, \infty)$ and X is solution of the equation:

$$\begin{cases} -X_{xx} - b = 0 & \text{on } \Gamma_0 \times (1, \infty) \\ X(0,t) = X(1,t) = 0 & \text{for } t \in (1, \infty). \end{cases}$$

Taking into account the uniqueness of solutions of the system (3.2) we obtain that $Z^0 = (\vec{u^0}, r^0, X^0, Y^0) = (\vec{0}, b, ba(x), 0)$. Hence Z^0 is an equilibrium. Therefore $E(t)$ is a strict Lyapunov function.

We are now in conditions to apply La Salle's Invariance Principle.

Let now $U^0 = (\vec{v^0}, p^0, W^0, W^1)$ be the initial data for (1.5). By La Salle's Invariance Principle it follows that the trajectory tends to the set of the equilibrium points when the times goes to infinity. Let us prove that, in fact, the trajectory converges to a unique point.

Integrating the second equation of (1.5) in Ω we deduce that the quantity $\displaystyle\int_\Omega p^0 + \int_{\Gamma_0} W^0$ is constant along the trajectory. Since the equilibrium points are of the form $(\vec{0}, b, ba(x), 0)$ it follows that the corresponding solution of (1.5) tends to an unique equilibrium point, the one for which $b = \dfrac{12}{13}\left(\displaystyle\int_\Omega p^0 + \int_{\Gamma_0} W^0\right)$. □

Remark 4. *We can decompose the space $\mathcal{X}_0$ as $\mathcal{X}_0 = \mathcal{X}_0^0 \oplus \mathcal{X}_0^1$, where:*

$$\mathcal{X}_0^0 = \left\{(\vec{v}^0, p^0, W^0, V^0) \in \mathcal{X}_0 : \textstyle\int_\Omega p^0 + \int_{\Gamma_0} W^0 = 0\right\},$$
$$\mathcal{X}_0^1 = \left\{(\vec{0}, b, b\, a(x), 0) \in \mathcal{X}_0,\ b \in \mathbb{R}\right\}.$$

The projection of the solution $U(t)$ of (1.5) on $\mathcal{X}_0^1$ is a constant function in time whereas, by Theorem 3.1, the projection on $\mathcal{X}_0^0$ tends to zero as t goes to infinity.

4. The lack of uniform decay

In this paragraph we prove that the rate of decay is not uniform. Results like this are typical for linear hybrid systems in which the dissipation is very weak: it can force the strong stabilization but it cannot ensure the uniform decay.

First of all we recall that a strongly continuous semigroup $\{S(t)\}_{t \geq 0}$ has exponential decay if there are two constants $\omega > 0$ and $M > 0$ such that

$$||S(t)|| \leq M \exp(-\omega t), \quad \forall t \geq 0. \tag{4.1}$$

We also remark that, in the case of linear semigroups, the exponential decay is equivalent to the uniform decay. Therefore, if a linear semigroup $\{S(t)\}_{t \geq 0}$ does not have exponential decay then there are initial data U^0 such that $S(t)U^0$ decays arbitrarily slowly to zero. More precisely, if $\psi : [0, \infty) \longrightarrow \infty$ is a continuous decreasing function such that $\psi(t) \to 0$ as $t \to \infty$ then there exist an initial data $U^0 \in \mathcal{X}_0$ and a sequence $(t_k)_{k \geq 0}$ tending to infinity such that $||S(t_k)U^0|| > \psi(t_k)$ (see [13]).

When the Dirichlet boundary conditions of W in (1.5) are replaced by the Neumann ones, i.e. if W is assumed to satisfy

$$W_x(0,t) = W_x(1,0) = 0, \quad t > 0$$

this result is easy to show. Indeed, under these boundary conditions one can find a sequence of solutions $\{(\vec{v_n}, p_n, W_n)\}_{n\in\mathbb{N}}$ of the type $(\vec{v_n}, p_n, W_n) = e^{-\lambda_n t}(\vec{v_n}, p_n, W_n)$ in separated variables such that $\mathcal{R}e\,\lambda_n \to 0$ as $n \to \infty$ (see [17]). However, the separation of variables does not apply with the boundary conditions we are considering.

In order to prove that, for our system, there is no uniform decay we analyze first a conservative problem. Next, using the fact that these two systems are very close one from another (in a way that we shall make precise later on) we prove the desired property.

We consider now the following undamped system in $\vec{v}$, p and W:

$$\left\{\begin{array}{ll} \vec{v}' + \nabla p = 0 & \text{in } \ \Omega \times (0,\infty) \\ p' + \operatorname{div} \vec{v} = 0 & \text{in } \ \Omega \times (0,\infty) \\ \vec{v}\cdot\nu = 0 & \text{on } \ \Gamma_1 \times (0,\infty) \\ \vec{v}\cdot\nu = W' & \text{on } \ \Gamma_0 \times (0,\infty) \\ W'' - W_{xx} - p = 0 & \text{on } \ \Gamma_0 \times (0,\infty) \\ W_x(0,t) = W_x(1,t) = 0 & \text{for } \ t \in (0,\infty) \\ \vec{v}(0) = \vec{v^0}, \quad p(0) = p^0 & \text{in } \ \Omega \\ W(0) = W^0, \quad W'(0) = W^1 & \text{on } \ \Gamma_0. \end{array}\right. \tag{4.2}$$

Remark 5. *Since we have dropped the dissipative term W' in the string equation the system (4.2) is conservative. On the other hand we remark that the Dirichlet boundary conditions for the string have been replaced by Neumann boundary conditions. This will allow us to use the separation of variables and to obtain useful informations about the eigenvalues and eigenfunctions of the system. We do this in Lemmas* 4.1 *and* 4.2.

The initial data $(\vec{v^0}, p^0, W^0, W^1)$ is considered in the space of finite energy:

$$\mathcal{X} = \mathcal{L} \times L^2(\Omega) \times H^1(\Gamma_0) \times L^2(\Gamma_0), \tag{4.3}$$

We define the energy associated to this system in the same way as in (1.6). We also define in $\mathcal{X}$ the unbounded operator $(\mathcal{D}(B), \mathcal{B})$:

$$\begin{gathered} \mathcal{B}(\vec{v}, p, W, V) = (\nabla p, \operatorname{div} \vec{v}, -V, -W_{xx} - p), \\ \mathcal{D}(B) = \{U = (\vec{v}, p, W, V) \in \mathcal{X} : \ \mathcal{B}(U) \in \mathcal{X}, \, \vec{v}\cdot\nu = 0 \text{ on } \Gamma_1, \\ \vec{v}\cdot\nu = V \text{ on } \Gamma_0, W_x(0) = W_x(1) = 0\}\,. \end{gathered}$$

Lemma 4.1. *The operator $\mathcal{B}$ has a sequence of purely imaginary eigenvalues $(\lambda_n i)_{n\in\mathbb{N}}$ where λ_n are the roots of the equation:*

$$\zeta \ \ tan \zeta = 1. \tag{4.4}$$

Proof. We look for a sequence of solutions $\{(\vec{v_n}, p_n, W_n)\}_{n\in\mathbb{N}}$ for (4.2) of the type $(\vec{v_n}, p_n, W_n) = e^{-\lambda_n i t}(\vec{u_n}, r_n, v_n)$ where $\vec{u_n} = \vec{u_n}(y)$, $r_n = r_n(y)$ and $v_n \in \mathbb{R}$.

We can see that (4.2) has solutions of this form if $(\vec{u_n}, r_n, v_n)$ satisfies

$$(4.5)\qquad \begin{cases} -\lambda_n\, i\, \vec{u_n} + \nabla r_n = 0 \text{ for } y \in (0,1) \\ -(\lambda_n)^2 r_n - (r_n)_{yy} = 0 \text{ for } y \in (0,1) \\ (r_n)_y(1) = 0,\ (r_n)_y(0) = -(\lambda_n)^2 v_n \\ (\lambda_n)^2 v_n + r_n(0) = 0. \end{cases}$$

It follows that $r_n(y) = \cos(\lambda_n(y-1))$, $\vec{u_n} = \dfrac{1}{\lambda_n\, i}\nabla r_n$ and $v_n = -\dfrac{1}{\lambda_n}\sin\lambda_n$ solves (4.5) if λ_n is solution of the algebraic equation (4.4).

It is well known that, for each $n \in \mathbb{N}$, there is a root of this equation which belongs to the interval $\left(n\pi - \dfrac{\pi}{2},\, n\pi + \dfrac{\pi}{2}\right)$. This concludes the proof. □

Remark 6. *A very similar proof allows us to show that, if in the string equation in system (4.2) we introduce the dissipative term W', there is a sequence of eigenvalues such that $\mathcal{R}e\,(\lambda_n) \to 0$ as $n \to \infty$ as we mentioned before. This implies that the decay rate of the associated semigroup is not uniform. In the case of system (1.5) under consideration it is difficult to show directly the existence of such solutions since we can not use separation of variables.*

Remark 7. *The roots $(\lambda_n)_n$ of the equation (4.4) have the following asymptotic development:*

$$\lambda_n = n\pi + \frac{1}{n\pi} + \mathcal{O}\left(\frac{1}{n^3}\right), \quad as\ n \to \infty.$$

For details see [19], p. 12.

To each eigenvalue $\lambda_n i$ given by Lemma 4.1 it corresponds an eigenfunction ξ_n defined by:

$$(4.6)\qquad \xi_n = \begin{pmatrix} \dfrac{1}{\lambda_n\, i}\,\nabla\cos(\lambda_n(y-1)) \\ \cos(\lambda_n(y-1)) \\ -\dfrac{\sin\lambda_n}{\lambda_n} \\ i\sin\lambda_n \end{pmatrix}.$$

We shall denote by ξ_n^j, $j \in \{1,\ldots,4\}$, the components of ξ_n.

Lemma 4.2. *If $(\xi_n)_n$ is the sequence of eigenfunctions of system (4.2) corresponding to the eigenvalues $(\lambda_n\, i)_n$ given by Lemma 4.1 then:*

i) The last two components of ξ_n tend to zero when n tends to infinity.

ii) The sequence $(\xi_n)_n$ does not tend to zero in $\mathcal{X}$ when n tends to infinity.

Proof. i) Since $\lambda_n = n\pi + \dfrac{1}{n\pi} + \mathcal{O}\left(\dfrac{1}{n^3}\right)$ it follows that $(\sin\lambda_n)_n$ tends to zero in $\mathbb{R}$ when n tends to infinity.

ii) We simply remark that

$$||\xi_n||^2_{\mathcal{X}_0} \geq ||\xi_n^2||^2_{L^2(\Omega)} = \frac{1}{2} - \frac{\sin 2\lambda_n}{4\lambda_n} \longrightarrow \frac{1}{2} \text{ as } n \to \infty. \quad \square$$

Remark 8. *Lemma 4.2 shows that there are solutions of (4.2) in which the effect of the vibrating string vanishes asymptotically. This indicates that the boundary conditions for the string are not very important at high frequencies. Since the system with Neumann boundary conditions does not have an exponential decay (see Remark 6) we can expect that this will be the case for system (1.5) too. Indeed, as the proof of the following Theorem shows, the solutions of (4.2) can be slightly modified in order to obtain solutions of (1.5) with arbitrarily small exponential decay rate.*

The main result of this paper is the following:

Theorem 4.3. *The decay rate of the semigroup $\{S(t)\}_{t\geq 0}$ is not exponential in the space $\mathcal{X}_0^0$.*

Proof. We shall prove the theorem by contradiction. Suppose that $\{S(t)\}_{t\geq 0}$ has exponential decay in $\mathcal{X}_0^0$, i.e. there are two constants $\omega > 0$ and $M > 0$ such that:

$$||S(t)||_{\mathcal{X}_0^0} \leq M \exp(-\omega t), \quad \forall t \geq 0.$$

Let $\mathcal{R}(\mathcal{A} : \mu)$ be the resolvent of $\mathcal{A}$ in μ, $\mathcal{R}(\mathcal{A} : \mu) = (\mathcal{A} - \mu\mathcal{I})^{-1}$, where μ is a complex number in the resolvent set of $\mathcal{A}$. We recall that $\mathcal{R}(\mathcal{A} : \mu) = \int_0^\infty e^{\mu t} S(t)\, dt$ (see [20], Theorem 3.1, p. 8). Hence

$$(4.7) \qquad ||\mathcal{R}(\mathcal{A} : \mu)||_{\mathcal{X}_0^0} \leq \int_0^\infty e^{\mathcal{R}e\,\mu\, t} ||S(t)||_{\mathcal{X}_0^0}\, dt \leq \int_0^\infty M\, e^{(\mathcal{R}e\,\mu - \omega)\,t}\, dt.$$

Since the operator $\mathcal{A}$ is dissipative we have that the resolvent is well defined from $\mathcal{X}_0^0$ to $\mathcal{D}(A)$ for all imaginary numbers μ (with $\mathcal{R}e\,\mu = 0$). In this case we obtain from (4.7) that the resolvents are uniformly bounded:

$$(4.8) \qquad ||\mathcal{R}(\mathcal{A} : \mu)||_{\mathcal{X}_0^0} \leq \frac{M}{\omega} \quad \text{for all } \mu \text{ with } \mathcal{R}e\,\mu = 0.$$

We shall prove that there exist a sequence of imaginary numbers $(\lambda_n\, i)_{n\in\mathbb{N}}$, $\lambda_n \in \mathbb{R}$, and a sequence of functions $(\Phi_n)_{n\in\mathbb{N}} \subset \mathcal{X}_0^0$, $||\Phi_n||_{\mathcal{X}_0^0} = 1$, such that

$$(4.9) \qquad ||\mathcal{R}(\mathcal{A} : \mu)\Phi_n||_{\mathcal{X}_0^0} \longrightarrow \infty \text{ when } n \longrightarrow \infty.$$

This contradicts (4.8) and the proof is completed.

In order to do this let $(\lambda_n\, i)_{n\in\mathbb{N}}$ be the sequence of eigenvalues of the problem (4.2) given in Lemma 4.1 and let $(\xi_n)_{n\in\mathbb{N}}$ be the corresponding eigenfunctions given by (4.6).

Observe that $\xi_n \notin \mathcal{X}_0^0$ because the third component, which is a constant, does not belong to $H_0^1(\Gamma_0)$. We shall "cut-off" this constant function in order to get a slightly modified one in $H_0^1(\Gamma_0)$.

For each $n \in \mathbb{N}$ we define the function $u_n : [0,1] \longrightarrow [-1,1]$ by:

$$(4.10) \qquad u_n = \begin{cases} (-|\lambda_n|x + 1)e^{\frac{|\lambda_n|x}{|\lambda_n|x - 1}}, & \text{if } x \in \left[0, \frac{1}{|\lambda_n|}\right) \\ (|\lambda_n|x - |\lambda_n| + 1)e^{\frac{|\lambda_n|(1-x)}{|\lambda_n|(1-x)-1}}, & \text{if } x \in \left(1 - \frac{1}{|\lambda_n|}, 1\right] \\ 0 & \text{otherwise,} \end{cases}$$

and the function h_n as the solution of:

$$(4.11) \qquad \begin{cases} -\Delta h_n + h_n = 0 & \text{in } \Omega \\ \dfrac{\partial h_n}{\partial \nu} = 0 & \text{on } \Gamma_1 \\ \dfrac{\partial h_n}{\partial \nu} = (\sin \lambda_n)\, u_n & \text{on } \Gamma_0. \end{cases}$$

Let now

$$\psi_n = \begin{pmatrix} \nabla h_n \\ \frac{\sin \lambda_n}{\lambda_n} \int_0^1 u_n \\ \frac{\sin \lambda_n}{\lambda_n} u_n \\ -i \sin \lambda_n \, u_n \end{pmatrix} \qquad \text{and} \qquad \varphi_n = \xi_n - \psi_n.$$

From the definitions of the functions u_n and h_n it follows that $\varphi_n \in \mathcal{D}(A) \cap \mathcal{X}_0^0$ for all $n \in \mathbb{N}$.

Finally let $\Phi_n = \dfrac{(\mathcal{A} - \lambda_n \, i \, \mathcal{I})\varphi_n}{||(\mathcal{A} - \lambda_n \, i \, \mathcal{I})\varphi_n||_{\mathcal{X}_0^0}}$.

We obtain that:

$$\mathcal{R}(\mathcal{A} : \lambda_n \, i)\, \Phi_n = \frac{\varphi_n}{||(\mathcal{A} - \lambda_n \, i \, \mathcal{I})\varphi_n||_{\mathcal{X}_0^0}},$$

and we want to prove that (4.9) holds.

Since we need more information about the norms of φ_n and $(\mathcal{A} - \lambda_n \, i\mathcal{I})\varphi_n$ we shall prove first some properties of the functions u_n and h_n.

Lemma 4.4. *The functions u_n and h_n defined by (4.10) and (4.11) respectively have the following properties:*

i) $||u_n||_{L^2}^2 = \mathcal{O}\left(\dfrac{1}{\lambda_n}\right)$.

ii) $||(u_n)_{xx}||_{L^2}^2 = \mathcal{O}\left((\lambda_n)^3\right)$.

iii) $||h_n||_{H^1}^2 = \mathcal{O}\left(\dfrac{1}{(\lambda_n)^3}\right)$.

iv) $||\Delta h_n||_{L^2}^2 = ||h_n||_{L^2}^2 = \mathcal{O}\left(\dfrac{1}{(\lambda_n)^3}\right)$.

Proof. i) From (4.10) we obtain

$$||u_n||_{L^2(0,1)}^2 = 2\int_0^{\frac{1}{|\lambda_n|}} (-|\lambda_n|x + 1)^2 \, e^{\frac{2|\lambda_n|x}{|\lambda_n|x-1}} \leq 2\int_0^{\frac{1}{|\lambda_n|}} (-|\lambda_n|x + 1)^2 = \mathcal{O}\left(\frac{1}{\lambda_n}\right).$$

ii) We have that

$$\begin{aligned} ||(u_n)_{xx}||_{L^2(0,1)}^2 &= 2|\lambda_n|^4 \int_0^{\frac{1}{|\lambda_n|}} \frac{1}{(|\lambda_n|x - 1)^4} \, e^{\frac{2|\lambda - n|x}{|\lambda_n|x-1}} \\ &= 2|\lambda_n|^3 \int_0^1 \frac{1}{(s-1)^4} \, e^{\frac{2s}{s-1}} \, ds \leq 2c\,|\lambda_n|^3, \end{aligned}$$

where $c = \displaystyle\int_0^1 \frac{1}{(s-1)^4} \, e^{\frac{2s}{s-1}} \, ds$ is a constant which does not depend on λ_n.

iii) From (4.11) we deduce that, for all $\delta > 0$,

$$\begin{aligned} ||h_n||^2_{H^1(\Omega)} + ||h_n||^2_{L^2(\Omega)} &= \left| -\sin\lambda_n \int_{\Gamma_0} u_n\, h_n \right| \leq \frac{|\sin\lambda_n|}{2}\left(\frac{1}{\delta}\int_{\Gamma_0} |u_n|^2 + \delta \int_{\Gamma_0} |h_n|^2 \right) \\ &\leq \frac{|\sin\lambda_n|}{2}\left(\frac{1}{\delta}\int_{\Gamma_0} |u_n|^2 + \delta c ||h_n||^2_{H^1(\Omega)} \right). \end{aligned}$$

Taking $\delta = \frac{1}{|\sin\lambda_n| c}$ we obtain that

$$||h_n||^2_{H^1(\Omega)} \leq c|\sin\lambda_n|^2 \int_{\Gamma_0} |u_n|^2.$$

Here c is a generic positive constant that may vary from line to line.
Since $\sin\lambda_n = \mathcal{O}\left(\frac{1}{\lambda_n}\right)$ and $\int_{\Gamma_0} |u_n|^2 = \mathcal{O}\left(\frac{1}{\lambda_n}\right)$ by i), iii) follows.
iv) We simply observe that

$$|| - \Delta h_n||_{L^2(\Omega)} = ||h_n||_{L^2(\Omega)} \leq ||h_n||_{H^1(\Omega)}$$

and use iii). The proof of the Lemma is now completed. □

In order to complete the proof of the theorem we estimate $||(\mathcal{A} - \lambda_n\, i\,\mathcal{I})\varphi_n||_{\mathcal{X}_0^0}$ and $||\varphi_n||_{\mathcal{X}_0^0}$ when n tends to infinity.

Observe first that, by Lemma 4.4 i), we have

$$||\varphi_n||_{\mathcal{X}_0^0} \geq ||\xi_n^2||_{L^2(\Omega)} - \left| \frac{\sin\lambda_n}{\lambda_n} \int_0^1 u_n \right| \longrightarrow \frac{1}{2} \text{ as } n \longrightarrow \infty \tag{4.12}$$

On the other hand

$$(\mathcal{A} - \lambda_n\, i\,\mathcal{I})\varphi_n = \begin{pmatrix} -\nabla h_n \\ -\Delta h_n + \sin\lambda_n \displaystyle\int_0^1 u_n \\ 0 \\ \dfrac{\sin\lambda_n}{\lambda_n}(u_n)_{xx} + \dfrac{\sin\lambda_n}{\lambda_n}\displaystyle\int_0^1 u_n + \sin\lambda_n\, u_n \end{pmatrix}.$$

We obtain that

$$\begin{aligned} ||(\mathcal{A} - \lambda_n\, i\,\mathcal{I})\varphi_n||^2_{\mathcal{X}_0^0} &\leq ||h_n||^2_{H^1(\Omega)} + 2||\Delta h_n||^2_{L^2(\Omega)} \\ &\quad + \left|\frac{2\sin\lambda_n}{\lambda_n}\right|^2 \left(||(u_n)_{xx}||^2_{L^2(\Gamma_0)} + ||u_n||^2_{L^2(\Gamma_0)} \right) + 4|\sin\lambda_n|^2 ||u_n||^2_{L^2(\Gamma_0)}. \end{aligned}$$

Taking into account the results of Lemma 4.4 and the fact that $\sin\lambda_n = \mathcal{O}\left(\frac{1}{\lambda_n}\right)$ we obtain that

$$||(\mathcal{A} - \lambda_n\, i\,\mathcal{I})\varphi_n||^2_{\mathcal{X}_0^0} \longrightarrow 0 \text{ when } n \longrightarrow \infty. \tag{4.13}$$

The last result together with (4.12) contradicts (4.8). So the assumption that $\{S(t)\}_{t\geq 0}$ has exponential decay must be false and the proof is completed. □

Remark 9. *Analyzing the exponential stability of the classical wave equation with dissipation on the boundary*

$$
\begin{cases} u'' - \Delta u = 0 & \text{in } \Omega \times (0,\infty) \\ \dfrac{\partial u}{\partial \nu} + u' = 0 & \text{on } \Gamma_0 \times (0,\infty) \\ u = 0 & \text{on } \Gamma_1 \times (0,\infty) \end{cases} \tag{4.14}
$$

Bardos, Lebeau and Rauch in [6] prove that if one characteristic ray escapes to the dissipative region Γ_0 we can construct solutions with an arbitrary decay rate and with the energy concentrated along this ray. In our case every segment $\{(x, y_0) : x \in (0,1)\}$, for any $y_0 \in (0,1)$, constitutes a ray with such a property and their argument could be applied as well.

Nevertheless the proof of Theorem 4.3 shows that we can find a sequence of solutions of (1.5) with the energy uniformly distributed in all Ω and with arbitrarily small exponential decay rate. Indeed, if $(\Phi_n)_n$ is the sequence considered in the proof, let $(S(t)\Phi_n)_n$ be the sequence of corresponding solutions of (1.5). By (4.7) we have that

$$
||\mathcal{R}(\mathcal{A} : \lambda_n i)\Phi_n||_{\mathcal{X}_0^0} \leq \int_0^\infty ||S(t)\Phi_n||_{\mathcal{X}_0^0}\, dt.
$$

If $(S(t)\Phi_n)_n$ had an uniform exponential decay rate, for example, $||S(t)\Phi_n||_{\mathcal{X}_0^0} \leq M exp(-\omega t)$, then

$$
||\mathcal{R}(\mathcal{A} : \lambda_n i)\Phi_n||_{\mathcal{X}_0^0} \leq \frac{M}{\omega}
$$

which is not true since (4.9) holds.

Therefore the lack of uniform decay of our system is of a different nature and is related not only to the support of the dissipative mechanism but also to the nature of the boundary conditions or of the coupling between the different components of the system.

Remark 10. *We mention that in the proof of Theorem 4.3 we may start with solutions $(\vec{v_n}, p_n, W_n)_{n\in\mathbb{N}}$ of (4.2) of the type $(\vec{v_n}, p_n, W_n) = e^{-\lambda_n i t}(\vec{u_n}, r_n, v_n)\cos(m\pi x)$ where $\vec{u_n} = \vec{u_n}(y)$, $r_n = r_n(y)$, $v_n \in \mathbb{R}$ and an arbitrary $m \in \mathbb{N}$. Therefore we can find a sequence of solutions of (1.5) with arbirary exponential decay rate and with a fixed frequency of vibration in the $x-$direction ($m \in \mathbb{N}$ fixed). This is due to the fact that the one-dimensional problems obtained by separating the variable x do not have an exponential decay for m fixed. This is an important difference with respect to system (4.14) in which the exponential decay holds if the frequency of vibration in the $x-$direction is fixed, but with a decay rate that vanishes as $m \to \infty$.*

5. Comments

In [3] a two-dimensional model is presented in which, on the subset Γ_0 of the boundary, an Euler-Bernoulli beam with fixed ends is considered. The methods developped in this paper can be adapted to this type of problems too.

The results of Sections 2 and 3 can be generalized to similar models in other domains. For instance, if Ω is a bounded open set in $\mathbb{R}^2$ with smooth boundary and

Γ_0 is an open subset of the boundary of the domain, one can replace in (1.5) the wave equation satisfied by W by

$$W'' - \frac{d^2W}{d\tau^2} + W' - p = 0 \quad \text{on} \quad \Gamma_0 \times (0, \infty)$$

where $\frac{d}{d\tau}$ is the derivative in the tangential direction.

The results of Section 4 may be extended to some particular geometries. For instance, in [15] we analyze the case in which Ω is a ball of $\mathbb{R}^2$ and the dissipative term acts on the whole boundary of Ω. We obtain that the corresponding system does not have exponential decay. This indicates something we already pointed out in Remark 9: the lack of uniform decay in this type of systems is due to the hybrid structure and not to the localization of the dissipation in a relatively small part of the boundary. Although this model does not have much physical meaning, all the techniques we used there can be adapted to the case of a cavity enclosed by a thin cylindrical shell which is much more realistic (see [4]).

Acknowledgments: The first author wishes to thank all organizers of the Project MATAROU TEMPUS JEP 2797 and especially Professor Doina Cioranescu for their support and dedication to this programme.

References

1. G. Avalos, *The Exponential Stability of a a Coupled Hyperbolic/Parabolic System Arising in Structural Acoustics*, Abstract Appl. Analysis, to appear.
2. G. Avalos and I. Lasiecka, *The Strong Stability of a Semigroup Arising from a Coupled Hyperbolic/Parabolic System*, Semigroup Forum, to appear.
3. H. T. Banks, W. Fang, R. J. Silcox and R. C. Smith, *Approximation Methods for Control of Acustic/Structure Models with Piezoceramic Actuators*, Journal of Intelligent Material Systems and Structures, 4 (1993), pp. 98–116.
4. H. T. Banks and R. C. Smith, *Well-Posedness of a Model for Structural Acoustic Coupling in a Cavity Enclosed by a Thin Cylindrical Shell*, J. Math. Analysis and Applications, 191 (1995), pp. 1–25.
5. H. T. Banks, R. C. Smith and Y. Wang, *Smart Material Structures. Modeling, estimation and control*, RAM, John Wiley & Sons, Masson, 1996.
6. C. Bardos, G. Lebeau and J. Rauch, *Sharp sufficient conditions for the observation, control and stabilization of waves from the boundary*, SIAM J. Control Optim., 30 (1992), pp. 1024–1065.
7. T. Cazenave and A. Haraux, *Introduction aux problèmes d'évolution semi-linéaires*, Mathématiques et Applications, 1, Ellipses, Paris, 1990.
8. P. Grisvard, *Elliptic Problems in Non-smooth Domains*, Pitman, 1985.
9. S. Hansen and E. Zuazua, *Exact Controllability and Stabilization of a Vibrating String with an Interior Point Mass*, SIAM J. Control Optim., 33, 5 (1995), pp. 1357–1391.
10. L. Hörmander, *The Analysis of Linear Partial Differential Operators I*, Springer-Verlag, 1990.
11. L. D. Landau and E. M. Lifshitz, *Fluid Mechanics*, Pergamon Press, 1987.
12. J. L. Lions, *Contrôlabilité exacte, perturbations et stabilisation de systèmes distribués, Tome 1: Contrôlabilité exacte*, Masson, RMA, Paris, 1988.
13. W. Littman and L. Marcus, *Some Recent Results on Control and Stabilization of Flexible Structures*, Univ. Minn., Mathematics Report 87–139.
14. W. Littman and L. Marcus, *Exact Boundary Controllability of a Hybrid System of Elasticity*, Archive Rat. Mech. Anal., 103, 3 (1988), pp. 193–236.

15. S. Micu, *Análisis de un modelo híbrido bidimensional fluido-estructura*, Ph. D. dissertation at Universidad Complutense de Madrid, 1996.
16. S. Micu and E. Zuazua, *Propriétés qualitatives d'un modèle hybride bi-dimensionnel intervenant dans le contrôle du bruit*, C. R. Acad. Sci. Paris, 319 (1994), pp. 1263–1268.
17. S. Micu and E. Zuazua, *Asymptotics for the spectrum of a fluid/structure hybrid system arising in the control of noise*, preprint.
18. S. Micu and E. Zuazua, *Stabilization and Periodic Solutions of a Hybrid System Arising in the Control of Noise*, Proceedings of the IFIP TC7/WG-7.2 International Conference, Laredo, España, Lecture Notes in Pure and Applied Mathematics, Vol. 174, Marcel Dekker, New York, 1996, pp. 219–230.
19. F. W. J. Olver, *Asymptotics and Special Functions*, Academic Press, 1974.
20. A. Pazy, *Semi-groups of Linear Operators and Applicatios to Partial Differential Equations*, Springer-Verlag, 1983.
21. J. Ralston, *Solutions of the wave equation with localized energy*, Comm. Pure Appl. Math. 22 (1969), pp. 807–823.
22. B. Rao. *Uniform Stabilization of a Hybrid System of Elasticity*, SIAM J. Cont. Optim., 33, 2 (1995), pp. 440–454.

Sorin Micu
Departamento de Matemática Aplicada
Facultad de Ciencias Matemáticas
Universidad Complutense
28040 Madrid, Spain

and

Facultatea de Matematica-Informatica
Universitatea din Craiova, 1100, Romania
e-mail:sorin@sunma4.mat.ucm.es

Enrique Zuazua
Departamento de Matemática Aplicada
Facultad de Ciencias Matemáticas
Universidad Complutense
28040 Madrid, Spain
e-mail:zuazua@sunma4.mat.ucm.es

International Series of Numerical Mathematics
Vol. 126, © 1998 Birkhäuser Verlag, Basel

Dirichlet Boundary Control of Parabolic Systems with Pointwise State Constraints*

BORIS S. MORDUKHOVICH AND KAIXIA ZHANG
Department of Mathematics
Wayne State University

ABSTRACT. In this paper we study an optimal control problem for linear parabolic systems with pointwise state constraints and measurable controls acting in the Dirichlet boundary conditions. Using the framework of mild solutions to parabolic systems with nonregular dynamics, we prove a general existence theorem of optimal controls and derive necessary optimality conditions for the state-constrained problem under consideration. Our variational analysis is based on a well-posed penalization procedure to approximate state constraints and then to study a parametric family of approximating problems. The final result establishes necessary optimality conditions for the original state-constrained problem by passing to the limit from approximating problems under a proper constraint qualification.

1991 *Mathematics Subject Classification.* 49K20, 49J20, 35K50, 93C20

Key words and phrases. Approximation, pointwise state constraints, constraint qualification, Dirichlet boundary controls, parabolic equations, and mild solutions.

1. Introduction

This paper is devoted to optimal control of parabolic systems with nonregular Dirichlet boundary conditions and pointwise state constraints. It is well known that the Dirichlet boundary control case is the most challenging and the least developed since such conditions offer the *lowest regularity properties* of the parabolic dynamics; cf. [1], [2], [5]–[12], [17], and references therein. The presence of pointwise state constraints brings an additional *nonsmoothness* to optimal control problems and requires the development of special methods for their variational analysis.

In this paper we provide such an analysis based on the theory of mild solutions to nonregular parabolic systems and well-posed smooth approximations. Crucial elements of this analysis and the corresponding results have been presented in [14]–[16] for certain special cases of the problem under consideration related to minimax control in uncertainty conditions.

In this paper we consider a general Dirichlet boundary control problem with a nonlinear integral cost functional involving the final state of the n-dimensional linear parabolic equation. Under natural assumptions we prove the existence of optimal controls and necessary optimality conditions in the presence of magnitude control and state constraints. To obtain necessary optimality conditions for the state-constrained problem we develop a constructive penalization procedure involving smooth approximations of multivalued maximal monotone operators. We establish

*This research was partly supported by the National Science Foundation grant DMS-9404128 and the USA-Israel BSF grant 94-00237.

the *well-posedness/strong convergence* of approximations in appropriate spaces and derive necessary optimality conditions for approximating solutions. Finally, necessary optimality conditions for the original state-constrained problem are established under a proper *constraint qualification* which is different from the standard Slater interiority type.

The paper is organized as follows. In Section 2 we formulate and discuss the Dirichlet boundary control problem of our study, present preliminary results from the theory of mild solutions, and prove a general existence theorem of optimal controls. Section 3 concerns with the development and justification of the main approximation procedure; it contains convergence results as well as necessary optimality conditions for approximating solutions. In the final Section 4 we furnish a limiting process to derive necessary optimality conditions for the original state-constrained problem under a proper constraint qualification.

2. Problem Setting and Existence of Optimal Solutions

Let $\Omega \subset \mathbb{R}^N$ be an open and bounded domain whose boundary Γ is an $(n-1)-$dimensional manifold. With $T>0$ we set $Q := (0,T)\times\Omega$ and $\Sigma := (0,T]\times\Gamma$. Let A be a second-order uniformly strongly elliptic operator on Ω given in the form

$$A := -\sum_{i,j=1}^{N} \frac{\partial}{\partial x_i}(a_{ij}(x)\frac{\partial}{\partial x_j}) + \sum_{i=1}^{N} a_i(x)\frac{\partial}{\partial x_i} + a_0(x)$$

with the smooth real-valued data $a_{ij}(x)$, $a_i(x)$, and $a_0(x)$.

We consider the following *Dirichlet boundary control system* for linear parabolic equations

$$(2.1) \qquad \begin{cases} y_t + Ay = f \ \text{ a.e. in } \ Q, \\ y(0,x) = y_0(x), \ x\in\Omega, \\ y(t,x) = u(t,x), \ (t,x)\in\Sigma \end{cases}$$

where y_t denotes the derivative of y with respect to time t, $f\in L^\infty(Q)$, and $y_0(x)\in H_0^1(\Omega)\cap H^2(\Omega)$. In what follows we impose *pointwise state and control constraints* of the magnitude type:

$$(2.2) \qquad \begin{aligned} a \le y(t,x) \le b \ \text{ a.e. } \ (t,x)\in Q, \\ c \le u(t,x) \le d \ \text{ a.e. } \ (t,x)\in \Sigma \end{aligned}$$

where both intervals $[a,b]$ and $[c,d]$ contain 0.

We say that u is a *feasible control* to system (2.1) if the corresponding trajectory y satisfies the state constraints (2.2). We always assume that system (2.1) admits at least one feasible control u.

Denote by

$$U_{ad} := \{u\in L^p(0,T;L^2(\Gamma)) \mid c\le u(t,x)\le d \text{ a.e. } (t,x)\in\Sigma\}$$

the set of *admissible* controls where p is a positive number that will be specified later. In the sequel the solution (trajectory) y of (2.1) corresponding to $u\in U_{ad}$ is understood in the *mild sense*; cf. [1], [8]–[11], [18]. This means that $y : [0,T] \to L^2(\Omega)$ is continuous and admits the following Cauchy-like representation:

$$
\begin{aligned}
y(t) &= S(t)y_0 + \int_0^t S(t-\tau)f(\tau)d\tau \\
&\quad + \int_0^t A^{3/4+\delta}S(t-\tau)A^{1/4-\delta}Du(\tau)d\tau \quad \forall\delta \in (0,1/4]
\end{aligned}
\tag{2.3}
$$

where $S(\cdot)$ is the strongly continuous analytic semigroup generated by the operator $-A$, $f(\cdot) \in L^\infty(Q)$, and

$$D : L^2(\Gamma) \to \mathcal{D}(A^{1/4-\delta}) = H^{1/2-2\delta}(\Omega)$$

is the so-called *Dirichlet map*. The latter operator is defined by $z = Du$ through the solution of the elliptic boundary-value problem

$$\begin{cases} -Az = 0 \ \text{ in } Q, \\ z(t,x) = u(t,x), \ (t,x) \in \Sigma. \end{cases}$$

It is well known that the Dirichlet map is continuous for $\delta \in (0,1/4]$ and, moreover, system (2.1) has a *unique mild solution* for each $u \in U_{ad}$ when p is sufficiently large; see, e.g., [11] and [16] for more discussion and references.

Note that, being a $L^2(\Omega)$-valued function, $y(\cdot) = y(t,x)$ is merely *measurable* with respect to (t,x). This lack of continuity creates certain technical difficulties to deal with nonregular Dirichlet boundary conditions. Nevertheless, mild solutions provide a reliable ground to study optimal control problems involving such conditions.

Let us consider the performance index (cost functional) given by

$$
\begin{aligned}
J(u,y) &:= \int_\Omega \varphi(y(T,x))dx + \iint_Q g(t,x,y(t,x))dtdx \\
&\quad + \iint_\Sigma h(t,x,u(t,x))dtd\sigma_x
\end{aligned}
\tag{2.4}
$$

where σ_x is the Lebesgue measure on Γ. Observe that the first term in (2.4) depends on the *final state* of (2.1) that creates additional difficulties in the framework of nonregular Dirichlet boundary conditions; see, e.g., Chapter 3 of [12].

Throughout the paper we impose the following hypotheses on the integrands in (2.4):

(i) $\varphi \in C^1(\mathbb{R})$ and there is a nonnegative function $k_1 \in L^2(\mathbb{R})$ as well as a constant $c_1 \geq 0$ such that

$$|\varphi'(z)| \leq k_1(z) + c_1|z| \ \ \forall z \in \mathbb{R}.$$

(ii) g is measurable in (t,x), continuous in y, and $|g(t,x,y)|$ is majorized by a $L^1(Q)$-function for all $y \in [a,b]$. In addition, $\dfrac{\partial g}{\partial y}$ is measurable in (t,x) for any $y \in \mathbb{R}$ and there is a nonnegative function $k_2 \in L^2(Q)$ as well as a constant $c_2 \geq 0$ such that

$$|\frac{\partial g}{\partial y}(t,x,y)| \leq k_2(t,x) + c_2|y| \ \text{ a.e. } (t,x) \in Q, \ \ \forall y \in \mathbb{R}.$$

(iii) h is measurable in (t, x), convex and continuous in u, and bounded from below by a $L^1(\Sigma)$-function for all $u \in [c, d]$. In addition, $\dfrac{\partial h}{\partial u}$ is measurable in (t, x) for any $u \in \mathbb{R}$ and there is a nonnegative function $k_3 \in L^q(\Sigma)$ $(1/p + 1/q = 1)$ such that

$$|\frac{\partial h}{\partial u}(t, x, u)| \leq k_3(t, x) \text{ a.e. } (t, x) \in \Sigma, \quad \forall u \in [c, d].$$

The main concern of this paper is the following *optimal control problem:*

(P) minimize the cost functional (2.4) over the Dirichlet boundary control system (2.1) subject to $u \in U_{ad}$ and the state constraints (2.2).

The first question we consider is the *existence of optimal controls* to problem (P). To establish a general theorem in this direction as well as other convergence results for mild solutions of (2.1) we are going to employ certain continuity properties of the linear operator

$$\mathcal{L}u = (\mathcal{L}u)(t) := \int_0^t A^{3/4+\delta} S(t - \tau) A^{1/4-\delta} Du(\tau) d\tau \tag{2.5}$$

from $L^p(0, T; L^2(\Gamma))$ into $L^r(0, T; H^{1/2-\varepsilon}(\Omega))$ where $p,\ r \in [1, \infty]$, $\delta \in (0, 1/4]$, and $\varepsilon \in (0, 1/2]$. Here $H^{1/2-\varepsilon}(\Omega) \subset L^2(\Omega)$ is the Sobolev space whose norm $\|y\|_{1/2-\varepsilon}$, being stronger than $\|y\|_{L^2(\Omega)}$, can be defined by $\|y\|_{1/2-\varepsilon} := \|A^{1/4-\varepsilon/2}\|_{L^2(\Omega)}$; cf. [12]. Note that $H^0(\Omega) = L^2(\Omega)$. When $t = T$, we use $\mathcal{L}_T$ to denote (2.5).

The following assertion was proved in [16], Proposition 3.1, based on estimates in Washburn [18] and Lasiecka–Triggiani [9]. Similar but somewhat different properties were established in [10].

Proposition 1. *Let $p > 4/\varepsilon$ with $\varepsilon \in (0, 1/2]$. Then the operator $\mathcal{L} : L^p(0, T; L^2(\Gamma)) \to C([0, T]; H^{1/2-\varepsilon}(\Omega))$ is linear and continuous. Moreover, the operator $\mathcal{L}_T : L^p(0, T; L^2(\Gamma)) \to H^{1/2-\varepsilon}(\Omega)$ is also continuous and its adjoint operator $\mathcal{L}_T^* : H^{-1/2+\varepsilon}(\Omega) \to L^q(0, T; L^2(\Gamma))$ $(1/p + 1/q = 1)$ is given by $\mathcal{L}_T^* = (AS(T - t)D)^*$.*

The next assertion, proved in Proposition 3.4 of [16], is crucial in passing to the limit in approximation procedures throughout the paper.

Proposition 2. *Let $p > 4/\varepsilon$ with $\varepsilon \in (0, 1/2)$. Then the weak convergence of $u_n \to u$ in $L^p(0, T; L^2(\Gamma))$ implies*

$$\mathcal{L}u_n \to \mathcal{L}u \text{ strongly in } L^2(Q) \text{ as } n \to \infty.$$

In what follows we always assume that p is *sufficiently large* to ensure the convergence property in Proposition 2 with some $\varepsilon \in (0, 1/2)$. Now we can formulate and prove the existence of optimal controls in (P).

Theorem 3. *Under the assumptions made above there exists an optimal solution $(\bar{u}, \bar{y}) \in U_{ad} \times C([0, T]; H^{1/2-\varepsilon}(\Omega))$ to the Dirichlet boundary control problem* (P).

Proof. Let (u_n, y_n), $n = 1, 2, \ldots$, be a minimizing sequence of feasible controls u_n in (P). For each $n = 1, 2, \ldots$ we consider the corresponding mild solution y_n of system (2.1) that is uniquely defined by u_n and belongs to the space $C([0, T]; H^{1/2-\varepsilon}(\Omega))$ where ε is any given number in $(0, 1/2]$ (this easily follows from Proposition 1). We always take $\varepsilon < 1/2$ to ensure the convergence property in Proposition 2 with large p. Since $\{u_n\} \subset U_{ad}$ is weakly compact in $L^p(0,T;L^2(\Gamma))$, there exist a control $\bar{u} \in U_{ad}$ and a subsequence of $\{u_n\}$, still labelled as $\{u_n\}$, such that

$$u_n \to \bar{u} \;\text{ weakly in }\; L^p(0,T;L^2(\Gamma)) \;\text{ as }\; n \to \infty.$$

Proposition 1 ensures that operator (2.5) acting from $L^p(0,T;L^2(\Gamma))$ into $C([0,T];$ $H^{1/2-\varepsilon}(\Omega))$ is *weakly continuous*. By (2.3) this implies that

$$y_n \to \bar{y} \;\text{ weakly in }\; C([0,T];H^{1/2-\varepsilon}(\Omega))$$

where $\bar{y}$ is a mild solution of (2.1) corresponding to $\bar{u}$. Now employing Proposition 2, we conclude that

$$y_n \to \bar{y} \;\text{ strongly in }\; L^2(Q) \;\text{ as }\; n \to \infty.$$

The latter ensures the existence of a subsequence $\{y_{n_k}\} \subset \{y_n\}$ with

$$y_{n_k}(t,x) \to \bar{y}(t,x) \;\text{ a.e. }\; (t,x) \in Q \;\text{ as }\; k \to \infty.$$

Such a pointwise convergence implies that the limiting trajectory $\bar{y}$ satisfies the state constraints (2.2) since each y_n has this property. Therefore, $\bar{u}$ is a feasible control to (P).

To prove the optimality of $\bar{u}$ in (P) we invoke the well-known fact that due to (iii) the last term in (2.4) is a *weakly lower semicontinuous* functional in the space $L^p(0,T;L^2(\Gamma))$. Furthermore, the Lebesgue dominated convergence theorem allows us to pass to the limit under the integral signs in the first and second terms of (2.4), due to the pointwise convergence of $y_{n_k} \to \bar{y}$ and assumptions (i) and (ii). Therefore,

$$J(\bar{u},\bar{y}) \le \liminf_{k\to\infty} J(u_{n_k}, y_{n_k})$$

that proves the optimality of $\bar{u}$ in (P). $\square$

Remark. We do not need smoothness assumptions on φ, g, and h to prove the existence of optimal controls in Theorem 3. The most essential requirements for this are the convexity of h in u and the right choice of p and ε ensuring the convergence/continuity properties in Propositions 1 and 2. However, we use the smoothness assumptions in the subsequent sections to derive necessary optimality conditions. To simplify the exposition we have combined all the assumptions together.

3. Necessary Optimality Conditions in Well-Posed Approximations

In this section we develop a well-posed approximation procedure allowing us to remove the state constraints in (P). We establish an appropriate *strong* convergence of approximations and derive necessary optimality conditions for approximating solutions. The latter results can be viewed as *suboptimality* conditions for the state-constrained problem (P) being the base to obtain necessary optimality conditions for (P) in the next section.

Let $\alpha : \mathbb{R} \Rightarrow \mathbb{R}$ be a multivalued *maximal monotone operator* of the form

$$\alpha(r) = \begin{cases} [0,\infty) & \text{if } r = b \\ (-\infty,0] & \text{if } r = a \\ 0 & \text{if } a < r < b \\ \emptyset & \text{if either } r < a \text{ or } r > b. \end{cases}$$

Using the Yosida approximation $\gamma^{-1}(r-(1+\gamma\alpha)^{-1}r)$ of $\alpha(\cdot)$ and then a C_0^∞-mollifier in $\mathbb{R}$, we may choose a *smooth* approximation of $\alpha(\cdot)$ as

$$\alpha_\gamma(r) = \begin{cases} \gamma^{-1}(r-b) - 1/2 & \text{if } r \geq b+\gamma \\ (2\gamma^2)^{-1}(r-b)^2 & \text{if } b \leq r < b+\gamma \\ \gamma^{-1}(r-a) + 1/2 & \text{if } r \leq a-\gamma \\ -(2\gamma^2)^{-1}(r-a)^2 & \text{if } a-\gamma < r \leq a \\ 0 & \text{if } a < r < b \end{cases} \tag{3.1}$$

with the property $|\gamma\alpha'_\gamma(r)| \leq 1$ for all $r \in \mathbb{R}$ and $\gamma > 0$; cf. [2], p. 322.

Let $(\bar{u}, \bar{y})$ be a given optimal solution to problem (P). We consider the following parametric family of boundary control problems without state constraints:

(P_γ) minimize $J_\gamma(u,y) := J(u,y) + \|u-\bar{u}\|^p_{L^p(0,T;L^2(\Gamma))} + \gamma\|\alpha_\gamma(y)\|^2_{L^2(0,T;L^2(\Omega))}$
over $u \in U_{ad}$ subject to system (2.1).

We are going to study problems (P_γ) from the three perspectives: existence of optimal solutions, their convergence to $(\bar{u},\bar{y})$ as $\gamma \to \infty$, and necessary optimality conditions for them as $\gamma > 0$. The next proposition answers the first question.

Proposition 4. *Let $p > 4/\varepsilon$ with $\varepsilon \in (0,1/2)$. For each $\gamma > 0$ problem (P_γ) has at least one optimal solution $(u_\gamma, y_\gamma) \in U_{ad} \times C([0,T]; H^{1/2-\varepsilon}(\Omega))$.*

Proof. The set of feasible solutions to (P_γ) is not empty since it obviously contains $(\bar{u},\bar{y})$ for any $\gamma > 0$. First we should check that the cost functional in (P_γ) is *proper*, i.e.,

$$J_\gamma(u,y) > -\infty \quad \forall \gamma > 0 \tag{3.2}$$

for all feasible solutions (u,y) to (P_γ). It easily follows from assumptions (i)–(iii) that

$$J(u,y) + \|u-\bar{u}\|^p_{L^p(0,T;L^2(\Gamma))} > -\infty.$$

To establish (3.2) it remains to show that

$$\|\alpha_\gamma(y)\|_{L^2(0,T;L^2(\Omega))} < \infty \quad \forall \gamma > 0. \tag{3.3}$$

Taking into account the definition of mild solutions (2.3) and estimates in [9] and [18], one gets

$$\|y(t)\|_{L^2(\Omega)} \leq M\Big(1 + \frac{\max\{|c|,d\}\sqrt{\mathrm{meas}(\Gamma)}}{1-4\delta} t^{\frac{1-4\delta}{4}}\Big) \text{ for any fixed } \delta \in (0,1/4)$$

with some constant $M > 0$. Due to (3.1) the latter implies (3.3) and hence (3.2); cf. the proof of Proposition 4.1 in [16] for more details.

Now arguing as in the proof of Theorem 3 and using Propositions 1 and 2, we conclude that the cost functional in (P_γ) is weakly lower semicontinuous on the set of feasible controls U_{ad} which is weakly compact in $L^p(0,T;L^2(\Gamma))$. Therefore, the existence of optimal controls in (P_γ) follows from the classical Weierstrass theorem. □

Next we establish the *well-posedness* of the approximation procedure under consideration proving the *strong convergence* of optimal solutions in (P_γ) to the given optimal pair $(\bar{u},\bar{y})$ in the original problem (P).

Theorem 5. *Let $(\bar{u},\bar{y})$ be the given optimal solution to* (P) *and let $\{(u_\gamma,y_\gamma)\}$ be a sequence of optimal solutions to the approximating problems* (P_γ)*. Then there is a subsequence $\{\gamma_k\}\subset\{\gamma\}$ such that*

$$u_{\gamma_k}\to\bar{u} \text{ strongly in } L^p(0,T;L^2(\Gamma)),\ y_{\gamma_k}\to\bar{y} \text{ strongly in } C([0,T];H^{1/2-\varepsilon}(\Omega)),$$
$$\text{and } J_{\gamma_k}(u_{\gamma_k},y_{\gamma_k})\to J(\bar{u},\bar{y}) \text{ as } k\to\infty.$$

Proof. Since $(\bar{u},\bar{y})$ is feasible to (P_γ) for each $\gamma>0$, one has

$$J_\gamma(u_\gamma,y_\gamma)\le J_\gamma(\bar{u},\bar{y})=J(\bar{u},\bar{y})\ \ \forall\gamma>0. \tag{3.4}$$

Due to (3.4) and assumptions (i)–(iii) we get

$$\gamma\|\alpha_\gamma(y_\gamma)\|^2_{L^2(0,T;L^2(\Omega))}\le M\ \ \forall\gamma>0$$

for some constant M. This yields

$$\gamma\|\alpha_\gamma(y_\gamma)\|_{L^2(0,T;L^2(\Omega))}\to 0 \text{ as } \gamma\to 0. \tag{3.5}$$

Since U_{ad} is weakly compact in the reflexive Banach space $L^p(0,T;L^2(\Gamma))$, there exists a subsequence of $\{u_\gamma\}$, still denoted by $\{u_\gamma\}$, such that

$$u_\gamma\to\tilde{u} \text{ weakly in } L^p(0,T;L^2(\Gamma)) \text{ as } \gamma\to 0 \tag{3.6}$$

for some $\tilde{u}\in U_{ad}$. Denote by $\tilde{y}$ a mild solution of (2.1) corresponding to $\tilde{u}$ and employing Proposition 2, one can find a subsequence $\{\gamma_k\}\subset\{\gamma\}$ such that

$$y_{\gamma_k}(t,x)\to\tilde{y}(t,x) \text{ a.e. in } Q \text{ as } k\to\infty. \tag{3.7}$$

To pass to the limit in (3.4) we need to show that $\tilde{y}$ satisfies the state constraints (2.2). For this purpose let us consider the following sets:

$$\Omega^t_{1a}:=\{x\in\Omega\mid a-\gamma<y(t,x)\le a\};\ \ \Omega^t_{2a}:=\{x\in\Omega\mid y(t,x)\le a-\gamma\};$$
$$\Omega^t_{1b}:=\{x\in\Omega\mid b\le y(t,x)<b+\gamma\};\ \ \Omega^t_{2b}:=\{x\in\Omega\mid y(t,x)\ge b+\gamma\}.$$

They are Lebesgue measurable due to the choice of $y\in C([0,T];H^{1/2-\varepsilon}(\Omega))$. Taking into account (3.5) and the structure of $\alpha_\gamma(\cdot)$ in (3.1), one has

$$\begin{aligned}&\int_0^T\int_{\Omega^t_{1a}}(2\gamma)^{-2}(y_\gamma(t,x)-a)^4dtdx+\int_0^T\int_{\Omega^t_{2a}}(y_\gamma(t,x)-a+\gamma/2)^2dtdx\\&\quad+\int_0^T\int_{\Omega^t_{1b}}(2\gamma)^{-2}(y_\gamma(t,x)-b)^4dtdx\\&\quad+\int_0^T\int_{\Omega^t_{2b}}(y_\gamma(t,x)-b-\gamma/2)^2dtdx\to 0 \text{ as } \gamma\to 0.\end{aligned}$$

Applying Lemma 4.2 of [16] similarly to the proof of Theorem 4.3 therein, we conclude that

$$a \le \tilde{y}(t,x) \le b \text{ a.e. in } Q,$$

i.e., $(\tilde{u},\tilde{y})$ is a feasible pair to the state-constrained problem (P). The latter yields

$$J(\tilde{u},\tilde{y}) \ge J(\bar{u},\bar{y}). \tag{3.8}$$

Now passing to the limit in (3.4) and taking into account (3.6)–(3.8) as well as the weak lower semicontinuity of the cost functional (2.4) in the control space $L^p(0,T;L^2(\Gamma))$, we arrive at

$$\lim_{k\to\infty} \|u_{\gamma_k} - \bar{u}\|^p_{L^p(0,T;L^2(\Gamma))} = 0 \text{ and } \lim_{k\to\infty} \gamma_k \|\alpha_{\gamma_k}(y_{\gamma_k})\|^2_{L^2(0,T;L^2(\Omega))} = 0. \tag{3.9}$$

The first equality in (3.9) means that $u_{\gamma_k} \to \bar{u}$ strongly in $L^p(0,T;L^2(\Gamma))$ as $k\to\infty$. By Proposition 1 this implies that $y_{\gamma_k} \to \bar{y}$ strongly in $C([0,T];H^{1/2-\varepsilon}(\Omega))$ as $k\to\infty$. Therefore, one has $\tilde{u}=\bar{u}$ and $\tilde{y}=\bar{y}$. Finally, the cost functional convergence in the theorem follows from the second equality in (3.9). □

The last result of this section provides a necessary condition for an optimal control u_γ to each approximating problem (P_γ). This condition is expressed in terms of the adjoint operators to $\mathcal{L}$ and $\mathcal{L}_T$ considered in Proposition 1.

Theorem 6. *Let (u_γ, y_γ) be an optimal pair to problem (P_γ). Then one has*

$$\begin{aligned} 0 \le & \iint_\Sigma [(\mathcal{L}_T^*\varphi'(y_\gamma))(t,x) + \mathcal{L}^*(\frac{\partial g}{\partial y}(t,x,y_\gamma) + 2\gamma\alpha'_\gamma(y_\gamma)\alpha_\gamma(y_\gamma)) \\ & + \frac{\partial h}{\partial u}(t,x,u_\gamma)]u\,dt\,d\sigma_x + 2p\int_0^T \|u_\gamma - \bar{u}\|^{p-2}_{L^2(\Gamma)}(\int_\Gamma (u_\gamma - \bar{u})u\,d\sigma_x)dt \end{aligned} \tag{3.10}$$

for any $u \in L^p(0,T;L^2(\Gamma))$ such that $u_\gamma + \theta u \in U_{ad}$ for all $\theta\in[0,\theta_0]$ with some $\theta_0>0$.

Proof. Consider variations of u_γ of the form $u_\gamma+\theta u \in U_{ad}$ with $u\in L^p(0,T;L^2(\Gamma))$ where $\theta\in[0,\theta_0]$ for some $\theta_0>0$. Denote by $y_{\gamma u}$ a mild solution of (2.1) corresponding to $u_\gamma+\theta u$ and consider a function $\psi:[0,\theta_0]\to\mathbb{R}$ defined by

$$\psi(\theta) := J_\gamma(u_\gamma+\theta u, y_{\gamma u}).$$

Clearly ψ attains its minimum at $\theta=0$. Moreover, Proposition 1 implies that

$$y_{\gamma u} \to y_\gamma \text{ strongly in } C([0,T];H^{1/2-\varepsilon}(\Omega)) \text{ as } \theta\to 0 \text{ and}$$

$$\frac{y_{\gamma u}(T,x)-y_\gamma(T,x)}{\theta} = \mathcal{L}_T u, \quad \frac{y_{\gamma u}(t,x)-y_\gamma(t,x)}{\theta} = \mathcal{L}u \ \ \forall\theta>0.$$

Employing these results and the classical mean value theorem, we come up with

$$\begin{aligned} 0 \le & \liminf_{\theta\to 0} \frac{\psi(\theta)-\psi(0)}{\theta} \\ = & \liminf_{\theta\to 0} \frac{1}{\theta}[\int_\Omega \varphi'(y_\gamma(T,x)+\theta_1(y_{\gamma u}(T,x)-y_\gamma(T,x)))(y_{\gamma u}(T,x)-y_\gamma(T,x))dx \\ & + \iint_Q \frac{\partial g}{\partial y}(t,x,y_\gamma+\theta_2(y_{\gamma u}-y_\gamma))(y_{\gamma u}-y_\gamma)dt\,dx \end{aligned}$$

$$
\begin{aligned}
&+\iint_{\Sigma}\frac{\partial h}{\partial u}(t,x,u_\gamma+\theta_3\theta u)\theta u dt d\sigma_x\\
&+\int_0^T(\|u_\gamma+\theta u-\bar u\|^{p-2}_{L^2(\Gamma)}+\ldots+\|u_\gamma-\bar u\|^{p-2}_{L^2(\Gamma)})(\int_\Gamma\theta u(2u_\gamma-2\bar u+\theta u)d\sigma_x)dt\\
&+\gamma\iint_Q(\alpha_\gamma(y_{\gamma u})+\alpha_\gamma(y_\gamma))\alpha'_\gamma(y_\gamma+\theta_4(y_{\gamma u}-y_\gamma))(y_{\gamma u}-y_\gamma)dtdx]
\end{aligned}
$$

where $\theta_i = \theta_i(t,x) \in [0,1]$ a.e. in Q for $i = 1,2,3,4$. Observe that $\theta_i(y_{\gamma u} - y_\gamma) \to 0$ *strongly* in $L^2(Q)$ as $\theta \to 0$ for $i = 1,2,3,4$ and that $\alpha_\gamma(y_{\gamma u}) + \alpha_\gamma(y_\gamma) \in L^2(0,T;L^2(\Omega))$. Then by using assumptions (i)–(iii) and the Lebesgue dominated convergence theorem, we obtain

$$
\begin{aligned}
0\le&\int_\Omega\varphi'(y_\gamma(T,x))\mathcal{L}_T u dx+\iint_Q(\frac{\partial g}{\partial y}(t,x,y_\gamma)+2\gamma\alpha'_\gamma(y_\gamma)\alpha_\gamma(y_\gamma))\mathcal{L}u dt dx\\
&+\iint_\Sigma\frac{\partial h}{\partial u}(t,x,u_\gamma)u dt d\sigma_x+2p\int_0^T\|u_\gamma-\bar u\|^{p-2}_{L^2(\Gamma)}(\int_\Gamma(u_\gamma-\bar u)u d\sigma_x)dt.
\end{aligned}
\tag{3.11}
$$

The latter implies (3.10) and ends the proof of the theorem. □

4. Necessary Optimality Conditions with State Constraints

In the last part of this paper we develop a limiting procedure to derive necessary optimality conditions for the original Dirichlet boundary control problem (P) with pointwise state constraints. This procedure is based on passing to the limit in necessary optimality conditions for the approximating problems (P_γ) by taking into account the strong convergence results established in Section 3. Analyzing these necessary optimality conditions (Theorem 6), we can observe that to pass to the limit therein one needs to get a *uniform bound* for the perturbation term $\gamma\alpha'_\gamma(\cdot)\alpha_\gamma(\cdot)$ in an appropriate space. Such a bound does not follow from the previous consideration without additional assumptions. To furnish this let us impose a *constraint qualification condition* (CQ) for the state constraints in problem (P). In what follows $\|\cdot\|_\infty$ and $\|\cdot\|_1$ denote the norms in $L^\infty(Q)$ and $L^1(Q)$, respectively.

(CQ) *There exist $\tilde u \in U_{ad}$ and $\eta > 0$ such that for all $\zeta \in L^\infty(Q)$ with $\|\zeta\|_\infty \le 1$ the mild solution $\tilde y$ of* (2.1) *corresponding to $\tilde u$ satisfies the condition*

$$a \le \tilde y(t,x) + \eta\zeta(t,x) \le b \ \text{a.e. in } Q.$$

Observe that this qualification condition is different from the classical Slater interiority one in the corresponding space; compare, e.g., [13]. In particular, (CQ) does *not* imply that the set of feasible trajectories y has nonempty interior in the space $C([0,T];H^{1/2-\varepsilon}(\Omega))$. We refer the reader to [3] and [4] for more discussions on the related qualification conditions for the case of parabolic systems with distributed controls.

The next lemma provides the desired uniform estimate that turns out to be crucial in our limiting procedure.

Proposition 7. *In addition to the assumptions made above we impose the qualification condition* (CQ). *Then there exists a constant $C > 0$ such that*

$$\|\gamma\alpha'_\gamma(y_\gamma)\alpha_\gamma(y_\gamma)\|_1 \le C \ \ \forall\gamma > 0. \tag{4.1}$$

Proof. Given $\tilde{u}$ in (CQ), let us substitute $u = \tilde{u} - u_\gamma$ into (3.11). Employing the monotonicity of $\alpha_\gamma(\cdot)$, one has

$$
\begin{aligned}
0 \le & \int_\Omega \varphi'(y_\gamma(T,x))\mathcal{L}_T(\tilde{u} - u_\gamma)dx + \iint_Q \frac{\partial g}{\partial y}(t,x,y_\gamma)\mathcal{L}(\tilde{u} - u_\gamma)dtdx \\
& + \iint_\Sigma \frac{\partial h}{\partial u}(t,x,u_\gamma)(\tilde{u} - u_\gamma)dtd\sigma_x \\
& + 2p\int_0^T \|u_\gamma(t) - \bar{u}(t)\|^{p-2}_{L^2(\Gamma)}\Big(\int_\Gamma (u_\gamma - \bar{u})(\tilde{u} - u_\gamma)d\sigma_x\Big)dt \\
& + 2\iint_Q \gamma\alpha'_\gamma(y_\gamma)\alpha_\gamma(y_\gamma)(\mathcal{L}\tilde{u} - \mathcal{L}u_\gamma)dtdx \\
\le & \int_\Omega \varphi'(y_\gamma(T,x))(\tilde{y}(T,x) - y_\gamma(T,x))dx + \iint_Q \frac{\partial g}{\partial y}(t,x,y_\gamma)(\tilde{y} - y_\gamma)dtdx \\
& + \iint_\Sigma \frac{\partial h}{\partial u}(t,x,u_\gamma)(\tilde{u} - u_\gamma)dtd\sigma_x \\
& + 2p\int_0^T \|u_\gamma(t) - \bar{u}(t)\|^{p-2}_{L^2(\Gamma)}\Big(\int_\Gamma (u_\gamma - \bar{u})(\tilde{u} - u_\gamma)d\sigma_x\Big)dt \\
& - 2\iint_Q \gamma\alpha'_\gamma(y_\gamma)(\alpha_\gamma(y_\gamma) - \alpha_\gamma(\tilde{y} + \eta\zeta))(y_\gamma - \tilde{y} - \eta\zeta)dtdx \\
& - 2\eta\iint_Q \gamma\alpha'_\gamma(y_\gamma)\alpha_\gamma(y_\gamma)\zeta dtdx \\
\le & \int_\Omega \varphi'(y_\gamma(T,x))(\tilde{y}(T,x) - y_\gamma(T,x))dx + \iint_Q \frac{\partial g}{\partial y}(t,x,y_\gamma)(\tilde{y} - y_\gamma)dtdx \\
& + \iint_\Sigma \frac{\partial h}{\partial u}(t,x,u_\gamma)(\tilde{u} - u_\gamma)dtd\sigma_x \\
& + 2p\int_0^T \|u_\gamma(t) - \bar{u}(t)\|^{p-2}_{L^2(\Gamma)}\Big(\int_\Gamma (u_\gamma - \bar{u})(\tilde{u} - u_\gamma)d\sigma_x\Big)dt \\
& - 2\eta\iint_Q \gamma\alpha'_\gamma(y_\gamma)\alpha_\gamma(y_\gamma)\zeta dtdx \;\; \forall\zeta \in L^\infty(Q) \text{ with } \|\zeta\|_\infty \le 1.
\end{aligned}
$$

Now taking into account Theorem 5, we can find a constant $C > 0$ independent of γ such that

$$\iint_Q \gamma\alpha'_\gamma(y_\gamma)\alpha_\gamma(y_\gamma)\zeta dtdx \le C \;\; \forall\gamma > 0 \;\; \forall\zeta \in L^\infty(Q) \text{ with } \|\zeta\|_\infty \le 1.$$

This estimate yields (4.1) and ends the proof. □

Let us denote by $ba(Q)$ the space of bounded additive functions (generalized measures) on subsets of Q that vanish on sets of the Lebesgue measure zero. It is well known that this space can be identified with the dual space to $L^\infty(Q)$ in the following sense: for each $\Lambda \in (L^\infty(Q))^*$ there is a unique $\lambda \in ba(Q)$ such that

$$\Lambda(w) = \iint_Q w\lambda(dtdx) \;\; \forall w \in L^\infty(Q).$$

In the sequel we do not distinguish between $\Lambda \in (L^\infty(Q))^*$ and its counterpart $\lambda \in ba(Q)$. Recall that (supp λ) means the *support set* for $\lambda \in (L^\infty(Q))^*$ where this measure is not zero. In what follows the convergence along a *generalized sequence* means the convergence of a *net* in the weak* topology of the space $(L^\infty(Q))^*$ where the topological and sequential limits are different.

For the optimal trajectory $\bar{y}(t,x)$ to problem (P) we define the set

$$Q_{ab} := \{(t,x) \in Q \mid \bar{y}(t,x) = a \text{ or } \bar{y}(t,x) = b\}$$

where the state constraints (2.2) are *active*. This set plays an essential role in the results below.

Proposition 8. *Under the assumptions made in Proposition* 7 *there exist* $\lambda \in (L^\infty(Q))^*$ *with* supp $\lambda \subset Q_{ab}$ *and a generalized sequence of* $\{\gamma\}$ *along which*

$$2\gamma\alpha'_\gamma(y_\gamma)\alpha_\gamma(y_\gamma) \to \lambda \ \textit{weakly}^* \ \textit{in} \ (L^\infty(Q))^* \ \textit{as} \ \gamma \to 0.$$

Proof. We just sketch the proof referring the reader to [16] for more details in a similar setting. Let us define

$$\Lambda_\gamma(w) := 2\iint_Q \gamma\alpha'_\gamma(y_\gamma)\alpha_\gamma(y_\gamma)w\,dtdx \ \ \forall w \in L^\infty(Q)$$

for each $\gamma > 0$. Proposition 7 ensures the *uniform boundedness* of $\{\Lambda_\gamma\}$ in $(L^\infty(Q))^*$. Due to weak* compactness of the unit ball in a dual space we find $\Lambda \in (L^\infty(Q))^*$ and a generalized sequence of $\{\gamma\}$ along which

$$\lim_{\gamma\to 0}\Lambda_\gamma(w) = \lim_{\gamma\to 0} 2\iint_Q \gamma\alpha'_\gamma(y_\gamma)\alpha_\gamma(y_\gamma)w\,dtdx = \Lambda(w) \ \ \forall w \in L^\infty(Q). \tag{4.2}$$

It remains to show that supp $\lambda \subset Q_{ab}$. To this end we observe that

$$\text{meas}(\{(t,x) \in Q \mid \bar{y}(t,x) < a \ \text{ or } \ \bar{y}(t,x) > b\}) = 0.$$

Thus assuming that supp $\lambda \not\subset Q_{ab}$, one has a set $\tilde{Q}$ with the properties

$$\text{meas}(\tilde{Q}) > 0, \ \ \lambda(\tilde{Q}) \neq 0, \ \text{ and } \ \tilde{Q} \subset \{(t,x) \in Q \mid a < \bar{y}(t,x) < b\}. \tag{4.3}$$

Now arguing in the same way as in [16], we find a nonnegative function $c(\rho)$ such that $c(\rho) \to 0$ when $\rho \to 0$ and

$$|\Lambda(w)| \le c(\rho) \ \ \forall w \in L^\infty(Q), \ \ \text{supp}\, w \subset \tilde{Q}$$

for all ρ sufficiently small. This yields

$$\Lambda(w) = 0 \ \ \forall w \in L^\infty(Q), \ \ \text{supp}\, w \subset \tilde{Q}$$

which contradicts (4.3) and ends the proof of the proposition. □

Now we are ready to derive necessary optimality conditions for optimal solutions to (P) by passing to the limit in (3.10). Due to the weak* convergence result of Proposition 8 we need to show that the operator $\mathcal{L}$ defined by (2.5) is continuous from $L^\infty(\Sigma)$ into $L^\infty(\Omega)$ (note that this is different from Proposition 1). The next theorem establishes this property and provides the desired necessary optimality conditions for the original state-constrained problem (P).

Theorem 9. *Let $(\bar{u},\bar{y})$ is an optimal solution to problem* (P) *under all the assumptions made above. Then there is a measure $\lambda \in (L^\infty(Q))^*$ with* supp $\lambda \subset Q_{ab}$ *such that*

$$
\begin{aligned}
0 \le & \iint_\Sigma [(\mathcal{L}_T^*\varphi'(\bar{y}))(t,x) + \mathcal{L}^*(\frac{\partial g}{\partial y}(t,x,\bar{y})) + \frac{\partial h}{\partial u}(t,x,\bar{u})](u-\bar{u})dtd\sigma_x \\
& + \iint_\Sigma (u-\bar{u})(\mathcal{L}^*\lambda)(dtd\sigma_x) \;\; \forall u \in U_{ad}.
\end{aligned} \tag{4.4}
$$

Proof. Let $\{(u_\gamma, y_\gamma)\}$ be a sequence of optimal solutions to problems (P_γ) that strongly converges to $(\bar{u},\bar{y})$ due to Theorem 5 satisfying the necessary optimality conditions in Theorem 6. It follows from (3.10) that

$$
\begin{aligned}
0 \le & \iint_\Sigma [(\mathcal{L}_T^*\varphi'(y_\gamma))(t,x) + \mathcal{L}^*(\frac{\partial g}{\partial y}(t,x,y_\gamma) + 2\gamma\alpha_\gamma'(y_\gamma)\alpha_\gamma(y_\gamma)) \\
& + \frac{\partial h}{\partial u}(t,x,u_\gamma)](u-u_\gamma)dtd\sigma_x \\
& + 2p\int_0^T \|u_\gamma - \bar{u}\|_{L^2(\Gamma)}^{p-2}(\int_\Gamma (u_\gamma-\bar{u})(u-u_\gamma)d\sigma_x)dt \;\; \forall u \in U_{ad}.
\end{aligned} \tag{4.5}
$$

Our purpose is to pass to the limit in (4.5) as $\gamma \to 0$ along a generalized subsequence. Due to Proposition 1, Theorem 5, and the well-known continuity of the operator $\mathcal{L}^* : L^2(0,T;L^2(\Omega)) \to L^2(0,T;L^2(\Gamma))$ (see, e.g., [11]) we have $\varphi'(y_\gamma(T,\cdot)) \in L^2(\Omega) \subset H^{-1/2+\varepsilon}(\Omega)$ for all $\gamma > 0$ and

$$
\begin{aligned}
& \iint_\Sigma [(\mathcal{L}_T^*\varphi'(y_\gamma))(t,x) + \mathcal{L}^*(\frac{\partial g}{\partial y}(t,x,y_\gamma)) + \frac{\partial h}{\partial u}(t,x,u_\gamma)](u-u_\gamma)dtd\sigma_x \to \\
& \iint_\Sigma [(\mathcal{L}_T^*\varphi'(\bar{y}))(t,x) + \mathcal{L}^*(\frac{\partial g}{\partial y}(t,x,\bar{y})) + \frac{\partial h}{\partial u}(t,x,\bar{u})](u-\bar{u})dtd\sigma_x \;\; \forall u \in U_{ad}.
\end{aligned}
$$

Since the last term in (4.5) converges to 0, it remains to show that

$$
\iint_\Sigma (u-u_\gamma)\mathcal{L}^*(2\gamma\alpha_\gamma'(y_\gamma)\alpha_\gamma(y_\gamma))dtd\sigma_x \to \iint_\Sigma (u-\bar{u})(\mathcal{L}^*\lambda)(dtd\sigma_x) \tag{4.6}
$$

as $\gamma \to 0$ for any $u \in U_{ad}$. Due to Proposition 8 property (4.6) immediately follows from the weak* continuity of the operator $\mathcal{L}^* : (L^\infty(Q))^* \to (L^\infty(\Sigma))^*$. In turn, this weak* continuity of the adjoint operator is a direct consequence of the strong continuity of the operator $\mathcal{L}$ in (2.5) considered from $L^\infty(\Sigma)$ into $L^\infty(Q)$. To justify the latter property we follow [16] and invoke some results from the theory of generalized solutions to parabolic equations.

Let $v \in L^2(\Sigma)$ be a boundary condition in (2.1). According to [12, Theorem 9.1], there is a unique $y(v) \in L^2(Q)$, called a *generalized solution* to (2.1), such that

$$
\begin{aligned}
& \iint_Q y(v)(-\frac{\partial z}{\partial t} + A^*z)dtdx = -\iint_\Sigma v\frac{\partial v}{\partial \nu_A}dtd\sigma_x \\
& \forall z \in \{z \in H^{2,1}(Q) \mid z(t,x) = 0,\; (t,x) \in \Sigma,\; z(T,x) = 0\}
\end{aligned} \tag{4.7}
$$

where ν_A is an outer normal to Γ associated with the operator A.

Let $v \in L^\infty(\Sigma)$ and let $y = \mathcal{L}v$ be the corresponding mild solution to system (2.1). We are going to show that such y coincides with the generalized solution to (2.1) in the sense of (4.7). Since $L^\infty(\Sigma) \subset L^p(0,T;L^2(\Gamma))$, we may consider v as an element of $L^p(0,T;L^2(\Gamma))$ and use the fact that $\mathcal{D}(\Sigma)$, the space of C^∞ functions on Σ with compact supports, is dense in $L^p(0,T;L^2(\Gamma))$. This gives a sequence $\{v_n\} \subset \mathcal{D}(\Sigma)$ such that

$$v_n \to v \text{ strongly in } L^p(0,T;L^2(\Gamma)) \text{ as } n \to \infty.$$

Since for each $v_n \in \mathcal{D}(\Sigma)$ system (2.1) has a unique *classical solution* y_n, we automatically get that $y_n = \mathcal{L}v_n$ and it satisfies (4.7). Moreover, it follows from Proposition 1 with $\varepsilon = 1/2$ that

$$\|\mathcal{L}v - y_n\|_{C([0,T];L^2(\Omega))} = \|\mathcal{L}v - \mathcal{L}v_n\|_{C([0,T];L^2(\Omega))} \to 0 \text{ as } n \to \infty.$$

Taking into account all these facts, we have

$$\begin{aligned}
&|\iint_Q \mathcal{L}v(-\frac{\partial z}{\partial t} + A^*z)dtdx + \iint_\Sigma v\frac{\partial z}{\partial \nu_A}dtd\sigma_x| \\
&\le |\iint_Q (\mathcal{L}v - y_n)(-\frac{\partial z}{\partial t} + A^*z)dtdx| + |\iint_\Sigma (v - v_n)\frac{\partial z}{\partial \nu_A}dtd\sigma_x| \\
&\le \|\mathcal{L}v - y_n\|_{C([0,T];L^2(\Omega))}\| - \frac{\partial z}{\partial t} + A^*z\|_{L^2(0,T;L^2(\Omega))}T^{1/2} \\
&\quad + \|v - v_n\|_{L^p(0,T;L^2(\Gamma))}\|\frac{\partial z}{\partial \nu_A}\|_{L^2(0,T;L^2(\Gamma))}T^{1/\tilde{q}} \to 0 \text{ as } n \to \infty
\end{aligned}$$

where $\tilde{q} := \dfrac{2(p-1)}{p-2}$. Thus we obtain

$$\begin{aligned}
&\iint_Q \mathcal{L}v(-\frac{\partial z}{\partial t} + A^*z)dtdx = -\iint_\Sigma v\frac{\partial z}{\partial \nu_A}dtd\sigma_x \\
&\forall z \in \{z \in H^{2,1}(Q) \mid z(t,x) = 0,\ (t,x) \in \Sigma,\ z(T,x) = 0\}.
\end{aligned}$$

The latter means that the mild solution $y = \mathcal{L}v$ is also a generalized solution to (2.1) for any $v \in L^\infty(\Sigma)$. Using the uniqueness of generalized solutions and the fact that the generalized solution operator is a continuous map from $L^\infty(\Sigma)$ into $L^\infty(Q)$ (see, e.g., [12, pp. 205–206]), we conclude that the linear operator $\mathcal{L}$ is continuous from $L^\infty(\Sigma)$ into $L^\infty(Q)$. This completes the proof of the theorem. □

References

1. A. V. Balakrishnan, *Applied Functional Analysis*, 2nd edition, Springer-Verlag, New York, 1981.
2. V. Barbu, *Analysis and Control of Nonlinear Infinite Dimensional Systems*, Academic Press, Boston, 1993.
3. M. Bergounioux and D. Tiba, General optimality conditions for constrained convex control problems, *SIAM J. Contr. Optim.*, to appear.
4. M. Bergounioux and F. Tröltzsch, Optimality conditions and generalized bang-bang principle for a state-constrained semilinear parabolic problem, *Numer. Funct. Anal. Optim.* **17** (1996), 517–536.
5. H. O. Fattorini, Optimal control problems with state constraints for semilinear distributed parameter systems, *J. Optim. Theory Appl.* **88** (1996), 25–59.
6. H. O. Fattorini and H. Frankowska, Infinite- dimensional control problems with state constraints, *Lecture Notes Contr. Inform. Sci.* **154**, pp. 52–62, Springer-Verlag, Berlin, 1991.
7. H. O. Fattorini and T. Murphy, Optimal problems for nonlinear parabolic boundary control problems: the Dirichlet boundary conditions, *Diff. Integ. Eq.* **7** (1994), 1367–1388.

8. I. Lasiecka, State constrained control problems for parabolic systems: regularity of optimal solutions, *Appl. Math. Optim.* **6** (1980), 1–29.
9. I. Lasiecka and R. Triggiani, Dirichlet boundary control problems for parabolic equations with quadratic cost: analiticity and Riccati feedback synthesis, *SIAM J. Contr. Optim.* **21** (1983), 41–67.
10. I. Lasiecka and R. Triggiani, The regulator problem for parabolic equations with Dirichlet boundary controls, I: Riccati's feedback synthesis and regularity of optimal solutions, *Appl. Math. Optim.* **16** (1987), 147–168.
11. I. Lasiecka and R. Triggiani, *Differential and Algebraic Riccati Equations with Applications to Boundary/Point Control Problems*, Lecture Notes in Contr. Inform. Sci. **164**, Springer-Verlag, Berlin, 1991.
12. J.-L. Lions, *Optimal Control of Systems Governed by Partial Differential Equations*, Springer-Verlag, Berlin, 1971.
13. U. Mackenroth, Convex parabolic boundary control problems with pointwise state constraints, *J. Math. Anal. Appl.* **87** (1982), 256–277.
14. B. S. Mordukhovich and K. Zhang, Bang-bang principle for state-constrained parabolic systems with Dirichlet boundary controls, *Proc. 33rd. Conf. Dec. Contr.*, 1994, 3418–3423.
15. B. S. Mordukhovich and K. Zhang, Existence, approximation, and suboptimality conditions for minimax control of heat transfer systems with state constraints, *Lecture Notes in Pure and Applied Mathematics* **150**, pp. 251–270, Marcel Dekker, New York, 1994.
16. B. S. Mordukhovich and K. Zhang, Minimax control of parabolic systems with Dirichlet boundary conditions and state constraints, *Appl. Math. Optim.*, to appear.
17. F. Tröltzsch, *Optimality Conditions for Parabolic Control Problems and Applications*, Teubner Texte, Leipzig, 1984.
18. D. Washburn, A bound on the boundary input map for parabolic equations with applications to time optimal control, *SIAM J. Contr. Optim.* **17** (1979), 652–671.

Boris S. Mordukhovich
Department of Mathematics
Wayne State University
Detroit, MI 48202, USA
E-mail: boris@math.wayne.edu

Kaixia Zhang
Department of Mathematics
Wayne State University
Detroit, MI 48202, USA

International Series of Numerical Mathematics
Vol. 126, © 1998 Birkhäuser Verlag, Basel

Second Order Optimality Conditions and Stability Estimates for the Identification of Nonlinear Heat Transfer Laws

ARND RÖSCH
Department of Mathematics
TU Chemnitz–Zwickau

ABSTRACT. Consider the heat equation with a nonlinear function α in the boundary condition which depends only on the solution u of the initial-boundary value problem. The unknown function α belongs to a set of admissible functions. For this problem the existence of a second Fréchet derivative of the control-state mapping is proved. Based on this result a necessary second order optimality condition is formulated. For the investigated objective sufficient second order condition are closely connected with stability estimates. Using the knowledge about stability estimates, it is shown that already for simple cases the usual sufficient conditions can not be fulfilled.

1991 *Mathematics Subject Classification.* 35K05, 35K60, 35R30, 49K20

Key words and phrases. Identification, inverse problems, heat equation, nonlinear boundary conditions, Fréchet differentiability, second order optimality conditions, stability estimates.

1. Introduction

This paper deals with the connection between second order optimality conditions and stability estimates for the identification of nonlinear heat transfer laws. The identification problem is formulated as an optimal control problem where the unknown heat exchange function plays the part of the control. The unknown function α is assumed as a function of the temperature u. The control system under consideration is governed by a semilinear parabolic equation, hence the control problem belongs to the class of nonconvex optimization problems. In contrast to parabolic control problems with convex objective functionals and linear equations, where the list of references and optimality conditions is very extensive, only a few investigations have been devoted to the case of nonlinear parabolic equations. In nonconvex problems, sufficient second order optimality conditions at the optimal point are a substitute for the convexity. The theory of sufficient second order conditions for twice differentiable extremal problems in function spaces is known to be more rich and interesting than that for problems in finite-dimensional spaces. This is due to the so-called two-norm discrepancy expressing the noncompatibility of the norms needed for second order optimality condition. This difficulty was resolved successfully by Ioffe [6] and Maurer [7]. The theory was extended on a class of parabolic boundary control problems in the papers of Goldberg and Tröltzsch [4],[5]. But the investigating type of identification problems has an other complicate structure. Therefore it is one goal of this paper to show that it is impossible to get similar conditions for the identification problem.

To ensure a sufficient optimality condition it is often required that there exist terms in the objective which depend directly on the control. For that reason the usual way is using a Tikhonov-regularization. But the Tikhonov-regularization is also the most popular way to get well-posedness and stability of inverse problems, see Tikhonov/Arsenin [13], Tikhonov/Goncharskij/Stepanov/Yagola [14]. Hence there is a close connection between sufficient second order condition and stability in this well-investigated class of boundary control problems. But this approach has disadvantages for the application to the nonlinear identification problem. A natural way to handle the nonlinear identification problem consists in using a compact set of admissible controls α (that means "admissible laws"). For practical applications it is often easy to bound the maximal growth of the heat exchange coefficient with respect to the temperature. The compactness of the set of admissible controls is the main property to get statements about stability and well-posedness in suitable chosen function spaces, see Rösch [8]. In this approach we have no regularity term. By the way, it is difficult to find a proper function space for the regularity term for the identification of a heat exchange coefficient depending only on the boundary temperature. These facts generate an own specific of identification problems. The compactness of the admissible control set relieves the derivation of several properties. Otherwise, with the absence of the Tikhonov-term is also absent the natural quadratic convexity term of the objective with respect to the control. Another difference is caused in the non-standard structure of the control-state-mapping. Usually the optimality conditions need no derivatives of the control, but in our case every differentiation of the objective uses derivatives of the control α. For that reason it is necessary to require C^2-regularity of the control in order to get second order conditions.

In this paper we want to discuss second order optimality conditions for the identification problem and the connection to stability estimates. The *optimal control problem* we are going to investigate is to *minimize* the functional

$$\Psi(\alpha) = \int_0^T \int_\Gamma (u(t,x) - q(t,x))^2 dS_x dt, \tag{1.1}$$

subject to

$$\begin{aligned} \frac{\partial u}{\partial t}(t,x) &= \Delta_x u(t,x) \text{ on } (0,\mathrm{T}] \times \Omega \\ u(0,x) &= u^0(x) \text{ on } \Omega \\ \frac{\partial u}{\partial n}(t,x) &= \alpha(u(t,x))(\vartheta - u(t,x)) \text{ on } (0,\mathrm{T}] \times \Gamma \end{aligned} \tag{1.2}$$

where the *control* α is taken from the set

$$\begin{aligned} U_{ad} := &\{\alpha \in C^2[\vartheta_1, \vartheta_2], 0 < m_1 \le \alpha(u) \le M_1, m_2 \le \alpha'(u) \le M_2, \\ &m_3 \le \alpha''(u) \le M_3\} \end{aligned}$$

In this setting, $\Omega \subset R^m$ is a bounded domain with C^∞-boundary Γ, $T > 0$ a fixed time, ϑ a fixed temperature and $q \in L_2((0,T) \times \Gamma)$ is a given function of "measurements". ϑ_1 and ϑ_2 are defined by

$$\begin{aligned} \vartheta_1 &= \min(\vartheta, \inf_{x\in\Omega} u^0(x)) \\ \vartheta_2 &= \max(\vartheta, \sup_{x\in\Omega} u^0(x)). \end{aligned}$$

Interpreting the process as a heating problem, the variable u means the temperature of the material, u^0 the initial temperature, ϑ the constant temperature of the surrounding medium, and α the unknown heat transfer function, playing the part of the control.

2. Preliminary results

In this section we introduce some notations and recall known results on the behaviour of the parabolic system (1.2), which belongs to the class of semilinear problems. For convenience we shall apply the theory of analytic semigroups. We shall heavily rely on results by Amann [1],[2] for semilinear parabolic problems. The assumptions and preparations we shall need here are nearly the same as in Rösch/Tröltzsch [12], where well-posedness of the parabolic system is proved.

In all what follows we work in Sobolev–Slobodeckij spaces $W_p^{2\hat{\sigma}}(\Omega)$ and $W_p^{2\sigma}(\Omega)$ with

$$\frac{m}{p} < 2\sigma < 2\hat{\sigma} < 1 + \frac{1}{p}.$$

Note that this inequality ensures the continuity of the regarded functions. The solution of the heat equation u is looked upon in the Banach space $C([0,T], W_p^{2\sigma}(\Omega))$ provided that the initial value u^0 belongs to $W_p^{2\hat{\sigma}}(\Omega)$.

Let A be a linear, positive, and elliptic differential operator. Then the parabolic equation

$$\begin{aligned} \frac{\partial u}{\partial t} &= -Au \\ u(0) &= u^0 \end{aligned}$$

subject to homogeneous boundary conditions gives rise to an analytic semigroup of linear continuous operators denoted by $S(t)$.

Following [12] we define A: $L_p(\Omega) \supset D(A) \longrightarrow L_p(\Omega)$ by
$D(A) = \{w \in W_p^2(\Omega) : \quad \partial w/\partial n\,|_\Gamma = 0\}, \quad Aw = (-\Delta + I)w$ for $w \in D(A)$.
Then the initial value problem

$$\begin{aligned} u'(t) &= -Au(t) \\ u(0) &= u^0 \end{aligned}$$

has the unique solution $u(t) = S(t)u^0$. The semigroup $S(t)$ generated by $-A$, $S(t) =$ "$\exp(-At)$" is an analytic semigroup of linear continuous operators in $L_p(\Omega)$.

For solving an initial value problem with an *inhomogeneous* boundary condition a special solution of the corresponding elliptic boundary value problem is needed. Let

$g \in L_p(\Gamma)$. The mapping, which assigns to g the solution v of the elliptic boundary problem

$$\begin{aligned} \Delta v - v &= 0 \\ \frac{\partial v}{\partial n} &= g, \end{aligned} \tag{2.1}$$

is denoted by N, i.e. $v = Ng$. Transforming the heat equation with $u = we^t$, we obtain

$$\begin{aligned} \Delta w(t,x) - w(t,x) &= \frac{\partial w}{\partial t}(t,x) \quad \text{on } (0,\mathrm{T}] \times \Omega \\ w(0,x) &= u^0(x) \quad \text{on } \Omega \\ \frac{\partial w}{\partial n}(t,x) &= \alpha(w(t,x)e^t)(\vartheta - w(t,x)e^t)e^{-t} \quad \text{on } (0,\mathrm{T}] \times \Gamma. \end{aligned} \tag{2.2}$$

Now the semigroup approach can be applied. The operator $-A$ is known to generate a strongly continuous and analytic semigroup $\{S(t), \quad t \geq 0\}$ of linear continuous operators in $L_p(\Omega)$, see Friedman [3]. N is a continuous mapping from $L_p(\Gamma)$ to $W_p^s(\Omega)$ for all $s < 1 + 1/p$, cf. Triebel [15].

Regarding the function w as an abstract function $w = w(t)$ with values in the Banach space $W_p^{2\sigma}(\Omega)$ the nonlinear Bochner integral equation

$$w(t) = \int_0^t AS(t-s)NB(\tau w(s))\, ds \; + \; S(t)u^0 \tag{2.3}$$

is obtained. We refer to Amann [1],[2]. In this equation, τ denotes the trace operator and B is the Nemytskij operator defined by

$$B(v)(t,x) = \alpha(v(x)e^t)(\vartheta - v(x)e^t)e^{-t}, \quad v \in C(\Gamma). \tag{2.4}$$

Here the trace operator maps $W_p^{2\sigma}(\Omega)$ into $C(\Gamma)$. Inserting the backward transformation $w = e^{-t}u$, we get in turn

$$u(t) = \int_0^t AS(t-s)Ne^{(t-s)}\alpha(\tau u(s))(\vartheta - \tau u(s))ds + e^t S(t)u^0. \tag{2.5}$$

For each $\alpha \in U_{ad}$ we get a unique solution $u \in C^{0,\delta}([0,T], W_p^{2\sigma}(\Omega))$. This solution satisfies the maximum principle

$$\vartheta_1 \leq \quad u(t,x) \quad \leq \quad \vartheta_2. \tag{2.6}$$

Let us slightly simplify the notation. Denoting the kernel of the integral in (2.5) by $k(t-s) := AS(t-s)Ne^{(t-s)}$, we get

$$u(t) = \int_0^t k(t-s)\alpha(\tau u(s))(\vartheta - \tau u(s))ds + e^t S(t)u^0. \tag{2.7}$$

The right hand side of the Bochner integral equation (2.7) depends only on the boundary values of u. For that reason, we shall investigate this equation only on the boundary. Therefore, it is convenient to introduce the trace of u by $x = \tau u$

with $\tau: C^{0,\delta}([0,T],W_p^{2\sigma}(\Omega)) \mapsto C([0,T]\times\Gamma)$ and to consider the boundary integral equation

$$x(t) = \tau \int_0^t k(t-s)\alpha(x(s))(\vartheta - x(s))ds + \tau e^t S(t)u^0. \tag{2.8}$$

Next we define a mapping Φ by

$$(\Phi\alpha)(t) = x(t), \tag{2.9}$$

where $x(t)$ is the solution of the boundary integral equation (2.8). It is also possible to work with weak solutions under reasonable assumptions, but in this case the derivation of several results is much more complicated. For the reader who prefers to work with other techniques we will write all equations in the form of PDE. Nevertheless, the solutions are assumed as solutions of the corresponding Bochner integral equation.

In Rösch [9] it is proved that Φ is Fréchet differentiable from $C^1[\vartheta_1,\vartheta_2]$ to $C([0,T]\times\Gamma)$ at a point α_0. Furthermore, let v be the solution of the initial boundary value problem

$$\begin{aligned} \frac{\partial v}{\partial t}(t,x) &= \Delta_x v(t,x) \text{ on } (0,\mathrm{T}]\times\Omega \\ v(0,x) &= 0 \text{ on } \Omega \\ \frac{\partial v}{\partial n}(t,x) &= (\alpha_0'(u_0(t,x))(\vartheta - u_0(t,x)) - \alpha_0(u_0(t,x)))v \\ &\quad +\alpha(u_0(t,x))(\vartheta - u_0(t,x)) \text{ on } (0,\mathrm{T}]\times\Gamma. \end{aligned} \tag{2.10}$$

Then the Fréchet derivative $\Phi'(\alpha_0)\alpha$ is the trace of the solution v on the boundary Γ. We expose the dependence of v with respect to the direction and write for instance v_β which means v is the solution of (2.9) for $\alpha=\beta$. This notation is useful for derivation of the second derivative which follows in the next section.

3. Second Fréchet derivative

In this section we want to prove that the mapping Φ has a second Fréchet derivative which is the base of the second order optimality conditions.

Theorem 3.1 (Existence of a second Fréchet derivative). *Φ has a second Fréchet derivative as mapping from $C^2[\vartheta_1,\vartheta_2]$ to $C([0,T]\times\Gamma)$ in every point $\alpha_0\in U_{ad}$. Moreover, let w be the solution of*

$$\begin{aligned} &\frac{\partial w}{\partial t}(t,x) = \Delta_x w(t,x) \textit{ on } (0,T]\times\Omega \\ &w(0,x) = 0 \textit{ on } \Omega \\ &\frac{\partial w}{\partial n}(t,x) = (\alpha_0''(u_0(t,x))(\vartheta - u(t,x)) - 2\alpha_0'(u_0(t,x)))v_\beta(t,x)v_\gamma(t,x) \\ &\qquad +(\alpha_0'(u_0(t,x))(\vartheta - u(t,x)) - \alpha_0(u_0(t,x)))w(t,x) \\ &\qquad +(\gamma'(u_0(t,x))(\vartheta - u_0(t,x)) - \gamma(u_0(t,x)))v_\beta(t,x) \\ &\qquad +(\beta'(u_0(t,x))(\vartheta - u_0(t,x)) - \beta(u_0(t,x)))v_\gamma(t,x) \textit{on } (0,T]\times\Gamma. \end{aligned} \tag{3.1}$$

In this setting u_0 is the solution of (1.2) for $\alpha = \alpha_0$. Then the trace of the solution w is the second Fréchet derivative of Φ at the point α_0 in the directions β and γ, that means $\Phi''(\alpha_0)[\beta,\gamma] = \tau w$.

Proof: To prove this theorem we investigate two admissible controls α_0 and $\alpha_1 = \alpha_0 + \varepsilon \cdot \gamma$ with corresponding solution u_0 and u_1 of (1.2) and corresponding first Fréchet derivative v_0 and v_1 in direction β (see (2.10)). Without lost of generality let $\|\gamma\| = 1$. Then we have $\|\alpha_1 - \alpha_0\| = \varepsilon$. The goal of the next steps is to show that the remainder term has the property $v_1 - v_0 - \varepsilon \cdot w = o(\varepsilon)$. Now we discuss the difference $\delta v = v_1 - v_0$ which solves the parabolic problem

$$\begin{aligned}
\frac{\partial \delta v}{\partial t}(t,x) &= \Delta_x \delta v(t,x) \text{ on } (0,\mathrm{T}] \times \Omega \\
\delta v(0,x) &= 0 \text{ on } \Omega \qquad (3.2)\\
\frac{\partial \delta v}{\partial n}(t,x) &= (\alpha_1'(u_1(t,x))(\vartheta - u_1(t,x)) - \alpha_1(u_1(t,x)))v_1 \\
&\quad -(\alpha_0'(u_0(t,x))(\vartheta - u_0(t,x)) - \alpha_0(u_0(t,x)))v_0 \qquad (3.3)\\
&\quad +\beta(u_1(t,x))(\vartheta - u_1(t,x)) \\
&\quad -\beta(u_0(t,x))(\vartheta - u_0(t,x)) \text{ on } (0,\mathrm{T}] \times \Gamma. \qquad (3.4)
\end{aligned}$$

It is easy to see that only the boundary condition is interesting. For that reason we discuss only this part in the next. To shorten the notation we drop the arguments t and x. First we discuss term (3.4):

$$\begin{aligned}
\beta(u_1)(\vartheta-u_1)-\beta(u_0)(\vartheta-u_0) &= \beta(u_1)(\vartheta-u_1)-\beta(u_1)(\vartheta-u_0) \\
&\quad + \beta(u_1)(\vartheta - u_0) - \beta(u_0)(\vartheta - u_0) \qquad (3.5)\\
&= -\beta(u_1)(u_1-u_0)+(\beta(u_1)-\beta(u_0))(\vartheta-u_0) \\
&= (-\beta(u_0) + \beta'(u_0)(\vartheta - u_0))v_\gamma\varepsilon + o(\varepsilon)
\end{aligned}$$

Now we discuss term (3.3):

$$\begin{aligned}
&(\alpha_1'(u_1)(\vartheta - u_1) - \alpha_1(u_1))v_1 - (\alpha_0'(u_0)(\vartheta - u_0) - \alpha_0(u_0))v_0 \\
&= (\alpha_1'(u_1)(\vartheta - u_1) - \alpha_1(u_1))v_1 - (\alpha_1'(u_1)(\vartheta - u_0) - \alpha_1(u_0))v_1 \qquad (3.6)\\
&\quad +(\alpha_1'(u_1)(\vartheta - u_0))v_1 - (\alpha_1'(u_0)(\vartheta - u_0))v_1 \qquad (3.7)\\
&\quad +(\alpha_1'(u_0)(\vartheta - u_0) - \alpha_1(u_0))v_1 - (\alpha_0'(u_0)(\vartheta - u_0) - \alpha_0(u_0))v_1 \qquad (3.8)\\
&\quad +(\alpha_0'(u_0)(\vartheta - u_0) - \alpha_0(u_0))v_1 - (\alpha_0'(u_0)(\vartheta - u_0) - \alpha_0(u_0))v_0 \qquad (3.9)
\end{aligned}$$

Next we handle the term (3.6)-(3.9). Term (3.8) is easy:

$$\begin{aligned}
&(\alpha_1'(u_0)(\vartheta - u_0) - \alpha_1(u_0))v_1 - (\alpha_0'(u_0)(\vartheta - u_0) - \alpha_0(u_0))v_1 \\
&= \varepsilon(\gamma'(u_0)(\vartheta - u_0) - \gamma(u_0))v_1 \qquad (3.10)\\
&= \varepsilon(\gamma'(u_0)(\vartheta - u_0) - \gamma(u_0))v_\beta + o(\varepsilon)
\end{aligned}$$

For term (3.7) we get

$$\begin{aligned}
(\alpha_1'(u_1)(\vartheta - u_0))v_1 - (\alpha_1'(u_0)(\vartheta - u_0))v_1 &= (\alpha_1'(u_1) - \alpha_1'(u_0))(\vartheta - u_0)v_1 \\
&= \alpha_1''(u_0)(\vartheta - u_0)v_\gamma v_1\varepsilon + o(\varepsilon) \qquad (3.11)\\
&= \alpha_0''(u_0)(\vartheta - u_0)v_\gamma v_\beta\varepsilon + o(\varepsilon)
\end{aligned}$$

Similar we deal with term (3.6)

$$\begin{aligned} & (\alpha_1'(u_1)(\vartheta - u_1) - \alpha_1(u_1))v_1 - (\alpha_1'(u_1)(\vartheta - u_0) - \alpha_1(u_0))v_1 \\ & = (-\alpha_1'(u_1)(u_1 - u_0) - (\alpha_1(u_1) - \alpha_1(u_0)))v_1 \\ & = -2\alpha_1'(u_1)v_\gamma v_\beta \varepsilon + o(\varepsilon) \end{aligned} \tag{3.12}$$

Now we discuss the term

$$r = v_1 - v_0 - \varepsilon \cdot w \tag{3.13}$$

which is the remainder term. This remainder solves a parabolic problem. Using the equations (3.5) and (3.10)-(3.12) we get:

$$\begin{aligned} \frac{\partial r}{\partial t}(t,x) &= \Delta_x r(t,x) \text{ on } (0,\mathrm{T}] \times \Omega \\ r(0,x) &= 0 \text{ on } \Omega \\ \frac{\partial r}{\partial n}(t,x) &= (\alpha_0'(u_0(t,x))(\vartheta - u_0(t,x)) - \alpha_0(u_0(t,x)))r + o(\varepsilon) \text{ on } (0,\mathrm{T}] \times \Gamma. \end{aligned} \tag{3.14}$$

For that reason $r = o(\varepsilon)$ holds and Φ has a second Fréchet derivative. □

4. Second order conditions and stability estimates

The two times Fréchet differentiability of Φ is the key point to formulate second order conditions. First of all we recall the necessary first order condition. In Rösch [10] we find the two formulations

$$\Psi'(\alpha_0)[\beta] = 2\int_0^T \int_\Gamma v_\beta(t,x)(u_0(t,x) - q(t,x))dS_x dt \geq 0 \tag{4.1}$$

and

$$\Psi'(\alpha_0)[\beta] = 2\int_0^T \int_\Gamma \beta(u_0(t,x))(\vartheta - u_0(t,x))y_0(t,x)dS_x dt \geq 0. \tag{4.2}$$

where y_0 is the solution of the adjoint system for $\alpha = \alpha_0$, $u = u_0$

$$\begin{aligned} -\frac{\partial y}{\partial t}(t,x) &= \Delta_x y(t,x) \text{ on } (0,\mathrm{T}] \times \Omega \\ y(T,x) &= 0 \text{ on } \Omega \\ \frac{\partial y}{\partial n}(t,x) &= (\alpha'(u(t,x))(\vartheta - u(t,x)) - \alpha(u(t,x)))y \\ & \quad + u(t,x) - q(t,x) \text{ on } (0,\mathrm{T}] \times \Gamma. \end{aligned} \tag{4.3}$$

Based on theorem 3.1 about the second Fréchet derivative it is easy to formulate second order optimality condition.

Theorem 4.1 (Necessary second order optimality condition). *Let α_0 be an optimal control with associated state u_0. For all admissible directions $\beta = \alpha - \alpha_0$, $\alpha \in U_{ad}$ with $\Psi'(\alpha_0)[\beta] = 0$ the necessary second order condition*

$$\Psi''(\alpha_0)[\beta,\beta] = 2\int_0^T\int_\Gamma w_{\beta\beta}(t,x)(u_0(t,x) - q(t,x)) + v_\beta(t,x)^2 dS_x dt \geq 0 \tag{4.4}$$

holds.

In this setting $w_{\beta\beta}$ is the solution of (3.1) with and $\gamma = \beta$. The proof is very simple and for that reason we resign to sketch it here.

Proposition: *It makes no sense to formulate sufficient second order conditions. For very simple cases such properties can not hold.*

Let us now discuss sufficient second order optimality conditions. Usually sufficient optimality conditions require one or both of the conditions

$$\Psi'(\alpha_0)[\beta] \geq \delta\|\beta\| \tag{4.5}$$

$$\Psi''(\alpha_0)[\beta,\beta] \geq \delta\|\beta\|^2 \tag{4.6}$$

for all admissible direction β and a positive δ. First the norm of β is the norm of the differentiation but in several applications it can be weakened with a other norm.

We will see now that such conditions cannot hold for the simplest case. For that purpose we choose an example for which is $\Psi(\alpha_0) = 0$, that means there exists an exact solution. thus we can replace the "measurements" q by the corresponding state u_0. Using formula (4.1) we get easy $\Psi'(\alpha_0)[\beta] = 0$ for all directions β. For that reason a condition of type (4.5) cannot work here.

We assume now that a condition of type (4.6) holds which a suitable norm.

$$\Psi''(\alpha_0)[\beta,\beta] \geq \delta\|\beta\|^2 \tag{4.7}$$

We investigate an admissible control $\alpha = \alpha_0 + \beta$ with corresponding state u. Using the Taylor expansion we have

$$\begin{aligned}\Psi(\alpha) &= \Psi(\alpha_0) + \Psi'(\alpha_0)[\beta] + \frac{1}{2}\Psi''(\alpha_0)[\beta,\beta] + o(\|\beta\|^2_{C^2}) \\ &= \frac{1}{2}\Psi''(\alpha_0)[\beta,\beta] + o(\|\beta\|^2_{C^2})\end{aligned} \tag{4.8}$$

If we have in (4.6) the C^2–norm, then we get

$$\Psi(\alpha_0) \geq \delta\|\beta\|^2_{C^2} \tag{4.9}$$

for a positive δ. Using the definition of Ψ,

$$\|u - u_0\|^2_{L_2([0,T]\times\Gamma)} \geq \delta\|\beta\|^2_{C^2} \tag{4.10}$$

holds. With a new $\delta > 0$ we get

$$\|u - u_0\|_{L_2([0,T]\times\Gamma)} \geq \delta\|\alpha - \alpha_0\|_{C^2} \tag{4.11}$$

We can interpret inequality (4.11) as a stability estimate. For the simple case $\Omega = [0,1]$ we find several stability estimates in Rösch [11]. An essential point in this paper is the discussion under which assumptions estimates of the more general type

$$\|u-u_0\| \geq c\|\alpha-\alpha_0\|^\nu \tag{4.12}$$

can be hold. First of all the norm of the difference of the controls has to be evaluated only on a smaller set as $[\vartheta_1, \vartheta_2]$ the so-called reference set. Variations outside of this set have no influence of the state. This fact compiles several investigation but it is not a crucial point. We only recall here the definition of reference sets.

Definition: Let $\alpha \in U_{ad}$ and u the corresponding solution of (1.2).
The set $M := \{\bar{u} : \bar{u} = u(t,x),\ t \in [0,T]\ x \in \Gamma\}$ is called *reference set* of α.

In the illustrative example and the concluding remark of this paper we can see that for a very simple example we can not expect Lipschitz type estimates

$$\|u-u_0\| \geq c\|\alpha-\alpha_0\| \tag{4.13}$$

with the same norms for u and α (for instance C or L_2).

Inequality (4.11) requires much more than that namely a Lipschitz estimate of the L_2-norm of u with respect to the C^2-norm of α. May be it is possible to weak the norm on the right-hand side of inequality (4.11). To ensure that all terms of the second Fréchet derivative make sense we must necessarily require that the increment $\alpha-\alpha_0$ belongs to $C^{0,1}$. Thus the L_2-norm or a weaker norm for $\alpha-\alpha_0$ is unimaginable.

Summarizing these deliberations, we notice that already for simply cases neither the sufficient estimates of type (4.5) nor sufficient estimates of type (4.6) hold. Nevertheless, the investigation of the second derivative delivers useful results. First of all we have the necessary second order condition. Second we are able to construct second order descent algorithm. In this context second order means using second order information not quadratical convergence.

In this context it seems to be a way out to use an additional Tikhonov-term. But this way has disadvantages. Because of the dependence of the temperature of α we have to discuss what means $\|\alpha\|^2_{L_2}$. If we choose the norm over the reference set (for instance $L_2(M)$) then we get the problem that every α generates its own reference set. For that reason the norm is not differentiable. The use of $\|\alpha(u)\|_{L_2((0,T)\times\Gamma)}$ geneates additional differentiation term because of the dependence from α. Possibilities to overcome this problems could be the use of $\|\alpha(q)\|_{L_2((0,T)\times\Gamma)}$ or an iterative Tikhonov regularization $\|\alpha(u_n)\|_{L_2((0,T)\times\Gamma)}$.

The investigations of this identification problems shows that for such problems it is necessary to find new ways. It is imaginable that the function of the sufficient second order condition can be fulfilled by a stability estimate of the type (4.12). In [11] we find several estimates of this form for the one-dimensional case. For instance it is possible to get the inequality

$$\|u-u_0\|_{L_2} \geq c\|\alpha-\alpha_0\|^\nu_{C(M)} \tag{4.14}$$

where M denotes the reference set of α_0. By means of such properties and the knowledge of the second order information it should be possible to construct numerical algorithms and to prove convergence rates. These are the goals of the future. The investigation of this papers show that the standard way can not be gone for this type of identification problems.

References

1. H. Amann: Parabolic evolution equations with nonlinear boundary conditions. In *Proc. Sympos. Pure Math., Vol. 45, Part I "Nonlinear Functional Analysis"*, pages 17–27. F.E. Browder, Ed., 1986.
2. H. Amann: Parabolic evolution equations with nonlinear boundary conditions. *J. Diff. Equations*, 72:201–269, 1988.
3. A. Friedman: Optimal control for parabolic equations. *J. Math. Anal. Appl.*, 18:479–491, 1967.
4. H. Goldberg and F. Tröltzsch: Second order optimality conditions for nonlinear parabolic boundary control problems. In *Lecture Notes Contr. Inf. Sci.*, pages 93–103. Proc. Int. Conference on optimal control of partial differential equations, 1991.
5. H. Goldberg and F. Tröltzsch: Second order sufficient optimality conditions for a class of non–linear parabolic boundary control problems. *SIAM J. Contr. Opt.*, 31(4):1007–1025, 1993.
6. A. D. Ioffe: Necessary and sufficient conditions for a local minimum 3: Second order conditions and augmented duality. *SIAM J. Control Opt.*, 17:266–288, 1979.
7. H. Maurer: First and second order sufficient optimality conditions in mathematical programming and optimal control. *Math. Programming Study*, 14:163–177, 1981.
8. A. Rösch: Identification of nonlinear heat transfer laws by optimal control. *Num. Funct. Analysis and Optimization*, 15(3&4):417–434, 1994.
9. A. Rösch: Fréchet differentiability of the solution of the heat equation with respect to a nonlinear boundary condition. *Z. Anal. u. Anw.*, 15(3):603–618, 1996.
10. A. Rösch: Identification of nonlinear heat transfer laws by means of boundary data. In *Progress in Industry (at ECMI 94)*, pages 405–412. Wiley–Teubner, 1996.
11. A. Rösch: Stability estimates for the identification of nonlinear heat transfer laws. *Inverse Problems*, 12:743–756, 1996.
12. A. Rösch and F. Tröltzsch: An optimal control problem arising from the identification of nonlinear heat transfer laws. *Archives of Control Sciences*, 1(3–4):183–195, 1992.
13. A. N. Tikhonov and V. Y. Arsenin: *Methods of Solution for Ill-posed Problems (in Russian)*. Nauka, Moscow, 1974.
14. A. N. Tikhonov, A. V. Goncharskij, V. V. Stepanov, and A. G. Yagola: *Regularization algorithm and a-priori information (in Russian)*. Nauka, Moscow, 1983.
15. H Triebel: *Interpolation theory, function spaces, differential operators*. J. A. Barth Verlag, Heidelberg-Leipzig, 1995.

Arnd Rösch
Department of Mathematics
TU Chemnitz–Zwickau
D-09107 Chemnitz, Germany

International Series of Numerical Mathematics
Vol. 126, © 1998 Birkhäuser Verlag, Basel

LQR Control of Shell Vibrations via Piezoceramic Actuators*

R.C.H. DEL ROSARIO AND R.C. SMITH

Center for Research in Scientific Computation
North Carolina State University

Department of Mathematics
Iowa State University

ABSTRACT. A model-based LQR method for controlling vibrations in cylindrical shells is presented. Surface-mounted piezoceramic patches are employed as actuators which leads to unbounded control input operators. Modified Donnell-Mushtari shell equations incorporating strong or Kelvin-Voigt damping are used to model the system. The model is then abstractly formulated in terms of sesquilinear forms. This provides a framework amenable for proving model well-posedness and convergence of LQR gains using analytic semigroup results combined with LQR theory for unbounded input operators. Finally, numerical examples demonstrating the effectiveness of the method are presented.

1991 *Mathematics Subject Classification.* Primary 49J20, 73K50; Secondary 93C95

Key words and phrases. LQR feedback control, thin shell model, Galerkin approximation.

1. Introduction

The use of shell models to describe structural dynamics is pervasive in applications ranging from noise reduction in aircraft to flow control in flexible pipes. While general shell equations can be used in a variety of geometries, they all share the property that component displacements are coupled due to the geometry. This leads to significant challenges when developing appropriate models and approximation techniques, and constructing effective controllers.

In this paper, we consider cylindrical shells due to their prevalence in applications. Control is provided by piezoceramic patches bonded in pairs to the surface of the shell. These transducers provide significant actuating capabilities due to the piezoelectric effect in which input voltages generate strains in the patches. Utilization of the converse piezoelectric effect (strains produce voltages) also allows the patches to be employed as sensors. When combined with their light weight, space efficiency and reasonable cost, these properties make the patches highly effective control elements in a variety of applications. From a mathematical perspective, the use of surface-mounted piezoceramic patches leads to unbounded control input operators.

*This research was supported in part by the National Aeronautics and Space Administration under NASA Contract Number NAS1-19480 while RCS was a visiting scientist at the Institute for Computer Applications in Science and Engineering (ICASE), NASA Langley Research Center, Hampton, VA 23681. Additional support was also provided in part under NASA grant NAG-1-1600.

Experimental work has already demonstrated the potential for success when employing the patches as actuators in applications involving cylindrical shells [8, 13]. However, these initial investigations have not, in general, utilized the full potential of the patches due to limitations in hardware, models, approximation techniques and control laws. For example, a common means of approximating shell dynamics to yield systems which are ultimately used to calculate control gains is through modal expansions [10]. However, closed form expressions for the modes can be determined only for a limited set of models with severely restrictive boundary conditions. The use of incorrect modes when calculating control gains can lead to loss of control authority and possible controller instabilities.

In this paper, we present a model-based method for controlling shell vibrations. For simplicity, the Donnell-Mushtari shells equations with Kelvin-Voigt damping are used as a model (the assumption of strong or Kelvin-Voigt damping is reasonable and typical for many shell materials such as aluminum). The methods are general, however, and can be applied to higher-order models (e.g., Byrne-Flügge-Lur'ye model) if the application warrants. A general Galerkin method based on splines is then used to discretize the infinite dimensional system. Through the choice of basis, the method is significantly more flexible than general modal methods when considering the boundary conditions and material nonhomogeneities which arise in typical applications.

The model and approximate system are then employed in an LQR full state feedback theory to obtain feedback gains and, ultimately, controlling voltages to the patches. While full state measurements are not available using current instrumentation, and hence the techniques cannot directly be implemented in experiments, they provide an important first step in the design of effective compensators based on state estimates calculated using a limited number of observations (see [5]). The consideration of the LQR performance also illustrates properties of the system and model-based control techniques and facilitates investigations regarding issues such as patch number and configuration. Finally, the consideration of the problem provides a step toward the development of model-based controllers for fully coupled structural acoustic and fluid/structure systems involving cylindrical shells.

The strong and weak forms of the Donnell-Mushtari shells equations are outlined in Section 2. In presenting this model, care is taken to include both passive (material) and active (actuator) contributions due to the patches. An abstract form of the model, based on sesquilinear forms, is also presented. This provides a natural setting to prove model well-posedness and convergence properties of the LQR control law. LQR full state feedback laws for systems with no exogenous force or forces which are periodic in time are presented in Section 3. In the former case, convergence of the approximate suboptimal gains to the optimal gains for the infinite dimensional system is proven using analytic semigroup theory in combination with LQR results for unbounded control input operators. A Fourier-Galerkin method for approximating the system dynamics is outlined in Section 4, and the effectiveness of the LQR method for periodic forces is demonstrated through a numerical example in Section 5. This example demonstrates that through the use of the model-based methodology with general Galerkin approximations, significant attenuation in shell vibrations can be obtained using piezoceramic patches.

2. PDE Model

The system under consideration consists of a thin cylindrical shell with surface-mounted piezoceramic patches. It is assumed that the patches are mounted in pairs with edges aligned with the circumferential and longitudinal axes of the shell. The edges of the shell are taken to be fixed in accordance with common experimental clamping techniques.

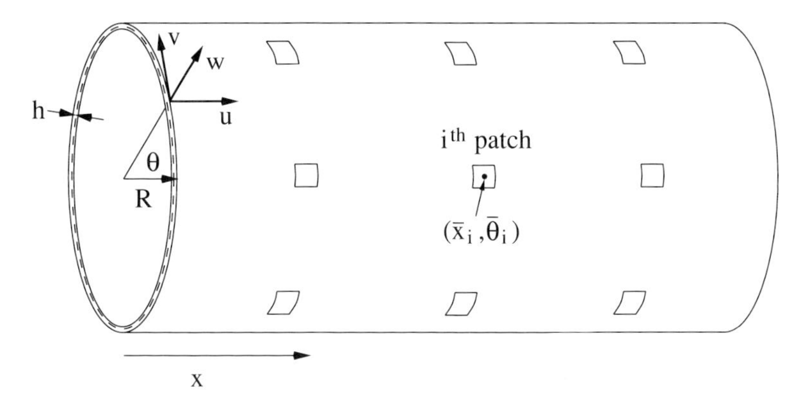

Figure 1. Thin cylindrical shell with surface mounted piezoceramic patches.

To specify the geometry for the corresponding model, we consider the longitudinal direction to be aligned along the x-axis as depicted in Figure 1. The displacements of the middle surface in the longitudinal, circumferential and transverse directions are denoted by u, v and w, respectively while the length, thickness and radius of the shell are denoted by ℓ, h, R. The region occupied by the middle surface is denoted by Γ_0. Finally, the shell is assumed to have mass density ρ, Young's modulus E, Poisson ratio ν, Kelvin-Voigt damping coefficient c_D and air damping coefficient μ.

Actuator and/or sensor capabilities are provided by s pairs of surface-mounted piezoceramic patches. It is assumed that all the patches have thickness h_{pe}, Young's modulus E_{pe}, Poisson ratio ν_{pe} and Kelvin-Voigt damping coefficient $c_{D_{pe}}$. Furthermore, it is assumed that the glue bonding layer provides negligible contribution to the structural dynamics (the reader is referred to [3, 5] for details concerning the incorporation of differing patch characteristics and bonding layers in the ensuing models). The region covered by the ith patch pair, with edges at x_{1i}, x_{2i}, θ_{1i}, θ_{2i}, is delineated by the characteristic function

$$\chi_{pe_i}(x,\theta) = \begin{cases} 1\ , & x_{1i} \le x \le x_{2i}\ ,\ \theta_{1i} \le \theta \le \theta_{2i} \\ 0\ , & \text{otherwise}\ . \end{cases}$$

The indicator function $\mathcal{S}_{pe_i}(x,\theta) \equiv S_{1,2}(x)\hat{S}_{1,2}(\theta)$, where

$$S_{1,2}(x) = \begin{cases} 1 & ,\ x < (x_{1i}+x_{2i})/2 \\ 0 & ,\ x = (x_{1i}+x_{2i})/2 \\ -1 & ,\ x > (x_{1i}+x_{2i})/2 \end{cases} \quad , \quad \hat{S}_{1,2}(\theta) = \begin{cases} 1 & ,\ \theta < (\theta_{1i}+\theta_{2i})/2 \\ 0 & ,\ \theta = (\theta_{1i}+\theta_{2i})/2 \\ -1 & ,\ \theta > (\theta_{1i}+\theta_{2i})/2 \end{cases}$$

delineates the sense of the forces generated by the ith pair. The symmetry of the function arises from the property that for homogeneous patches having uniform thickness, equal but opposite strains are generated about the point $\left(\bar{x}_i, \bar{\theta}_i\right) = ((x_{1i}+x_{2i})/2, (\theta_{1i}+\theta_{2i})/2)$.

2.1. Strong Form of the Modeling Equation. We consider here the modified Donnell-Mushtari equations

$$
\begin{aligned}
&R\rho h\frac{\partial^2 u}{\partial t^2} - R\frac{\partial N_x}{\partial x} - \frac{\partial N_{\theta x}}{\partial \theta} = R\hat{q}_x - R\sum_{i=1}^{s}\frac{\partial (N_x)_{pe_i}}{\partial x}\mathcal{S}_{pe_i}(x,\theta)\\
&R\rho h\frac{\partial^2 v}{\partial t^2} - \frac{\partial N_\theta}{\partial \theta} - R\frac{N_{x\theta}}{\partial x} = R\hat{q}_\theta - \sum_{i=1}^{s}\frac{\partial (N_\theta)_{pe_i}}{\partial \theta}\mathcal{S}_{pe_i}(x,\theta)\\
&R\rho h\frac{\partial^2 w}{\partial t^2} + R\mu\frac{\partial w}{\partial t} - R\frac{\partial^2 M_x}{\partial x^2} - \frac{1}{R}\frac{\partial^2 M_\theta}{\partial \theta^2} - 2\frac{\partial^2 M_{x\theta}}{\partial x \partial \theta} + N_\theta\\
&\qquad = R\hat{q}_n - \sum_{i=1}^{s}\left[R\frac{\partial^2 (M_x)_{pe_i}}{\partial x^2} + \frac{1}{R}\frac{\partial^2 (M_\theta)_{pe_i}}{\partial \theta^2}\right]
\end{aligned}
\tag{2.1}
$$

as a model for the thin shell dynamics. As detailed in [3, 5, 9], these equations are obtained through force and moment balancing with only low order terms retained. Here $M_x, M_\theta, M_{\theta x}$ and $M_{x\theta}$ are internal moments while $N_x, N_\theta, N_{\theta x}$ and $N_{x\theta}$ denote internal force resultants. External surface forces are denoted by $\hat{q}_x, \hat{q}_\theta, \hat{q}_n$ whereas the external resultants (line moments and forces) generated by the ith patch pair are designated by $(M_x)_{pe_i}, (M_\theta)_{pe_i}, (N_x)_{pe_i}, (N_\theta)_{pe_i}$.

Expressions for the internal force and moment resultants are derived under the assumption that stress is proportional to a linear combination of strain and strain rate. This yields a model which incorporates Kelvin-Voigt or strong internal damping. As detailed in [3, 5], the resultants $N_x, N_{x\theta}, N_{\theta x}, M_x, M_{x\theta}, M_{\theta x}$ derived under this assumption are

$$
\begin{aligned}
N_x &= \frac{Eh}{1-\nu^2}\left(\varepsilon_x + \nu\varepsilon_\theta\right) + \sum_{i=1}^{s}\frac{2E_{pe}h_{pe}}{1-\nu_{pe}^2}\left(\varepsilon_x + \nu_{pe}\varepsilon_\theta\right)\chi_{pe_i}(x,\theta)\\
&\quad + \frac{c_D h}{1-\nu^2}\left(\dot{\varepsilon}_x + \nu\dot{\varepsilon}_\theta\right) + \sum_{i=1}^{s}\frac{2c_{D_{pe}}h_{pe}}{1-\nu_{pe}^2}\left(\dot{\varepsilon}_x + \nu_{pe}\dot{\varepsilon}_\theta\right)\chi_{pe_i}(x,\theta)\\
N_{x\theta} &= N_{\theta x} = \frac{Eh}{2(1+\nu)}\varepsilon_{x\theta} + \sum_{i=1}^{s}\frac{E_{pe}h_{pe}}{(1+\nu_{pe})}\varepsilon_{x\theta}\chi_{pe_i}(x,\theta)\\
&\quad + \frac{c_D h}{2(1+\nu)}\dot{\varepsilon}_{x\theta} + \sum_{i=1}^{s}\frac{c_{D_{pe}}h_{pe}}{(1+\nu_{pe})}\dot{\varepsilon}_{x\theta}\chi_{pe_i}(x,\theta)\\
M_x &= \frac{Eh^3}{12(1-\nu^2)}\left(\kappa_x + \nu\kappa_\theta\right) + \sum_{i=1}^{s}\frac{2E_{pe}a_3}{3(1-\nu_{pe}^2)}\left(\kappa_x + \nu_{pe}\kappa_\theta\right)\chi_{pe_i}(x,\theta)\\
&\quad + \frac{c_D h^3}{12(1-\nu^2)}\left(\dot{\kappa}_x + \nu\dot{\kappa}_\theta\right) + \sum_{i=1}^{s}\frac{2c_{D_{pe}}a_3}{3(1-\nu_{pe}^2)}\left(\dot{\kappa}_x + \nu_{pe}\dot{\kappa}_\theta\right)\chi_{pe_i}(x,\theta)
\end{aligned}
\tag{2.2}
$$

$$M_{x\theta} = M_{\theta x} = \frac{Eh^3}{24(1+\nu)}\tau + \sum_{i=1}^{s} \frac{E_{pe}a_3}{3(1+\nu_{pe})}\tau\chi_{pe_i}(x,\theta) + \frac{c_D h^3}{24(1+\nu)}\dot{\tau} + \sum_{i=1}^{s} \frac{c_{D_{pe}}a_3}{3(1+\nu_{pe})}\dot{\tau}\chi_{pe_i}(x,\theta)$$

where the constant $a_3 \equiv (h/2 + h_{pe})^3 - h^3/8$ results from integration through the thickness of the patch. Expressions for the resultants N_θ and M_θ can be obtained by replacing $\varepsilon_x, \varepsilon_\theta, \kappa_x, \kappa_\theta$ in the expressions for N_x and M_x by $\varepsilon_\theta, \varepsilon_x, \kappa_\theta, \kappa_x$, respectively. The midsurface strains and changes in curvature for the Donnell-Mushtari model are

$$\begin{aligned} &\varepsilon_x = \frac{\partial u}{\partial x} \quad , \quad \varepsilon_\theta = \frac{1}{R}\frac{\partial v}{\partial \theta} + \frac{w}{R} \quad , \quad \varepsilon_{x\theta} = \frac{\partial v}{\partial x} + \frac{1}{R}\frac{\partial u}{\partial \theta} \\ &\kappa_x = -\frac{\partial^2 w}{\partial x^2} \quad , \quad \kappa_\theta = -\frac{1}{R^2}\frac{\partial^2 w}{\partial \theta^2} \quad , \quad \tau = -\frac{2}{R}\frac{\partial^2 w}{\partial x \partial \theta} \, . \end{aligned} \tag{2.3}$$

Note that for the undamped shell which is devoid of patches, the resultant equations (2.2) reduce to the classical Donnell-Mushtari expressions

$$N_x = \frac{Eh}{(1-\nu^2)}\left[\frac{\partial u}{\partial x} + \frac{\nu}{R}\left(\frac{\partial v}{\partial \theta} + w\right)\right] , \quad M_x = \frac{-Eh^3}{12(1-\nu^2)}\left[\frac{\partial^2 w}{\partial x^2} + \frac{\nu}{R^2}\frac{\partial^2 w}{\partial \theta^2}\right]$$

$$N_\theta = \frac{Eh}{(1-\nu^2)}\left[\frac{1}{R}\frac{\partial v}{\partial \theta} + \frac{w}{R} + \nu\frac{\partial u}{\partial x}\right] , \quad M_\theta = \frac{-Eh^3}{12(1-\nu^2)}\left[\frac{1}{R^2}\frac{\partial^2 w}{\partial \theta^2} + \nu\frac{\partial^2 w}{\partial x^2}\right]$$

$$N_{x\theta} = N_{\theta x} = \frac{Eh}{2(1+\nu)}\left[\frac{\partial v}{\partial x} + \frac{1}{R}\frac{\partial u}{\partial \theta}\right] , \quad M_{x\theta} = M_{\theta x} = \frac{-Eh^3}{12R(1+\nu)}\frac{\partial^2 w}{\partial x \partial \theta}$$

(e.g., see [9]).

To characterize the external or active patch contributions, it is typical to start with the assumption that the strains generated by a patch are proportional to the applied voltage [3]. Since differing voltages can be applied to the outer and inner patches in the pair, we will differentiate between the two with $V_{i1}(t)$ and $V_{i2}(t)$ used to denote the voltages to the outer and inner patches in the ith pair, respectively. The proportionality constant relating the generated strain to the input voltage is designated by d_{31}. As detailed in [3], the total external moments and forces generated by the patches are

$$\begin{aligned} (M_x)_{pe_i} &= \frac{-E_{pe}}{1-\nu_{pe}} \cdot \frac{d_{31}}{h_{pe}}\chi_{pe_i}(x,\theta)\left[\left(\frac{a_2}{2} + \frac{a_3}{3R}\right)V_{i1} - \left(\frac{a_2}{2} - \frac{a_3}{3R}\right)V_{i2}\right] \\ (M_\theta)_{pe_i} &= \frac{-E_{pe}}{1-\nu_{pe}} \cdot \frac{d_{31}a_2}{2h_{pe}}\chi_{pe_i}(x,\theta)[V_{i1} - V_{i2}] \\ (N_x)_{pe_i} &= \frac{-E_{pe}}{1-\nu_{pe}} \cdot \frac{d_{31}}{h_{pe}}\chi_{pe_i}(x,\theta)\mathcal{S}_{pe_i}(x,\theta)\left[\left(h_{pe} + \frac{a_2}{2R}\right)V_{i1} + \left(h_{pe} - \frac{a_2}{2R}\right)V_{i2}\right] \\ (N_\theta)_{pe_i} &= \frac{-E_{pe}}{1-\nu_{pe}} d_{31}\chi_{pe_i}(x,\theta)\mathcal{S}_{pe_i}(x,\theta)[V_{i1} - V_{i2}] \end{aligned} \tag{2.4}$$

where $a_2 = (h/2 + h_{pe})^2 - h^2/4$ and $a_3 = (h/2 + h_{pe})^3 - h^3/8$. When substituted into (2.1), the expressions (2.4) provide the input from the patches when voltages are applied.

Finally, the fixed-edge boundary conditions

$$u = v = w = \frac{dw}{\partial x} = 0 \quad , \quad x = 0, \ell \tag{2.5}$$

are used to model the end behavior of the shell. These boundary conditions are appropriate for experimental setups in which heavy endcaps prevent edge movement. Note that alternative boundary conditions such as simply supported or "almost fixed" (see [4]) can be employed if edge movement is suspected.

2.2. Weak Form of Modeling Equations. The strong form (2.1) of the modeling equations requires first and second derivatives of the moment and force resultants. As noted in (2.2) and (2.4), both the internal and external moment and force resultants are discontinuous due to the piezoceramic patches. Hence formal analysis and approximation using the strong form of the modeling equations lead to difficulties due to differentiation of Dirac distributions.

To alleviate these difficulties, it is advantageous to consider a weak form of the modeling equations which can be derived from Hamilton's principle (energy considerations). While equivalent to the strong form under suitable smoothness assumptions, the weak form provides a more natural setting for analysis and approximation.

The state variables for the problem in second-order form are taken to be $y = (u, v, w)$ in the state space $H = L^2(\Gamma_0) \times L^2(\Gamma_0) \times L^2(\Gamma_0)$. For the fixed-edge boundary conditions (2.5), the space of test functions is taken to be $V = H_0^1(\Gamma_0) \times H_0^1(\Gamma_0) \times H_0^2(\Gamma_0)$ where

$$\begin{aligned} H_0^1(\Gamma_0) &= \{\eta \in H^1(\Gamma_0) \,|\, \eta(0) = \eta(\ell) = 0\} \\ H_0^2(\Gamma_0) &= \{\eta \in H^2(\Gamma_0) \,|\, \eta(0) = \eta_x(0) = \eta(\ell) = \eta_x(\ell) = 0\}\,. \end{aligned}$$

For $\Phi = (u, v, w)$ and $\Psi = (\eta_1, \eta_2, \eta_3)$, the H and V inner products are taken to be

$$\langle \Phi, \Psi \rangle_H = \int_{\Gamma_0} \rho h u \overline{\eta}_1 d\gamma + \int_{\Gamma_0} \rho h v \overline{\eta}_2 d\gamma + \int_{\Gamma_0} \rho h w \overline{\eta}_3 d\gamma$$

and

$$\begin{aligned} \langle (E, E_{pe})\Phi, \Psi \rangle_V \;=\; & \int_{\Gamma_0} \left\{ \frac{Eh}{1-\nu^2} \left[(\varepsilon_x + \nu\varepsilon_\theta) \overline{\frac{\partial \eta_1}{\partial x}} + \frac{1}{2R}(1-\nu)\varepsilon_{x\theta} \overline{\frac{\partial \eta_1}{\partial \theta}} \right] \right. \\ & \left. + \sum_{i=1}^{s} \frac{2E_{pe}h_{pe}}{1-\nu_{pe}^2} \chi_{pe_i}(x,\theta) \left[(\varepsilon_x + \nu_{pe}\varepsilon_\theta) \overline{\frac{\partial \eta_1}{\partial x}} + \frac{1}{2R}(1-\nu_{pe})\varepsilon_{x\theta} \overline{\frac{\partial \eta_1}{\partial \theta}} \right] \right\} d\gamma \\ + & \int_{\Gamma_0} \left\{ \frac{Eh}{1-\nu^2} \left[(\varepsilon_\theta + \nu\varepsilon_x) \overline{\frac{\partial \eta_1}{\partial \theta}} + \frac{1}{2R}(1-\nu)\varepsilon_{x\theta} \overline{\frac{\partial \eta_2}{\partial x}} \right] \right. \\ & \left. + \sum_{i=1}^{s} \frac{2E_{pe}h_{pe}}{1-\nu_{pe}^2} \chi_{pe_i}(x,\theta) \left[(\varepsilon_\theta + \nu_{pe}\varepsilon_x) \overline{\frac{\partial \eta_2}{\partial \theta}} + \frac{1}{2R}(1-\nu_{pe})\varepsilon_{x\theta} \overline{\frac{\partial \eta_2}{\partial x}} \right] \right\} d\gamma \end{aligned}$$

$$
\begin{aligned}
&+ \int_{\Gamma_0} \left\{ \frac{Eh}{1-\nu^2} \left[\frac{1}{R}(\varepsilon_\theta + \nu\varepsilon_x)\overline{\eta_3} - \frac{h^2}{12}(\kappa_x + \nu\kappa_\theta)\overline{\frac{\partial^2 \eta_3}{\partial x^2}} \right.\right. \\
&\qquad \left. - \frac{h^2}{12R^2}(\kappa_\theta + \nu\kappa_x)\overline{\frac{\partial^2 \eta_3}{\partial \theta^2}} - \frac{h^2}{12R}(1-\nu)\tau\overline{\frac{\partial^2 \eta_3}{\partial x \partial \theta}} \right] \\
&\qquad + \sum_{i=1}^{s} \frac{2E_{pe}}{3(1-\nu_{pe}^2)}\chi_{pe_i}(x,\theta) \left[\frac{3h_{pe}}{R}(\varepsilon_\theta + \nu_{pe}\varepsilon_x)\overline{\eta_3} - a_3(\kappa_x + \nu_{pe}\kappa_\theta)\overline{\frac{\partial^2 \eta_3}{\partial x^2}} \right. \\
&\qquad \left.\left. - \frac{a_3}{R^2}(\kappa_\theta + \nu_{pe}\kappa_x)\overline{\frac{\partial^2 \eta_3}{\partial \theta^2}} - \frac{a_3}{R}(1-\nu_{pe})\tau\overline{\frac{\partial^2 \eta_3}{\partial x \partial \theta}} \right] \right\} d\gamma
\end{aligned}
$$

where $\varepsilon_x, \varepsilon_\theta, \varepsilon_{x\theta}, \kappa_x, \kappa_\theta, \tau$ are defined in (2.3) and $d\gamma = Rd\theta dx$. The dependence of the inner product on the Young's moduli is explicitly included in the definition to provide a notation for defining analogous damping expressions later in this work.

The weak form of (2.1), as derived in [5] from energy principles, is given by

$$
\begin{aligned}
&\int_{\Gamma_0} \left\{ R\rho h \frac{\partial^2 u}{\partial t^2}\overline{\eta_1} + RN_x\overline{\frac{\partial \eta_1}{\partial x}} + N_{\theta x}\overline{\frac{\partial \eta_1}{\partial \theta}} - R\hat{q}_x\overline{\eta_1} - R\sum_{i=1}^{s}(N_x)_{pe_i}\overline{\frac{\partial \eta_1}{\partial x}} \right\} d\gamma = 0 \\
&\int_{\Gamma_0} \left\{ R\rho h \frac{\partial^2 v}{\partial t^2}\overline{\eta_2} + N_\theta\overline{\frac{\partial \eta_2}{\partial \theta}} + RN_{x\theta}\overline{\frac{\partial \eta_2}{\partial x}} - R\hat{q}_\theta\overline{\eta_2} - \sum_{i=1}^{s}(N_\theta)_{pe_i}\overline{\frac{\partial \eta_2}{\partial \theta}} \right\} d\gamma = 0 \\
&\int_{\Gamma_0} \left\{ R\rho h \frac{\partial^2 w}{\partial t^2}\overline{\eta_3} + R\mu\frac{\partial w}{\partial t}\overline{\eta_3} + N_\theta\overline{\eta_3} - RM_x\overline{\frac{\partial^2 \eta_3}{\partial x^2}} - \frac{1}{R}M_\theta\overline{\frac{\partial^2 \eta_3}{\partial \theta^2}} \right. \\
&\qquad \left. -2M_{x\theta}\overline{\frac{\partial^2 \eta_3}{\partial x \partial \theta}} - R\hat{q}_n\overline{\eta_3} + \sum_{i=1}^{s}\left[R(M_x)_{pe_i}\overline{\frac{\partial^2 \eta_3}{\partial x^2}} + \frac{1}{R}(M_\theta)_{pe_i}\overline{\frac{\partial^2 \eta_3}{\partial \theta^2}} \right] \right\} d\gamma = 0
\end{aligned}
\tag{2.6}
$$

for all $\Psi = (\eta_1, \eta_2, \eta_3) \in V$. A comparison between (2.6) and (2.1) illustrates that in the weak form, derivatives are transferred from the discontinuous resultants onto suitably smooth test functions. This alleviates the difficulties associated with the discontinuities and reduces smoothness requirements on approximate solutions.

2.3. Abstract Formulation. To define appropriate sesquilinear forms, we group stiffness components separately from damping components. To this end, we define $\sigma_i : V \times V \to \mathbb{C}$, $i = 1, 2$ by

$$
\begin{aligned}
\sigma_1(\Phi, \Psi) &= \langle (E, E_{pe})\Phi, \Psi \rangle_V \\
\sigma_2(\Phi, \Psi) &= \Big\langle (c_D, c_{D_{pe}})\Phi, \Psi \Big\rangle_V + \int_{\Gamma_0} \mu w \overline{\eta}_3 d\gamma \,.
\end{aligned}
\tag{2.7}
$$

Note that $\langle (c_D, c_{D_{pe}})\Phi, \Psi \rangle_V$ differs from $\langle (E, E_{pe})\Phi, \Psi \rangle_V$ only in that Young's moduli are replaced by Kelvin-Voigt damping coefficients. It can be directly verified that the stiffness form σ_1 satisfies

(H1) $|\sigma_1(\Phi,\Psi)| \leq c_1|\Phi|_V|\Psi|_V$, for some $c_1 \in \mathbb{R}$ (Bounded)

(H2) $\mathrm{Re}\,\sigma_1(\Phi,\Phi) \geq c_2|\Phi|_V^2$, for some $c_2 > 0$ (V-Elliptic)

(H3) $\sigma_1(\Phi,\Psi) = \overline{\sigma_1(\Psi,\Phi)}$ (Symmetric)

for all $\Phi, \Psi \in V$. Moreover, the damping term σ_2 satisfies

(H4) $|\sigma_2(\Phi,\Psi)| \leq c_3|\Phi|_V|\Psi|_V$, for some $c_3 \in \mathbb{R}$ (Bounded)

(H5) $\mathrm{Re}\,\sigma_2(\Phi,\Phi) \geq c_4|\Phi|_V^2$, for some $c_4 > 0$ (V-Elliptic).

Remark 1. The symmetry of σ_1 is dependent upon the choice of shell model and ultimately reflects the Maxwell-Betti reciprocity theorem. While the Donnell-Mushtari model yields a symmetric sesquilinear form σ_1, other models such as the Timoshenko shell model will not yield a symmetric form.

To represent control contributions, let $U = \mathbb{R}^s$ denote the Hilbert space of control inputs and define $B \in \mathcal{L}(U, V^*)$ by

$$\langle Bu(t), \Psi \rangle_{V^*,V} = \int_{\Gamma_0} \sum_{i=1}^{s} \left\{ (N_x)_{pe_i} \overline{\frac{\partial \eta_1}{\partial x}} + \frac{1}{R}(N_\theta)_{pe_i} \overline{\frac{\partial \eta_2}{\partial \theta}} \right.$$
$$\left. -(M_x)_{pe_i} \overline{\frac{\partial^2 \eta_3}{\partial x^2}} - \frac{1}{R^2}(M_\theta)_{pe_i} \overline{\frac{\partial^2 \eta_3}{\partial \theta^2}} \right\} d\gamma$$

for $\Psi \in V$. Here $\langle \cdot, \cdot \rangle_{V^*,V}$ denotes the usual duality product. Finally, with the definition $\tilde{g} = (1/\rho h)[\hat{q}_x, \hat{q}_\theta, \hat{q}_n]$, we can write the weak form (2.6) in the abstract variational form

$$\langle \ddot{y}(t), \Psi \rangle_{V^*,V} + \sigma_2(\dot{y}(t), \Psi) + \sigma_1(y(t), \Psi) = \langle Bu(t) + \tilde{g}(t), \Psi \rangle_{V^*,V} . \tag{2.8}$$

To pose the problem in a first-order form amenable for control applications, we define the product spaces $\mathcal{H} = V \times H$ and $\mathcal{V} = V \times V$ with the norms

$$|(\phi_1, \phi_2)|_{\mathcal{H}}^2 = |\phi_1|_V^2 + |\phi_2|_H^2$$
$$|(\phi_1, \phi_2)|_{\mathcal{V}}^2 = |\phi_1|_V^2 + |\phi_2|_V^2 .$$

The state is taken to be $z(t) = (y(t), \dot{y}(t)) \in \mathcal{H}$. Finally, the product space forcing terms are formulated as

$$g(t) = \begin{bmatrix} 0 \\ \tilde{g}(t) \end{bmatrix} \quad , \quad \mathcal{B}u(t) = \begin{bmatrix} 0 \\ Bu(t) \end{bmatrix} . \tag{2.9}$$

The weak form (2.8) can then be rewritten as

$$\begin{aligned} &\langle \dot{z}(t), \Lambda \rangle_{\mathcal{V}^*,\mathcal{V}} + \sigma(z(t), \Lambda) = \langle \mathcal{B}u(t) + g(t), \Lambda \rangle_{\mathcal{V}^*,\mathcal{V}} \quad \text{for } \Lambda \in \mathcal{V} \\ &z(0) = z_0 = (y_0, y_1) \end{aligned} \tag{2.10}$$

where $\sigma : \mathcal{V} \times \mathcal{V} \to \mathbb{C}$ is given by

$$\sigma(\phi, \psi) = - \langle \phi_2, \psi_1 \rangle_V + \sigma_1(\phi_1, \psi_2) + \sigma_2(\phi_2, \psi_2)$$

for $\phi = (\phi_1, \phi_2), \psi = (\psi_1, \psi_2) \in \mathcal{V}$. As proven in [5, page 109], σ is $\mathcal{V}$ continuous and for $\lambda > 0$, $\sigma(\cdot, \cdot) + \lambda \langle \cdot, \cdot \rangle_{\mathcal{H}}$ is $\mathcal{V}$-elliptic. From the continuity of σ, it follows that one can define an operator $\widetilde{\mathcal{A}} \in \mathcal{L}(\mathcal{V}, \mathcal{V}^*)$ by $\sigma(\Upsilon, \Lambda) = \langle \widetilde{\mathcal{A}}\Upsilon, \Lambda \rangle_{\mathcal{V}^*,\mathcal{V}}$.

To obtain a strong form of the first-order system which is appropriate for control purposes, consider the system operator

$$\text{dom}\mathcal{A} = \{(\phi_1, \phi_2) \in \mathcal{H} | \phi_2 \in V, A_1\phi_1 + A_2\phi_2 \in H\}$$

$$\mathcal{A} = \begin{bmatrix} 0 & I \\ -A_1 & -A_2 \end{bmatrix} \tag{2.11}$$

with $A_1, A_2 \in \mathcal{L}(V, V^*)$ defined by

$$\langle A_i\phi_1, \phi_2 \rangle_{V^*,V} = \sigma_i(\phi_1, \phi_2) \quad , \quad i = 1, 2 .$$

It should be notated that $\mathcal{A}$ is the negative of the restriction to dom$\mathcal{A}$ of $\widetilde{\mathcal{A}} \in \mathcal{L}(\mathcal{V}, \mathcal{V}^*)$ so that $\sigma(\Upsilon, \Lambda) = \langle -\mathcal{A}\Upsilon, \Lambda \rangle_{\mathcal{H}}$ for $\Upsilon \in \text{dom}\mathcal{A}, \Lambda \in \mathcal{V}$.

A strong form of the abstract system model is then given by

$$\begin{aligned} \dot{z}(t) &= \mathcal{A}z(t) + \mathcal{B}u(t) + g(t) \quad \text{in } \mathcal{V}^* = V \times V^* \\ z(0) &= z_0 . \end{aligned} \tag{2.12}$$

The rigorous equivalence of solutions is established through the following theorems.

Theorem 1. Under Hypotheses (H1)–(H5) on σ_1 and σ_2, $\widetilde{\mathcal{A}}$ generates an analytic semigroup $\mathcal{T}(t)$ on $\mathcal{V}, \mathcal{H}$ and $\mathcal{V}^*$. In terms of this semigroup, the representation

$$z(t) = \mathcal{T}(t)z_0 + \int_0^t \mathcal{T}(t-s)[\mathcal{B}u(s) + g(s)]ds \tag{2.13}$$

defines a mild solution to (2.12) for $z_0 \in \mathcal{V}^*$ and $\mathcal{B}u + g \in L^2((0,T); \mathcal{V}^*)$. Furthermore, this semigroup is (uniformly) exponentially stable on $\mathcal{V}, \mathcal{H}$ and $\mathcal{V}^*$.

Theorem 2. Let z_{sg} denote the semigroup solution to (2.12) given by (2.13) and let v_{var} denote the weak solution to (2.8). Under hypotheses (H1)-(H5), it follows that $z_{sg}(z_0, \mathcal{F}) = z_{var}(z_0, \mathcal{F})$ for $z_0 \in \mathcal{H}$ and $\mathcal{F} \equiv \mathcal{B}u + g \in L^2((0,T); \mathcal{V}^*)$.

Following the convention of [14], we will use the same notation for the semigroups defined on $\mathcal{V}, \mathcal{H}$ and $\mathcal{V}^*$ since each semigroup is an extension or restriction of the others. Note that dom$\mathcal{A}$ defined in (2.11) is actually $\text{dom}_{\mathcal{H}}\widetilde{\mathcal{A}}$, the domain of $\widetilde{\mathcal{A}}$ as a generator of $\mathcal{T}(t)$ in $\mathcal{H}$. As detailed in Lemma 3.6.1 and Theorem 3.6.1 of [14] (see also Section IV.6 of [12] and Chapter 2, Theorem 5.2 of [11]), the property that $\widetilde{\mathcal{A}}$ generates an analytic semigroup on $\mathcal{V}, \mathcal{H}$ and $\mathcal{V}^*$ results from the continuity and $\mathcal{V}$-ellipticity of σ. The exponential stability of $\mathcal{T}(t)$ on $\mathcal{H}$ for second-order systems with strong damping is demonstrated in [1] while the exponential stability of $\mathcal{T}(t)$ on $\mathcal{V}$ and $\mathcal{V}^*$ in this setting is proven in Lemma 3.3 of [2]. Finally, Theorem 2 is a reformulation of Theorem 4.14 of [5] and details can be found therein.

3. LQR Control Problem

In the last section, the PDE system modeling the dynamics of the thin shell with surface-mounted piezoceramic actuators was written in the abstract first-order form

$$\begin{aligned}\dot{z}(t) &= \mathcal{A}z(t) + \mathcal{B}u(t) + g(t)\\ z(0) &= z_0\end{aligned}$$

in $\mathcal{V}^*$. In this section, LQR control results for both the original infinite dimensional problem and approximating finite dimensional problems will be discussed. Two cases will be considered, namely when $g \equiv 0$ and g is periodic in time. In both cases, it is assumed that state observations in an observation space Y have the form

$$z_{ob}(t) = \mathcal{C}z(t) \tag{3.1}$$

where $\mathcal{C} \in \mathcal{L}(\mathcal{H}, Y)$ is bounded. The assumption that $\mathcal{C}$ is bounded is made to simplify the exposition and the reader is referred to [2] for arguments pertaining to the case of unbounded observation operators.

3.1. No Exogenous Input. For the case in which $g \equiv 0$, the infinite horizon problem concerns the determination of a control u which minimizes the quadratic cost functional

$$J(u, z_0) = \int_0^\infty \left\{|\mathcal{C}z(t)|_Y^2 + |\mathcal{R}^{1/2}u(t)|_U^2\right\} dt \tag{3.2}$$

subject to

$$\begin{aligned}\dot{z}(t) &= \mathcal{A}z(t) + \mathcal{B}u(t)\\ z(0) &= z_0\,.\end{aligned}$$

The positive, self-adjoint operator $\mathcal{R}$ is used to weight various components of the control.

As detailed in [2, 5], if $(\mathcal{A}, \mathcal{B})$ is stabilizable and $(\mathcal{A}, \mathcal{C})$ is detectable, then the optimal control minimizing (3.2) is given by

$$\bar{u}(t) = -\mathcal{R}^{-1}\mathcal{B}^*\Pi\bar{z}(t)$$

where Π solves the algebraic Riccati equation

$$(\mathcal{A}^*\Pi + \Pi\mathcal{A} - \Pi\mathcal{B}\mathcal{R}^{-1}\mathcal{B}^*\Pi + \mathcal{C}^*\mathcal{C})z = 0 \quad \text{for all } z \in \mathcal{V}$$

and $\bar{z}(t) = \mathcal{S}(t)z_0$. Here $\mathcal{S}(t)$ is the closed loop semigroup generated by $\mathcal{A}-\mathcal{B}\mathcal{R}^{-1}\mathcal{B}^*\Pi$.

For implementation purposes, it is necessary to define an approximate system and controls, and determine convergence criteria for these approximate controls when fed back into the infinite dimensional system. The approximations are considered in a Galerkin framework with trajectories evolving in the finite dimensional subspaces $\mathcal{V}^N \subset \mathcal{V} \subset \mathcal{H}$. It is assumed that the approximation method satisfies the standard convergence conditions

(H1N) For any $z \in \mathcal{V}$, there exists a sequence $\tilde{z}^N \in \mathcal{V}^N$ such that $|z - \tilde{z}^N|_{\mathcal{V}} \to 0$ as $N \to \infty$.

The finite dimensional operators and approximating system are then determined as follows. The operator $\mathcal{A}^N : \mathcal{V}^N \to \mathcal{V}^N$ which approximates $\mathcal{A}$ is defined by restricting σ to $\mathcal{V}^N \times \mathcal{V}^N$; this yields

$$\left\langle -\mathcal{A}^N \Upsilon, \Lambda \right\rangle_{\mathcal{H}} = \sigma(\Upsilon, \Lambda) \qquad \text{for all } \ \Upsilon, \Lambda \in \mathcal{V}^N \,. \tag{3.3}$$

For each N, the C_0 semigroup on $\mathcal{V}^N$ which is generated by $\mathcal{A}^N$ is denoted by $\mathcal{T}^N(t)$. The control operator is approximated by $\mathcal{B}^N \in \mathcal{L}(U, \mathcal{V}^N)$ given by

$$\left\langle \mathcal{B}^N u, \Lambda \right\rangle_{\mathcal{H}} = \langle u, \mathcal{B}^* \Lambda \rangle_{\mathcal{H}} \qquad \text{for all } u \in U \,,\, \Lambda \in \mathcal{V}^N \tag{3.4}$$

while $\mathcal{C}^N$ denotes the restriction of the observation operator $\mathcal{C}$ to $\mathcal{V}^N$. Finally, we let P^N denote the usual orthogonal projection of $\mathcal{H}$ onto $\mathcal{V}^N$ which by definition satisfies

$$\text{(i) } P^N \Upsilon \in \mathcal{V}^N \quad \text{for } \Upsilon \in \mathcal{H}$$

$$\text{(ii) } \left\langle P^N \Upsilon - \Upsilon, \Lambda \right\rangle_{\mathcal{H}} = 0 \quad \text{for all } \Lambda \in \mathcal{V}^N \,.$$

This projection can be extended to $P^N \in \mathcal{L}(\mathcal{V}^*, \mathcal{V}^N)$ by replacing the $\mathcal{H}$-inner product $\langle \Upsilon, \Lambda \rangle_{\mathcal{H}}$ by the duality product $\langle \Upsilon, \Lambda \rangle_{\mathcal{V}^*, \mathcal{V}}$ and considering $\Upsilon \in \mathcal{V}^*$.

The approximate problem corresponding to (2.10) with $g \equiv 0$ can then be formulated as

$$\begin{aligned} &\frac{d}{dt} \left\langle z^N(t), \Lambda \right\rangle_{\mathcal{H}} + \sigma(z^N(t), \Lambda) = \left\langle \mathcal{B}^N u(t), \Lambda \right\rangle_{\mathcal{H}} \qquad \text{for all } \Lambda \in \mathcal{V}^N \\ &z^N(0) = P^N z_0 \,. \end{aligned}$$

This has the solution

$$z^N(t) = \mathcal{T}^N(t) P^N z_0 + \int_0^t \mathcal{T}^N(t-s) P^N \mathcal{B}^N u(s) ds \,.$$

The following theorems taken from [2, 5] can be used to establish the convergence of the approximate gains to their infinite dimensional counterparts for certain classes of shell models (see specifically Theorem 7.10 and Lemma 7.13 of [5]).

Theorem 3. Assume that the injection $i : V \hookrightarrow H$ is compact. Moreover, suppose that the damping sesquilinear form can be decomposed as $\sigma_2 = \delta\sigma_1 + \hat{\sigma}_2$, for some $\delta > 0$, where the continuous sesquilinear form $\hat{\sigma}_2$ satisfies for some $\lambda \in \mathbb{R}$

$$Re\ \hat{\sigma}_2(\phi, \phi) \geq -\frac{\delta}{2} |\phi|_V^2 - \lambda |\phi|_H^2 \quad \text{for all } \phi \in V \,.$$

Finally, suppose that the operator $A_1^{-1} \widehat{A}_2$, where $\widehat{A}_2 \in \mathcal{L}(V, V^*)$ is defined by $\left\langle \widehat{A}_2 \phi, \eta \right\rangle_{V^*, V} = \hat{\sigma}_2(\phi, \eta)$, is compact on V.

If for some $\omega \in \mathbb{R}$ and $M \geq 1$, $\mathcal{T}(t)$ satisfies

$$|\mathcal{T}(t)|_{\mathcal{L}(\mathcal{H})} \leq M e^{\omega t} \quad , \quad t \geq 0 \,,$$

then for any $\varepsilon > 0$ there exists an integer N_ε such that for $N \geq N_\varepsilon$,

$$|\mathcal{T}^N(t) P^N|_{\mathcal{L}(\mathcal{H})} \leq \widetilde{M} e^{(\omega + \varepsilon)t} \quad , \quad t \geq 0$$

for some constant $\widetilde{M} > 0$ independent of N.

Theorem 4. Assume that the injection $i : V \hookrightarrow H$ is compact. Let the sesquilinear form σ associated with the first-order system (2.10) be continuous and $\mathcal{V}$-elliptic. Assume that the operators $\mathcal{A}, \mathcal{B}, \mathcal{C}$ of (2.11), (2.9), (3.1), respectively, satisfy: $(\mathcal{A}, \mathcal{B})$ is stabilizable and $(\mathcal{A}, \mathcal{C})$ is detectable where $\mathcal{B} \in \mathcal{L}(U, \mathcal{V}^*)$ is unbounded and $\mathcal{C} \in \mathcal{L}(\mathcal{H}, Y)$ is bounded. Consider an approximation method which satisfies (H1N). Finally, suppose that for fixed N_0 and $N > N_0$, the pair $(\mathcal{A}^N, \mathcal{B}^N)$ is uniformly stabilizable and $(\mathcal{A}^N, \mathcal{C}^N)$ is uniformly detectable.

Then for N sufficiently large, there exists a unique nonnegative self-adjoint solution $\Pi^N \in \mathcal{L}(\mathcal{V}^*, \mathcal{V})$ to the Nth approximate algebraic Riccati equation

$$\mathcal{A}^{N^*}\Pi^N + \Pi^N \mathcal{A}^N - \Pi^N \mathcal{B}^N \mathcal{R}^{-1} \mathcal{B}^{N^*} \Pi^N + \mathcal{C}^{N^*}\mathcal{C}^N = 0$$

in $\mathcal{V}^N$. There also exist constants $M_3 \geq 1$ and $\omega_3 > 0$ independent of N such that $\mathcal{S}^N(t) = e^{(\mathcal{A}^N - \mathcal{B}^N \mathcal{R}^{-1} \mathcal{B}^{N^*} \Pi^N)t}$ satisfies

$$\left|\mathcal{S}^N(t)\right|_{\mathcal{V}^N} \leq M_3 e^{-\omega_3 t} \quad , \quad t > 0\,.$$

Moreover, the convergence of the Riccati and control operators

$$\Pi^N P^N z \xrightarrow{s} \Pi z \ \text{ in } \mathcal{V} \ \text{ for every } z \in \mathcal{V}^*$$

$$\left|\mathcal{B}^{N^*}\Pi^N P^N - \mathcal{B}^*\Pi\right|_{\mathcal{L}(\mathcal{H},U)} \to 0\ ,$$

as $N \to \infty$, is obtained.

Example 1. We consider in this example a shell with constant parameters ρ, E, ν, c_D. Such a case would arise if modeling a homogeneous shell or a shell in which the variance of material properties across regions with actuators is negligible. The sesquilinear forms for this model are specified in (2.7). Due to the constant coefficients, σ_2 can be written as $\sigma_2 = \delta\sigma_1 + \hat{\sigma}_2$ where $\delta = \frac{c_D}{E}$ and $\hat{\sigma}_2(\Phi, \Psi) = \mu \int_{\Gamma_0} w\eta_3 d\gamma$. It follows immediately that

$$Re\hat{\sigma}_2(\phi, \phi) = \mu \int_{\Gamma_0} \phi^2 d\gamma \geq -\frac{\delta}{2}|\phi|_V^2$$

for all $\phi \in V$. The boundedness of the operator $\widehat{A}_2$ generated by $\hat{\sigma}_2$ follows directly from the boundedness of $\hat{\sigma}_2$. Furthermore, it is noted that $A_1^{-1} \in \mathcal{L}(V^*, V)$ can be written as an operator on $V \to V$ by $A_1^{-1} = A_1^{-1} i^* i$ where the injections $i : V \hookrightarrow H, i^* : H \hookrightarrow V^*$ are compact. Thus A_1^{-1} is compact on V which implies that $A_1^{-1}\widehat{A}_2$ is compact on V since it is formed from the product of compact and bounded linear operators. Finally, the exponential stability of $\mathcal{T}(t)$, the stabilizability of $(\mathcal{A}, \mathcal{B})$ and the detectability of $(\mathcal{A}, \mathcal{B})$ are guaranteed by Theorem 1. The hypotheses of Theorem 3 are then satisfied for this system and one obtains uniform bounds on the approximating semigroups. The convergence of the Riccati and control operators is then obtained from Theorem 4.

3.2. Periodic Exogenous Input. A reasonable assumption in many mechanical systems is that g is periodic in time with period τ. The system to be controlled in this case is

$$\begin{aligned} \dot{z}(t) &= \mathcal{A}z(t) + \mathcal{B}u(t) + g(t) \\ z(0) &= z(\tau) \end{aligned} \tag{3.5}$$

and an appropriate quadratic functional to be minimized is

$$J_\tau(u) = \frac{1}{2}\int_0^\tau \{|\mathcal{C}z(t)|_Y^2 + |\mathcal{R}^{1/2}u(t)|_U^2\}\, dt\,.$$

Note that the periodic exogenous term g can be used to model inputs such as noise generated by rotating engine components (e.g., propellers or turbines) or periodic electromagnetic disturbances.

To guarantee the existence of a unique Riccati solution and control for the system (3.5), it is assumed that $(\mathcal{A},\mathcal{B})$ is stabilizable and $(\mathcal{A},\mathcal{C})$ is detectable. Furthermore, it is assumed that $g \in L^2(0,\tau;\mathcal{H})$ and that $\mathcal{B}$ is bounded. Under these conditions, it is verified in [6] that the Riccati equation

$$\mathcal{A}^*\Pi + \Pi\mathcal{A} + \Pi\mathcal{B}\mathcal{R}^{-1}\mathcal{B}^*\Pi + \mathcal{C}^*\mathcal{C} = 0$$

has a unique solution. Furthermore, if r denotes the τ-periodic solution of the adjoint or tracking equation

$$\begin{aligned}\dot r(t) &= -[\mathcal{A} - \mathcal{B}\mathcal{R}^{-1}\mathcal{B}^*\Pi]^* r(t) + \Pi g(t)\\ r(0) &= r(\tau)\end{aligned}$$

and $\bar z$ is the closed loop solution of

$$\begin{aligned}\dot{\bar z}(t) &= [\mathcal{A} - \mathcal{B}\mathcal{R}^{-1}\mathcal{B}^*\Pi]\bar z(t) - \mathcal{B}\mathcal{R}^{-1}\mathcal{B}^* r(t) + g(t)\\ \bar z(0) &= \bar z(\tau)\,,\end{aligned}$$

then the optimal control is given by

$$\bar u(t) = -\mathcal{R}^{-1}\mathcal{B}^*[\Pi\bar z(t) - r(t)]\,. \tag{3.6}$$

The LQR theory for this case is less complete than that for systems with no exogenous input and is currently limited to bounded control inputs $\mathcal{B}$. The synthesis of the theory for unbounded input operators and periodic exogenous forces is currently under investigation. The effectiveness of the method is illustrated in the final example of this work.

4. Approximation Method

A Galerkin method was used to approximate the solutions u, v, w to the system (2.6), or equivalently, (2.10). The approximating subspaces were taken of the form

$$\mathcal{V}^N = \text{span}\{\mathcal{B}_{u_k}\} \times \text{span}\{\mathcal{B}_{v_k}\} \times \text{span}\{\mathcal{B}_{w_k}\}$$

where $\mathcal{B}_{u_k}, \mathcal{B}_{v_k}, \mathcal{B}_{w_k}$ denote bases for the u, v and w displacements, respectively. To exploit the tensor nature of the shell domain Γ_0 and periodicity in θ, the bases were constructed with Fourier components in θ and cubic splines in x (see [7] for details).

The approximate displacements were then given by the expansions

$$u^N(t,\theta,x) = \sum_{k=1}^{\mathcal{N}_u} u_k(t)\mathcal{B}_{u_k}(\theta,x)$$

$$v^N(t,\theta,x) = \sum_{k=1}^{\mathcal{N}_v} v_k(t)\mathcal{B}_{v_k}(\theta,x)$$

$$w^N(t,\theta,x) = \sum_{k=1}^{\mathcal{N}_w} w_k(t)\mathcal{B}_{w_k}(\theta,x)\,.$$

To obtain a finite dimensional system with matrices corresponding to the finite dimensional operators in (3.3) and (3.4), the sesquilinear forms σ_1 and σ_2 were restricted to $\mathcal{V}^N$. This yields the matrix system

$$\begin{bmatrix} K_E^{\mathcal{N}} & 0 \\ 0 & M^{\mathcal{N}} \end{bmatrix} \begin{bmatrix} \dot{\vartheta}^{\mathcal{N}}(t) \\ \ddot{\vartheta}^{\mathcal{N}}(t) \end{bmatrix} = \begin{bmatrix} 0 & K_E^{\mathcal{N}} \\ -K_E^{\mathcal{N}} & -K_{c_D}^{\mathcal{N}} \end{bmatrix} \begin{bmatrix} \vartheta^{\mathcal{N}}(t) \\ \dot{\vartheta}^{\mathcal{N}}(t) \end{bmatrix} + \begin{bmatrix} 0 \\ \tilde{B}^{\mathcal{N}} \end{bmatrix} [u(t)] + \begin{bmatrix} 0 \\ \tilde{g}^{\mathcal{N}}(t) \end{bmatrix}$$

$$\begin{bmatrix} K_E^{\mathcal{N}} & 0 \\ 0 & M^{\mathcal{N}} \end{bmatrix} \begin{bmatrix} \vartheta^{\mathcal{N}}(0) \\ \dot{\vartheta}^{\mathcal{N}}(0) \end{bmatrix} = \begin{bmatrix} y_1^{\mathcal{N}} \\ y_2^{\mathcal{N}} \end{bmatrix}$$

where $\vartheta^{\mathcal{N}}(t) = [u_1(t),\ldots,u_{\mathcal{N}_u}, v_1(t),\ldots,v_{\mathcal{N}_v}, w_1(t),\ldots,w_{\mathcal{N}_w}]^T$ contains the $\mathcal{N} = \mathcal{N}_u + \mathcal{N}_v + \mathcal{N}_w$ generalized Fourier coefficients. The s patch inputs are contained in $u(t) = [u_1(t),\ldots,u_s(t)]^T$. The reader is referred to [7] for details concerning the construction of the mass, stiffness and damping matrices $M^{\mathcal{N}}, K_E^{\mathcal{N}}, K_{c_D}^{\mathcal{N}}$, the inputs $\tilde{B}^{\mathcal{N}}, \tilde{g}^{\mathcal{N}}(t)$ and the initial conditions $y_1^{\mathcal{N}}, y_2^{\mathcal{N}}$.

Multiplication by the inverted mass matrix yields the Cauchy equation

$$\begin{aligned} \dot{z}^N(t) &= A^N z^N(t) + B^N u(t) + g^N(t) \\ z^N(0) &= z_0^N\,, \end{aligned} \tag{4.1}$$

where $z^N \in \mathbb{R}^{2\mathcal{N}} = [\vartheta^{\mathcal{N}}(t), \dot{\vartheta}^{\mathcal{N}}(t)]^T$. This system forms the constraint equations used in the finite dimensional LQR theory discussed in Section 3.

5. Numerical Example

We consider here an exogenous force g which is periodic in time with period $\tau = 1000\pi$ (500 Hz). The distribution of the force was taken to be binormal in the transverse and longitudinal directions and centered at $(x,\theta) = (\ell/2, 0)$ and $(x,\theta) = (\ell/2,\pi)$ as depicted in Figure 2. The magnitude of the transverse component $\hat{q}_n$ was one hundred times that of the longitudinal component $\hat{q}_x$ so as to model an input consisting primarily of acoustic sources located adjacent to $(\ell/2, 0)$ and $(\ell/2,\pi)$.

Six pairs of piezoceramic patches of length $1\,cm$ and radial measure $\pi/3$ were employed as actuators. The locations and material properties of the patches along with the dimensions and physical parameters for the shell are summarized in Table 1.

To accommodate the periodic exogenous force g, control inputs to the twelve patches were computed using the feedback law (3.6). Note that in this formulation, independent voltages are determined for the individual patches. This provides the capability of generating both inplane forces and bending moments in the regions covered by the patches so that longitudinal, circumferential and transverse vibrations can be controlled.

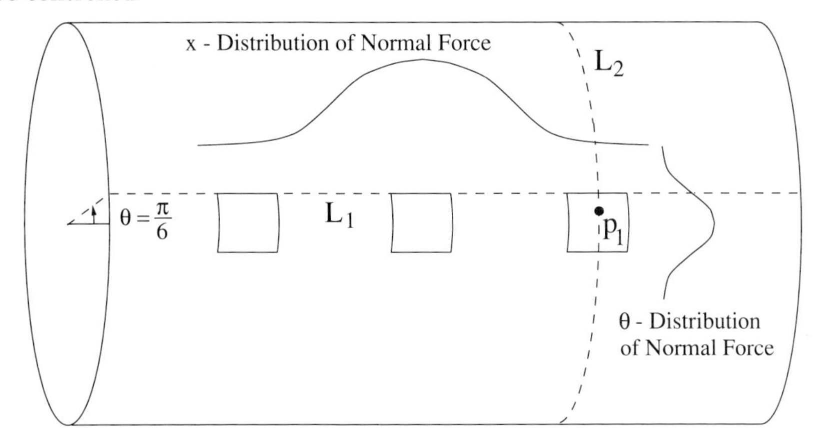

Figure 2. Distribution of normal forcing function at $\theta = 0$ and $\theta = \pi$. Observation lines $L_1 = \{(x,\theta)|0 \leq x \leq \ell, \theta = \pi/6\}$, $L_2 = \{(x,\theta)|x = 3\ell/4, 0 \leq \theta \leq 2\pi\}$ and observation point $p_1 = (3\ell/4, \pi/32)$.

Time histories of the uncontrolled and controlled shell displacements at the point $p_1 = (3\ell/4, \pi/32)$, depicted in Figure 2, are plotted in Figure 3. The open loop trajectories exhibit both a transient response settling into steady state and a beat phenomenon due to the close proximity of the driving frequency and natural frequencies for the shell. At this observation point, all three displacement components are reduced by more than 90% when controlling voltages are fed back to the patches.

	Dimensions	Parameters
Shell	$h = .00127\, m$ $R = .4\, m$ $\ell = 1\, m$	$\rho = 2700\, kg/m^3$ $E = 7.1 \times 10^{10}\, N/m^2$ $c_D = 2.816 \times 10^{-5}\, Nms$ $\nu = .33$ $\mu = 58.97 Ns/m^2$
Patches	$h_{pe} = .0001778\, m$ Centers (x,θ): $(.25,0), (.5,0), (.75,0)$ $(.25,\pi), (.5,\pi), (.75,\pi)$ Dimensions: $x : 0.1\, cm$, $\theta : \pi/3$	$\rho_{pe} = 7600\, kg/m^3$ $E_{pe} = 6.3 \times 10^{10}\, N/m^2$ $c_{D_{pe}} = 3.211 \times 10^{-5}\, Nms$ $\nu_{pe} = .31$ $d_{31} = 190 \times 10^{-12}\, m/V$

Table 1. Dimensions and physical parameters for the shell and patches.

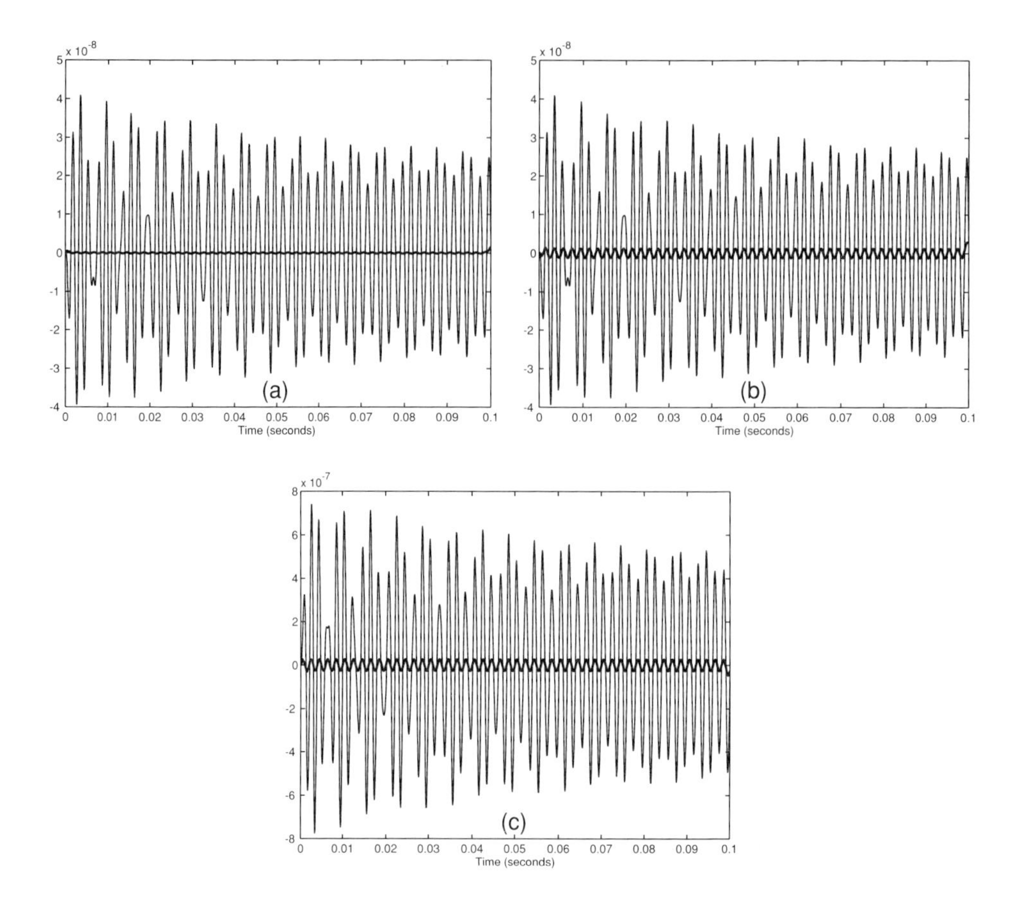

Figure 3. Uncontrolled and controlled shell displacements at the point

$$p_1 = (3\ell/4, \pi/32)\ ;$$

(a) longitudinal u,
(b) circumferential v,
(c) transverse w displacements;
—— (uncontrolled), —— (controlled).

To illustrate the spatial attenuation due to the feedback of voltages to the patches, root mean square (rms) plots of the uncontrolled and controlled trajectories along the axial line L_1 and circumferential line L_2 (see Figure 2) are plotted in Figure 4 and 5, respectively. For the open loop case, these plots illustrate a standing wave in all three components of the displacement. The figures also demonstrate significant reductions in all three displacement levels, even in regions not covered by patches. This further illustrates the effectiveness through which the model-based control law can be used to attenuate shell vibrations.

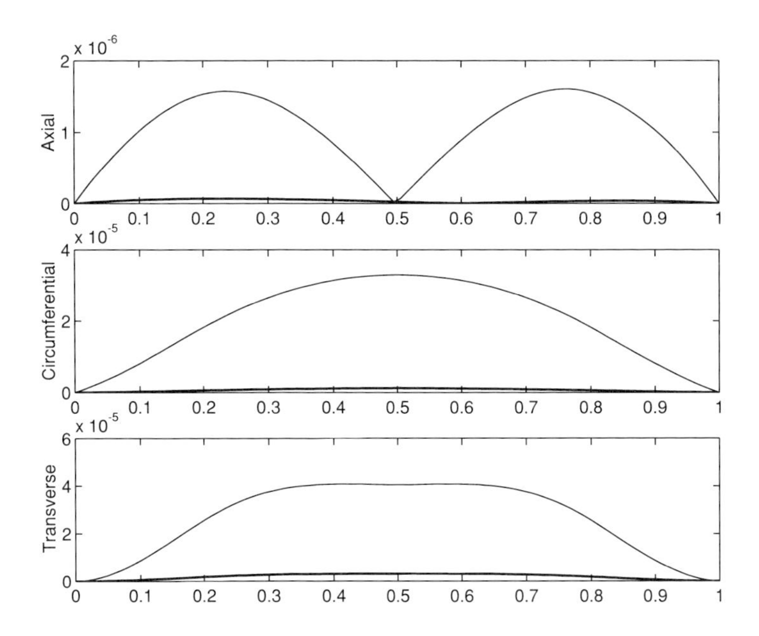

Figure 4. Root mean square (rms) displacements along the axial line L_1; —— (uncontrolled), —— (controlled).

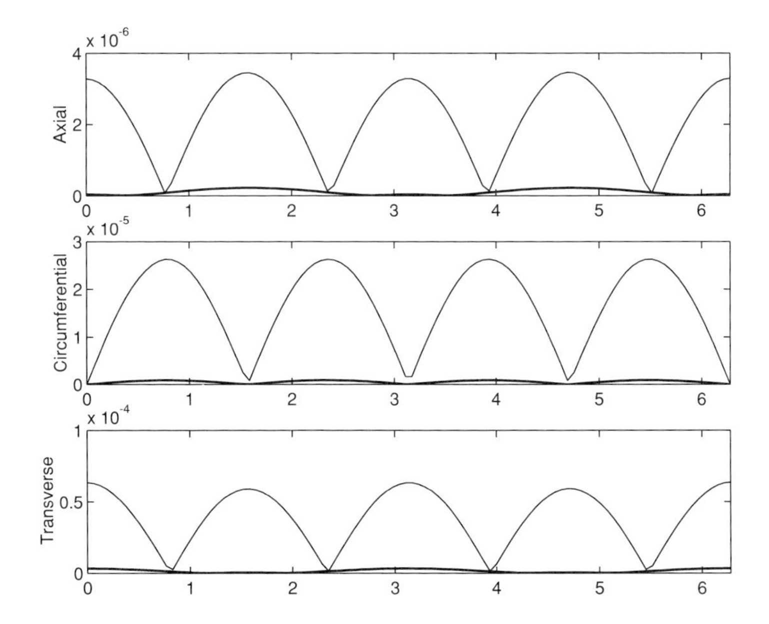

Figure 5. Root mean square (rms) displacements along the circumferential line L_2; —— (uncontrolled), —— (controlled).

6. Conclusions

A model-based LQR method for controlling shell vibrations has been presented here. While developed in the context of a modified Donnell-Mushtari cylindrical shell model, the method is quite general and can be directly extended to other models and geometries. Under the assumption of strong or Kelvin-Voigt damping (a reasonable and typical assumption for many shell materials), model well-posedness and convergence of control gains is obtained using analytic semigroup theory combined with LQR results for unbounded input operators.

The Galerkin method used to approximate the system dynamics utilizes bases constructed from tensored Fourier polynomials and modified cubic splines. As discussed in [7], case must be taken when developing methods for approximating shell dynamics so as to avoid shear or membrane locking. One manifestation of locking is the existence of model dynamics which are incorrectly approximated by the numerical method. The use of a numerical method which exhibits locking can lead to a loss of control authority and potential controller destabilization if the approximations are sufficiently inaccurate. Further details regarding issues concerning the approximation of shell dynamics and convergence properties of the numerical method can be found in [7].

The numerical example demonstrates the effectiveness of the model-based control method for attenuating all three components of the shell displacement in the presence of both transient and steady state dynamics. Furthermore, by modeling the global shell dynamics and patch interactions through coupled PDE and constructing the control law in terms of these PDE, significant reductions in displacement levels throughout the shell are obtained, even in regions devoid of patches. Numerical implementation of the LQR method in this manner provides a first step toward the development of model-based state estimators and compensators which can be experimentally implemented in shell applications.

Acknowledgements: The authors thank H.T. Banks, CRSC, North Carolina State University, for input regarding various aspects of this investigation.

References

1. H.T. Banks and K. Ito, "A Unified Framework for Approximation in Inverse Problems for Distributed Parameter Systems," *Control-Theory and Advanced Technology*, 4, pp. 73–90, 1988.
2. H.T. Banks and K. Ito, "Approximation in LQR Problems for Infinite Dimensional Systems with Unbounded Input Operators," CRSC Technical Report CRSC-TR94-22, November 1994; *Journal of Mathematical Systems, Estimation and Control*, to appear.
3. H.T. Banks, R.C. Smith and Y. Wang, "The Modeling of Piezoceramic Patch Interactions with Shells, Plates, and Beams," *Quarterly of Applied Mathematics*, 53(2), pp. 353–381, 1995.
4. H.T. Banks, R.C. Smith and Y. Wang, "Modeling and Parameter Estimation for an Imperfectly Clamped Plate," CRSC Technical Report CRSC-TR95-2, 1995; *Computation and Control IV*, (K.L. Bowers and J. Lund, eds.), Birkhäuser, Boston, 1995, pp. 23–42.
5. H.T. Banks, R.C. Smith and Y. Wang, *Smart Material Structures: Modeling, Estimation and Control*, Masson/John Wiley, Paris/Chichester, 1996.
6. G. Da Prato, "Synthesis of Optimal Control for an Infinite Dimensional Periodic Problem," *SIAM J. Control and Optimization*, 25(3), 1987, pp. 706–714.
7. R. del Rosario and R.C. Smith, "Spline Approximation of Thin Shell Dynamics," ICASE Report 96-26, March 1996; *International Journal for Numerical Methods in Engineering*, to appear.

8. C.R. Fuller, S.D. Snyder, C.H. Hansen and R.J. Silcox, "Active Control of Interior Noise in Model Aircraft Fuselages Using Piezoceramic Actuators," Paper 90-3922, AIAA 13th Aeroacoustics Conference, Tallahassee, FL, October 1990.
9. A.W. Leissa, *Vibration of Shells,* NASA SP-288, 1973; Reprinted by the Acoustical Society of America through the American Institute of Physics, 1993.
10. Lester and Lefebvre, "Piezoelectric Actuator Models for Active Sound and Vibration Control of Cylinders," Proceedings of the Conference on Recent Advance in Active Control of Sound and Vibration, Blacksburg, VA, 1991, pp. 3–26.
11. A. Pazy, *Semigroups of Linear Operators and Applications to Partial Differential Equations,* Springer-Verlag, New York, 1983.
12. R.E. Showalter, *Hilbert Space Methods for Partial Differential Equations,* Pitman Publishing Ltd., London, 1977.
13. R.J. Silcox, S. Lefebvre, V.L. Metcalf, T.B. Beyer and C.R. Fuller, "Evaluation of Piezoceramic Actuators for Control of Aircraft Interior Noise," Proceedings of the DGLR/AIAA 14th Aeroacoustics Conference, Aachen, Germany, May 11–14, 1992.
14. H. Tanabe, *Equations of Evolution,* Pitman Publishing Ltd., London, 1979.

R.C.H. del Rosario
Center for Research in Scientific Computation
North Carolina State University
Raleigh, NC 27695, USA
e-mail:rcdelros@eos.ncsu.edu

R.C. Smith
Department of Mathematics
Iowa State University
Ames, IA 50011, USA
e-mail:rsmith@iastate.edu

International Series of Numerical Mathematics
Vol. 126, © 1998 Birkhäuser Verlag, Basel

The Algebraic Riccati Equation in Discrete and Continuous Time

OLOF J. STAFFANS
Department of Mathematics
Åbo Akademi University

ABSTRACT. The standard state space solution of the finite-dimensional continuous time quadratic cost minimization problem has a straightforward extension to infinite-dimensional problems with bounded or moderately unbounded control and observation operators. However, if these operators are allowed to be sufficiently unbounded, then a strange change takes place in one of the coefficients of the algebraic Riccati equation, and the continuous time Riccati equation begins to resemble the discrete time Riccati equation. To explain why this phenomenon must occur we discuss a delay equation of difference type that can be formulated both as a discrete time system and as a continuous time system, and show that in this example the continuous time Riccati equation can be recovered from the discrete time Riccati equation. A particular feature of this example is that the Riccati operator does not map the domain of the generator into the domain of the adjoint generator, as it does in the standard case.

1991 *Mathematics Subject Classification.* Primary 49J27, 93A05, 93C55, 93C60

Key words and phrases. Algebraic Riccati equation, linear quadratic optimal control, well-posed linear systems, spectral factorization.

1. The Discrete and Continuous Time Riccati Equations

We begin by comparing two different optimization problems, namely the discrete time and the continuous time quadratic cost minimization problems.

In the discrete time quadratic cost minimization problem we study the discrete time system

$$\begin{aligned} x_{n+1} &= Ax_n + Bu_n, \\ y_n &= Cx_n + Du_n, \quad n \in \mathbb{N} = \{0,1,2,\dots\}. \end{aligned} \tag{1.1}$$

Here u_n belongs to the input space U, x_n to the state space H, and y_n to the output space Y. These are Hilbert spaces, and A, B, C and D are bounded linear operators between the appropriate spaces. The problem is to find a sequence $u_n \in l^2(\mathbb{N};U)$ that minimizes the cost functional

$$W(x_0,u) = \sum_{n=0}^{\infty} \left(\langle y_n, y_n\rangle + \langle u_n, Ru_n\rangle\right), \tag{1.2}$$

where R is a given positive definite operator on U. Under mild assumptions (stabilizability, detectability, and coercivity of the cost function), the optimal control u^{opt} is of state feedback type, i.e., there is a bounded linear operator K such that

$u_n^{opt} = Kx_n^{opt}$ for all $n \in \mathbb{N}$, and the optimal cost $W(x_0, u^{opt})$ can be written in the form

$$W(x_0, u^{opt}) = \langle x_0, Px_0 \rangle, \qquad x_0 \in H,$$

where P is a positive definite operator on H, the Riccati operator. Moreover, the feedback operator K and the Riccati operator P satisfy the equations

$$SK = -\left(B^*PA + D^*C\right), \tag{1.3}$$

$$A^*PA - P + C^*C = K^*SK, \tag{1.4}$$

$$S = R + D^*D + B^*PB. \tag{1.5}$$

We shall refer to these equations as the *discrete time Lure equations.* We call S the *sensitivity operator* of the discrete time problem, due to the fact that it describes the sensitivity of the optimal solution with respect to a nonzero closed loop control signal; cf. [Malinen(1997)]. In the standard case the sensitivity operator S is invertible, and by eliminating K and S we get the discrete time algebraic Riccati equation

$$A^*PA - P + C^*C = (B^*PA + D^*C)^*(R + D^*D + B^*PB)^{-1}(B^*PA + D^*C)\,. \tag{1.6}$$

See, for example, [Curtain and Zwart(1995), pages 329–332] or [Malinen(1997)].

In the continuous time quadratic cost minimization problem we study the continuous time system

$$\begin{aligned} z'(t) &= \mathcal{A}z(t) + \mathcal{B}u(t), \\ y(t) &= \mathcal{C}z(t) + Du(t), \quad t \in \mathbb{R}^+ = (0, \infty), \\ z(0) &= z_0. \end{aligned} \tag{1.7}$$

Here $u(t)$ belongs to the input space U, $z(t)$ to the state space H, and $y(t)$ to the output space Y, still Hilbert spaces. We suppose that $\mathcal{A}$ generates a strongly continuous semigroup A on H and that D is bounded, and, for the moment, we also take the operators $\mathcal{B}$ and $\mathcal{C}$ to be bounded. Naturally, we interpret (1.7) in the strong sense, i.e., z is given by $z(t) = A(t)x_0 + \int_0^t A(t-s)\mathcal{B}u(s)\,ds$ for $t \in \mathbb{R}^+$. This time the problem is to find a control $u \in L^2(\mathbb{R}^+; U)$ that minimizes the cost functional

$$\mathcal{W}(x_0, u) = \int_0^\infty \left(\langle y(t), y(t)\rangle + \langle u(t), Ru(t)\rangle\right) dt, \tag{1.8}$$

with a positive definite R. Again, under mild assumptions (stabilizability, detectability, and coercivity of the cost function), the optimal control u^{opt} is of state feedback type, i.e., there is a bounded linear operator $\mathcal{K}$ such that $u^{opt}(t) = \mathcal{K}z^{opt}(t)$ for all $t \in \mathbb{R}^+$, and the optimal cost $\mathcal{W}(x_0, u^{opt})$ can be written in the form

$$\mathcal{W}(x_0, u^{opt}) = \langle x_0, \mathcal{P}x_0 \rangle, \qquad x_0 \in H,$$

where $\mathcal{P}$ is the positive definite continuous time Riccati operator. Moreover, the feedback operator $\mathcal{K}$ and the Riccati operator $\mathcal{P}$ satisfy the continuous time Lure equations

$$\mathcal{S}\mathcal{K} = -\left(\mathcal{B}^*\mathcal{P} + D^*\mathcal{C}\right), \tag{1.9}$$

$$\mathcal{A}^*\mathcal{P} + \mathcal{P}\mathcal{A} + \mathcal{C}^*\mathcal{C} = \mathcal{K}^*\mathcal{S}\mathcal{K}, \tag{1.10}$$

$$\mathcal{S} = R + D^*D. \tag{1.11}$$

Equations (1.9) and (1.11) hold on H and U, respectively, and (1.10) is valid on the domain $dom(\mathcal{A})$ of $\mathcal{A}$; in particular, $\mathcal{P}$ maps $dom(\mathcal{A})$ into $dom(\mathcal{A}^*)$. Again we call $\mathcal{S}$ the *sensitivity operator* of the continuous time problem, since it describes the sensitivity of the cost function with respect to a nonzero closed loop input. In the standard case the sensitivity operator $\mathcal{S}$ is invertible, and we can eliminate $\mathcal{K}$ and $\mathcal{S}$ to get the continuous time algebraic Riccati equation

$$\mathcal{A}^*\mathcal{P} + \mathcal{P}\mathcal{A} + \mathcal{C}^*\mathcal{C} = (\mathcal{B}^*\mathcal{P} + D^*\mathcal{C})^* (R + D^*D)^{-1} (\mathcal{B}^*\mathcal{P} + D^*\mathcal{C}), \tag{1.12}$$

valid on $dom(\mathcal{A})$. See, for example, [Curtain and Zwart(1995), pages 316–317] or [Staffans(1997b)].

There are some striking similarities and differences between the discrete and continuous time Lure equations (1.3)–(1.5) and (1.9)–(1.11) and Riccati equations (1.6) and (1.12). Maybe the most important difference is that the discrete time sensitivity operator S depends on the discrete time Riccati operator P, but that the continuous time sensitivity operator $\mathcal{S}$ does not depend on the continuous time Riccati operator $\mathcal{P}$, and, if we ignore the difficulties caused by the unbounded operator $\mathcal{A}$, the structure of the discrete time Lure equations (1.3)–(1.5) is more complicated than the structure of the continuous time equations (1.9)–(1.11).

2. The Discrete and Continuous Time Closed Loop Riccati Equations

Above we have written the discrete and continuous time optimality conditions and Riccati equations in "open loop" form, i.e., they are written in terms of the original operators A, B, C, and D in (1.1), and $\mathcal{A}$, $\mathcal{B}$, $\mathcal{C}$, and D in (1.7). It is also possible to give "closed loop" versions of the same equations.

If we in (1.1) replace u_n by a new control v_n according to the formula

$$u_n = Kx_n + v_n, \qquad n \in \mathbb{N},$$

then (1.1) becomes

$$\begin{aligned} x_{n+1} &= A_{\circlearrowleft}x_n + Bv_n, \\ y_n &= C_{\circlearrowleft}x_n + Dv_n, \quad n \in \mathbb{N}, \end{aligned}$$

where

$$A_{\circlearrowleft} = A + BK, \qquad C_{\circlearrowleft} = C + DK.$$

Replacing A and C in (1.3)–(1.5) by $A = A_{\circlearrowleft} - BK$ and $C = C_{\circlearrowleft} - DK$ we get the discrete time closed loop Lure equations

$$\begin{aligned} RK &= -(B^*PA_{\circlearrowleft} + D^*C_{\circlearrowleft}), \\ A_{\circlearrowleft}^*PA_{\circlearrowleft} - P + C_{\circlearrowleft}^*C_{\circlearrowleft} &= -K^*RK, \end{aligned}$$

The operator R need not be invertible, but if it is, then we can eliminate K to get the discrete time closed loop algebraic Riccati equation

$$A_{\circlearrowleft}^*PA_{\circlearrowleft} - P + C_{\circlearrowleft}^*C_{\circlearrowleft} = -(B^*PA_{\circlearrowleft} + D^*C_{\circlearrowleft})^* R^{-1} (B^*PA_{\circlearrowleft} + D^*C_{\circlearrowleft}).$$

The difference compared to (1.3)–(1.6) consists in a change of sign in the quadratic term, and the fact that the operator S has been replaced by R and no longer enters the equations. Clearly, the invertibility of R is a stronger condition than the invertibility of S.

In the continuous time case we can proceed in the same way. We separate the feedback contribution to the control from the external control, and write

$$u(t) = \mathcal{K}x(t) + v(t), \qquad t \in \mathbb{R},$$

to get the closed loop system

$$\begin{aligned} z'(t) &= \mathcal{A}_{\circlearrowleft} z(t) + \mathcal{B}v(t), \\ y(t) &= \mathcal{C}_{\circlearrowleft} z(t) + Dv(t), \quad t \in \mathbb{R}^+, \\ z(0) &= z_0, \end{aligned}$$

where

$$\mathcal{A}_{\circlearrowleft} = \mathcal{A} + \mathcal{B}\mathcal{K}, \qquad \mathcal{C}_{\circlearrowleft} = \mathcal{C} + D\mathcal{K}.$$

Replacing $\mathcal{A}$ and $\mathcal{C}$ in (1.9)–(1.11) by $\mathcal{A} = \mathcal{A}_{\circlearrowleft} - \mathcal{B}\mathcal{K}$ and $\mathcal{C} = \mathcal{C}_{\circlearrowleft} - D\mathcal{K}$ we get the continuous time closed loop Lure equations

$$\begin{aligned} R\mathcal{K} &= -\left(\mathcal{B}^*\mathcal{P} + D^*\mathcal{C}_{\circlearrowleft}\right), \\ \mathcal{A}_{\circlearrowleft}^*\mathcal{P} + \mathcal{P}\mathcal{A}_{\circlearrowleft} + \mathcal{C}_{\circlearrowleft}^*\mathcal{C}_{\circlearrowleft} &= -\mathcal{K}^*R\mathcal{K}. \end{aligned} \tag{2.1}$$

If R is invertible, then we can eliminating $\mathcal{K}$ to get the continuous time closed loop algebraic Riccati equation

$$\mathcal{A}_{\circlearrowleft}^*\mathcal{P} + \mathcal{P}\mathcal{A}_{\circlearrowleft} + \mathcal{C}_{\circlearrowleft}^*\mathcal{C}_{\circlearrowleft} = -\left(\mathcal{B}^*\mathcal{P} + D^*\mathcal{C}_{\circlearrowleft}\right)^* R^{-1}\left(\mathcal{B}^*\mathcal{P} + D^*\mathcal{C}_{\circlearrowleft}\right). \tag{2.2}$$

Again, the invertibility condition on R is a stronger one than the invertibility condition on $\mathcal{S}$ (whenever D is nonzero). Comparing these equations to the corresponding open loop equations we see the same changes as in the discrete time case.

The closed loop discrete and continuous time Lure equations and Riccati equations resemble each other more than the corresponding open loop equations, due to the fact that the operators S and $\mathcal{S}$ have disappeared. However, observe that the closed loop equations contain an extra implicit dependence on the feedback operators K and $\mathcal{K}$, hidden in the definitions of $A_{\circlearrowleft}$, $C_{\circlearrowleft}$, $\mathcal{A}_{\circlearrowleft}$, and $\mathcal{C}_{\circlearrowleft}$, and that they are less general in the sense that we need an invertibility condition on R instead of invertibility conditions on S and $\mathcal{S}$, respectively.

3. Unbounded Control and Observation Operators

Up to now we have assumed the continuous time control operator $\mathcal{B}$ and observation operator $\mathcal{C}$ to be bounded. They can be allowed to be somewhat unbounded without any significant nontechnical additions to the theory. This applies, in particular, to the class of smooth Pritchard-Salamon systems studied in, e.g., [Pritchard and Salamon(1985), Pritchard and Salamon(1987)] and [van Keulen(1993)]. However, if $\mathcal{B}$ and $\mathcal{C}$ are sufficiently unbounded then the structure of the continuous time Lure equations (1.9)–(1.11) changes, and they become even more similar to the discrete time Lure equations (1.3)–(1.5).

The main problem is how to define the term $\mathcal{B}^*\mathcal{P}$ in (1.9) when $\mathcal{B}$ is unbounded. The largest class of systems that we are able to cope with is the class of well-posed and regular Salamon-Weiss systems; see [Salamon(1987), Salamon(1989)] and [Weiss(1994a), Weiss(1994b)] for the relevant theory. In this theory the natural domain for $\mathcal{B}^*$ is $dom(\mathcal{A}^*)$. As before, we want (1.10) to hold on $dom(\mathcal{A})$, hence (1.9) should also hold at least on $dom(A)$. Thus, the operator $\mathcal{B}^*\mathcal{P}$ should be defined at least on $dom(\mathcal{A})$. Since the natural domain of $\mathcal{B}^*$ is $dom(\mathcal{A}^*)$, we would like $\mathcal{P}$ to map $dom(\mathcal{A})$ into $dom(\mathcal{A}^*)$ (as it does in the case of bounded $\mathcal{B}$ and $\mathcal{C}$). However, this will not be true in general, and in particular, it is not true in the example that we present below. Thus, we are forced to extend $\mathcal{B}^*$ to a larger domain. This extension is not unique, due to the fact that $dom(\mathcal{A}^*)$ need not be dense in the larger domain (this will be the case in the example given below).

The necessary extension of $\mathcal{B}^*$ can be carried out in at least two different ways. Instead of extending $\mathcal{B}^*$, [Flandoli et al.(1988)] show that in the case where $D = 0$ and $\mathcal{C}$ is bounded it is possible to find *some* extension of $\mathcal{B}^*\mathcal{P}$ such that the Riccati equation (1.12) holds on $dom(\mathcal{A})$ (this result applies to some non-regular systems as well). However, the definition of the extended $\mathcal{B}^*\mathcal{P}$ given by [Flandoli et al.(1988)] is quite implicit (it is part of the proof of [Flandoli et al.(1988), Corollary 4.9]), and it far from obvious how to compute this extension from the original data. Moreover, it is not clear to what extent that result applies when $D \neq 0$ or $\mathcal{C}$ is unbounded (as is the case in the example that we present below).

Our solution, found in [Staffans(1997a), Staffans(1997b)], is quite different. We impose an extra "regular spectral factorization assumption", the content of which is that both the input/output map of the original system and a particular spectral factor should be regular together with their adjoints in the sense of [Weiss(1994a)]. See [Staffans(1997a), Staffans(1997b)] for details. In order to verify this assumption for a particular system one needs good information about its input/output behavior. This type of information is readily avaliable for delay equations but not for general PDEs. In particular, it follows from [Staffans(1995), Lemma 2.1] that this assumption is satisfied in the example presented below, but it is still an open question whether or not it is satisfied in most of the really interesting PDE examples.

The regular spectral factorization assumption enables us to replace the extension of $\mathcal{B}^*\mathcal{P}$ used in [Flandoli et al.(1988)] by $\overline{\mathcal{B}}^*\mathcal{P}$, where $\overline{\mathcal{B}}^*$ stands for the straightforward *Weiss extension* [Weiss(1994a)]

$$\overline{\mathcal{B}}^* x = \lim_{\beta\to+\infty} \mathcal{B}^*\beta\left(\beta I - \mathcal{A}^*\right)^{-1} x \tag{3.1}$$

of $\mathcal{B}^*$. As shown in [Staffans(1997b)], if we use this extension, then we must add a correction term to the continuous time sensitivity operator $\mathcal{S}$ and replace the definition (1.11) of $\mathcal{S}$ by

$$\mathcal{S} = R + D^*D + \lim_{\alpha\to+\infty} \overline{\mathcal{B}}^*\mathcal{P}\left(\alpha I - \mathcal{A}\right)^{-1}\mathcal{B} \tag{3.2}$$

(this limit exists in the strong sense whenever the regular spectral factorization assumption holds). Equations (1.9) and (1.10) remain valid (with $\mathcal{B}^*$ replaced by $\overline{\mathcal{B}}^*$). Observe that (3.2) agrees with (1.11) whenever $\mathcal{B}$ is bounded. As in the discrete time case, it can be shown [Staffans(1997a)] that $\mathcal{S} \geq R + D^*D$, and that $\mathcal{S}$ depends only on the weight R and the transfer function of the system, i.e., $\mathcal{S}$ is independent of the particular realization $\mathcal{A}$, $\mathcal{B}$, and $\mathcal{C}$. The physical interpretations of S and $\mathcal{S}$ are identical [Malinen(1997), Staffans(1997b)].

Since the sensitivity operator S does not show up in the closed loop Lure and Riccati equations, it is to be expected that these should still remain the same as in the case of bounded control operator $\mathcal{B}$ and observation operator $\mathcal{C}$. Indeed, this is the case, as shown in [Staffans(1997a)].

The purpose of this paper is to present an example where the change from (1.11) to (3.2) takes place. This example is a delay equation of difference type. It can be formulated both as a discrete time system and as a continuous time system. In this example the continuous time sensitivity operator $\mathcal{S}$ is the same as the discrete time sensitivity operator S, and the two Riccati equations (1.6) and (1.12) (with $(R + D^*D)^{-1}$ replaced by $\mathcal{S}^{-1}$) become more or less equivalent. We remark that this example has been discovered independently by [Weiss and Weiss(1997)]. Some additional details of this example are presented in [Staffans(1996a)] and [Weiss and Weiss(1997)].

Another example illuminating the difference between the two extensions of $\mathcal{B}^*$ used in [Staffans(1997b)] and [Flandoli et al.(1988)] is found in a recent preprint by [Weiss and Zwart(1996)]. In that example $D = 0$ and $\mathcal{C}$ is bounded.

For completeness, let us point out the fact that the present theory says nothing about the solvability of the system (1.9), (1.10), and (3.2): Is the solution unique, and can these equations be used to actually compute Π, $\mathcal{K}$ and $\mathcal{S}$? In other words, the converse part of the theory is still missing.

4. The Delay Equation

In the rest of this note we consider the following delay equation of difference type:

$$
\begin{aligned}
x(t) &= Ax(t-T) + Bu(t), \quad && t \in [0,\infty), \\
y(t) &= Cx(t-T) + Du(t), \quad && t \in [0,\infty), \\
x(t) &= \text{given}, && t = [-T, 0).
\end{aligned}
\tag{4.1}
$$

Here $u(t) \in U$, $x(t) \in H$, $y(t) \in Y$ (all Hilbert spaces), and A, B, C, and D are bounded linear operators between the appropriate spaces. For simplicity, we assume, in addition, that the system (4.1) is exponentially stable, but this assumption can be replaced by a stabilizability and detectability assumption. Moreover, we assume that the data have been chosen in such a way that the discrete time Lure equations (1.3)–(1.5) have a unique solution, with P positive definite and S strictly positive definite, i.e., $S \geq \epsilon I$ for some $\epsilon > 0$. The cost function $\mathcal{W}$ that we want to minimize is given by (1.8).

5. Two Discrete Time Formulations

It is easy to reformulate (4.1) as a discrete time system. This can even be done in two conceptually different ways. In both cases we start with the observation that if we define

$$\begin{aligned} u_n(t) &= u(t+nT), \\ x_n(t) &= x(t+(n-1)T), \\ y_n(t) &= y(t+nT), \qquad n \in \mathbb{N}, \qquad t \in [0,T), \end{aligned}$$

then (4.1) becomes

$$\begin{aligned} x_{n+1}(t) &= Ax_n(t) + Bu_n(t), \\ y_n(t) &= Cx_n(t) + Du_n(t), \qquad n \in \mathbb{N}, \qquad t \in [0,T), \\ x_0(t) &= \text{given}, \qquad t \in [0,T), \end{aligned} \tag{5.1}$$

and the cost function $\mathcal{W}$ can be written in the form

$$\begin{aligned} \mathcal{W}(x_0, u) &= \int_0^\infty (\langle y(t), y(t)\rangle + \langle u(t), Ru(t)\rangle)\, dt \\ &= \sum_{n=0}^{\infty} \left(\int_0^T (\langle y_n(t), y_n(t)\rangle + \langle u_n(t), Ru_n(t)\rangle)\, dt \right) \\ &= \int_0^T \left(\sum_{n=0}^{\infty} (\langle y_n(t), y_n(t)\rangle + \langle u_n(t), Ru_n(t)\rangle) \right) dt. \end{aligned} \tag{5.2}$$

The two different expressions given above for the cost function gives rise to two different interpretations. In the first interpretation we take the input, state, and output spaces to be

$$\mathcal{U} = L^2(0,T;U), \qquad \mathcal{H} = L^2(0,T;H), \qquad \mathcal{Y} = L^2(0,T;Y),$$

and we have a standard discrete time minimization problem.

In the second interpretation we observe that, for each fixed $t \in [0,T)$, the sequences $x_n(t)$ and $y_n(t)$ depend only on $x_0(t)$ and $u_n(t)$, and not on $x_0(s)$ and $u_n(s)$ for $s \neq t$. This means that the system (5.1) is really a collection of independent equations, parametrized by the real parameter $t \in [0,T)$. Moreover, it follows from the last line in (5.2) that in order to minimize the total cost it suffices to minimize each t-parametrized problem separately. Thus, in this interpretation, we have an infinite number (parametrized by $t \in [0,T)$) of problems that are otherwise identical, but have different initial states $x_0 = x_0(t)$. Each subproblem is a discrete time minimization problem of the type described in Section 1, with input space U, state space H, and output space Y. Let us in the sequel denote the common optimal feedback and Riccati operators for these subproblems by K and P, respectively. Then, for all $t \in [0,T)$ and $n \in \mathbb{N}$, we have $u_n^{opt}(t) = Kx_n^{opt}(t)$, or if we recall the definitions of $x_n(t)$ and $u_n(t)$,

$$u^{opt}(t) = Kx^{opt}(t-T), \qquad t \geq 0. \tag{5.3}$$

Moreover, by (5.2), the total optimal cost will be

$$\begin{aligned}\mathcal{W}(x_0, u^{opt}) &= \int_0^T W(x_0(t), u^{opt}(t))\, dt \\ &= \int_0^T \langle x_0(t), P x_0(t)\rangle\, dt \\ &= \int_{-T}^0 \langle x(t), P x(t)\rangle\, dt.\end{aligned} \tag{5.4}$$

6. A Continuous Time Formulation

Although the two discrete time formulations given above are very natural, the most common approach is to formulate (4.1) as a continuous time problem rather than a discrete time problem. Equation (4.1) is a special case of what is usually called a "difference equation"; see [Hale(1977), Section 12.3]. The standard method to rewrite this into a continuous time system is to solve (4.1) to get x, then to translate x to the left, and to restrict x to $[-T, 0)$ to get a new initial function given on $[-T, 0)$ for the same equation. In this setting the input, state, and output spaces become U, $\mathcal{H} = L^2(-T, 0; H)$, and Y, respectively, and the state $z(t)$ at time t is given by

$$z(t) = (s \mapsto x(t+s)), \qquad s \in [-T, 0).$$

The generator $\mathcal{A}$ of the semigroup that we get in this way is the differentiation operator

$$\mathcal{A}z = z', \quad dom(\mathcal{A}) = \left\{ z \in W^{1,2}(-T, 0; H) \;\middle|\; z(0) = Az(-T) \right\}. \tag{6.1}$$

Its adjoint is the differentiation operator

$$\mathcal{A}^* z = -z', \quad dom(\mathcal{A}^*) = \left\{ z \in W^{1,2}(-T, 0; H) \;\middle|\; z(-T) = A^* z(0) \right\}. \tag{6.2}$$

The input and output operators $\mathcal{B}$ and $\mathcal{C}$ are unbounded, and they are defined through the equations

$$\mathcal{B}^* z = B^* z(0) \text{ for } z \in dom(\mathcal{A}^*), \qquad \mathcal{C}z = Cz(-T) \text{ for } z \in dom(\mathcal{A}). \tag{6.3}$$

The resulting system is well-posed and regular. For details, see [Staffans(1996b), Theorem 6.1].

From the discussion in the previous section we know the optimal solution to the quadratic cost minimization problem. By (5.3), $u^{opt}(t) = Kx^{opt}(t - T)$, hence the continuous time state feedback operator $\mathcal{K}$ is unbounded, and it is given by

$$\mathcal{K}z = Kz(-T), \quad z \in dom(\mathcal{A}), \tag{6.4}$$

and by (5.4), the continuous time Riccati operator is

$$\mathcal{P}z = Pz, \quad z \in L^2(-T, 0; H). \tag{6.5}$$

We claim that $\mathcal{P}$ does not in general map $dom(\mathcal{A})$ into $dom(\mathcal{A}^*)$. This can be seen as follows. By (6.1) and (6.2), $\mathcal{P}$ maps $dom(\mathcal{A})$ into $dom(\mathcal{A}^*)$ if and only if $P = A^*PA$. However, if this is the case, then we can iterate the equation $P = A^*PA$ to get $P = (A^*)^k PA^k$ for every $k \in \mathbb{N}$, and letting $k \to \infty$ we find that $P = 0$. Thus, the only case in which $\mathcal{P}$ maps $dom(\mathcal{A})$ into $dom(\mathcal{A}^*)$ is when $P = 0$, i.e., the optimal cost is zero and also $\mathcal{P} = 0$. In all other cases, in order to give a meaning to the term $\mathcal{A}^*\mathcal{P}$ in (1.12) we have to extend the domain of $\mathcal{B}^*$. Since $\mathcal{P}$ maps $dom(\mathcal{A})$ into $W^{1,2}(-T, 0; H)$, it suffices to define $\mathcal{B}^*x$ for all $x \in W^{1,2}(-T, 0; H)$ (see the discussion in Section 2). Equation (6.3) does not define $\mathcal{B}^*$ uniquely on $W^{1,2}(-T, 0; H)$ since $dom(\mathcal{A}^*)$ is not dense in this space. The Weiss extension of $\mathcal{B}^*$ (cf. (3.1)) is given by

$$\overline{\mathcal{B}}^* x = B^* x(0), \tag{6.6}$$

for all x in $W^{1,2}(-T, 0; H)$.

7. Computation of the Operator $\mathcal{S}$

Since we know $\mathcal{P}$, we can compute $\mathcal{S}$ from (3.2). For each $f \in \mathcal{H} = L^2(-T, 0; H)$ and $\alpha \in \rho(\mathcal{A})$ we have

$$\left((\alpha I - \mathcal{A})^{-1} f\right)(t) = \left(I - e^{-\alpha T} A\right)^{-1} \int_{-T}^{0} e^{\alpha(t-s)} f(s)\, ds - \int_{-T}^{t} e^{\alpha(t-s)} f(s)\, ds.$$

By letting f tend to $\mathcal{B}u = B\delta_0 u$ (where δ_0 is the unit atom at zero) in the distribution sense we get

$$\left((\alpha I - \mathcal{A})^{-1} \mathcal{B}u\right)(t) = e^{\alpha t} \left(I - e^{-\alpha T} A\right)^{-1} Bu, \tag{7.1}$$

hence, by (3.2) and (6.6),

$$\begin{aligned}
\mathcal{S}u &= (R + D^*D)u + \lim_{\alpha\to\infty} \overline{\mathcal{B}}^* \mathcal{P} (\alpha I - \mathcal{A})^{-1} \mathcal{B}u \\
&= (R + D^*D)u + \lim_{\alpha\to\infty} B^* P \left(I - e^{-\alpha T} A\right)^{-1} Bu \\
&= (R + D^*D)u + B^*PBu = Su.
\end{aligned}$$

Thus $\mathcal{S} = S$.

8. Verification of the Modified Continuous Time Lure Equations

Above we have solved the quadratic cost minimization problem for equation (1.1) with cost function (1.2) by appealing to the discrete time theory. Here we shall show that the continuous time feedback operator $\mathcal{K}$ and Riccati operator $\mathcal{P}$ satisfy (1.9) and (1.10), where $\mathcal{S}$ is the operator that we computed above, i.e., $\mathcal{S} = S$.

Let us start with the verification of (1.9). Take $z \in dom(\mathcal{A})$. By (6.4), $\mathcal{K}z = Kz(-T)$, and by (6.5) and (6.6),

$$(\overline{\mathcal{B}}^* \mathcal{P} + D^*\mathcal{C})z = B^*Pz(0) + D^*Cz(-T).$$

Replacing $z(0)$ by $Az(-T)$ we get

$$(\overline{\mathcal{B}}^* \mathcal{P} + D^*\mathcal{C})z = (B^*PA + D^*C)z(-T).$$

Thus, (1.9) follows from (1.3), (6.4), and the fact that $\mathcal{S} = S$.

It remains to verify (1.10). Take z_0, $z_1 \in dom(\mathcal{A})$. Since $\mathcal{P}$ maps $W^{1,2}(-T,0;H)$ into itself we can integrate by parts to get

$$\begin{aligned}
&\langle \mathcal{A}z_0, \mathcal{P}z_1\rangle + \langle z_0, \mathcal{P}\mathcal{A}z_1\rangle \\
&\quad = \int_{-T}^{0} \langle z_0'(t), (\mathcal{P}z_1)(t)\rangle \, dt + \int_{-T}^{0} \langle z_0(t), (\mathcal{P}z_1')(t)\rangle \, dt \\
&\quad = \langle z_0(0), (\mathcal{P}z_1)(0)\rangle - \langle z_0(-T), (\mathcal{P}z_1)(-T)\rangle \\
&\qquad + \int_{-T}^{0} \langle z_0(t), (\mathcal{P}z_1')(t) - (\mathcal{P}z_1)'(t)\rangle \, dt \\
&\quad = \langle Az_0(-T), PAz_1(-T)\rangle - z_0(-T)Pz_1(-T),
\end{aligned}$$

where the last equation follows from the facts that $z_0(0) = Az_0(-T)$, $z_1(0) = Az_1(-T)$, and $(\mathcal{P}z)(t) = Pz(t)$. Thus

$$\langle \mathcal{A}z_0, \mathcal{P}z_1\rangle + \langle z_0, \mathcal{P}\mathcal{A}z_1\rangle = \langle z_0(-T), (A^*PA - P)z_1(-T)\rangle \, .$$

This equation, together with (1.4), (6.3), and (6.4), gives (1.10).

References

[Curtain and Zwart(1995)] R. F. Curtain and H. Zwart. *An Introduction to Infinite-Dimensional Linear Systems Theory.* Springer-Verlag, New York, 1995.

[Flandoli et al.(1988)] F. Flandoli, I. Lasiecka, and R. Triggiani. Algebraic Riccati equations with non-smoothing observation arising in hyperbolic and Euler-Bernoulli boundary control problems. *Annali Mat. pure appl.*, CLIII:307–382, 1988.

[Hale(1977)] J. K. Hale. *Theory of Functional Differential Equations.* Springer-Verlag, Berlin and New York, 1977.

[Malinen(1997)] J. Malinen. Nonstandard discrete time cost optimization problem: the spectral factorization approach. Preprint, 1997.

[Pritchard and Salamon(1985)] A. J. Pritchard and D. Salamon. The linear-quadratic control problem for retarded systems with delays in control and observation. *IMA Journal of Mathematical Control & Information*, 2:335–362, 1985.

[Pritchard and Salamon(1987)] A. J. Pritchard and D. Salamon. The linear quadratic control problem for infinite dimensional systems with unbounded input and output operators. *SIAM Journal on Control and Optimization*, 25:121–144, 1987.

[Salamon(1987)] D. Salamon. Infinite dimensional linear systems with unbounded control and observation: a functional analytic approach. *Transactions of the American Mathematical Society*, 300:383–431, 1987.

[Salamon(1989)] D. Salamon. Realization theory in Hilbert space. *Mathematical Systems Theory*, 21:147–164, 1989.

[Staffans(1995)] O. J. Staffans. Quadratic optimal control of stable systems through spectral factorization. *Mathematics of Control, Signals, and Systems*, 8:167–197, 1995.

[Staffans(1996a)] O. J. Staffans. On the discrete and continuous time infinite-dimensional algebraic Riccati equations. *Systems and Control Letters*, 29:131–138, 1996a.

[Staffans(1996b)] O. J. Staffans. Quadratic optimal control through coprime and spectral factorizations. Reports on Computer Science & Mathematics 178, Åbo Akademi University, Åbo, Finland, 1996b.

[Staffans(1997a)] O. J. Staffans. Quadratic optimal control of stable well-posed linear systems. To appear in Transactions of American Mathematical Society, 1997a.

[Staffans(1997b)] O. J. Staffans. Quadratic optimal control of well-posed linear systems. Submitted to SIAM Journal on Control and Optimization, 1997b.

[van Keulen(1993)] B. van Keulen. *H_∞-Control for Distributed Parameter Systems: A State Space Approach.* Birkhäuser Verlag, Basel Boston Berlin, 1993.

[Weiss(1994a)] G. Weiss. Transfer functions of regular linear systems. Part I: Characterizations of regularity. *Transactions of American Mathematical Society*, 342:827–854, 1994a.

[Weiss(1994b)] G. Weiss. Regular linear systems with feedback. *Mathematics of Control, Signals, and Systems*, 7:23–57, 1994b.

[Weiss and Zwart(1996)] G. Weiss and H. Zwart. An example in LQ optimal control. Preprint, 1996.

[Weiss and Weiss(1997)] M. Weiss and G. Weiss. Optimal control of stable weakly regular linear systems. To appear in Mathematics of Control, Signals, and Systems, 1997.

Olof J. Staffans
Department of Mathematics
Åbo Akademi University
FIN-20500 Åbo, Finland
e-mail: Olof.Staffans@abo.fi

International Series of Numerical Mathematics
Vol. 126, © 1998 Birkhäuser Verlag, Basel

The Wave Equation with Neuman Controls: On Lions's F Space

DANIEL TATARU*

Department of Mathematics
Princeton University

ABSTRACT. A crucial issue in boundary controllability problems for partial differential equations is choosing the appropriate regularity for the space of controls. This space should satisfy two criteria: it should be large enough, so that controllability is possible; and it should be small enough, so that the initial regularity of solutions is preserved. Reconciling these two opposite criteria is sometimes easy and sometimes very difficult. The aim of this article is to provide a solution for this problem for the wave equation with Neuman controls.

1991 *Mathematics Subject Classification.* 93C20

Key words and phrases. Exact zero controllability, boundary control, wave equation.

1. The controllability problem

We denote the coordinates in $\mathbb{R} \times \mathbb{R}^n$ by $(x_0 = t, x_1, \ldots, x_n)$. For convenience we call t the time coordinate and x the space coordinate. ξ stands in the sequel for the corresponding Fourier variable. Set

$$D_j = \frac{1}{i}\frac{\partial}{\partial x_j}.$$

Let Ω be a bounded set in $\mathbb{R}^n$ with C^2 boundary $\partial\Omega$. Then we use the notations $\langle \cdot, \cdot \rangle$, respectively $\langle \cdot, \cdot \rangle_\partial$ for the L^2 inner product in $\Omega \times [0,T]$, respectively $\partial\Omega \times [0,T]$.

Consider a second order hyperbolic partial differential operator

$$P(t, x, D) = \partial_j a^{jk} \partial_k$$

in $\Omega \times [0,T]$ so that the surfaces $t = \text{const}$ are space-like. The corresponding inhomogeneous Neuman problem is

$$\left\{ \begin{array}{ll} Pu = 0 & \text{in } \Omega \times [0,T], \\ \partial_\nu u = g & \text{in } \partial\Omega \times [0,T], \\ u(0) = u_0 & \text{in } \Omega, \\ \partial_\mu u(0) = u_1 & \text{in } \Omega. \end{array} \right. \tag{1.1}$$

Here ∂_μ is the conormal derivative with respect to the surfaces $t = \text{const}$,

$$\partial_\mu = a^{0k} \partial_k$$

*Research partially supported by NSF grant DMS9622942 and by an Alfred P. Sloan fellowship.

and ∂_ν is the conormal derivative with respect to the lateral boundary $\partial\Omega \times [0,T]$,

$$\partial_\nu = \nu^j a^{jk} \partial_k$$

where ν is the unit outer normal to $\partial\Omega$.

If the coefficients are smooth then the homogeneous problem (i.e. with $g = 0$) is well-posed in $H^s(\Omega) \times H^{s-1}(\Omega)$ at least if $-1/2 < s < 3/2$. For other values of s one cannot simply use the H^s spaces and the appropriate compatibility conditions are required on the boundary. If the coefficients of P are only C^1 then the homogeneous problem is well-posed in $H^s(\Omega) \times H^{s-1}(\Omega)$ for $0 \leq s \leq 1$.

Let Γ be an open subset of $\partial\Omega \times [0,T]$. Then the exact boundary controllability problem with Neuman boundary controls localized in Γ can be stated as follows:

Given any initial data $(u_0, u_1) \in H^s \times H^{s-1}$ find a boundary control $g \in G_s$, supported in Γ, so that the solution u to (1.1) satisfies $u(T) = \partial_\mu u(T) = 0$.

The allowable range of s depends on the regularity of the coefficients (see the comments above). Since the Neuman problem is backwards well-posed, one can also start at time T with 0 Cauchy data, solve the equation backwards and try to find g so that at time $t = 0$ the Cauchy data is exactly (u_0, u_1).

The problem we are interested in here is that of the choice of the space of controls G_s. A good choice (see [4]) should have the following properties:

(C) G_s should be large enough, so that controllability is possible.

(CR) G_s should also be sufficiently small, so that the solutions to (1.1) stay in H^s between the times 0 and T.

The usual choice in the literature is the obvious one, $G_s = H^{s-1}$. Under appropriate assumptions on the coefficients and on the size of the set Γ (see e.g. [2], [3], [1], [6]) this space is large enough to satisfy (C). However, if $n > 1$ then it fails to satisfy (CR). Hence, one can infer that the good choice of G_s is a space slightly smaller than H^{s-1}. The first idea one is tempted to try is to substitute H^{s-1} by H^{q-1}, $q > s$; unfortunately, in this case (C) fails. The explanation is that in the hyperbolic region of the cotangent bundle of the boundary (which correspond to the singularities that hit the boundary transversally and are reflected) the microlocal H^{s-1} regularity for functions in G^s is the correct one; it is only in the glancing region (which corresponds to singularities propagating in directions tangent to the boundary) that one needs some better microlocal regularity for G^s functions.

We conclude the discussion of the controllability problem with remarks about some of its features which are only indirectly related to the problem considered here.

Remark 1.1. *(On suitable geometric assumptions) Obtaining exact boundary controllability results requires certain assumptions on the geometry of the controlled region Γ relative to Ω and on the regularity of the coefficients of the hyperbolic operator. The two most important such sets of conditions are*

i) The geometric optics condition (see [1]), based on the idea of propagating information along rays. This requires any generalized bicharacteristic of P in $\Omega \times [0,T]$ to hit Γ in a nondiffractive point. A nondiffractive point is a point where the ray would leave the domain Ω if there were no boundary.

ii) The pseudoconvexity condition (see [7]), based on the idea of propagating information across pseudoconvex surfaces. This requires the existence of a strongly pseudoconvex function ϕ in $\Omega \times [0,T]$ which is negative at times $0, T$ but positive *at some intermediate time t and which satisfies $\partial_\nu \phi < 0$ outside Γ. (the result was proved in [6] for the case when the control is taken on the entire boundary; to obtain the more precise geometrical condition on Γ one needs to use more refined Carleman estimates for solutions to boundary value problems as in [7].)*

Remark 1.2. *(The regularity of the coefficients) Well-posedness of the hyperbolic problem requires essentially C^1 coefficients. Many of the features of the problem depend on this regularity. The geometric optics method requires C^2 coefficients. The Carleman estimates method requires only C^1 coefficients. The range of admissible values for s also depends on the regularity of the coefficients. For instance, with C^1 coefficients and the equation in divergence form the admissible range is $0 \leq s \leq 1$.*

2. Duality and the observability problem

Consider now the dual homogeneous problem

$$\begin{cases} P^* v = 0 & \text{in } \Omega \times [0,T], \\ \partial_\nu v = 0 & \text{in } \partial\Omega \times [0,T], \\ v(0) = v_0 & \text{in } \Omega, \\ \partial_\mu v(0) = v_1 & \text{in } \Omega, \end{cases} \tag{2.1}$$

which is well-posed in the space $H^q(\Omega) \times H^{q-1}(\Omega)$.

To it we associate the stable observability problem

Given the observation $v_{|\Gamma}$ in a space $F^q = F^q_\Gamma$, determine the initial data $(v_0, v_1) \in H^q \times H^{q-1}$.

Now the question is how to choose the observation space F^q. It should be small enough, so that observability can hold; but it should also be large enough, so that it contains $v_{|\Gamma}$ for any initial data $(v_0, v_1) \in H^q \times H^{q-1}$. This can be summarized in the following two inequalities (O), which guarantees that stable observability holds, and (OR), which gives the regularity of the observation.

$$\textbf{(O)} \qquad |v_0|_q + |v_1|_{q-1} \leq c|v|_{F^q_\Gamma}$$

$$\textbf{(OR)} \qquad |v_0|_q + |v_1|_{q-1} \geq c|v|_{F^q_\Gamma}$$

The controllability and observability problems are dual. To make this more precise start with the following integration by parts,

$$\langle u, \partial_\mu v\rangle - \langle \partial_\mu u, v\rangle \,|_{t=T}^{t=0} = \langle u, Pv\rangle - \langle v, Pu\rangle + \langle u, \partial_\nu v\rangle_\partial - \langle v, \partial_\nu u\rangle_\partial \tag{2.2}$$

for any smooth functions u, v. Suppose now that u solves the inhomogeneous problem (1.1) with zero Cauchy data at time T and that v solves the homogeneous problem (2.1). Then (2.2) becomes

$$\langle u_0, v_1\rangle - \langle u_1, v_0\rangle = -\langle g, v\rangle_\partial \tag{2.3}$$

which is the duality relation which connects the controllability problem with the corresponding observability problem. For the controllability problem one needs to look at the map

$$T : G_s(\Gamma) \to H^s(\Omega) \times H^{s-1}(\Omega), \qquad Tg = (u(0), \partial_\mu u(0))$$

(where u is assumed to solve (1.1) with 0 Cauchy data at time T). Then the controllability and regularity statements (C) and (CR) are equivalent to saying that T is a surjective, respectively a bounded operator.

For the observability problem, on the other hand, consider the map

$$S : H^q(\Omega) \times H^{q-1}(\Omega) \to F_q(\Gamma), \qquad T(v_0, v_1) = v_{|\Gamma}$$

(where v is assumed to solve (2.1)). Then (O) and (OR) are equivalent to saying that S is bounded from below, respectively that S is bounded.

By (2.3) the duality $S = T'$ is achieved if we take $q + s = 1$ and $G_s = F'_q$. The controllability and regularity statements (C), (CR) are then equivalent to the observability and regularity estimates (O), (OR). Hence, determining the correct spaces G^s for the boundary controls reduces to finding the appropriate spaces F^q for controllability.

Lions's idea is the following. Suppose we know the uniqueness result $v = 0$ in Γ implies $v_0 = v_1 = 0$. Then define the norm of the space F^q exactly by

$$|v|_{F_q(\Gamma)} = |v_0|_q + |v_1|_{q-1}$$

Such a space will have the right properties, and its dual is the good space for controllability.

Thus, it remains to characterize the space F_q. Intuitively, the dependence of the F_q norms on q should be fairly simple. One would expect that F_q and F_r differ by exactly $q - r$ derivatives, $F_q = D^{r-q} F_r$. Then it is best to characterize F_q for a given value of q. The simplest choice, which is used in the sequel, is $q = 1$. Then we want to identify the space $F = F_1$, defined by

$$|v|_{F_\Gamma} = |v_0|_{H^1} + |v_1|_{L^2}$$

3. The F space

Define the "tangential" component R of P on the boundary $\partial\Omega \times [0, T]$ by

$$Ru = Pu \quad \text{whenever} \quad \partial_\nu u = \partial_\nu^2 u = 0$$

This is equivalent to saying that local coordinates can be chosen in which $\partial\Omega = \{x_n = 0\}$ and

$$p(x, \xi) = \xi_n^2 - r(x, \xi') \quad \text{on } \{x_n = 0\} \tag{3.1}$$

Since the boundary $\partial\Omega \times [0, T]$ is time-like, it follows that R is also hyperbolic. If P has C^1 coefficients then R also has C^1 coefficients.

Introduce now the spaces $X^{s,\theta}$ associated to R by

$$X^{s,0} = H^s \quad X^{s,1} = \{u \in H^s; \ \ Ru \in H^{s-1}\} \quad X^{s,-1} = H^s + R^* H^{s+1}$$

$$X^{s,\theta} = [X^{s,0}, X^{s,1}]_\theta \quad 0 < \theta < 1 \qquad X^{s,-\theta} = [X^{s,0}, X^{s,-1}]_\theta \quad 0 < \theta < 1$$

(complex interpolation). These spaces are L^2 type Sobolev spaces which have a special structure near the characteristic set of R. It is easier to understand these spaces in the constant coefficient case. Then the above definitions are equivalent to

$$X^{s,\theta} = \{u | \ (1 + |\xi|)^s (1 + \frac{|r(\xi)|}{|\xi|})^\theta \hat{u} \in L^2 \ \}$$

One can see that the index s corresponds to classical derivatives, while the index θ corresponds to "derivatives" away from the characteristic set of R. Such spaces have been investigated in detail in [8]. By interpolation one can easily prove that they have some good properties such as microlocalization and the expected mapping properties for pseudodifferential operators and R:

$$OPS^m : X^{s,\theta} \to X^{s-m,\theta} \tag{3.2}$$

$$R : X^{s,\theta} \to X^{s-1,\theta-1} \tag{3.3}$$

On a bounded open set Γ we define the $X^{s,\theta}$ space in the standard way, as the restriction to Γ of $X^{s,\theta}$ functions.

Remark 3.1. *If the coefficients of R are only C^1 then the $X^{s,\theta}$ spaces are well-defined and have the above properties only for $|s| \leq 1$, $|s-\theta| \leq 1$. This is exactly the range which can be obtained by interpolation from the spaces $X^{1,1}$, $X^{0,1}$, H^1 and their duals.*

Then our main result is

Theorem 3.1. *Suppose that P has C^1 coefficients and that the boundary $\partial\Omega$ is of class C^2. Assume that a suitable set of geometric assumptions (see Remark 1.1) is fulfilled. Then the F norm is equivalent to the $X^{1/2,1/2}$ norm.*

Note that $X^{0,0} = L^2$ and

$$X^{1,1} = \{w \in H^1 \mid Rw \in L^2\}$$

Hence, if we interpret R as a selfadjoint operator in L^2 then the space $X^{1/2,1/2}$ can be locally characterized as

$$X^{1/2,1/2} = H^{1/2} \cap D(R^{1/2})$$

This result leads to the following optimal choices for the observability and controllability problems:

Corollary 3.1. *Suppose that P has C^1 coefficients and that the boundary $\partial\Omega$ is of class C^2. Then the F_q norm is equivalent to the $X^{q-1/2,1/2}$ norm and the optimal space of controls G_s is $X^{s-1/2,-1/2}$, $0 \leq s, q \leq 1$. If both the coefficients of P and the boundary are smooth then the same result holds for all real s, q.*

Before we prove the theorem, let us make one important observation. The F norm, as we have defined it, applies only to traces of solutions to the Neuman problem, which is not a dense set in any Sobolev space. Hence, there is definitely more than one norm say, on smooth functions, which extends F. A striking example of that is given by the following version of the above theorem:

Theorem 3.2. *Suppose that P has C^1 coefficients and that the boundary $\partial\Omega$ is of class C^2. Assume that a suitable set of geometric assumptions is fulfilled. Then the F norm is equivalent to the $H^{2/3} \cap X^{1/2,1/2}$ norm.*

The $X^{1/2,1/2}$ and $H^{2/3} \cap X^{1/2,1/2}$ norms are certainly not equivalent in general. However, it turns out that they are equivalent on the function space where F is defined.

Proof. We need to prove the following two estimates

$$|v_0|_1 + |v_1|_0 \leq c|v|_{X^{1/2,1/2}(\Gamma)} \tag{3.4}$$

$$|v_0|_1 + |v_1|_0 \geq c|v|_{H^{2/3} \cap X^{1/2,1/2}(\partial\Omega\times[0,T])} \tag{3.5}$$

The second one is a trace regularity result which was proved in [9]. The assumptions in [9] are that the coefficients are sufficiently smooth. It now appears that the same result holds even for C^1 coefficients; however, this will be proved elsewhere.

To give a complete proof to (3.4) we would need to redo the appropriate observability estimates for various sets of geometric conditions (see e.g. [1],[6]). Fortunately, in all these cases the boundary traces appear in the same way. The common computation done in all these works is local, near the observed part of the boundary Γ. It goes like this:

One starts with a quadratic form of the form $\langle Pu, Qu\rangle$, where Q is of order 1 and is either a differential operator or an operator of the form

$$Q = Q_0 D_n + Q_1$$

where Q_0, Q_1 are tangential pseudodifferential operators of order 0, respectively 1, with purely imaginary symbol. The main step is then to compute by integration by parts (commuting)

$$2Re\langle Pv, Qv\rangle = \langle Pv, Qv\rangle + \langle Qv, Pv\rangle = A(v,v) + B(v,v)$$

where A is an second order interior quadratic form and B is a second order boundary quadratic form. A is then used to estimate the H^1 interior norm of v, and B is bounded by

$$B(v,v) \leq c|v|^2_{H^1(\Gamma)}$$

Our aim, therefore, is to refine this to

$$B(v,v) \leq c|v|^2_{X^{1/2,1/2}}$$

To do that, we need to determine what is B. It is easier to do that in local coordinates where (3.1) holds. Then

$$P = D_n^2 - R, \quad Q = Q_0 D_n + Q_1$$

When we integrate by parts to commute P and Q we obtain

$$2Re\langle Pv, Qv\rangle = \text{interior terms} + \langle D_n v, iQ_1 v\rangle_\partial + \langle D_n v, iQ_0 D_n v\rangle_\partial + \langle iQ_0 v, Rv\rangle_\partial$$

Recalling the boundary condition $D_n v = 0$ it follows that

$$B(v,v) = \langle iQ_0 v, Rv\rangle_\partial$$

By (3.2), (3.3) this can be bounded by

$$|B(v,v)| \leq c|Rv|_{X^{-1/2,-1/2}}|Q_0 v|_{X^{1/2,1/2}} \leq c|v|^2_{X^{1/2,1/2}}$$

This concludes the proof. Note that the above argument requires only C^1 regularity of the coefficients of P. □

References

1. C. Bardos, G. Lebeau and R. Rauch *Sharp sufficient conditions for the observation, control, and stabilization of waves from the boundary.* SIAM J. Control Optim. 30 (1992), no. 5, 1024–1065
2. I. Lasiecka, R. Triggiani *Exact controllability of the wave equation with Neumann boundary control.* Appl. Math. Optim. 19 (1989), no. 3, 243–290
3. W. Littman *Remarks on boundary control for polyhedral domains and related results*, Proc. Conf. On boundary Control, Sophia Antipolis, oct. 1990
4. J.L. Lions *Controlabilite Exacte. Perturbations et Stabilisation des Systemes Distribue Vol 1 et 2* Masson, Paris (1988)
5. D. Tataru *A-priori estimates of Carleman's type in domains with boundary* Journal de Mathematiques Pure et Appl., 73, 1994, 355–387
6. D. Tataru *Boundary controllability for conservative P.D.E.*, Applied Math. and Optimization, 31, 1995, 257–295
7. D. Tataru *Carleman estimates and unique continuation near the boundary for P.D.E.'s* Journal de Math. Pure et Appl. 75, 1996 367–408
8. D. Tataru *On the X_θ^s spaces and unique continuation for semilinear hyperbolic equations* Comm. PDE, 21 (1996), no 5–6
9. D. Tataru *On the regularity of the boundary traces for the wave equation*, Annali de la Scuola Normale Superiore di Pisa, to appear

Daniel Tataru
Department of Mathematics
Princeton University
Princeton, NJ 08540, USA

International Series of Numerical Mathematics
Vol. 126, © 1998 Birkhäuser Verlag, Basel

On the Pointwise Stabilization of a String

M. TUCSNAK

Ecole Polytechnique
Centre de Mathématiques Appliquées
and Université de Versailles

ABSTRACT. We consider an initial and boundary value problem the one dimensional wave equation with damping concentrated at an interior point. Our main results assert that the decay rate is uniform for regular initial data and give lower estimates of the decay rate. An essential intermediate step is the description of the spectrum of the associate dissipative operator.

1991 *Mathematics Subject Classification.* 93D15, 93C20, 35B37

Key words and phrases. Invariance principle, unique continuation, strong stabilization, decay rates.

1. Introduction and statement of the main results

The main goal of the present paper is to study the asymptotic behaviour of solutions for the following initial and boundary value problem:

$$u''(x,t) - \frac{\partial^2 u}{\partial x^2}(x,t) + u'(a,t)\delta_a = 0, \ \forall x \in (0,1), \ \forall t \in (0,\infty) \tag{1.1}$$

$$u(0,t) = u(1,t) = 0, \ \forall t \in (0,\infty) \tag{1.2}$$

$$u(x,0) = u^0(x), \ u'(x,0) = u^1(x), \ \forall x \in (0,1), \tag{1.3}$$

where δ_a is the Dirac mass concentrated in the point $a \in (0,1)$ and by u', u'' we denoted the time derivatives of u. Equations (1.1)–(1.3) are dissipative since

$$E'(t) = -|u'(a,t)|^2, \tag{1.4}$$

where $E = E(t)$ is the energy

$$E(t) = \frac{1}{2}\int_0^1 \left[|u'(x,t)|^2 + \left|\frac{\partial u}{\partial x}(x,t)\right|^2 \right] dx.$$

The main known results concerning the asymptotic behaviour of solutions of (1.1)–(1.3) can be summarized as follows:

Theorem 1.1. *1. For any $a \in (0,1)$ the problem (1.1)–(1.3) admits a unique solution u satisfying*

$$u \in C([0,\infty), H_0^1(0,1)) \cap C^1([0,\infty), L^2(0,1)).$$

2. The solution u of (1.1)–(1.3) has the decay property

$$\lim_{t\to\infty}(\|u(t)\|_{H^1(0,1)} + \|u'(t)\|_{L^2(0,1)}) = 0. \tag{1.5}$$

if and only if

$$a \in (0,1) \cap (\mathbb{R} - \mathbb{Q}). \tag{1.6}$$

3. For any a satisfying (1.6) the decay of u to zero is not uniform in the energy space. More precisely, for any function $\psi : [0,\infty) \to \mathbb{R}$ with $\lim_{t\to\infty}\psi(t) = 0$ there exists a sequence (t_n), $t_n \to \infty$ and a solution u_ψ of (1.1)–(1.3) such that

$$\frac{\|\{u_\psi(t_n), u'_\psi(t_n)\}\|_{H_0^1(0,1)\times L^2(0,1)}}{\|\{u_\psi(0), u'_\psi(0)\}\|_{H_0^1(0,1)\times L^2(0,1)}} \geq \psi(t_n), \ \forall n \geq 1. \tag{1.7}$$

The results in Theorem 1.1 were essentially proved in [2] and [7]. However, for the sake of completeness we shall sketch the proof in section 2.

In the present paper we shall prove that if we assume some additional smoothness of the initial data we can achieve a definite rate of decay for the solutions of (1.1)–(1.3). Moreover we shall show that the energy decays at most as a negative power of time. In order to state the precise result we shall consider a subspace of $H_0^1(0,1) \times L^2(0,1)$ defined by

$$\begin{aligned}(1.8)\quad \mathcal{D}(\mathcal{A}) &= \\ &= \{(u,v) \in [H_0^1(0,1)]^2 | u \in H^2(0,a) \cap H^2(a,1), \frac{\partial u}{\partial x}(a+) - \frac{\partial u}{\partial x}(a-) = v(a)\},\end{aligned}$$

endowed with the norm

$$||(u,v)||^2_{\mathcal{D}(\mathcal{A})} = \|u\|^2_{H^2(0,a)} + \|u\|^2_{H^2(a,1)} + \|v\|^2_{H^1(0,1)}. \tag{1.9}$$

Our first result on the uniform decay of solutions of (1.1)–(1.3) is

Proposition 1.1. *For any a satisfying (1.6) there exists a function $h_a : [0,\infty) \to \mathbb{R}$ with $\lim_{t\to\infty} h_a(t) = 0$, such that the solution u of (1.1)–(1.3) satisfies*

$$\|\{u(t), u'(t)\}\|_{H_0^1(0,1)\times L^2(0,1)} < h_a(t)||(u^0,u^1)||_{\mathcal{D}(\mathcal{A})}, \tag{1.10}$$

$$\forall (u^0,u^1) \in \mathcal{D}(\mathcal{A}), \ \forall t \geq 0.$$

Our main result shows that for any irrational a the function h_a in Proposition 1.1 tends to zero at most as $\frac{1}{\sqrt{t}}$. More precisely we have

Theorem 1.2. *For any a satisfying (1.6) there exists a sequence $t_n \to \infty$ and a sequence (u_n) of solution of (1.1)–(1.2) such that*

$$\lim_{n\to\infty}\sqrt{t_n}\frac{\|\{u_n(t_n), u'_n(t_n)\}\|_{H_0^1(0,1)\times L^2(0,1)}}{\|\{u_n(0), u'_n(0)\}\|_{\mathcal{D}(\mathcal{A})}} = C \in (0,\infty). \tag{1.11}$$

Moreover for any $\delta > 0$ there exists a constant a satisfying (1.6) such that

$$\lim_{n\to\infty} t_n^\delta \frac{\|\{u_n(t_n), u_n'(t_n)\}\|_{H_0^1(0,1)\times L^2(0,1)}}{\|\{u_n(0), u_n'(0)\}\|_{\mathcal{D}(\mathcal{A})}} > 0, \tag{1.12}$$

for some u_n and t_n as above.

The plan of this paper is as follows: in the second section we prove some preliminary results including Theorem 1.1; in the third section we prove the main results; we end up with a section devoted to further comments and other related questions.

2. Wellposedness and strong stabilization results

We shall first study the wellposedness of (1.1)–(1.3) by using the theory of semigroups. With $\mathcal{D}(\mathcal{A})$ defined by (1.8) and $X = H_0^1(0,1) \times L^2(0,1)$ we consider the operator $\mathcal{A} : \mathcal{D}(\mathcal{A}) \to X$ defined by

$$\mathcal{A}\begin{pmatrix} u \\ v \end{pmatrix} = \begin{pmatrix} -v \\ v(a)\delta_a - \frac{d^2u}{dx^2} \end{pmatrix}. \tag{2.1}$$

The wellposedness theorem for (1.1)–(1.3) will be a simple consequence of the following result:

Lemma 2.1. *The space $\mathcal{D}(\mathcal{A})$ is dense in X and $-\mathcal{A}$ is the generator of a continuous semigroup of contractions in X.*

Proof. The proof of the density of $\mathcal{D}(\mathcal{A})$ in X is a simple exercise so we shall skip it.

If we suppose that X is endowed with the scalar product

$$\left(\begin{pmatrix} u_1 \\ v_1 \end{pmatrix}, \begin{pmatrix} u_2 \\ v_2 \end{pmatrix}\right)_X = \int_0^1 \left(\frac{du_1}{dx}\frac{du_2}{dx} + v_1 v_2\right) dx,$$

a simple calculation shows that

$$\left(\mathcal{A}\begin{pmatrix} u \\ v \end{pmatrix}, \begin{pmatrix} u \\ v \end{pmatrix}\right)_X = 2v^2(a) \geq 0, \forall \begin{pmatrix} u \\ v \end{pmatrix} \in \mathcal{D}(\mathcal{A}),$$

so $\mathcal{A}$ is monotone. In order to prove that $-\mathcal{A}$ generates a semigroup of linear contractions on X it suffices to show that $\mathcal{A}$ is onto, i.e. $\mathcal{A}(\mathcal{D}(\mathcal{A})) = X$. Let $\begin{pmatrix} f \\ h \end{pmatrix} \in X$ and consider the equation

$$\mathcal{A}\begin{pmatrix} u \\ v \end{pmatrix} = \begin{pmatrix} f \\ h \end{pmatrix}, \tag{2.2}$$

which can be written as

$$v = -f \in H_0^1(0,1), \tag{2.3}$$

$$-\frac{d^2u}{dx^2} + v(a)\delta_a = h \in L^2(0,1). \tag{2.4}$$

From (2.3), (2.4) and a simple elliptic regularity result it follows that equation (2.2) admits a unique solution $\begin{pmatrix} u \\ v \end{pmatrix} \in \mathcal{D}(\mathcal{A})$, so $\mathcal{A}$ is onto. □

In order to study the asymptotic behaviour of the solutions of (1.1)–(1.3) we shall need the following compactness result.

Lemma 2.2. *The space $\mathcal{D}(\mathcal{A})$ is compactly embedded in X and the operator $\mathcal{A}^{-1}$ is compact from X into X.*

Proof. We first notice that $\mathcal{A}^{-1}$ is a linear continuous isomorphism from X onto $\mathcal{D}(\mathcal{A})$ and from $L^2(\Omega) \times H^{-1}(\Omega)$ onto X. It suffices then to use the fact that X is compactly embedded in $L^2(\Omega) \times H^{-1}(\Omega)$. □

Proof of Theorem 1.1. In order to prove the first assertion it suffices to notice that (1.1)–(1.3) can be written

$$\begin{pmatrix} u \\ u' \end{pmatrix}' + \mathcal{A} \begin{pmatrix} u \\ u' \end{pmatrix} = 0,$$

and to apply Lemma 2.1.

In order to prove (1.5) consider the problem

$$w'' - \frac{\partial w}{\partial x^2} = 0, \text{ in } (0,1) \times (0,\infty) \tag{2.5}$$

$$w(0,t) = w(1,t) = 0, \ \forall t \in (0,\infty), \tag{2.6}$$

$$w'(a,t) = 0, \ \forall t \in (0,\infty). \tag{2.7}$$

A simple Fourier expansion of w combined with the independence of complex exponentials show that the only function w satisfying (2.5)–(2.7) is $w \equiv 0$. By applying a version of the invariance principle of LaSalle (cf.[3] and [4]) we obtain now that u satisfies (1.5).

We still have to prove the third assertion of Theorem (1.1). We first notice that if a is rational then $\mathcal{A}$ admits purely imaginary eigenvalues. If a is irrational then it can be approached by a sequence of rationals and one can show (see Lemma 3.1 below for details) that there exists a sequence μ_n of eigenvalues of $\mathcal{A}$ such that

$$\mathcal{R}e\mu_n \to 0, \ \mathcal{I}m \ \mu_n \to \infty. \tag{2.8}$$

Denote now by $S(t)$ the semigroup generated by $\mathcal{A}$ and by $\mathcal{L}(X,X)$ the space of linear operators from X into X. Relation (2.8) implies the estimate

$$\|S(t)\| \leq Me^{-\omega t}, \ \forall t \geq 0,$$

is false for any $\omega, M > 0$. It suffices now to apply a result from [9] to obtain assertion 3 of Theorem 1.1.

3. Decay estimates

In this section we shall prove Proposition 1.1 and Theorem 1.2.

Proof of Proposition 1.1. It suffices to show that

$$\lim_{t\to\infty} \|S(t)\|_{\mathcal{L}(\mathcal{D}(\mathcal{A}),X)} = 0, \tag{3.1}$$

where $S(t)$ is the semigroup generated by $\mathcal{A}$ and $\mathcal{L}(\mathcal{D}(\mathcal{A}), X)$ is the space of linear bounded operators from $\mathcal{D}(\mathcal{A})$ into X. In this case estimate (1.10) holds true with

$h_a(t) = \|S(t)\|_{\mathcal{L}(\mathcal{D}(\mathcal{A}),X)}$. Let us suppose that (3.1) is false, i.e. that there exists $\epsilon > 0$, $t_n \to \infty$ and $(U_n) \subset \mathcal{D}(\mathcal{A})$, $\|U_n\|_{\mathcal{D}(\mathcal{A})} = 1$ such that

$$\|S(t_n)U_n\|_X \geq \epsilon, \ \forall n \geq 1. \tag{3.2}$$

As by Lemma 2.2 $\mathcal{D}(\mathcal{A})$ is compactly embedded in X there exists a subsequence (U_{n_k}) of (U_n) and $U \in X$, $U \neq 0$ such that

$$\lim_{k\to\infty} \|U_{n_k} - U\|_X = 0. \tag{3.3}$$

As $S(t)$ is a semigroup of contractions relation above and (3.3) imply that

$$\lim_{k\to\infty} \|S(t_{n_k})U_{n_k} - S(t_{n_k})U\|_X = 0. \tag{3.4}$$

From (3.2) and (3.4) we obtain that

$$\|S(t_{n_k})U\|_X \geq \frac{\epsilon}{2},$$

for k large enough, which obviously contradicts (1.5). □

An essential intermediate step in the proof of Theorem 1.2 is the study of the eigenvalues and of the eigenvectors of the operator $\mathcal{A} : \mathcal{D}(\mathcal{A}) \to H_0^1(0,1) \times L^2(0,1)$, where $\mathcal{D}(\mathcal{A})$ is defined by (1.8) and A is defined by (2.1). The eigenvectors and eigenvalues of $\mathcal{A}$ are characterized by the following result:

Lemma 3.1. *If $a \in (0,1)$, $a \notin \mathbb{Q}$ then a complex number λ is an eigenvalue of $\mathcal{A}$ if and only if λ satisfies the equation*

$$3 - e^{2\lambda} - e^{2a\lambda} - e^{2(1-a)\lambda} = 0. \tag{3.5}$$

All the eigenvalues of $\mathcal{A}$ are simple and the eigenvector $\begin{pmatrix}\phi_\lambda \\ \psi_\lambda\end{pmatrix}$ corresponding to the eigenvalue $-\lambda$ is given by

$$\phi_\lambda(x) = \begin{cases} \dfrac{e^{\lambda x} - e^{-\lambda x}}{e^{\lambda\xi} - e^{-\lambda\xi}}, & 0 < x < a, \\[2ex] \dfrac{e^{\lambda(x-1)} - e^{-\lambda(x-1)}}{e^{\lambda(\xi-1)} - e^{-\lambda(\xi-1)}}, & a < x < 1. \end{cases}, \quad \psi_\lambda(x) = -\lambda\phi_\lambda(x). \tag{3.6}$$

Proof. A simple calculation shows that relation

$$\mathcal{A}\begin{pmatrix}\phi_\lambda \\ \psi_\lambda\end{pmatrix} = \lambda\begin{pmatrix}\phi_\lambda \\ \psi_\lambda\end{pmatrix},$$

holds true if and only if

$$\psi_\lambda(x) = -\lambda\phi_\lambda(x), \tag{3.7}$$

and

$$-\frac{d^2\phi_\lambda}{dx^2}(x) + \lambda^2\phi_\lambda(x) = 0, \ x \in (0,a) \cup (a,1), \tag{3.8}$$

$$\phi_\lambda(a+) = \phi_\lambda(a-) = \phi_\lambda(a), \tag{3.9}$$

$$\frac{d\phi_\lambda}{dx}(a+) - \frac{d\phi_\lambda}{dx}(a-) = -\lambda\phi_\lambda(a), \tag{3.10}$$

$$\phi_\lambda(0) = \phi_\lambda(1) = 0. \tag{3.11}$$

If we suppose that $\phi_\lambda(a) = 0$ and we use that $a \notin \mathbb{Q}$ relations (3.8)–(3.11) imply that $\phi_\lambda(x) = 0$, $\forall x \in (0,1)$. This is why we shall admit that $\phi_\lambda(a) = 1$. From (3.8), (3.9), (3.11) we can then obtain that $\phi_\lambda, \psi_\lambda$ satisfy (3.6). Condition (3.10) (with $\phi(a) = 1$) implies now that λ satisfies (3.5). □

We can give the main proof of this section.

Proof of Theorem 1.2. Let us make the notation

$$G(a,\lambda) = 3 - e^{2\lambda} - e^{2a\lambda} - e^{2(1-a)\lambda},$$

with a satisfying (1.6) We first notice that the first assertion of Theorem 1.2 follows from the existence of a sequence $(z_n) \subset \mathbb{C}$ such that z_n is an eigenvalue of $\mathcal{A}$ and $0 < \mathcal{R}e z_n < \dfrac{K}{|\mathcal{I}m z_n|^2}$, $|\mathcal{I}m z_n| \to \infty$. In this case the sequence of solutions of (1.1)–(1.2) with initial data $\{\phi_{z_n}(x), 0\}$ satisfies (1.11) with $t_n = |\mathcal{I}m z_n|^2$. By Lemma 3.1 a sequence (z_n) satisfies conditions above if and only if

$$z_n = x_n + iy_n,\ y_n \to \infty,\ 0 < x_n \le \frac{K}{y_n^2},\ G(a, z_n) = 0. \tag{3.12}$$

In order to prove the existence of a sequence (z_n) satisfying (3.12) we first notice that by Theorem 5 in [1] there exist two sequences (p_n), $(q_n) \subset \mathbb{Z}$ such that

$$\left| a - \frac{p_n}{q_n} \right| \le \frac{5^{-\frac{1}{2}}}{q_n^2},\ \forall n \ge 1,\ q_n \to \infty. \tag{3.13}$$

By Rouché's theorem (cf. [10, p. 243]) it suffices to prove the existence of a constant $K > 0$ such that

$$\begin{aligned} &\left| G(a,\lambda) - \frac{\partial G}{\partial \lambda}(a, q_n\pi i)(\lambda - q_n\pi i) \right| \le \\ &\le \left| \frac{\partial G}{\partial \lambda}(a, q_n\pi i)(\lambda - q_n\pi i) \right|, \quad \text{if } |\lambda - q_n\pi i| = \frac{K}{q_n^2}. \end{aligned} \tag{3.14}$$

We shall first estimate the left hand side of (3.14) by writing the Taylor expansion

$$\begin{aligned} &G(a,\lambda) - \frac{\partial G}{\partial \lambda}(a, q_n\pi i)(\lambda - q_n\pi i) = \\ &= G(a, q_n\pi i) + \sum_{m\ge 2} \frac{1}{m!}\frac{\partial^m G}{\partial \lambda^m}(a, q_n\pi i)(\lambda - q_n\pi i)^m. \end{aligned} \tag{3.15}$$

Concerning the first term in the right hand side of (3.15) a simple calculation gives

$$\begin{aligned} G(a, q_n\pi i) &= 2 - e^{2\left(a - \frac{p_n}{q_n}\right)q_n\pi i} - e^{2\left(\frac{p_n}{q_n} - a\right)q_n\pi i} = \\ &= 4\sin^2\left[\left(a - \frac{p_n}{q_n}\right) q_n\pi\right]. \end{aligned}$$

By using (3.13) relation above implies the existence of $n_1 > 0$ such that

$$|G(a, q_n\pi i)| \leq \frac{4\pi^2}{5q_n^2}, \ \forall n \geq n_1. \tag{3.16}$$

We shall now estimate the infinite sum in the right hand side of (3.15). We can easily check that

$$\frac{\partial^m G}{\partial \lambda^m}(a, \lambda) = -2^m[e^{2\lambda} + a^m e^{2a\lambda} + (1-a)^m e^{2(1-a)\lambda}], \ \forall m \geq 1, \tag{3.17}$$

which implies that

$$\left|\frac{\partial^m G}{\partial \lambda^m}(a, q_n\pi i)\right| \leq 3 \cdot 2^m, \ \forall m \geq 1.$$

If $|\lambda - q_n\pi i| = \dfrac{K}{q_n^2}$ inequality above implies the existence of $n_2 > 0$ such that

$$\begin{aligned} &\sum_{m\geq 2} \frac{1}{m!} \frac{\partial^m G}{\partial \lambda^m}(a, q_n\pi i)(\lambda - q_n\pi i)^m \leq \\ &\leq 3(e^{\frac{2K}{q_n^2}} - \frac{2K}{q_n^2} - 1) \leq \frac{3K}{4q_n^4}, \ \forall n \geq n_2. \end{aligned} \tag{3.18}$$

By combining (3.15), (3.16) and (3.18) we obtain

$$\begin{aligned} &\left|G(a, \lambda) - \frac{\partial G}{\partial \lambda}(a, q_n\pi i)(\lambda - q_n\pi i)\right| \leq \frac{4\pi^2}{5q_n^2} + \frac{3K}{4q_n^4}, \\ &\forall n \geq \max(n_1, n_2), \ |\lambda - q_n\pi i| = \frac{K}{q_n^2}. \end{aligned} \tag{3.19}$$

On the other hand from (3.17) it follows that

$$\frac{\partial G}{\partial \lambda}(a, q_n\pi i) = -2\left[1 + ae^{2\left(a - \frac{p_n}{q_n}\right)q_n\pi i} + (1-a)e^{2\left(\frac{p_n}{q_n} - a\right)q_n\pi i}\right],$$

which implies that

$$\left|\frac{\partial G}{\partial \lambda}(a, q_n\pi i)\right| \geq 2\left\{1 + 2\cos\left[\left(a - \frac{p_n}{q_n}\right)q_n\pi\right]\right\}.$$

From relation above and (3.13) we obtain the existence of $n_3 > 0$ such that

$$\begin{aligned} &\left|\frac{\partial G}{\partial \lambda}(a, q_n\pi i)(\lambda - q_n\pi i)\right| \geq \frac{2K}{q_n^2}, \\ &\forall n \geq n_3, \ |\lambda - q_n\pi i| = \frac{K}{q_n^2}. \end{aligned} \tag{3.20}$$

Finally by combining (3.19) and (3.20) we obtain that (3.14) holds true for all $K > \frac{4\pi^2}{5}$, provided that n is large enough, so we obtain the first assertion of Theorem 1.2.

Let us now suppose that $\delta > 0$. According to [14] we can find $a \in (0,1)$ which is transcendental and such that there exist two sequences (p_n), $(q_n) \subset \mathbb{Z}$ with (q_n) strictly increasing such that

$$\left| a - \frac{p_n}{q_n} \right| < \frac{C}{q_n^\delta}, \quad \forall n \geq 1. \tag{3.21}$$

By using (3.21) and the estimates already proved in the first part of this proof one can easily show that the second assertion of 1.2 also holds true. □

4. Comments and related questions

The results in this paper can be generalized to the case of n-space dimensions, for the following problem:

$$u'' - \Delta u + g(u')\delta_\gamma = 0, \text{ in } \Omega \times (0, \infty) \tag{4.1}$$

$$u = 0, \text{ on } \Gamma \times (0, \infty) \tag{4.2}$$

$$u(x,0) = u^0(x), \; u'(x,0) = u^1(x), \text{ in } \Omega, \tag{4.3}$$

where:

(a) Ω is an open bounded subset of $\mathbb{R}^n$ with regular boundary Γ.
(b) $\gamma = \partial\omega$, where $\omega \subset \bar{\omega} \subset \Omega$ is an open set with regular boundary.
(c) δ_γ is the Dirac mass concentrated on γ and the function $g : \mathbb{R} \to \mathbb{R}$ is supposed to be continuous and strictly monotone, with $g(0) = 0$.

The two dimensional version of (4.1)–(4.3) was treated in [8] where it was proved that the solutions decay to zero for almost all $\omega \subset \Omega$. In the same paper it was proved that, in the one dimensional case, there exist $a \in (0,1)$ such that the solution of (1.1)–(1.3) goes to zero like $\sqrt{\frac{1}{t}}$.

A question related to the problem studied in this paper is the stabilization of elastic plates by the use of piezoelectric actuators (see [5] for appropriate models and [11], [12], [13] for the associated control problems). In this case new difficulties arise as the control function is scalar valued, so one may hope that strong stabilization holds only in the case of simple eigenvalues.

In general, the methods used in this paper apply for a large class of equations of the form $u'' + Au + g(u')\delta_\gamma = 0$ including the plate equation with various boundary conditions.

References

1. J. W. Cassals, *An introduction to diophantine approximation*, Cambridge University Press, Cambridge, 1966.
2. C. Chen, M. Coleman and H. H. West, *Pointwise stabilization in the middle of the span for second order systems, nonuniform and uniform decay of solutions*, SIAM J.Appl. Math., 47(1987), 751–780.
3. C. Dafermos, *Asymptotic behaviour of solutions of evolution equations* in "Nonlinear evolution equations" (M.G. Crandall, Ed.), 103–123, Academic Press, New York, 1978.
4. C. Dafermos and M. Slemrod, *Asymptotic behaviour of nonlinear contraction semigroups*, J. Diff. Eq., 13(1973), 97–106
5. Ph. Destuynder, I. Legrain, L. Castel, N. Richard, *Theoretical, numerical and experimental discussion of the use of piezoelectric devices for control-structure interaction* Eur. J. Mech., A/Solids, 11(1992), 181–213.
6. A. Haraux, *Stabilization of trajectories for some weakly damped hyperbolic equations*, J. Diff. Eq., 59(1985), 145–154.
7. L. F. Ho, *Controllability and stabilizability of coupled strings with control applied at the coupled points*, SIAM J. Control and Optimization, 31(1993), 1416–1436
8. S. Jaffard, M. Tucsnak and E. Zuazua, *Singular internal stabilization of the wave equation*, preprint
9. W. Littman, *Some recent results on control and stabilization of flexible structures*, Proceedings of the ComCon workshop on stabilization of flexible structures, Montpellier, France, December 1987.
10. W. Rudin, *Real and complex analysis*, McGraw-Hill, New York, 1974
11. M. Tucsnak,*Contrôle d'une poutre avec actionneur piézoélectrique,* Comptes Rendus de l'Acad. Sci. 319(1994), 697–702
12. M. Tucsnak,*Regularity and exact controllability for a beam with piezoelectric actuator*, SIAM J. on Control, 34(1996), 922–930
13. M. Tucsnak, *Control of plates vibrations by means of piezoelectric actuators*, Discrete and continuous dynamical systems, 2(1996), 281–293
14. G. Valiron, *Théorie des fonctions*, Masson, Paris, 1990.

M. Tucsnak
Ecole Polytechnique
Centre de Mathématiques Appliquées
F-91128 Palaiseau Cedex, France
and Université de Versailles

International Series of Numerical Mathematics
Vol. 126, © 1998 Birkhäuser Verlag, Basel

Exact Controllability of the Generalized Boussinesq Equation

BING-YU ZHANG
Department of Mathematical Sciences
University of Cincinnati

ABSTRACT. In this paper we consider distributed control of the system described by the generalized Boussinesq equation

$$u_{tt} = u_{xx} - (a(u) + u_{xx})_{xx} + f$$

on the periodic domain S, the unit circle in the plane.

In the case of local control, if the control f is allowed to act on the whole domain S, it is shown that the system is globally exactly controllable. In the case of local control where the control f is only allowed to act on a sub-domain of S, we show that the same result holds if the initial and terminal states have "small amplitude" in a certain sense.

1991 *Mathematics Subject Classification.* Primary 35Q20, 93C20

Key words and phrases. Boussinesq equation, exact controllability, contraction principle.

1. Introduction

In the present work we consider distributed control of a class of equations which may be described as being of *generalized Boussinesq* type. They have the general form

$$u_{tt} - u_{xx} = (a(u) - u_{xx})_{xx} + f \tag{1.1}$$

in which $u \equiv u(x,t)$, x, $t \in R$ and the subscripts denote the corresponding partial derivatives, $a : R \to R$ is a smooth function with $a(0) = 0$. The equation (1.1) is a perturbation of the linear wave equation which takes into account effects of weak nonlinearity and dispersion, and appears in the theory of nonlinear strings.

The classical Boussinesq equation is of the form

$$u_{tt} - u_{xx} + \frac{3}{2}(u^2)_{xx} + bu_{xxxx} = 0, \tag{1.2}$$

and was derived by Boussinesq [2] in 1872 as a model for the propagation of small amplitude, long waves on the surface of water. It possesses special, traveling-wave solutions called solitary waves. Historically, Boussinesq's theory [2] was the first to give a satisfactory, scientific explanation of the phenomenon of solitary waves described by Scott-Russell thirty years earlier. Depending on whether the coefficient b in the equation (1.2) is positive or negative, the equation (1.1) is called the "good" Boussinesq equation or the "bad" Boussinesq equation. The "bad" version is used to describe a two-dimensional flow of a body of water over a flat bottom with air above the water, assuming that the water waves have small amplitudes and the water is shallow. It also appeared in a posterior study of the Fermi-Pasta-Ulam (FPU) problem, which was performed to show that the finiteness of the thermal conductivity of an anharmonic lattice was related to nonlinear forces in the springs. However, the "bad" Boussinesq equation is notorious for its initial value problem (IVP) being not well-posed even locally (in time). Of the "bad" Boussinesq equation only solutions of soliton type, which can be found using the inverse scattering method, are known. For this reason, we only consider a generalized version of the "good" Boussinesq equation (1.1).

Our main concern is the study of equation (1.1) from a control point of view. In particular, we consider the equation posed on a periodic domain S, the unit circle in the complex plane

$$u_{tt} - u_{xx} = (a(u) - u_{xx})_{xx} + f, \quad x \in S,\ t \in R \tag{1.3}$$

with the forcing function $f \equiv f(x,t)$ as a control input. The goal is to influence the system by choosing an appropriate input f.

The control theory of Boussinesq-type equations was initiated by Liu and Russell [6], [7] and [8]. Both distributed control and boundary control of the Boussinesq equation have been considered. Some dissipative mechanism is introduced into the systems through appropriate feedback control laws. They showed that the small amplitude solutions of the resulting closed loop system are then exponentially stable.

In this paper we consider the exact control problem: choose an appropriate control input $f(x,t)$ to guide the system described by (1.3), during time interval $[0,T]$, from a given initial state to another preassigned terminal state in an appropriate function space of system states.

Note that for an appropriately smooth solution $u(x,t)$ of the unforced equation ($f(x,t) \equiv 0$) it is easy to see that any smooth solution u satisfies

$$\frac{d}{dt}\int_S u_t(x,t)dx = 0$$

for any $t \in R$. Therefore

$$\int_S u_t(x,t) = \int_S u_t(x,0)dx$$

and

$$\int_S u(x,t) = \int_S u(x,0)dx + t\int_S u_t(x,0)dx$$

for any $t \in R$. Usually one chooses the initial value $u_t(x,0)$ with $\int_S u_t(x,0)dx = 0$ (cf. [1]) so that both $\int_S u_t(x,t)dx$ and $\int_S u(x,t)dx$ are conserved for the unforced system. In order to keep these quantities conserved while conducting control we require that the control input f in system (1.3) satisfies

$$\int_S f(x,t)dx = 0, \quad \forall t \in R. \tag{1.4}$$

A more interesting case is obtained if some further *a priori* restrictions are imposed on the applied control $f(x,t)$. Let us suppose that $g(x)$ be a smooth function defined on S with its support contained in S satisfying

$$[g] := \int_S g(x)\, dx = 1$$

where $[g]$ denotes the mean value of the function g over the circle S. We restrict our attention to a control of the form

$$f(x,t) = Gh := g(x)\left(h(x,t) - \int_S g(y)h(y,t)dy\right). \tag{1.5}$$

Thus $h(x,t)$ may be considered as a new control input. It is easy to check that

$$\int_S f(x,t) \equiv 0$$

with f given by (1.5); therefore the restriction (1.4) is satisfied. Depending on the support of the function g in the domain S, there are two different control situations. If the support of g is the whole domain S then the control acts on the whole domain and we refer to it as global control. If the support of the function g is a proper subset of S, the control acts only on a sub-domain and we refer to it as local control. Obviously we have more control power in the global control situation than in the local control case. On the other hand, the local control situation includes more cases of practical interests and is therefore more relevant in general.

Now we describe the main results of this paper. Let $H^s(S)$ $(s \geq 0)$ be the space of all functions of the form

$$v(x) = \sum_{-\infty}^{\infty} v_k e^{ikx}$$

such that

$$\left\{\sum_{-\infty}^{\infty} |v_k|^2 (1+|k|)^{2s}\right\}^{1/2} < +\infty. \tag{1.6}$$

The left hand side of (1.6) is a Hilbert norm for $H^s(S)$; we denote it by $\|v\|_s$.

For the control problem just introduced the *exact control problem* consists in using the indicated control function to guide the system, during $[0,T]$, between a given pair of initial states $u(x,0) = \phi_0(x)$ and $u_t(x,0) = \psi_0(x)$ and a given pair of terminal states $u(x,T) = \phi_T(x)$ and $u_t(x,T) = \psi_T(x)$, in an appropriate function space of system states, necessarily, in view of the conserving control actions under consideration, such that

$$\int_S \psi_0(x)dx = \int_S \psi_T(x)dx = 0, \qquad \int_S \phi_0(x)dx = \int_S \phi_T(x)dx. \tag{1.7}$$

In the global control case, the control h acts on the whole domain S and we have the following strong controllability result.

Theorem 1.1. *Let $T > 0$ and $s \geq 0$ be given and assume that the function g in (1.5) satisfies*

$$|g(x)| > \beta > 0, \qquad \forall x \in S \tag{1.8}$$

Then for any $(\phi_0, \psi_0), (\phi_T, \psi_T) \in H^{s+2}(S) \times H^s(S)$ *satisfying (1.7) there exists a control function* $h \in L^2(0, T; H^s(S)$ *such that the equation*

$$u_{tt} - u_{xx} = (a(u) - u_{xx})_{xx} + Gh$$

has a solution $u \in C([0, T]; H^{s+2}(S)) \cap C^1([0, T]; H^s(S))$ *satisfying*

$$u(x, 0) = \phi_0(x), \qquad u_t(x, 0) = \psi_0(x),$$
$$u(x, T) = \phi_T(x), \qquad u_t(x, T) = \psi_T(x).$$

In other words, we have "global" exact controllability in the global control case.

In the *local control case*, the support of the function g may be a very small part of the domain S; thus our control power is quite limited. In this situation, we have the following "local" exact controllability result.

Theorem 1.2. *Let* $T > 0$ *and* $s \geq 0$ *be given. Then there exists a* $\delta > 0$ *such that for any* $(\phi_0, \psi_0), (\phi_T, \psi_T) \in H^{s+2}(S) \times H^s(S)$ *satisfying (1.7) and*

$$\|\phi_0\|_{s+2} + \|\phi_T\|_{s+2} < \delta, \qquad \|\psi_0\|_s + \|\psi_T\|_s < \delta,$$

there exists a control function $h \in L^2(0, T; H^s(S)$ *such that the equation*

$$u_{tt} - u_{xx} = (a(u) - u_{xx})_{xx} + Gh$$

has a solution $u \in C([0, T]; H^{s+2}(S)) \cap C^1([0, T]; H^s(S))$ *satisfying*

$$u(x, 0) = \phi_0(x), \qquad u_t(x, 0) = \psi_0(x),$$
$$u(x, T) = \phi_T(x), \qquad u_t(x, T) = \psi_T(x).$$

The paper is organized as follows. In section 2, we consider well-posedness of the initial value problem of the forced generalized Boussinesq equation posed on the periodic domain S:

$$\text{(1.9)} \qquad \begin{cases} u_{tt} - u_{xx} = (a(u) - u_{xx})_{xx} + f, & x \in S, \ t \in R \\ u(x, 0) = \phi(x), \quad u_t(x, 0) = \psi(x) \end{cases}$$

As it is known this is equivalent to considering periodic solutions of the equation posed on R. The local well-posedness of this problem could be established either by Kato's semigroup approach (see Bona and Sack [1]) or by Bona and Smith's regularization approach (see Liu and Russell [7]). But in this paper we provide a direct and simpler approach with the *contraction principle*. The advantage of this approach is that one not only obtains the well-posedness of the problem but is also able to show that the solution depends analytically on its initial data and the forcing term. It should be pointed out that one only expects a local well-posedness result for the IVP (1.9). Some solutions of the IVP (1.9) may blow up in finite time even though their initial data and the forcing term are smooth (see [5]). In section 3, we conduct a spectral analysis of the operator

$$A = \begin{pmatrix} 0 & 1 \\ \partial_x^2 - \partial_x^4 & 0 \end{pmatrix}$$

defined in the space $X_s \equiv H^{s+2}(S) \times H^s(S)$ for $s \geq 0$. We show that the operator A is a discrete spectral operator and its eigenvectors form a Riesz basis of the space X_s.

The result established in this section is the basis to obtain our main exact controllability results for the generalized Boussinesq equation (1.1). The proof of our main results of this paper, Theorem 1 and Theorem 2, are provided in section 4. As in our earlier joint paper with Russell [10] which dealt with the same control problem for the Korteweg-de Vries equation, we consider first a linear system associated with the nonlinear system (1.1):

$$\begin{cases} u_{tt} - u_{xx} + u_{xxxx} = Gh, & x \in S,\ t \in R \\ u(x,0) = \phi(x), & u_t(x,0) = \psi(x) \end{cases} \tag{1.10}$$

We will show that the system is exactly controllable in the space $H^{s+2}(S) \times H^s(S)$ for any $s \geq 0$. Moreover, we show that there exists a bounded linear operator K_T from the initial/terminal state pair (ϕ_0, ψ_0), (ϕ_T, ψ_T), each in the space $H^{s+2}(S) \times H^s(S)$, to the corresponding control h in the space $L^2([0,T]; H^s(S))$. Then the proofs of Theorem 1.1 and Theorem 1.2 follow from the same argument used in [10].

2. Well-posedness

In this section we establish the well-posedness of the initial value problem of the forcing general Boussinesq equation on a periodic domain S,

$$\begin{cases} u_{tt} = u_{xx} - (a(u) + u_{xx})_{xx} + f & x \in S,\ t \in R, \\ u(x,0) = \phi(x), & u_t(x,0) = \psi(x), \end{cases} \tag{2.1}$$

via the *contraction principle* approach.

Theorem 2.1. *Let $s \geq 0$ and $T > 0$ be given. Then for any $\phi \in H^{s+2}(S)$, $\psi \in H^s(S)$ and $f \in L^1(0,T; H^s(S))$, there exists a $T^* > 0$, depending only on $\|\phi\|_{s+2}$, $\|\psi\|$ and $\|f\|_{L^1(0.T;H^s(S))}$, such that the IVP (2.1) has a unique solution $u \in C([0,T^*]; H^{s+2}(S))$ with $u_t \in C([0,T^*]; H^s(S))$. In addition, the solution depends continuously on the initial data ϕ, ψ and the forcing term f in the respective spaces.*

Before we present the proof of the theorem, we rewrite (2.1) as the following equivalent first order evolution equation

$$\frac{d}{dt}\vec{u} = A\vec{u} + F(\vec{u}) + \vec{g}, \quad \vec{u}(0) = \vec{u}_0 \tag{2.2}$$

where

$$\vec{u} = \begin{pmatrix} u \\ u_t \end{pmatrix}, \qquad A = \begin{pmatrix} 0 & 1 \\ \partial_x^2 - \partial_x^4 & 0 \end{pmatrix} \tag{2.3}$$

and

$$F(\vec{u}) = \begin{pmatrix} 0 \\ (a(u))_{xx} \end{pmatrix}, \qquad \vec{g} = \begin{pmatrix} 0 \\ f \end{pmatrix}, \qquad \vec{u}_0 = \begin{pmatrix} \phi \\ \psi \end{pmatrix}. \tag{2.4}$$

For any $s \geq 0$ and $T > 0$, let X_s denote the Hilbert space

$$X_s = H^{s+2}(S) \times H^s(S)$$

equipped with the norm

$$\|\vec{u}\|_{X_s} = \left(\|u_x\|_s^2 + \|u_{xx}\|_s^2 + \|v\|_s^2\right)^{1/2}$$

for any $\vec{u} = \begin{pmatrix} u \\ v \end{pmatrix} \in X_s$. In addition, let

$$Y_{T,s} = C([0,T]; X_s).$$

It is easy to see that the operator A is a linear operator from X_s to X_s with $\mathcal{D}(A) = H^{s+4}(S) \times H^s(S)$. Besides, the adjoint operator of A

$$A^* = -A.$$

The operator A generates an isomorphic group $W(t)$ on the space X_s for any $s \geq 0$ and the standard semigroup theory gives us the following estimates.

Proposition 2.1. *Let $T > 0$ and $s \geq 0$ be given. There exists a constant $c > 0$ such that*

$$\sup_{t\in(0,T)} \|W(t)\vec{u}\|_{X_s} \leq \|\vec{u}\|_{X_s} \tag{2.5}$$

for any $\vec{u} \in X_s$ and

$$\sup_{t\in(0,T)} \|\int_0^t W(t-\tau)\vec{f}(\tau)d\tau\|_{X_s} \leq c\|\vec{f}\|_{L^1(0,T;X_s)} \tag{2.6}$$

for any $\vec{f} \in L^1(0,T;X_2)$.

Using the notations we just introduced, Theorem 2.1 can be restated as follows:

Theorem 2.1′. *Let $s \geq 0$ and $T > 0$ be given. Then for any $\vec{u}_0 \in X_s$ and $\vec{g} \in L^1(0,T;X_s)$, there exists a $T^* > 0$ depending only on $\|\vec{u}_0\|_{X_s}$ and $\|\vec{g}\|_{L^1(0,T;X_s)}$ such that the IVP (2.2) has a unique solution $\vec{u} \in Y_{T^*,s}$ and the corresponding solution map: $(\vec{u}_0, f) \in X_s \times L^1(0,T;X_s) \to u \in Y_{T^*,s}$ is continuous.*

Proof of Theorem 2.1′. Using the notation of the semigroup $W(t)$ we may write (2.2) in its integral form

$$\vec{u}(t) = W(t)\vec{u}_0 + \int_0^t W(t-\tau)(F(\vec{u}) + \vec{g})(\tau)d\tau \tag{2.7}$$

It suggests us to consider the map Γ defined on the space $C[0,T;X_s]$, for given $\vec{u}_0 \in X_s$ and $\vec{g} \in L^1(0,1;X_s)$, by

$$\Gamma(\vec{u}) = W(t)\vec{u}_0 + \int_0^t W(t-\tau)(F(\vec{u}) + \vec{g})(\tau)d\tau$$

for any $\vec{u} \in X_s$. For $M > 0$ and $T > 0$, let $S_{T,M}$ be a bounded subset of the space $C[0,T;X_s]$:

$$S_{T,M} = \left\{\vec{u} \in C[0,T;X_s];\ \sup_{t\in(0,T)} \|\vec{u}\|_{X_s} \leq M\right\}.$$

Applying (2.5)–(2.6) yields that for any $T > 0$ there exists a constant $c > 0$ such that

$$\sup_{t\in(0,T)} \|\vec{u}(t)\|_{X_s} \leq c\|\vec{u}_0\|_{X_s} + c\int_0^T (\|F(\vec{u}\|_{X_s} + \|\vec{g}\|_{X_s})d\tau.$$

Note that

$$\|F(\vec{u})\|_{X_s} = \|(a(u))_{xx}\|_s, \qquad \|\vec{g}(\cdot,t)\|_{X_s} = \|g(\cdot,t)\|_s,$$

for any $\vec{u} \in C[0,T;X_s]$ and

$$\begin{aligned}\|(a(u))_{xx}\|_s &\leq \|a''(u)u_x^2\|_s + \|a'(u)u_{xx}\|_s \\ &\leq \beta(\|u\|_{s+2})\|u\|_{s+2} \leq \beta(\vec{u}\|_{X_s})\|\vec{u}\|_{X_s}\end{aligned}$$

where $\beta(\cdot)$ is a continuous monotone increasing function only depending on a. One has that

$$\sup_{t\in(0,T)} \|\Gamma(\vec{u})\|_{X_s} \leq c(\|\vec{u}_0\|_{X_s} + \|f\|_{L^1(0,T;X_s)}) + cT\beta(\sup_{t\in(0,T)} \|\vec{u}(t))\|_{X_s} \sup_{t\in(0,T)} \|\vec{u}(t)\|_{X_s}.$$

Choose M and T^* such that

$$M = 2c(\|\vec{u}_0\|_{X_s} + \|f\|_{L^1(0,T;X_s)}), \quad cT^*\beta(2M) < 1/2. \tag{2.8}$$

Then

$$\sup_{t\in(0,T^*)} \|\Gamma(\vec{u})\|_{X_s} \leq M/2 + cT^*\beta(M)M < M$$

for any $\vec{u} \in S_{T^*,M}$. In addition, for any $\vec{u}$, $\vec{v} \in S_M$,

$$\Gamma(\vec{u}) - \Gamma(\vec{v}) = \int_0^t W(t-\tau)\left(F(\vec{u}) - F(\vec{v})\right)d\tau$$

and

$$F(\vec{u}) - F(\vec{v}) = \left(\int_0^1 a'(u + \lambda(v-u))d\lambda(u-v)\right)_{xx}.$$

Thus

$$\begin{aligned}\sup_{t\in(0,T^*)} \|\Gamma(\vec{u}) - F(\vec{v})\|_{X_s} &\leq cT^*\beta(\sup_{t\in(0,T^*)} \|\vec{u}+\vec{v}\|_{X_s}) \sup_{t\in(0,T)} \|\vec{u}-\vec{v}\|_{X_s} \\ &\leq cT^*\beta(2M) \sup_{t\in(0,T^*)} \|\vec{u}-\vec{v}\|_{X_s} \\ &\leq \sup_{t\in(0,T^*)} \|\vec{u}-\vec{v}\|_{X_s}/2.\end{aligned}$$

So Γ is a contraction in $S_{T^*,M}$. Its fixed point is the desired solution. □

As an advantage of this contraction principle approach, one can show that the solution map is not only continuous but also infinitely many times Fréchet differentiable if a is a C^∞ function and is analytic if a is a polynomial.

Corollary 2.1. *If the function a is C^∞ smooth, then the solution map:*

$$(\vec{u}_0, f) \in X_s \times L^1(0,T;H^s(S)) \to Y_{T,s}$$

is C^∞ smooth, i.e., it has any order of Fréchet derivatives. Furthermore, if a is a polynomial, then the solution map is analytic.

Proof. see [11].

3. Spectral analysis

In this section we conduct a spectral analysis of the operator A defined by (2.3). The result obtained in this analysis will be the basis to transfer the exact controllability problem of the associated linear system (1.10) to a corresponding moment problem.

Let us define

$$E_{1,0} = \begin{pmatrix} 1 \\ 0 \end{pmatrix}, \qquad E_{2,0} = \begin{pmatrix} 0 \\ 1 \end{pmatrix}$$

and

$$E_{1,k} = \frac{1}{k^2}\begin{pmatrix} e^{ikx} \\ 0 \end{pmatrix}, \qquad E_{2,k} = \begin{pmatrix} 0 \\ e^{ikx} \end{pmatrix}$$

for $k = \pm 1, \ \pm 2, \ldots$. An easy computation leads to

$$A(E_{1,k}, E_{2,k}) = (E_{1,k}, E_{2,k})\Sigma_K \quad \text{with} \quad \Sigma_k = \begin{pmatrix} 0 & k^2 \\ -(k^2+1) & 0 \end{pmatrix}.$$

The matrix Σ_k has the eigenvalues

$$\lambda_{k,1} = i\sqrt{k^2(k^2+1)}$$
(3.1)
$$\lambda_{k,2} = -i\sqrt{k^2(k^2+1)}$$

with the corresponding eigenvectors

$$\vec{e}_{k,1} = \begin{pmatrix} 1 \\ \frac{\lambda_{k,1}}{k^2} \end{pmatrix}, \qquad \vec{e}_{k,2} = \begin{pmatrix} 1 \\ \frac{\lambda_{k,2}}{k^2} \end{pmatrix} \tag{3.2}$$

for $k = \pm 1, \pm 2, \ldots$. In addition, $\lambda_{0,1} = \lambda_{0,2} = 0$ is also an eigenvalue with the corresponding eigenvector

$$\vec{e}_{0,1} = \vec{e}_{0,2} = \begin{pmatrix} 0 \\ 1 \end{pmatrix}.$$

Thus

$$\begin{aligned} A(E_{1,k}, E_{2,k})(\vec{e}_{k,1}, \vec{e}_{k,2}) &= (E_{1,k}, E_{2,k})\Sigma_k(\vec{e}_{k,1}, \vec{e}_{k,2}) \\ &= (\lambda_{1,k}(E_{1,k}, E_{2,k})\vec{e}_{k,1}, \lambda_{k,2}(E_{1,k}, E_{2,k})\vec{e}_{k,2}). \end{aligned}$$

So $\lambda_{k,1}$ and $\lambda_{k,2}$ are eigenvalues of the operator A with the corresponding eigenvectors

$$\vec{\eta}_{k,1} = (E_{1,k}, E_{2,k})\vec{e}_{k,1}, \qquad \vec{\eta}_{k,2} = (E_{1,k}, E_{2,k})\vec{e}_{k,2}.$$

for $k = \pm 1, \ \pm 2, \ \ldots$. In addition, $\lambda_0 = 0$ is the eigenvalue of the operator A with the corresponding eigenvector

$$\vec{\eta}_0 = \begin{pmatrix} 0 \\ 1 \end{pmatrix}.$$

Note that

$$M_k = (e_{k,1}, e_{k,2}) \to \begin{pmatrix} 1 & 1 \\ i & -i \end{pmatrix}$$

as $k \to \infty$. Thus

$$\lim_{k\to\infty} \det M_k = -2i \neq 0.$$

Since $\{E_{1,k}, E_{2,k}\},\ k = 0, \pm 1, \pm 2, \dots$ form an orthogonal basis for the space X_s, by [3] we know that

$$\{\vec{\eta}_0, \vec{\eta}_{k,1}, \vec{\eta}_{k,2},\ k = \pm 1, \pm 2, \dots\}$$

forms a Riesz basis for the space X_s.

Note that $A^* = -A$. The spectrum of A^* consists of the eigenvalues $\mu_0 = \lambda_0 = 0$ and

$$\mu_{k,j} = \overline{\lambda}_{k,j} = -\lambda_{k,j}$$

for $j = 1, 2$ and $k = \pm 1, \pm 2, \dots$ with the corresponding eigenvectors $\vec{\nu}_0 = \vec{\eta}_0$ and $\vec{\nu}_{k,j} = \vec{\eta}_{k,j}$. Furthermore, we define

$$m_{k,j} = \|\vec{\eta}_{k,j}\|_{X_s}$$

and

$$\hat{\phi}_{k,j} = \vec{\eta}_{k,j}/m_{k,j}$$

for $j = 1, 2$ and $k = \pm 1, \pm 2, \dots$. Then

$$\left\{\hat{\phi}_0 = \vec{\eta}_0, \quad \hat{\phi}_{k,j}, j = 1, 2 \text{ and } k = \pm 1, \pm 2, \dots\right\}$$

forms an orthonormal basis for the space X_s:

$$\langle \hat{\phi}_{k,j}, \hat{\phi}_{m,l}\rangle_{X_s} = \begin{cases} 1 & \text{if } k = m,\ j = l \\ 0 & \text{otherwise} \end{cases}$$

The above discussion can be summarized as the following theorem.

Theorem 3.1. *Let*

$$\lambda_n = \begin{cases} i\sqrt{n^2(n^2+1)} & n = 0, 1, 2, \dots, \\ -i\sqrt{n^2(n^2+1)} & n = -1, -2, \dots \end{cases}$$

$$\phi_{n,1} = \begin{cases} \hat{\phi}_{n,1} & n = 0, 1, 2, \dots \\ \hat{\phi}_{n,2} & n = -1, -2, -3, \dots \end{cases}$$

and

$$\phi_{n,2} = \begin{cases} \hat{\phi}_{-n,1} & n = 0, 1, 2, \dots \\ hat\phi_{-n,2} & n = -1, -2, \dots \end{cases}$$

Then

(a) *The spectrum of the operator A consists of eigenvalues $\{\lambda_n\}_{n=-\infty}^{\infty}$ in which λ_0 is a simple eigenvalue with the corresponding eigenvector $\phi_0 = \phi_{0,1} = \phi_{0,2}$ and each λ_n, $n = \pm 1, \pm 2, \dots$ is a double eigenvalue with the corresponding eigenvector $\phi_{n,j}$, $j = 1, 2$.*

(b) $\{\phi_0, \ \phi_{n,j}, \ j = 1, 2, \ n = \pm 1, \pm 2, \dots\}$ *forms an orthonormal basis for the space* X_s *and any* $\vec{w} \in X_s$ *has the following Fourier series expansion:*

$$\vec{w} = \alpha_0\phi_0 + \sum_{n=-\infty, \neq 0}^{+\infty} (\alpha_{n,1}\phi_{n,1} + \alpha_{n,2}\phi_{n,2})$$

with $\alpha_0 =< \vec{w}, \phi_0 >_{X_s}$, *and*

$$\alpha_{n,1} =< \vec{w}, \phi_{n,1} >_{X_s}, \quad \alpha_{n,2} =< \vec{w}, \phi_{n,2} >_{X_s}$$

for $n = \pm 1, \pm 2, \dots$

4. Exact controllability

In this section we prove the main results of this paper. Following the argument in [10] we first consider the corresponding linear control system

$$\text{(4.1)} \qquad \begin{cases} u_{tt} = u_{xx} - u_{xxxx} + Gh & x \in S, \ t \in R, \\ u(x,0) = \phi(x), & u_t(x,0) = \psi(x), \end{cases}$$

which can be written as the following abstract linear system

$$\text{(4.2)} \qquad \begin{cases} \frac{d\vec{y}}{dt} = A\vec{y} + Bh, \\ \vec{y}(0) = \vec{y}_0 \end{cases}$$

on the space X_s where

$$\vec{y} = \begin{pmatrix} u \\ u_t \end{pmatrix}, \quad Bh = \begin{pmatrix} 0 \\ Gh \end{pmatrix}$$

with $h \in L^2(0, T; H^s(S))$. The solution $\vec{y}(t)$ of this system can be written as

$$\text{(4.3)} \qquad \vec{y}(t) = W(t)\vec{y}_0 + \int_0^t W(t-\tau)Bh(\tau)d\tau.$$

Based on the spectral analysis of the operator A conducted earlier,

$$\text{(4.4)} \qquad \begin{aligned} \vec{y}(t) &= e^{\lambda_0 t}\alpha_0\phi_0 + \sum_{n\neq 0} e^{\lambda_n t}(\alpha_n\phi_{n,1} + \alpha_{n,2}\phi_{n,2}) \\ &+ \int_0^t e^{\lambda_0(t-\tau)}\beta_0(\tau)d\tau + \sum_{n\neq 0}\int_0^t e^{\lambda_n(t-\tau)}\left(\beta_{n,1}(\tau)\phi_{n,1} + \beta_{n,2}(\tau)\phi_{n,2}\right)d\tau \end{aligned}$$

where $\alpha_0 = \langle \vec{y}_0, \phi_0 \rangle_{X_s}$, $\beta_0 = \langle Bh, \phi_0 \rangle_{X_s}$ and

$$\alpha_{n,j} = \langle \vec{y}_0, \phi_{n,j} \rangle_{X_s}, \quad \beta_{n,j} = \langle Bh, \phi_{n,j} \rangle_{X_s} = \langle h, B^*\phi_{n,j} \rangle_{H^s(S)}$$

for $j = 1, 2$ and $n = \pm 1, \pm 2, \dots$ Note that

$$\begin{aligned} \langle Bh, \begin{pmatrix} u \\ v \end{pmatrix} \rangle_{X_s} &= \langle \begin{pmatrix} 0 \\ Gh \end{pmatrix}, \begin{pmatrix} u \\ v \end{pmatrix} \rangle_{X_s} \\ &= \langle h, Gv \rangle_{H^s(S)}. \end{aligned}$$

Thus

$$B^*\begin{pmatrix} u \\ v \end{pmatrix} = Gv \in H^s(S) \tag{4.5}$$

for any $\begin{pmatrix} u \\ v \end{pmatrix} \in X_s$.

Let

$$\mathcal{X}_s = \left\{ (\vec{y}_0, \vec{y}_T) \in X_s \times X_s, \quad \int_S (\vec{y}_0)_1 dx = \int_S (\vec{y}_T)_2 dx, \quad \int_S (\vec{y}_0)_2 dx = \int_S (\vec{y}_T)_2 dx = 0 \right\}.$$

Here, $(\vec{y})_j$ denotes the j–th component of $\vec{y}$ $(j = 1, 2)$. We show that the linear system (4.3) is exactly controllable. In fact we show that the following stronger result holds which is the key for obtaining exact controllability of the nonlinear system.

Theorem 4.1. *Let $T > 0$ be given. There exists a bounded linear operator*

$$K_T : \mathcal{X}_s \to L^2(0, T; H^s(S))$$

such that for any $(\vec{y}_0, \vec{y}_T) \in \mathcal{X}_s$, the solution of

$$\dot{\vec{y}}(t) = A\vec{y}(t) + BK_T(\vec{y}_0, \vec{y}_T), \quad \vec{y}(0) = \vec{y}_0$$

satisfies

$$\vec{y}(T) = \vec{y}_T$$

and

$$\|K_T(\vec{y}_0, \vec{y}_T)\|_{L^2(0,T;H^s(S))} \le c_T \left(\|\vec{y}_0\|^2_{X_s} + \|\vec{y}_T\|^2_{X_s} \right)^{1/2}$$

for some constant $c_T > 0$ independent of $(\vec{y}_0, \vec{y}_T)$.

Proof. Since the system (4.3) is time reversible we may assume, without loss of generality, that $\vec{y}_T = 0$. Then the exact control problem, letting $t = T$ in (4.4) and $\vec{y}(T) = 0$, consists of finding a $h \in L^2(0, T; H^s(S))$ such that

$$(\alpha_0 + T\beta_0)\phi_0 + \sum_{n\neq 0} e^{\lambda_n T}(\alpha_n \phi_{n,1} + \alpha_{n,2}\phi_{n,2})$$
$$+ \sum_{n\neq 0} \int_0^T e^{\lambda_n (T-\tau)} \left(\beta_{n,1}(\tau)\phi_{n,1} + \beta_{n,2}(\tau)\phi_{n,2}\right) d\tau = 0.$$

This leads to the solvability of the following moment problem

$$\alpha_0 + T\beta_0 = 0, \qquad \begin{cases} -e^{\lambda_n T}\alpha_{n,1} = \int_0^T e^{\lambda_n (T_\tau)}\beta_{n,1}(\tau)d\tau, \\ \\ -e^{\lambda_n T}\alpha_{n,2} = \int_0^T e^{\lambda_n (T_\tau)}\beta_{n,2}(\tau)d\tau \end{cases}$$

for $n = \pm 1, \pm 2, \ldots$ Here $(\phi_{n,j})_2$ denotes the second component of $\phi_{n,j}$ and according to the previous computation

$$(\phi_{nj})_2 = \begin{cases} b_k e^{ikx} & \text{if } j = 1 \\ \\ b_k e^{-ikx} & \text{if } j = 2. \end{cases}$$

where $0 < m < |b_k| < M$ for any k.

By denoting $p_k = e^{\lambda_k t}$, $\mathcal{P} \equiv \{p_k \mid -\infty < k < \infty\}$ forms a Riesz basis for its closed span, P_T in $L^2(0,T)$ (see [4]). We let $\mathcal{L} = \{q_k \mid -\infty < k < \infty\}$ be the unique dual Riesz basis for $\mathcal{P}$ in P_T, i.e., the functions in $\mathcal{L}$ are the unique elements of P_T such that

$$\int_0^T q_l(t)\overline{p_k(t)}dt = \delta_{kl}, \qquad -\infty < l,\, k < \infty. \tag{4.6}$$

We take the control h in (4.3) to have the form

$$h(x,t) = c_0 q_0(t) + \sum_{l\neq 0} q_l(t)(c_{l,1}G((\phi_{l,1})_2) + c_{l,2}G((\phi_{l,2})_2) \tag{4.7}$$

where the coefficients c_0, $c_{l,1}$ and $c_{l,2}$ are to be determined so that, among other things, the series (4.7) is appropriately convergent. Substituting (4.7) into (4.4) yields, by using the biorthogonality (4.6), that we have $\alpha_0 = c_0 < G(\phi_0), G(\phi_0) >_{X_s}$ and

$$\begin{cases} e^{-\lambda_n T}\alpha_{n,1} = c_{n,1}(G(\phi_{n,1})_2, G(\phi_{n,1})_2)_{H^s(S)} + c_{n,2}(G(\phi_{n,1})_2, G(\phi_{n,2})_2)_{H^s(S)} \\ e^{-\lambda_n T}\alpha_{n,1} = c_{n,1}(G(\phi_{n,2})_2, G(\phi_{n,1})_2)_{H^s(S)} + c_{n,2}(G(\phi_{n,2})_2, G(\phi_{n,2})_2)_{H^s(S)} \end{cases} \tag{4.8}$$

for $n = \pm 1, \pm 2, \ldots$ Let

$$\begin{aligned} \Delta_n &= \begin{vmatrix} (G(\phi_{n,1})_2, G(\phi_{n,1})_2)_{H^s(S)} & (G(\phi_{n,1})_2 G(\phi_{n,2})_2)_{H^s(S)} \\ (G(\phi_{n,2})_2, G(\phi_{n,1})_2)_{H^s(S)} & (G(\phi_{n,2})_2 G(\phi_{n,2})_2)_{H^s(S)} \end{vmatrix} \\ &= \|G(\phi_{n,1})_2\|_s^2 \|G(\phi_{n,2})_2\|_s^2 - \left|(G(\phi_{n,1})_2, G(\phi_{n,2})_{H^s(S)}\right|^2. \end{aligned}$$

Note that $\Delta_n \neq 0$ for any n since $G(\phi_{n,2})_2$ and $G(\phi_{n,1})_2$ are linearly independent. In addition, as $n \to \infty$, $(G(\phi_{n,1})_2, G(\phi_{n,2})_{H^s(S)} \to 0$ and $\|G(\phi_{n,j})_2\|^2 \sim b_k^2$, $j = 1, 2$. Hence there exists a $\epsilon > 0$ such that

$$|\Delta_n| > \epsilon$$

for any n and $c_{n,1}$ and $c_{n,2}$ are uniquely determined by (4.8). By Cramer's rule,

$$c_{n,1} = \frac{\Delta_{n,1}}{\Delta_n}, \qquad c_{n,2} = \frac{\Delta_{n,2}}{\Delta_n} \tag{4.9}$$

with

$$\Delta_{n,1} = \begin{vmatrix} e^{-\lambda_n T}\alpha_{n,1} & (G(\phi_{n,1})_2 G(\phi_{n,2})_2)_{H^s(S)} \\ e^{-\lambda_n T}\alpha_{n,2} & (G(\phi_{n,2})_2 G(\phi_{n,2})_2)_{H^s(S)} \end{vmatrix}$$

and

$$\Delta_{n,2} = \begin{vmatrix} (G(\phi_{n,1})_2, G(\phi_{n,1})_2)_{H^s(S)} & e^{-\lambda_n T}\alpha_{n,1} \\ (G(\phi_{n,2})_2, G(\phi_{n,1})_2)_{H^s(S)} & e^{-\lambda_n T}\alpha_{n,2} \end{vmatrix}.$$

It remains to show that h defined by (4.7) and (4.9) $\in L^2([0,T]; H^s(S)$. It follows from the same argument as in the proof of Theorem 1.1 in [10] and is therefore omitted here. The proof is complete. □

Now we turn to the nonlinear system

$$(4.10)\qquad \begin{cases} u_{tt} - u_{xx} = (a(u) - u_{xx})_{xx} + Gh \\ u(x,0) = \phi(x), \quad u_t(x,0) = \psi(x) \end{cases}$$

and prove the main results of this paper.

Proof of Theorem 1.1. According to Theorem 4.1, there exists $h_1 \in L^2(0,T;H^s(S))$ for which one may find $u \in C([0,T];H^{s+2}(S)) \cap C^1([0,T];H^s(S))$ satisfying

$$\begin{cases} u_{tt} - u_{xx} + u_{xxxx} = Gh_1, \\ u(x,0) = \phi_0(x), \quad u_t(x,0) = \psi_0(x), \\ u(x,T) = \phi_T(x), \quad u_t(x,T) = \psi_T(x) \end{cases}$$

for given (ϕ_0, ψ_0) and (ϕ_T, ψ_T) in the space $\mathcal{X}_s$. Adding $-a(u)_{xx}$ to both sides of the above equation one obtains

$$\begin{cases} u_{tt} - u_{xx} - (a(u))_{xx} + u_{xxxx} = Gh_1 - a(u)_{xx}, \\ u(x,0) = \phi_0(x), \quad u_t(x,0) = \psi_0(x), \\ u(x,T) = \phi_T(x), \quad u_t(x,T) = \psi_T(x). \end{cases}$$

Thus it suffices to show that there exists $h_2 \in C([0,T];H^s(S))$ such that

$$(Gh_2)(x,t) = -(a(u))_{xx}.$$

Note that $a(u)_{xx} = a'(u)u_{xx} + a''(u)u_x^2 \in C([0,T];H^s(S))$ since a is a smooth function and $u \in C([0,T];H^{s+2}(S))$. The existence of h_2 follows from exactly the same argument as in the proof of [10, Theorem 1.1]. The proof is complete. □

Proof of Theorem 1.2. We first write (4.10) as the first order evolution system

$$\begin{cases} \frac{d\vec{u}}{dt} = A\vec{u} + F(\vec{u}) + Bh, \\ \vec{u}(0) = \vec{u}_0 \end{cases}$$

which can be rewritten in its equivalent integral equation form

$$(4.11)\qquad \vec{u}(t) = W(t)\vec{u}_0 + \int_0^t W(t-\tau)(Bh((\tau) + \int_0^t W(t-\tau)F(\vec{u})(\tau)d\tau.$$

We define

$$\omega(T,\vec{u}) \equiv \int_0^T W(T-\tau)F(\vec{u})(\tau)d\tau.$$

According to Theorem 4.1, for given $\vec{u}_0,\ \vec{u}_T \in \mathcal{X}_s$, if one chooses

$$h = K_T(\vec{u}_0, \vec{u}_T + \omega(T,\vec{u}))$$

in equation (4.11), then

$$\vec{u}(t) = W(t)\vec{u}_0 + \int_0^t W(t-\tau)(BK_T(\vec{u}_0, \vec{u}_T + \omega(T,\vec{u}))(\tau)d\tau + \int_0^t W(t-\tau)F(\vec{u})(\tau)d\tau$$

and $\vec{u}(0) = \vec{u}_0$, $\vec{u}(T) = \vec{u}_T$ by virtue of the definition of the operator K_T. This suggests that we consider the map

$$\Gamma(\vec{u}) = W(t)\vec{u}_0 + \int_0^t W(t-\tau)(BK_T(\vec{u}_0, \vec{u}_T + \omega(T, \vec{u}))(\tau)d\tau + \int_0^t W(t-\tau)F(\vec{u})(\tau)d\tau.$$

If the map Γ is shown to be a contraction in an appropriate space, then its fixed point $\vec{u}$ is a solution of (4.11) with $h = K_T(\vec{u}_0, \vec{u}_T + \omega(T, \vec{u})$ and satisfies $\vec{u}(T) = \vec{u}_T$. An argument similar to the proof of Theorem 2.1 in section 2 shows that the map Γ is a contraction in the space X_s. The only modification one needs is that instead of choosing a small T, one chooses a small $\delta > 0$ and requires

$$\|\vec{u}_0\|_{X_s} \le \delta, \qquad \|\vec{u}_T\|_{X_s} \le \delta$$

so that the map Γ becomes a contraction (see also [11]). The proof is complete. $\square$

References

1. J. L. Bona and R. L. Sachs, Global existence of smooth solutions and stability of solitary waves for a generalized Boussinesq equation, *Comm. Math. Physics*, **118** (1988), 15–29.
2. J. Boussinesq, Théorie des ondes et de remous qui se propagent ..., *J. Math. Pures Appl.*, **17** (1872), 55–108.
3. S. W. Hansen, Bounds on functions biorthogonal to sets of complex exponentials; control of damped elastic systems, J. Math. Anal. Appl., **158** (1991), 487–509.
4. A. E. Ingham, Some trigonometrical inequalities with application to the theory of series, *Math. A.*, **41** (1936), 367–369
5. V. K. Kalantarov and O. A. Ladyzhenskaya, The occurrence of collapse for quasi-linear equations of parabolic and hyperbolic types, *J. Sov. Math.*, **10** (1978), 53–70.
6. F. L. Liu and D. L. Russell, Exponential decay rates for Boussinesq equation with boundary control, preprint.
7. F. L. Liu and D. L. Russell, Boussinesq equation on a periodic domains, preprint.
8. F. L. Liu and D. L. Russell, Stabilization of Boussinesq equation with periodic boundary conditions.
9. D. L. Russell and B.-Y. Zhang, Controllability and stabilizability of the third order linear dispersion equation on a periodic domain, *SIAM J. Control and Optim.*, **31** (1993), 659–676.
10. D. L. Russell and B.-Y. Zhang, Exact controllability and stabilizability of the Korteweg-de Vries equation, *Transaction of the American mathematical Society*, **348** (1996), 3643–3672.
11. B.-Y. Zhang, Analyticity of solutions of the generalized Korteweg-de Vries equation with respect to their initial values, *SIAM J. Math. Anal.*, **26** (1995), 1488–1513.

Bing-Yu Zhang
Department of Mathematical Sciences
University of Cincinnati
Cincinnati, Ohio 45221-0025, USA